Lecture Notes in Mechanical Engineering

Lecture Notes in Mechanical Engineering (LNME) publishes the latest developments in Mechanical Engineering—quickly, informally and with high quality. Original research reported in proceedings and post-proceedings represents the core of LNME. Volumes published in LNME embrace all aspects, subfields and new challenges of mechanical engineering. Topics in the series include:

- Engineering Design
- Machinery and Machine Elements
- Mechanical Structures and Stress Analysis
- Automotive Engineering
- Engine Technology
- Aerospace Technology and Astronautics
- Nanotechnology and Microengineering
- Control, Robotics, Mechatronics
- MEMS
- Theoretical and Applied Mechanics
- Dynamical Systems, Control
- Fluid Mechanics
- Engineering Thermodynamics, Heat and Mass Transfer
- Manufacturing
- Precision Engineering, Instrumentation, Measurement
- Materials Engineering
- Tribology and Surface Technology

To submit a proposal or request further information, please contact the Springer Editor of your location:

China: Dr. Mengchu Huang at mengchu.huang@springer.com
India: Priya Vyas at priya.vyas@springer.com
Rest of Asia, Australia, New Zealand: Swati Meherishi at swati.meherishi@springer.com
All other countries: Dr. Leontina Di Cecco at Leontina.dicecco@springer.com

To submit a proposal for a monograph, please check our Springer Tracts in Mechanical Engineering at http://www.springer.com/series/11693 or contact Leontina.dicecco@springer.com

Indexed by SCOPUS. All books published in the series are submitted for consideration in Web of Science.

More information about this series at http://www.springer.com/series/11236

Muthukumar Palanisamy · Velraj Ramalingam ·
Murugan Sivalingam
Editors

Theoretical, Computational, and Experimental Solutions to Thermo-Fluid Systems

Select Proceedings of ICITFES 2020

Springer

Editors
Muthukumar Palanisamy
Department of Mechanical Engineering
Indian Institute of Technology Guwahati
Guwahati, India

Velraj Ramalingam
Department of Mechanical Engineering
Anna University
Chennai, India

Murugan Sivalingam
Department of Mechanical Engineering
National Institute of Technology Rourkela
Rourkela, India

ISSN 2195-4356 ISSN 2195-4364 (electronic)
Lecture Notes in Mechanical Engineering
ISBN 978-981-33-4164-7 ISBN 978-981-33-4165-4 (eBook)
https://doi.org/10.1007/978-981-33-4165-4

This Springer imprint is published by the registered company Springer Nature Singapore Pte Ltd.
The registered company address is: 152 Beach Road, #21-01/04 Gateway East, Singapore 189721, Singapore

Contents

Numerical Investigation on Heat Transfer Characteristics of a Triangular Slotted Pin Fin Heat Sink 1
Rout Jagannath and Yamala Muralikrishna

Heat Transfer Analysis of an Equilateral Triangular Slotted Pin Fin: An Experimental Work 13
Chappa Bhaskara Rao and Yamala Muralikrishna

Thermo-Fluid Performance Evaluation of an Elliptical Tube Type Fin-and-Tube Heat Exchanger Supported with Winglets 23
S. K. Sarangi and D. P. Mishra

Experimental Analysis of a Crossflow Heat Exchanger Using Elliptical Shape Tube for Air Side Heat Recovery 35
Amruta Sonawane, Neelam S. Gohel, R. S. Jha, and Digvijay Shelar

An Experimental Investigation of Tube-in-Tube Type Heat Exchanger to Enhance Heat Transfer Using Titanium Oxide TiO_2-Water Nanofluid 43
K. Soma Sekhar, P. Vijay Kumar, and B. Kamala Priya

Earth Air Tube Heat Exchanger—A Parametric Study 53
Saif Nawaz Ahmad and Om Prakash

Natural Convection in a Square Porous Enclosure with a Pair of Fluid-Pockets 63
Jayesh Subhash Chordiya and Ram Vinoy Sharma

Numerical Analysis of Natural Convection in a Partially Open Square Cavity with Multiple Heat Sources 73
Gloria Biswal and Aurovinda Mohanty

Simulation Studies of Heat Transfer by Natural Convection from an Isothermal Rectangular Cylinder Using ANSYS Fluent 81
M. Naveen, B. Venkata Sai Raghu Vamsi, M. R. Ch. Sastry, and T. Siva Krishna

Forced Convective Wake Dynamics Past a Semicircular Cylinder at Incidence with a Downstream Circular Cylinder in Crossflow 93
Chitrak Mondal, Sandip Sarkar, and Nirmal Kumar Manna

Electro-thermal Convection in a Rectangular Enclosure 103
Nirmalendu Biswas, Aparesh Datta, and Nirmal Kumar Manna

Numerical Investigation of Aspect Ratio Effect on Biomimetic Heat Sink Model with Two Inlet-Outlet Pairs, for Cooling Rectangular Shaped Electronic Circuits 115
K. Kandassamy and B. Prabu

Effect of Filling Ratio on Performance of Two Loop Pulsating Heat Pipe 127
Est Dev Patel and Subrata Kumar

Optimization of Process Parameters of Sintered Copper Wick Heat Pipe Using Response Surface Methodology 137
B. Ch Nookaraju, P. S. V. Kurmarao, and S. Nagasarada

A Numerical Study on the Flapping Dynamics of a Heaving Flexible Foil in a Uniform Flow 149
Kuntal Patel, K. Supradeepan, and P. S. Gurugubelli

Effect of Heat Shields on Performance of Rolls Integrated with Internal Heaters and Its Simulation 159
Nalla Shivaprasad and U. S. Jyothi

Influence of Geometric and Operating Parameters on the Heat Transfer Enhancement of Synthetic Jet 171
Aswini Kumar Khuntia, Pandaba Patro, and Sanjoy Ghoshal

A Numerical Investigation of Effect of Shape of Orifice on the Heat Transfer Characteristics of a Synthetic Jet 183
Aswini Kumar Khuntia, Pandaba Patro, and Sanjoy Ghoshal

Atomization Characteristics of a Porous Injector 195
M. Jegan and M. Vadivukkarasan

Numerical Simulation of PCM-Based Heat Sink with Plate Fins for Thermal Management of Electronic Components 207
Anuj Kumar, Rohit Kothari, Pushpanjay K. Singh, M. P. Paulraj, Santosh K. Sahu, and Shailesh I. Kundalwal

Mixed Convective Power-Law Fluid Flow and Heat Transfer Characteristics Past a Semi-circular Cylinder Mounted with a Splitter Plate 219
Souparna Banerjee, Samrat Banik, Chitrak Mondal, Sandip Sarkar, and Nirmal Kumar Manna

Design and Analysis of Cold Box for Helium Liquefier/Refrigerator . . . 227
Kamlesh Kumar, Sandip Pal, and Murugan Sivalingam

Effect of Porosity on the Performance of an GM-Type DIPTR Using CFD 239
Pankaj Kumar, Ajay Kumar Gupta, Sanjay Kumar Gupta, and R. K. Sahoo

Numerical Analysis of Pulse Tube Cryocooler and Optimization Using Taguchi Method 253
Ajay Kumar Gupta, Pankaj Kumar, and R. K. Sahoo

Entropy Generation Minimization of Vapour Absorption Heat Transformer 265
S. Sekar, P. Chandrasekar, S. Kumar, Abraham J. S. Jospher, R. Sheeja, T. N. Valarmathi, and G. M. Lionus Leo

Exergy Studies on a Hybrid Desalination and Cooling Plant 275
Chalasani Chiranjeevi, Yendaluru Raja Sekhar, Muthuswamy Natarajan, Tangellapalli Srinivas, Mehran Hashemian, and Vinod Aditya

Performance Analysis of a Desert Cooler for Different Climatic Conditions in India 285
Prabhat Ranjan Mishra, Aneesh Somwanshi, and Vivek Kumar Gaba

Energy and Environmental Analyses of Active Solar Dryer for Medicinal Herbs Drying 297
D. V. N. Lakshmi and Muthukumar Palanisamy

Modelling and Trade-Off Analysis of Performance Parameters for Counter Flow Packed Bed Liquid Desiccant air Dehumidifier 309
Mrinal Bhowmik, Muthukumar Palanisamy, and R. Anandalakshmi

Design of Solar Photovoltaic/Thermal System (PVT) with Thermal Energy Storage for Air-Conditioning Applications 325
A. Sai Kaushik, Pulla Rao Muvvala, and Satya Sekhar Bhogilla

Measurement of Temperature Distribution Using Liquid Crystal Thermography Technique Over the Absorber Plate of Solar Air Heater 335
Dheeraj Kumar, Amit Kumar, and Apurba Layek

Compressible Flow Through Convergent–Divergent Nozzle 345
Buddha Dev Das, Rajdeep Sardar, Sandip Sarkar, and Nirmal Kumar Manna

A Comparative Study of Fluid Flow Characteristics of Dual Jet Using Different RANS-Based Turbulence Models . . . 355
Sanjay Singh Rathore and Suresh Kant Verma

Fluid Structure Interaction Study of Damper for Swish and Rattle Noise Refinement . . . 367
Swapnil S. Kulkarni, A. Satheesh, B. Ravi, and M. R. Saraf

A Numerical Model to Study the Fluid–Structure Interaction at a Bridge Pier . . . 379
Kanchibhotla Hima Teja, G. Srivalli, and V. Balakrishna Murthy

Numerical Investigation on Performance Enhancement of Rectangular Microchannel by Using Quadrilateral Fins Embedded on the Channel Bed . . . 391
Arindam Santra, Anirban Bose, and Arunabha Chanda

Investigation on Centrifugal Pump Performance for Various Nose-Cap Geometries . . . 401
Neeta Amol Mandhare and K. Karunamurthy

Determination of Coefficient of Contraction of Orifice with Variation of Geometrical Parameter . . . 413
Santosh Kumar Panda and Alok Patra

Solution of Incompressible Navier–Stokes Equation Using Upwind Scheme . . . 423
Banamali Dalai and Manas Kumar Laha

An Efficient Physically Adjusted Central Euler Solver . . . 433
Souren Misra

Experimental Study on Vertical Axis Wind Turbine to Harness Wind Power from Rapidly Moving Railway Locomotives . . . 445
Yendaluru Raja Sekhar, Muthuswamy Natarajan, Chalasani Chiranjeevi, Roy Sukanta, and Patil Yugandhar

Effect of Mach Number on the Rarefied Gas Flow Over a Forward-Facing Step . . . 451
Deepak Nabapure, Arjun Singh, and K. Ram Chandra Murthy

CFD Analysis of Twisted Gas Turbine Blade with Different Cooling Hole Geometries on Leading Edge . . . 461
Moughbul Basha and Mithilesh Kumar Sahu

Validation of Blade Failure of a Kaplan Turbine Under Adverse Conditions Using Numerical Analysis . . . 471
Atul S. Tumane, K. Kumar, Abhijeet Kulkarni, R. A. Kubde, and S. Ajai

Determination of Nusselt Number Over Artificially Roughened Solar Air Heater Using Numerical Approach 483
Amit Kumar, Dheeraj Kumar, and Apurba Layek

Unsteady Wake Dynamics Past a Triangular Cylinder at Incidence with a Downstream Semi-circular Cylinder at Re = 100 493
Akram Hossain Middya, Amartya Samanta, Chitrak Mondal, Sandip Sarkar, and Nirmal Kumar Manna

About the Editors

Dr. Muthukumar Palanisamy is a Professor in the Department of Mechanical Engineering at Indian Institute of Technology, Guwahati, India. He received PhD degree in Mechanical Engineering from IIT Madras during 2005. He received DAAD research fellowships during September–December 2000, June–July 2008, and June–July 2010. He is the recipient of IEI Young Engineer Award - 2010 in Mechanical Engineering from Institute of Engineers (India). He received Bhaskara Advanced Solar Energy Fellowship (BASE Fellowship) from Indo - U.S. Science and Technology Forum (USIEF), Jan 2014 and also received Er. M.P. Baya National Award 2015 by the Institution of Engineers (India), Udaipur. He served as visiting Professor in many Universities in the USA and Germany and also having research collaborations with over 20 International Universities. He has published over 250 research articles in various International Journals and conference proceedings. He is having seven national patents for his credit. He has supervised 15 PhD and 45 M.Tech student's theses. He has successfully completed 5 sponsored research projects worth of 192 lakhs and five consultancy projects worth of 35 lakhs. Currently he is handling seven research projects worth of 760 lakhs. His area of research includes hydrogen energy storage, metal hydride based thermal machines, coupled heat and mass transfer in porous medium, porous medium combustion, sorption heating and cooling systems, etc.

Dr. Velraj Ramalingam is a Professor, Institute for Energy Studies, and specialized in Heat Transfer, Computational Fluid Dynamics and Energy Storage. As a part of his doctoral work he carried out his research in Solar institute, Juelich, Germany, for a period of 20 months (1995–1997) under DAAD Fellowship by Germany. He has 30 years of teaching experience, and held additional responsibilities as Deputy Director, Centre for Affiliation of Institutions, for 6 years from 2004 to 2010; Director, AU – FRG CAD/CAM for 3 years from 2010 to 2013; and Director, Institute for Energy Studies for 5 years from 2013 to 2018. He has to his credit, 33 consultancy projects as principal coordinator, and 15 sponsored research projects as Principal Investigator which includes collaborative projects with Germany and UK. Under his Guidance 40 candidates have completed and 12 scholars are pursuing their Ph.D. research. He has published over 175 research papers in international journals of repute. His H - Index is 41 as per Google scholar citations. He has received Tamil Nadu Scientist Award – 2014 by TANSA in the year 2017, Active consultant award in the year 2011, Best innovation award in 2013 by Anna University, and Achiever Award by Appadurai Chair for contribution to TANGEDCO, Tamil Nadu in the year 2017. He is an expert member in various international and national forums in his field of research, reviewer for several international journals of repute, and member of governing council/academic council/board of studies for various universities and autonomous institutions.

Dr. Murugan Sivalingam is a Professor in the Department of Mechanical Engineering, National Institute of Technology, Rourkela, India. He obtained his master and PhD degree from Anna University, Chennai. His areas of research are in energy/renewable energy, waste management, refrigeration and air conditioning. He has 25 years of teaching experience. He was also a visiting researcher at Cranfield University, the UK. He has carried out his postdoctoral research work at VSB Technical University, Ostrava, The Czech Republic, in 2014–15. He was a Principal Investigator for two major sponsored research projects sponsored by the Government of India, and a sponsored research project funded by the Tamil Nadu State Council for Science and Technology, Tamil Nadu, India. He has published more than 75 research articles in reputed peer-reviewed international journals scoring his h-index 30 in google scholar citations. He has guided 8 PhD research scholars and 25 postgraduate and more than 50 undergraduate engineering students.

Numerical Investigation on Heat Transfer Characteristics of a Triangular Slotted Pin Fin Heat Sink

Rout Jagannath and Yamala Muralikrishna

Nomenclature

a	Size of perforation (mm)
d	Diameter of pin fin (mm)
l_f	Height of fin (mm)
N	Number of fins
N_p	Number of slots
Nu	Nusselt number
p	Pitch of perforations (m)
Re	Reynolds number
T_b	Base temperature of model (K)
Δp	Pressure drop (Pa)
u_i	Inlet velocity of air (m/s)
η	System performance parameter
ρ	Density of air (kg/m^3)

1 Introduction

Thermal execution of perforated pin fin heat sink relies upon different factors, for example, the pitch of aperture and size of puncturing, material properties of the heat

R. Jagannath · Y. Muralikrishna (✉)
Department of Mechanical Engineering, Gayatri Vidya Parishad College of Engineering (Autonomous), Visakhapatnam 530048, A.P., India
e-mail: muralikrishna.yamala@gvpce.ac.in

M. Palanisamy et al. (eds.), *Theoretical, Computational, and Experimental Solutions to Thermo-Fluid Systems*, Lecture Notes in Mechanical Engineering,
https://doi.org/10.1007/978-981-33-4165-4_1

sink. Literature concerning various geometries of heat sinks has been collected, and the details are furnished below. Shaeri et al. [1] examined, numerically, the liquid stream, and conjugate conduction-convective heat transfer from rectangular perforated fins with square windows. The outcomes demonstrate that new perforated fins have higher absolute heat transfer and extensive weight decrease in correlation with solid fins. An experimental and numerical study on finned metal foam and metal foam heat sinks under impinging air jet cooling was carried out by Feng et al. [2]. Fan et al. [3] explored the heat transfer and pressure drop of a novel cylindrical oblique fin heat sink fitted over cylindrical heat sources. Jeng et al. [4] contemplated tentatively heat transfer estimation of the cylindrical heat sink with sintered-metal-bead-layers fins and an implicit motor fan. The study showed that the influence of sintered-bronze-bead layers is negligible in free convection heat transfer. Kim et al. [5] analyzed the thermal performance of optimized plate-fin and pin–fin heat sinks with a vertically situated base plate in natural convection by defining the objective function. Numerical examination on geometric optimization of PCM-based pin fin heat sink is done by Pakrouh et al. [6]. The primary objective of the investigation is to get the designs that augment the heat sink operational time. Singh et al. [7] investigated the heat transfer attributes of an embossed heat sink having rehashed impacts on the fin surface exposed to natural convection. The heat dispersal capacity of the naturally cooled heat sink has been found to be increased by the use of impacts on the fin body. Sajedi et al. [8] broke down numerically the impact of splitter plate on the hydrothermal conduct of pin fin heat sink. To stay away from or debilitate the stream detachment and diminish the pressure drop through the heat sink, a thin plate is situated on the back of the pin. Damook et al. [9] examined utilizing corresponding trial and computational fluid dynamics strategies to find the advantages of utilizing the pin fin heat sinks with different perforations. Li et al. [10] performed numerical examinations on natural convection heat transfer from radial heat sinks with a perforated ring. Shin et al. [11] broke down numerically the qualities of the heat sink with the iconic breeze utilizing wire to parallel plate electrodes utilizing Computational Fluid Dynamics strategy for another cooling gadget of a light-emitting diode. Li et al. [12] examined tentatively the heat transfer of pin fin heat sinks cooled by double piezoelectric fans by infrared thermography. The impacts of the phase difference, the arrangement, the elevation of the piezoelectric fans, and the components of the heat sinks on the thermal performance of the heat sinks are examined. Al-Sallami et al. [13] researched numerically the warm wind currents over strip fin heat sinks utilizing a conjugate heat transfer model; they investigated computationally the advantages of utilizing strip fin heat sinks. Maji et al. [14] explored the heat transfer improvement of heat sink utilizing perforated pin fins with the straight and staggered course of action. Pin fins of different shapes with various perforation geometries to be specific roundabout, diamond-shaped and elliptical sort are considered in their investigation. Their calculations are done by differing inlet speeds, their numerical model is approved with test studies, and great understanding was watched. Ali et al. [15] examined tentatively for the optimization of heat transfer in electronic coordinated circuits utilizing close-packed PCMs filled pin fin heat sinks.

2 Problem Definition and Method of Solution

The present work is proposed to perform numerical heat transfer simulation over a rectangular solid and slotted pin fin heat sink for an inline and staggered fin array. The problem geometry, as shown in Fig. 1, comprises a rectangular base with a fin array protruded from it. Figure 1a–d shows the heat sink with inline slotted fin array, inline solid fin array, staggered slotted fin array, and staggered solid fin array, respectively.

Fins are provided with equilateral triangular slots on their lateral surfaces. A triangular slot offers more surface area than a circular slot for the same cross-section area of the slot. A heat sink is subjected to constant heat flux at its base, and the air is the fluid medium. A heat sink is placed in the fluid stream in such a way that the cross-section of the slot is normal to the direction of flow to ensure that the fluid passes through it.

The equations used for the simulation purpose are the continuity, momentum, and energy equations along with the equations for modeling the quantities of turbulence.

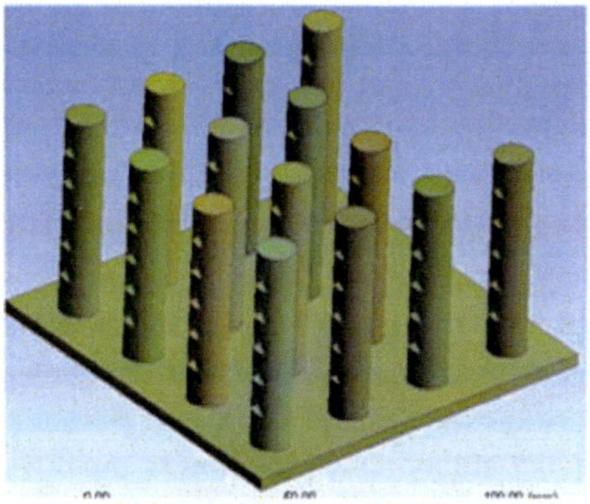

(a) Isometric view of Inline triangular perforated pin fin heat sink

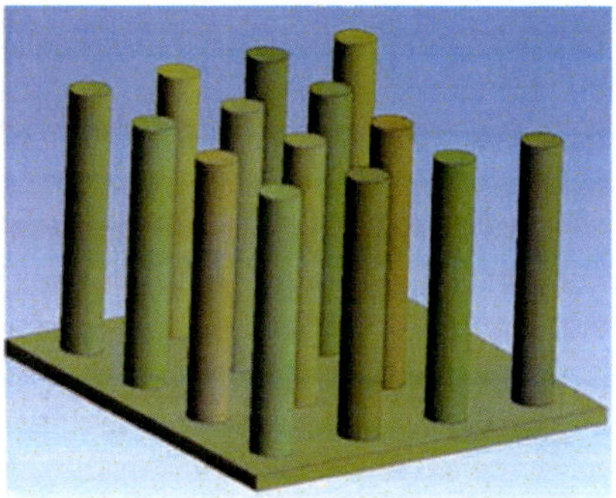

(b) Isometric view of Inline solid pin fin heat sink.

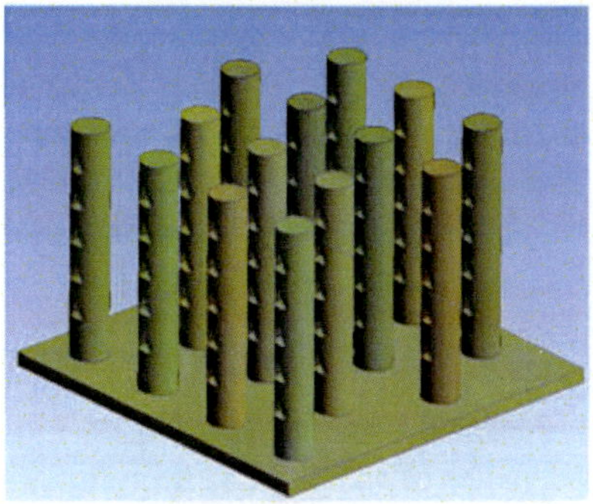

(c) Isometric view of Staggered triangular perforated pin fin heat sink.

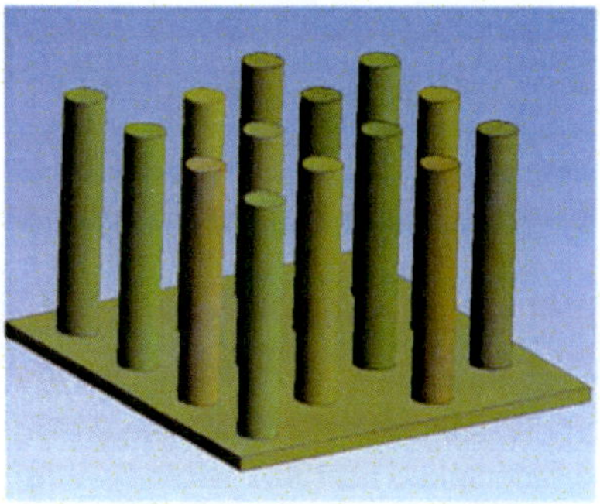

(d) Isometric view of staggered solid pin fin heat sink.

Fig. 1 Rectangular heat sink with slotted pin fin array

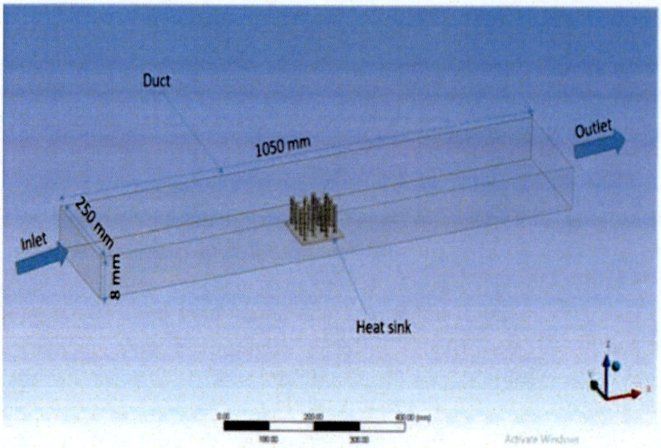

Fig. 2 Computational domain of pin fin heat sink

The assumptions considered in formulating these equations for the current investigations are incompressible flow, no viscous dissipation, steady and buoyancy effects. Standard k- ε model is utilized for the current simulation among a range of turbulence models existing in the ANSYS FLUENT.

Figure 2 shows the computational domain of the present problem, along with the boundary conditions. Airflow passes through the duct inlet, and the heat sink is placed exactly at the mid location of the duct. A heat sink is placed in such a way that the slots are normal to the direction of flow. The base of the heat sink is subjected to constant heat flux, and the remaining walls of the duct are adiabatic and zero heat flux is applied to the four walls of the duct except at the inlet and outlet of the duct. The no-slip wall condition is applied to the four walls of the duct. The boundary condition at the outlet of the duct is applied as pressure outlet conditions.

3 Result and Discussion

3.1 Grid Convergence Test

Grid independence test has been performed on one of the four model models to find the optimum number of elements for the analysis and to reduce the computation time. While doing the grid convergence test, the inlet velocity is taken as 4 m/s, and the applied heat flux is 6000 W/m^2. These values of velocity and heat flux are constant. The test is made by varying the element size by body sizing operation, and results are obtained for the base temperature of the model. Results show that the base temperature of the model decreases with an increase of mesh elements initially and reaches an asymptotic value for 79,014 number of elements. Therefore, the value is zoned to 79,014 mesh elements, and the computational analysis is carried out.

3.2 Validation of Results

In order to validate the results, the work performed by Maji et al. [14] is considered. A. Maji et al. performed the numerical investigation on a circular slotted pin fin model for both inline and staggered fin array. Convection heat transfer coefficient (*h*) is taken as a parameter for validating the results. The results of the present work have a decent agreement with benchmark results with a maximum deviation of 13.09%.

3.3 Local Temperature Profiles

Figures 3, 4, 5 and 6 show the local temperature profiles of all four geometries of the heat sink. The present study is made for a fixed input of heat flux, Reynolds number (Re) as shown in Figure. Figure 6 demonstrates that the temperatures along with the fin decrease from base to fin and is maximum at the base. It is evident to note that the local fin temperatures, at any location on the fin, increases in the downstream of flow. The above is true due to a fresh air stream is in contact with the first row. A similar pattern of temperature profiles has been observed for all four heat sink geometries. Further, the local temperatures of the fin, at any given location, for a model with triangular slotted fins are less than that of a solid pin fin model. This is due to an increased rate of convection with slots in fins. For a comparison between inline fin array and staggered fin array, Figs. 3 and 5 show that, the staggered fin array observes

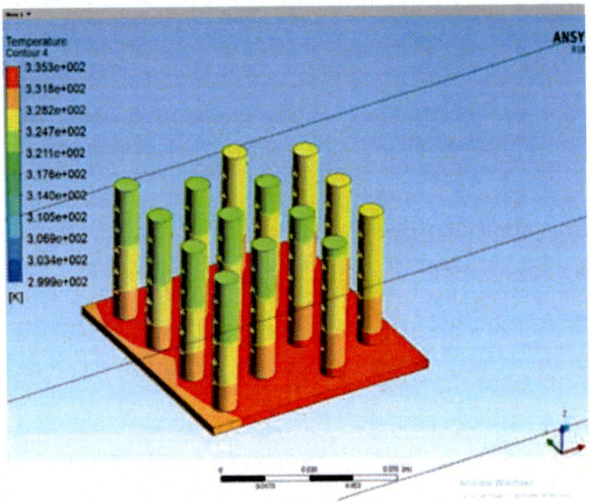

Fig. 3 Temperature distribution of staggered triangular slotted pin fin model

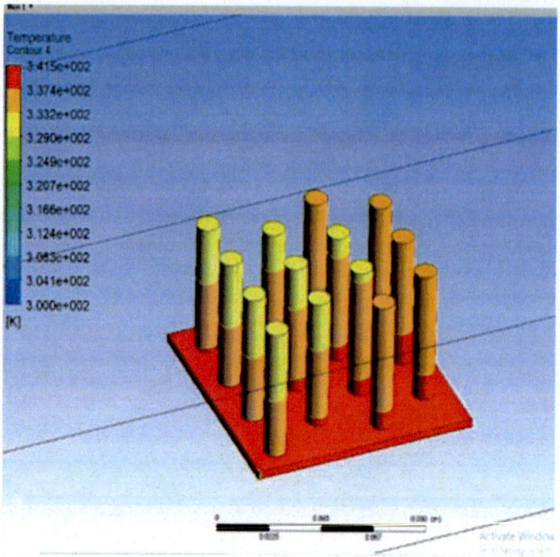

Fig. 4 Temperature distribution of solid staggered pin fin model

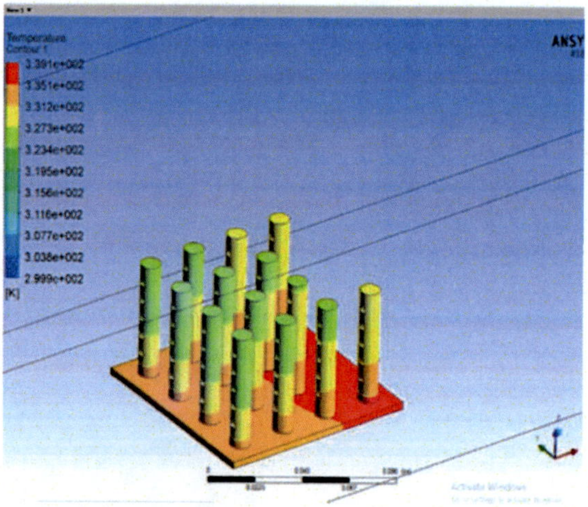

Fig. 5 Temperature distribution of inline triangular slotted pin fin model

low-temperature value than an inline array for the identical location of the fin and on the fin. Because the flow pattern of staggered fin array is more advantageous in comparison to the flow across an inline fin array. The above study concludes that the local heat sink temperatures are less for the triangular slotted pin fin model and are further dropped down by making use of staggered arrangement for fin array.

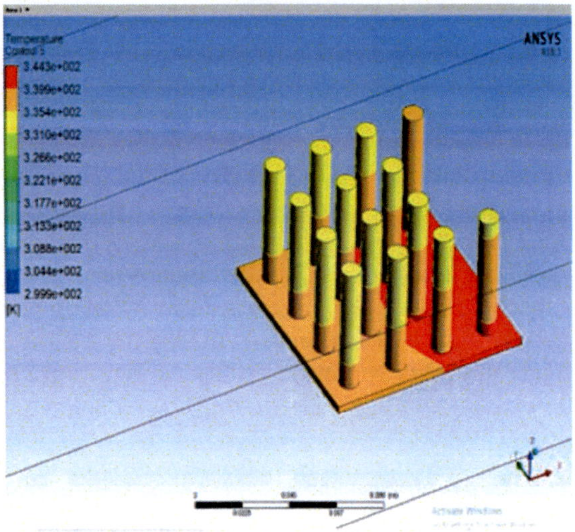

Fig. 6 Temperature distribution of inline solid pin fin model

3.4 Base Temperature of Heat Sink

The following studies are performed to understand the effect of different variables on the base temperature of the heat sink. Investigations have been performed to obtain the results for the base temperature of the heat sink by varying the inlet velocities from 4 to 12 m/s and by keeping the applied heat flux as constant for all the models. Figure 7 demonstrates the variation in the base temperature of the model with the inlet velocity of air for the inline and staggered arrangement for both solid and triangular slotted pin fins. The base temperatures decrease with an increase in velocity due to

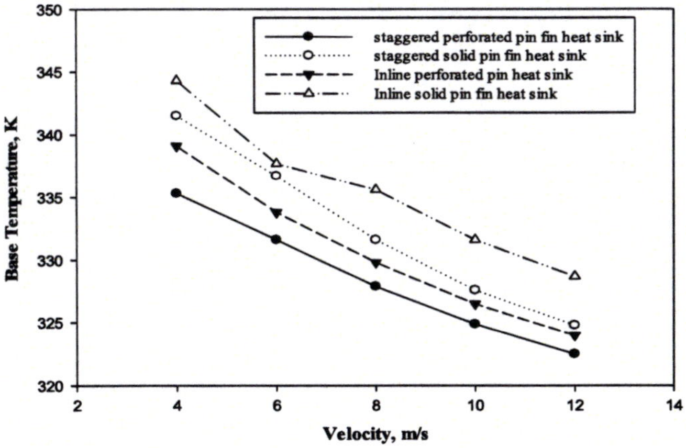

Fig. 7 Base temperature of various models by varying inlet velocity

an increased rate of convective heat dissipation. At all velocities, results show that the base temperature for the triangular slotted pin fin model is lower than that of the solid pin fin model in both the arrangements.

This is due to, an expected, the slots in fins enhances the rate of convective heat dissipation by providing more surface area of contact between the fin and cooling medium. For example, at a flow velocity 6 m/s, for an inline arrangement, the base temperature drops down by 4 °C by providing slots in fins. Further, the base temperature is even tested for two different kinds of fin array, namely (i) inline and (ii) staggered arrangement. Results disclose that the model with staggered fin array experiences less base temperature and is clearly obvious from Fig. 7. This is due to bifurcation of the mainstream at the front leading edge of each pin fin, while, for inline arrangement, most of the fins are in the wake region of preceding cylinders.

3.5 *Nusselt Number*

Nusselt number (Nu) is a non-dimensional convective heat flux that depends on flow properties, fluid properties and flow geometry. The following study has been made to outline the variation of Nusselt number with flow geometry for different flow velocities. Figure 8 describes the variation in Nusselt number of the model with the Reynolds number for an inline and staggered arrangement for both solid and slotted pin–fin models. For any given geometry, the Reynolds number is varied from 27,056 to 81,168 by keeping heat flux constant at 6000 W/m^2. The Nusselt number increases with the increase in Reynolds number due to the increased rate of convective heat transfer. At all Reynolds number, results show that the Nusselt number for the triangular slotted pin fin model is higher than that of the solid pin fin

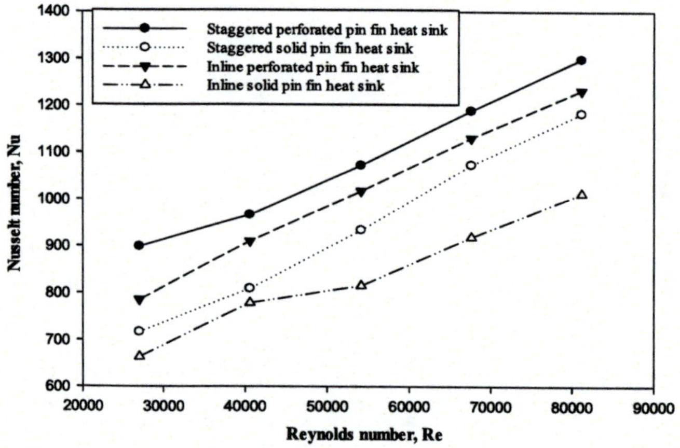

Fig. 8 Nusselt number of various models by varying Reynolds number

model in both the arrangements. Because, the slots provided in the fins, increase the apparent surface area for convective heat dissipation. A 35.38% increase in Nusselt number is observed by replacing an inline solid fin array with a staggered triangular slotted fin array.

3.6 *Pressure Drop Across the Heat Sink*

In fluid dynamics, it is equally important to estimate the pressure drop along with the Nusselt number. Because pressure drop (Δp) helps in determining the pumping power requirement. In most of the ones, the increased convective heat transfer, that is, increased Nusselt number, is always associated with increased pressure drop which demands more pumping power. So, in order to estimate the pressure drop across the heat sink with varying Reynolds number, the results are obtained. Figure 9 demonstrates the variation in pressure drop of the model with the Reynolds number for an inline and staggered arrangement for both solid and triangular slotted pin fin models. The pressure drop increases with an increase in Reynolds number and attains its highest value in the case of solid fins because the solid fins provide more obstruction to fluid flow than triangular slotted pin fins. At all Reynolds number, results show that the pressure drop for the solid pin fin model is higher than that of the triangular slotted pin fin model in both the arrangements. Substantial pressure drops have been observed by making use of a triangular slotted pin fin heat sink. For example, the pressure losses are 66.6% lower with an inline triangular slotted pin–fin arrangement in comparison to that of staggered solid pin fin array.

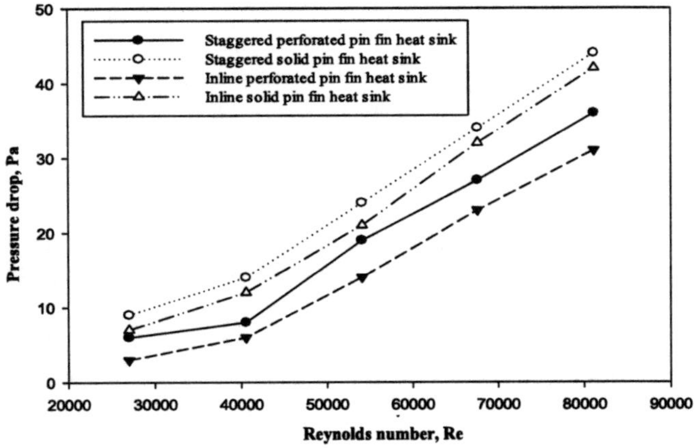

Fig. 9 Pressure drops of various models at different Reynolds number

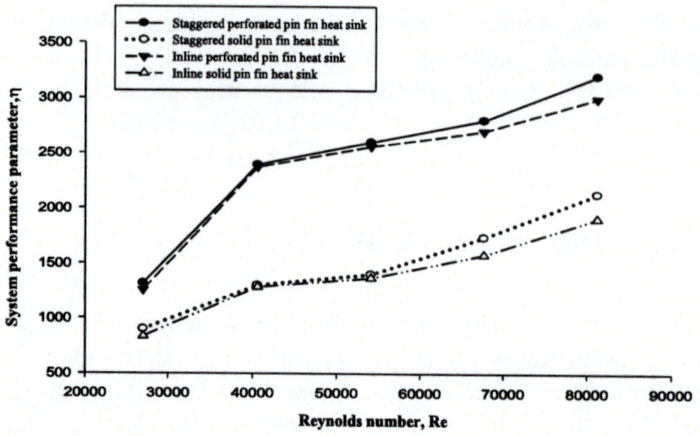

Fig. 10 System performance parameter of various models at different Reynolds number

3.7 Thermal Performance of Heat Sink

The thermal performance parameter (η) of the heat sink is defined as the ratio of the Nusselt number to the skin friction coefficient of the model. It signifies relative dominance of convective heat dissipation over the associated pressure drop, that is, pumping power requirements. Figure 10 depicts the variation in thermal performance parameter with Reynolds number for an inline and staggered arrangement for both solid and triangular slotted pin fin model geometries. The thermal performance parameter increases with an increase in Reynolds number due to an increased rate of thermal dissipation. At all Reynolds number, results show that the system performance parameter for the triangular slotted pin fin model is higher than that of the solid pin fin model in both the arrangements. This is due to that the slots in fins increase convection heat transfer in one way and reduces the pressure losses, and thus the pumping power required on the other way. The thermal performance of the system was enhanced by 57.9% with a staggered triangular slotted pin–fin arrangement in place of an inline solid pin–fin array. From the above results, it is evident that the Nusselt number for staggered triangular slotted pin fin model is best, and the pressure drop for an inline triangular slotted pin fin model is low. So to obtain the overall best configuration, the system performance parameter is used. In addition, the graph shows that the staggered triangular slotted pin fin model is best among all the models.

4 Conclusions

A numerical investigation over a triangular slotted pin fin and solid pin fin model with inline and staggered arrangement by varying inlet velocities has been made.

The results are validated for the convective heat transfer coefficient with benchmark solutions Maji et al. [14] and found to be in decent agreement with a maximum percentage of deviation of 13.09%. The average heat sink temperature is less for the triangular slotted pin fin model in comparison with the solid pin fin model and is further less in staggered arrangement. The maximum temperature of the heat sink, base temperature, is 9 °C less with a staggered triangular slotted pin–fin arrangement in comparison to that of inline solid pin fin array. The Nusselt number of staggered arrangement is more than that of inline arrangement at any given Reynolds number. Further, the triangular slotted pin fin model observes a high Nusselt number in comparison to that of the solid pin fin model. A 35.38% increase in Nusselt number is observed by replacing an inline solid fin array with a staggered triangular slotted fin array. The substantial drop in pressure losses has been observed with a triangular slotted fin. A 66.6% drop in pressure loss is achieved with a triangular slotted pin fin model in comparison to that of a solid pin fin model. The thermal performance parameter is 57.9% higher with a staggered triangular slotted pin–fin arrangement in place of inline solid pin fin array.

References

1. Shaeri MR, Yaghoubi M, Jafarpur K (2009) Heat transfer analysis of lateral equilateral triangular slotted fin models. Appl Energy 86:2019–2029
2. Feng SS, Kuang JJ, Wen T, Lu TJ, Ichimiya K (2014) An experimental and numerical study of finned metal foam models under impinging air jet cooling. Int J Heat Mass Transf 77:1063–1074
3. Fan Y, Lee P, Jin L-W, Chua B (2014) Experimental investigation on heat transfer and pressure drop of a novel cylindrical oblique fin model. Int J Therm Sci 76:1–10
4. Jeng T-M, Tzeng S-C (2014) Heat transfer measurement of the cylindrical model with sintered-metal-bead-layers fins and a built-in motor fan. Int Commun Heat Mass Transf 59:136–142
5. Joo Y, Kim SJ (2015) Comparison of thermal performance between plate-fin and pin-fin models in natural convection. Int J Heat Mass Transfer 83:345–356
6. Pakrouh R, Hosseini MJ, Ranjbar AA, Bahrampoury R (2015) A numerical method for PCM-based pin fin models optimization. Energy Convers Manage 103:542–552
7. Singh P, Patil AK (2015) Experimental investigation of heat transfer enhancement through embossed fin model under natural convection. Exp Thermal Fluid Sci 61:24–33
8. Sajedi R, Osanloo B, Talati F, Taghilou M (2016) Splitter plate application on the circular and square pin fin models. Microelectron Reliab 62:91–101
9. Al-Damook A, Kapur N, Summers JL, Thompson HM (2015) An experimental and computational investigation of thermal air flows through equilateral triangular slotted pin models. Appli Thermal Eng 89:365–376
10. Li B, Jeon S, Byon C (2016) Investigation of natural convection heat transfer around a radial model with a equilateral triangular slotted ring. Int J Heat Mass Transf 97:705–711
11. Shin DH, Baek SH, Ko HS (2016) Development of model with ionic wind for LED cooling. Int J Heat Mass Transf 93:516–528
12. Li H-Y, Wu Y-X (2016) Heat transfer characteristics of pin-fin models cooled by dual piezoelectric fans. Int J Therm Sci 110:26–35
13. Al-Sallami W, Al-Damook A, Thompson HM (2016) A numerical investigation of thermal airflows over strip fin models. Int Commun Heat Mass Transf 75:183–191
14. Maji A, Bhanja D, Patowari P (2017) Numerical investigation on heat transfer enhancement of model using equilateral triangular slotted pin fins with inline and staggered arrangement. Appl Therm Eng 125:596–616

15. Ali HM, Ashraf MJ, Giovannelli A, Irfan M, Irshad TB, Hamid HM, Hassan F, Arshad A (2018) Thermal management of electronics: an experimental analysis of triangular, rectangular and circular pin-fin models for various PCM's. Int J Heat Mass Transf 123:272–284

Heat Transfer Analysis of an Equilateral Triangular Slotted Pin Fin: An Experimental Work

Chappa Bhaskara Rao and Yamala Muralikrishna

Nomenclature

b	Side of an equilateral triangular slot (mm)
d	Diameter of the fin (mm)
h	Convective heat transfer coefficient (W/m^2 K)
l	Length of fin (mm)
$\dot{m}$	Mass flow rate of air (kg/min)
N	Number of slots
P	Power input (W)
Q	Amount of heat transfer (W)
q	Heat flux (W/ m^2)
T	Temperature (K)
T_{avg}	Average fin temperature (K)
T_{mf}	Mean film temperature (K)
T_{amb}	Ambient temperature (K)
T_{base}	Base temperature of fin (K)
T_x	Local Fin temperature (K)
x	Axial distance along the length of the fin (mm)
ε	Effectiveness
η	Efficiency

C. B. Rao · Y. Muralikrishna (✉)
Department of Mechanical Engineering, Gayatri Vidya Parishad College of Engineering (Autonomous), Visakhapatnam, Andhra Pradesh 48, India
e-mail: muralikrishna.yamala@gvpce.ac.in

M. Palanisamy et al. (eds.), *Theoretical, Computational, and Experimental Solutions to Thermo-Fluid Systems*, Lecture Notes in Mechanical Engineering,
https://doi.org/10.1007/978-981-33-4165-4_2

1 Introduction

Now a day's electronic components and automobiles are the major parameters in daily life. Heat generation and removal rate is the major problem faced by these devices. For the best performance of these devices, heat should be removed and maintain optimum temperature levels. So for this best way is mount fins and heat sinks to the heat-generating components. Based on this, so many studies are done to improve heat transfer characteristics. Naphon et al. [1] studied experimentally heat transfer characteristics of the inline and staggered taper pin fin heat sink. Reynolds numbers in 1000–9000 range and heat flux in 0.91–3.64 kW/m^2. It was found that the heat transfer rate increased by staggered configuration than the inline configuration. Yu et al. [2] conduct an experiment on the radial heat sink, with a horizontal base and rectangular fins under natural convection condition. The effect of the number of fins, fin length, height, and heat flux on thermal resistance and heat transfer coefficient and found optimal values for better performance. Pankaj et al. [3] examine the heat transfer characteristics of the embossed heat sink, having repeated impressions on the fin surface subjected to natural convection. The heat transfer characteristics of fin expressed in terms of Nusselt number improvement and fin effectiveness. Park et al. [4], by examining the heat transfer characteristics of the radial heat sink having a hollow cylinder, under natural convection condition, observed that the thermal performance of heat sink was improved with the hollow cylinder. Shen et al. [5] performed numerical simulation on natural convection heat transfer from vertical cylinder heat sinks with longitudinal fins and found that the average Nusselt number increased with the increase in the length of the fin. Numerical investigation on natural convection heat transfer around a radial heat sink with a slotted ring has been performed by Li et al. [6]. The diameter of the slot, number of slots, length of the slot, and orientation are examined. It was observed that a radial heat sink with a slotted ring has better thermal performance than a solid one and is further enhanced with upward orientation. Baik et al. [7] studied the heat transfer characteristics for chimney type radial heat sink under natural convection and observed that thermal performance of chimney-based heat sink is increased by 20% compared with radial heat sink without a chimney. Numerically and experimentally, an investigation on the thermal performance of the slotted pin fin heat sink under forced convection was performed by Tijani et al. [8]. Heat sink improves 1–4% of thermal efficiency to solid pin fin heat sink. Singh et al. [9] studied the heat transfer characteristics of slotted fin array and observed that the fin of slots with constant pitch and 4 mm diameter with 45° geometries dissipates more heat. The array of this fin with 10 mm spacing is best for a horizontal rectangular fin array. Feng et al. [10] conduct an experiment on a cross-fin heat sink consisting of a series of long fins and a series of perpendicularly arranged short fins under natural convection. It was demonstrated that fluid flow characteristics are improved with this cross fin heat sink. Compared to the reference plate-fin heat sink, the cross-fin heat sink enhanced the overall convective heat transfer coefficients by 11–15%. Menget al. [11] investigated the heat dissipation performance of a heat sink with the mounting angle under natural convection condition. It is found that the heat sink achieves the highest cooling power when its mounting angle is 900 and is 6.88%, lower than that of 900 for a 15° mounting angle. Chingulpitak et al. [12] investigated the thermal performance of lateral circular perforated plate-fin heat sinks (LAP-PFHSs). It is concluded that for all conditions, the LAP-PFHS with a diameter of the slot, number of perforations 3 mm, and 75 exhibits the highest heat transfer rate, about 11.6% higher than that of the SFHS.

It is obvious from the thorough literature review that most of the investigations pertaining to heat sinks and fins are numerically performed. The available experimental investigations for heat transfer over slotted pin fins are fewer and are inadequate. By keeping the above facts in mind, the present work deals with experimental heat transfer analysis from an equilateral triangular slotted pin fin with zero, five, and seven slots. The above analysis has been carried out at different mass flow rates of air.

2 Problem Description and Experimental Setup

Present work deals with experimental analysis of heat transfer from an equilateral triangular slotted pin fin for various heat inputs and at different mass flow rates of air. The problem geometry consists of an aluminum pin fin of length $l = 180$ mm and diameter $d = 12.5$ mm. Three different pin fin geometries are considered for the study and are pin fin with zero, five, and seven equilateral triangular slots of side $b = 6$ mm, as shown in Fig. 1. Fin is subjected to a constant heat flux at its base. Air is the cooling. The test is performed for two different heat inputs at various mass flow rates of air. The results are obtained to test the thermal performance of the slotted pin fin under various flow conditions and to compare with the solid fin.

The experimental setup as shown in Fig. 2a comprises a suction duct of size 1050 mm length, 150 mm wide, and 100 mm height connected to a centrifugal blower for air passage. In addition, it consists of a heater section, delivery section, and test panel. The delivery section is fitted with an orifice plate, to measure the mass flow rate of air, and a control valve, to regulate the airflow rates. The test panel consists of a voltmeter, ammeter, multi-channel temperature indicator, and dimmer stat. The heater section comprises a nichrome film heater wounded around a hollow cylinder made up of brass. The input to the heater can be regulated by making use

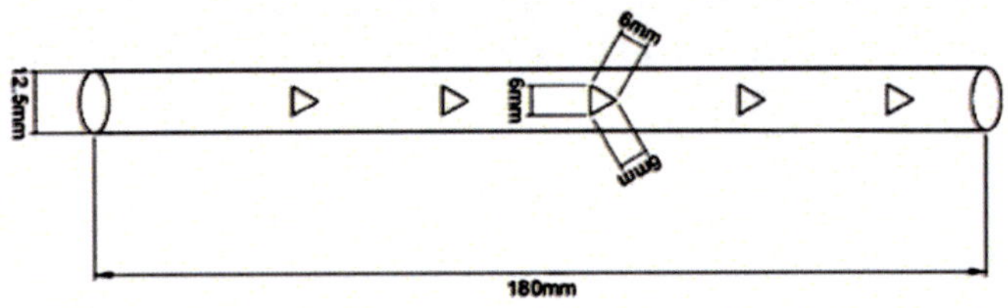

Fig. 1 Schematic of the slotted pin fin

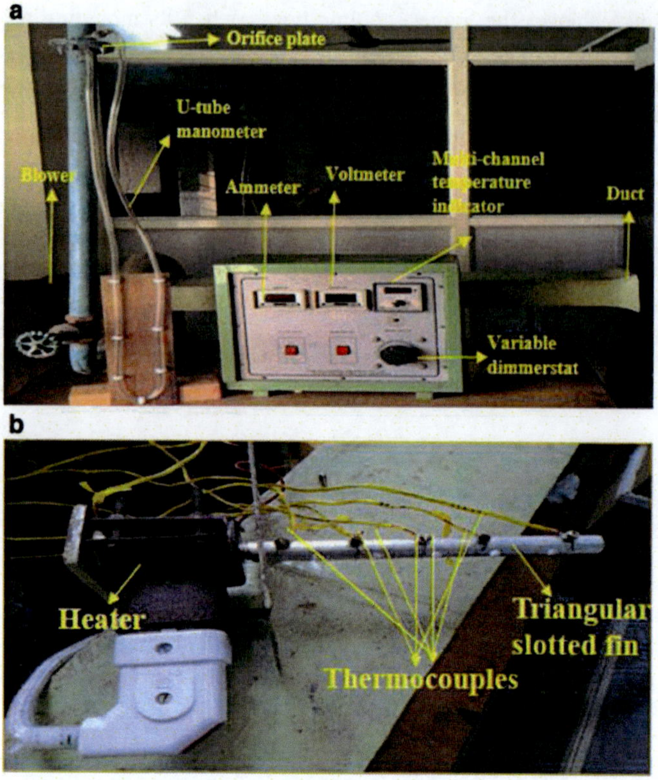

Fig. 2 **a** Experimental setup, **b** Heater and fin section with thermocouples

of dimmer stat. Fin base is inserted in the hollow portion of the heater section. So, that the heat generated in the film heater first conducts from the outer surface of the heater section to its inner surface and subsequently conducts along the fin attached to it. The heater-pin fin assembly is placed horizontally in the suction duct in such a way that the cross-section of the slots is in the normal direction to the airflow.

Five thermocouples are attached at equal distances along the fin to measure its temperature, and one more thermocouple is placed in the air stream to measure its temperature. The arrangement is shown in Fig. 2b. Figure 3 shows the different fin geometries used in the present study. Figure 3a–c are the three geometries provided with equilateral triangular slots with the side (b) 6 mm with a number of slots from (N) 0, 5, and 7, respectively.

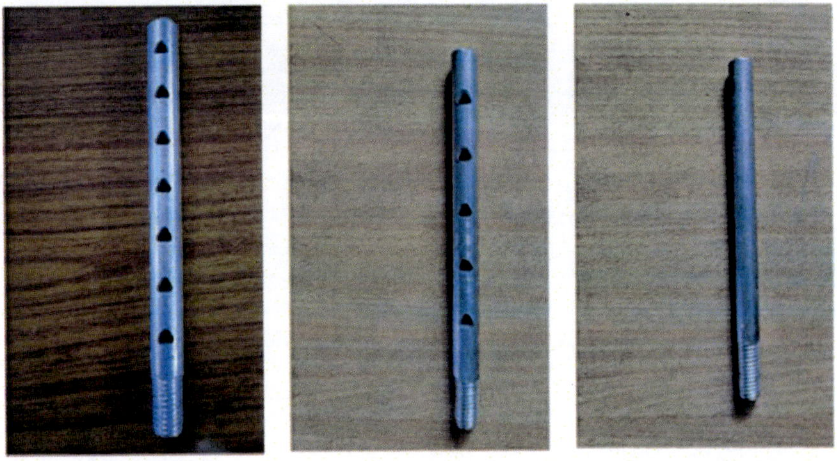

(a) 7 slots (b) 5 slots (c) 0 slots (solid)

Fig. 3 Equilateral triangular slotted pin fins

3 Result and Discussion

The current work deals with heat transfer analysis from an equilateral triangular slotted pin fin by conducting experiments on three different fin geometries at different mass flow rates of air and for different heat inputs. An experimental heat transfer analysis is performed on an equilateral triangular slotted pin fin with 0, 5, and 7 slots (N). A zero number of slots represent solid fin. Length (l) and diameter (d) of the fin are 180 mm and 12.5 mm. The side of an equilateral triangular slot (b) is 6 mm. The thermal conductivity of the fin material (k) is 200 W/mk. The mass flow rate of the air ($\dot{m}$) is varied from 0.285 kg/min to 0.513 kg/min. Results are obtained to establish the local fin temperature distribution, rate of heat flux (q), fin effectiveness (ϵ), fin efficiency (η), and weight reduction factor.

3.1 Local Fin Temperature Profiles

Figure 4a–c describes the local fin temperature profiles for a fin with 0, 5, and 7 slots for 0.513 kg/min, 0.457 kg/min, and 0.285 kg/min mass flow rate of air, respectively. The temperature decreases exponentially along the fin for all three geometries at all mass flow rates of air. It is observed from the above Figure that the local temperatures, at any location along the fin, decreases with an increasing number of slots. Because the fin with slots offers more area for heat transfer that increases convective heat dissipation.

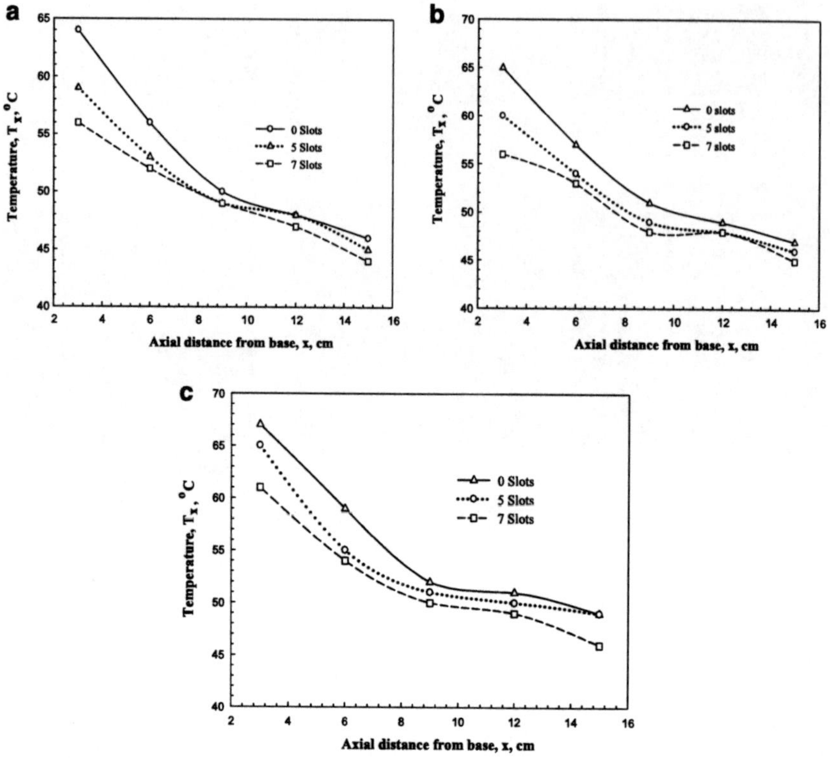

Fig. 4 **a** Local temperature distribution at 0.513 kg/min mass flow rate of air, **b**. Local temperature distribution at 0.415 kg/min mass flow rate of air, **c** Local temperature distribution at 0.285 kg/min mass flow rate of air

Due to the above fact, the thermal performance of a heat sink with slotted pin fins is better than that of a solid pin fin heat sink. The above said is true for all mass flow rates of air considered. Further, the impact of slots on local temperature values is more at the base and is mere at the end of the fin. A similar trend has been observed at different mass flow rates. For example, with the increase in slots from 0 to 7, the local temperature at the base is dropped down by 12.5%, 13.8%, and 8.9% at mass flow rates 0.285, 0.457, and 0.514 kg/min, respectively.

3.2 *Variation in Heat Dissipation Rate with the Number of Slots at Different Mass Flow Rates of Air*

It is known that a fin with slots increases the rate of heat dissipation by offering a more effective area. In order to estimate the increased rate of heat dissipation with

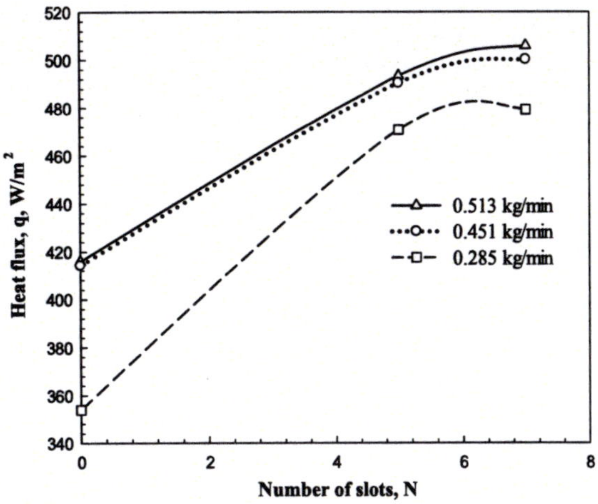

Fig. 5 Variation of heat flux at different mass flow rates and number of slots

increasing no of slots, the experiment has been conducted at three different mass flow rates, 0.285, 0.451, and 0.513 kg/min of air. Figure 5 shows an exponential enhancement in heat dissipation for a change in slots from 0 to 5, and the change is mere with further increase in slots to 7. A similar trend has been observed at all mass flow rates of air. As expected, for a given number of slots, the rate of heat dissipation is more for higher mass flow rates. For example, a 35% enhancement in heat in heat dissipation is obtained by making use of a fin with seven slots in place of a solid fin at 0.285 kg/min mass flow rate of air. The change in the mass flow rate of air from 0.285 to 0.513 kg/min increases the heat flux by 17% from a solid fin.

3.3 *Effectiveness of Equilateral Triangular Slotted Fin for Various Mass Flow Rates of Air*

Effectiveness is the performance parameter of a fin that defines the ratio of rate heat transfer with a fin to that of without a fin. Experiments have been carried out to elucidate the effect of the number of slots on fin effectiveness. Figure 6 demonstrates the results pertaining to the effectiveness of fin with 0, 5, and 7 slots for a range of mass flow rates of air from 0.285 to 0.513 kg/min. As discussed in the previous section, the increased rate of heat dissipation with the number of slots results in increased effectiveness. In addition, the change in mass flow rates of air from 0.285 to 0.513 kg/min improves the fin effectiveness for any number of slots. The above is true due to the increasing rate of heat dissipation with increased mass flow rates. The fin effectiveness is increased by 62%, at 0.513 kg/min mass flow rate of air, with the change in slots from 0 to 7.

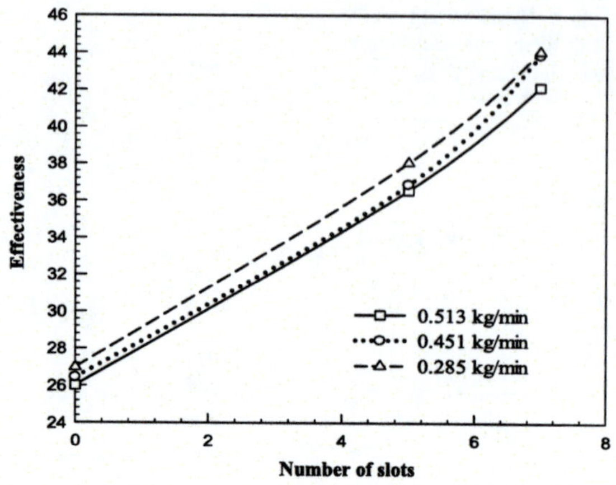

Fig. 6 Variation in fin effectiveness with different number of slots and mass flow rates of air

3.4 Weight Reduction Factor of Fin at Different Mass Flow Rates Air

Providing slots in a fin not only increase the effective surface area for convective heat dissipation but also decrease the weight of the fin. So for a given rate of heat dissipation, the weight of the slotted pin fin is less than that of the solid pin fin. Higher the numbers of slots lower weight of fin. The weight reduction factor signifies the percentage of reduction in weight by replacing a solid pin fin (0 slots) with a slotted pin fin for the same rate of heat dissipation. A test is conducted to evaluate the weight reduction factor for different mass flow rates of air. The above studies performed on the fin with 0, 5, and 7 slots with a change in mass flow rates of air from 0.285 to 0.513 kg/min. Figure 7 shows the concerning results of the above study. The weight of the fin required to dissipate the given rate of heat transfer at 0.285 kg/min with 0 slots (solid) is taken as the reference for comparison. Figure 5 demonstrates, for a given mass flow rate of air. The weight of the fin decreases with an increasing number

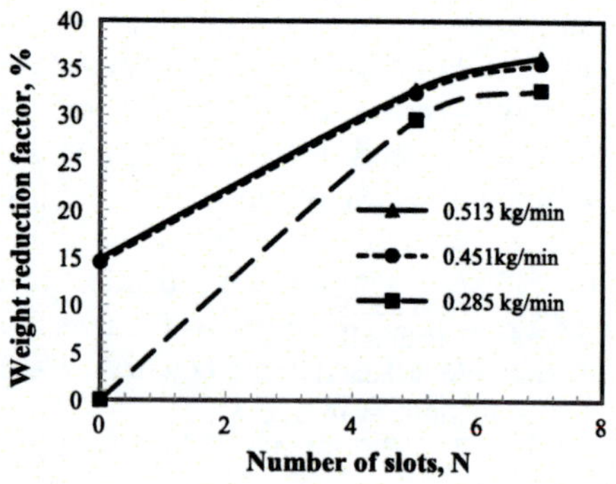

Fig. 7 Percentage of weight reduction for different mass flow rates and the number of slots

of slots. Further, for a given fin, the weight of the fin reduces with increasing mass flow rate of air. This is due to increased convective heat dissipation by providing more number of slots with a high mass flow rate of air. For a given mass flow rate of 0.285 kg/min. The weight of the fin is reduced by 32.70% by making use of a fin with seven slots in place of a solid fin. Also, 10.7% weight reduction is found in a seven slotted pin fin for a change in the mass flow rate of air from 0.285 to 0.513 kg/min. So, for a given mandated heat load, the weight of the heat sink is reduced drastically by making use of slotted pin fins instead of solid pin fin.

4 Conclusions

Experimental analysis on heat transfer from a pin fin with zero (solid), five, and seven equilateral triangular slots has been made. The results are obtained for three different mass flow rates ranging from 0.285 to 0.513 kg/min. The thermal performance of the fin is tested for local temperature distribution, rate of heat flux, effectiveness, and weight reduction. Further, comparative studies have been performed between slotted and solid pin fin at two different heat inputs. The local fin temperature of slotted fins is less in comparison with the solid fin. For example, at the base, for 0.513 kg/min, the temperature of the fin is reduced by 12.5% by replacing a zero slotted fin with seven slots. Heat flux from fin is increasing with an increase in the number of slots and mass flow rate of air. The rate of heat flux from a solid pin fin at 0.285 kg/min is 354 W/m^2 and is 479.28 W/m^2 for a fin with seven slots at the same mass flow rate of air. A 35% of increase in the heat flux by replacing a zero slotted fin with seven slots for 0.285 kg/min of the mass flow rate of air. Heat dissipated per unit surface area from the fin increases with an increase in the mass flow rate of air and number of slots owing to increased convection heat transfer coefficient and surface area for convection. The effectiveness of fin increases to 62% with the change in slots from zero to seven due to increased heat dissipation. A 36.23% weight reduction has been observed for a fin with seven slots in comparison to that of a solid fin for the same rate of heat dissipation.

References

1. Naphon P, Sookkasem A (2007) Investigation on heat transfer characteristics of tapered cylinder pin fin heat sinks. Energy Convers Manage 48:2671–2679
2. Yu S-H, Lee K-S, Yook S-J (2010) Natural convection around a radial heat sink. Int J Heat Mass Transf 53:2935–2938
3. Singh P, Patil AK (2015) Experimental investigation of heat transfer enhancement through embossed fin heat sink under natural convection. Exp Thermal Fluid Sci 61:24–33
4. Park S-J, Jang D, Lee K-S (2015) Thermal performance improvement of a radial heat sink with a hollow cylinder for LED down light applications. Int J Heat Mass Transf 89:1184–1189
5. Shen Q, Sun D, Xu Y, Jin T, Zhao X, Zhang N, Wu K, Huang Z (2016) Natural convection heat transfer along vertical cylinder heat sinks with longitudinal fins. Int J Thermal Sci 100:457–464

6. Li B, Jeon S, Byon C (2016) Investigation of natural convection heat transfer around a radial heat sink with a perforated ring. Int J Heat Mass Transf 97:705–711
7. Li B, Baik Y-J, Byon C (2016) Enhanced natural convection heat transfer of a chimney-based radial heat sink. Energy Convers Manage 108:422–428
8. Tijani AS, Jaffri NB (2018) Thermal analysis of perforated pin-fins heat sink under forced convection condition. Proc Manuf 24:290–298
9. Pankaj S, Santhosh B, Kulkarni JS (2018) Experimental investigation of heat transfer by natural convection with perforated pin fin array. Proc Manuf 20:311–317
10. Feng S, Shi M, Yan H, Sun S, Li F, Lu TJ (2018) Natural convection in a cross-fin heat sink. Appl Thermal Eng 132:30–37
11. Meng X, Zhu J, Wei X, Yan Y (2018) Natural convection heat transfer of a straight-fin heat sink. Int J Heat Mass Transf 123:561–568
12. Chingulpitak S, Ahn HS, Asirvatham LG, Wongwises S (2019) Fluid flow and heat transfer characteristics of heat sinks with laterally perforated plate fins. Int J Heat Mass Transf 138:293–303

Thermo-Fluid Performance Evaluation of an Elliptical Tube Type Fin-and-Tube Heat Exchanger Supported with Winglets

S. K. Sarangi and D. P. Mishra

1 Introduction

Fin-and-tube heat exchangers are in general used for the exchange of heat between a liquid and a gas (or air). The total thermal resistance of the heat exchanger comprises of airside thermal resistance, tube side convective resistance, and fin side conductive thermal resistance. Due to the very low heat transfer coefficient of the air, the airside convective thermal resistance in these heat exchangers is significantly higher. Also, in these heat exchangers, the wake region behind the tubes is larger where interaction between the hot and cold fluid is minimum. Thus these thermal and geometrical limits have encouraged researchers to opt for heat transfer enhancement techniques.

Vortex generation using wing and winglets is a passive heat transfer enhancement technique where these vortex generators are placed near to the surface of the tubes to guide the flow properly towards the wake region behind the tubes. Chen et al. [1] examined two to four staggered winglet rows of delta winglet type vortex generators. Winglets with staggered array were found to provide better heat transfer effect than the inline placed delta winglets. Biswas et al. [2] performed experiments with delta winglet vortex generator and found that compared to the plane fin-tube model, the winglet supported model can enhance heat transfer significantly in the tubes wake region. Torii et al. [3] proposed a "common flow up" orientation for the arrangement of winglets and compared the results with the "common flow down" orientation of delta winglets and found that the former provides greater heat transfer enhancement with a relatively lower pressure drop penalty [4]. Joardar and Jacobi [5, 6] performed experiments and numerical simulations by considering a 7-row fin-and-tube heat exchanger supported with delta winglets, and they confirmed that

S. K. Sarangi (✉) · D. P. Mishra
Birla Institute of Technology Mesra, Ranchi 835215, India
e-mail: shaileshjitu@gmail.com

M. Palanisamy et al. (eds.), *Theoretical, Computational, and Experimental Solutions to Thermo-Fluid Systems*, Lecture Notes in Mechanical Engineering,
https://doi.org/10.1007/978-981-33-4165-4_3

"common flow up" oriented winglet driven accelerated flow causes flow impingement and boundary layer modification. He et al. [7] performed numerical simulations to assess the performance of a rectangular winglet supported fin-and-tube heat exchanger. They found that for a seven-row tube geometry, the staggered winglet array provides heat transfer enhancement as good as the inline winglet array with further reduction in pressure drop penalty. Hu et al. [8] performed numerical simulations to explore the relationship between secondary flow intensity and the heat transfer characteristics of the heat transfer surface. Zhou and Feng [9] conducted experiments to analyze the performance of the punched hole type curved and plain winglets. They found that in both laminar and turbulent regions, the curved winglet geometries provided greater enhancement in heat transfer. Li et al. [10] proposed a symmetrical winglet configuration for heat transfer enhancement in a fin-and-tube heat transfer surface. They found that their proposed configuration for five-tube rows performs at par with six tube type wavy fin structure.

In recent years a lot of work has been done to find the optimum winglet location in fin-and-tube heat exchangers. Arora et al. [11] performed numerical simulations to optimize the winglet location in the fin-and-tube heat transfer surface with three inline rows of tubes. They considered the "common flow up" orientation of winglets. Sarangi and Mishra [12] opted a similar winglet orientation and investigated the winglet number, attack angle, and winglet location for best heat transfer performance. They found that placement of winglet pair over the central tube provides greater heat transfer enhancement with a relatively lower pressure drop. Similar numerical investigation but with "common flow down" winglet orientation was presented by Naik and Tiwari [13]. Sarangi et al. [14, 15] extended their work to five rows of tubes with plain and wavy rectangular winglets and presented optimum values of design parameters.

In the existing works of literature of fin-and-tube heat exchangers, most of the investigations have been performed by considering the circular tube and some with elliptical tube geometry. However, to the author's best knowledge, the two tube geometries have not been studied together. Also, the optimum winglet location for elliptical tube geometry has not been investigated. In this paper, we have compared the circular and elliptical tube geometries, and their combined effect on heat transfer and pressure drop performance have been presented, respectively, in terms of Nu and f. An enhancement factor η has been presented to summarize the combined effect of Nu and f.

2 Mathematical Formulation

2.1 Computational Domain

Figure 1a shows the top view of the fin-and-tube heat transfer surface. Due to the symmetrical arrangement, the region within the dashed lines as shown in Fig. 1a–c has been selected as the computational domain.

Rectangular winglet pairs with length equal to the tube diameter and height equal to fin pitch have been selected as the heat transfer enhancement tool. The winglets are arranged in a "common flow up" orientation to minimize the wake region and delay the flow separation. The winglet and fin thickness is the same and is equal to 0.18 mm.

For all the simulations in the present study, a streamwise and spanwise distance of 0 and 8.6 mm has been considered. For comparative analysis, the length of the elliptical tube semi-major axis (6.52 mm) and semi-minor axis (4 mm) has been selected in such a way that its perimeter is equal to the perimeter of the circular tube. All the solid bodies of the computational domain are assumed to be made up of aluminum, and air has been selected as the fluid. Air with an inlet temperature of 310.6 K flows across the inline arranged three rows of tubes having tube wall temperature of 291.77 K. The tube wall temperature has been assumed constant owing to the high heat transfer coefficient of fluid flowing inside the tubes and the high thermal conductivity of aluminum.

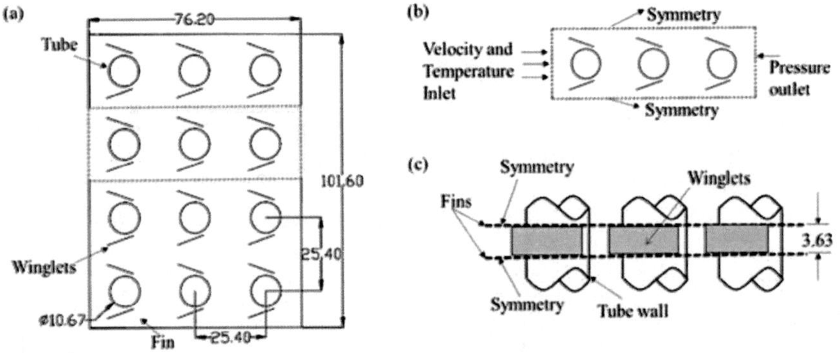

Fig. 1 **a** Side view of the fin-and-tube heat transfer surface supported with winglets, **b** side view of computational domain with imposed boundary conditions, and **c** top view of computational domain with extended tubes and imposed boundary conditions (all dimensions are in mm)

2.2 Governing Equation

For a laminar, steady, and incompressible flow following governing equations are required to be solved:

Continuity equation:

$$\frac{\partial}{\partial x_i}(\rho u_i) = 0 \tag{1}$$

Momentum equation:

$$\frac{D}{Dt}(\rho u_i) = \frac{\partial}{\partial x_i}\left(\mu \frac{\partial u_k}{\partial x_i}\right) - \frac{\partial p}{\partial x_i} \tag{2}$$

μ is the dynamic viscosity and is assumed constant in the operating temperature range.

Energy equation:

$$\frac{D}{Dt}(\rho T) = \frac{\partial}{\partial x_i}\left(\frac{k_a}{C_p}\frac{\partial T}{\partial x_i}\right) \tag{3}$$

2.3 Boundary Conditions and Parameter Definitions

The boundary conditions imposed on the computational domain are shown in Fig. 1b, c. Uniform velocity and temperature are applied at the inlet. Pressure outlet boundary condition is imposed at the outlet boundary, and the top, bottom, and side boundaries are imposed with symmetry boundary conditions.

Parameter Definitions.

Following heat transfer and fluid flow parameters have been computed to access the performance of the fin-and-tube heat transfer surface [14, 15]:

Total heat transfer

$$Q = \dot{m}\, C_P(\overline{T}_o - \overline{T}_{\text{in}}) \tag{4}$$

LMTD,

$$\Delta T = \frac{(T_w - \overline{T}_{\text{in}}) - (T_w - \overline{T}_o)}{\ln[(T_w - \overline{T}_{\text{in}})/(T_w - \overline{T}_o)]} \tag{5}$$

Hydraulic diameter,

$$D_h = 4\left(\frac{A_{\min}.L}{A_T}\right) \tag{6}$$

Reynolds number,

$$\mathrm{Re} = \frac{\rho u_{\mathrm{in}}}{\mu} D_h \tag{7}$$

Nusselt number

$$\mathrm{Nu} = \frac{h D_h}{k_a} \tag{8}$$

Heat transfer coefficient,

$$h = \frac{Q}{A_T.\Delta T} \tag{9}$$

Friction factor,

$$f = \frac{\Delta p}{\frac{\rho u_{\mathrm{in}}^2}{2} \frac{A_T}{A_{\min}}} \tag{10}$$

Enhancement factor [14, 15],

$$\eta = \left(\frac{\mathrm{Nu}}{\mathrm{Nu}_c}\right)\left(\frac{f}{f_c}\right)^{-1/3} \tag{11}$$

where, the subscript c stands for plain circular tube type heat transfer surface.

2.4 Numerical Method

Computational fluid dynamics code Fluent-18 based on the finite volume method has been used for the simulation process. The entire computational domain is meshed with structured hexahedral meshing elements. SIMPLE algorithm is used to ensure pressure–velocity coupling. For the discretization of convective terms appearing in the momentum equation, a first-order upwind scheme is used. The second-order upwind scheme changes the results only within 0.5%. The under relaxation factors for pressure, momentum, and energy were set, respectively, as 0.3, 0.7, and 0.9. The solution was said to be converged when the residuals for energy equation fall below 10^{-6} and that for continuity and momentum falls below 10^{-3}.

Fig. 2 Validation of the present CFD results with existing experimental results

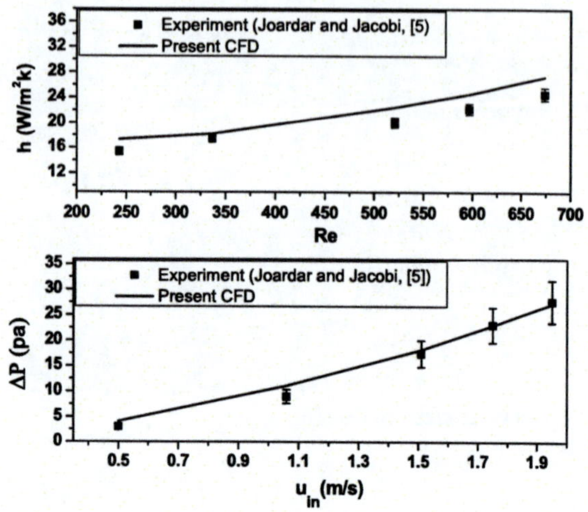

3 Results and Discussions

3.1 Grid Independency Test and Validation of Numerical Model

To achieve a balance between the computational space and accuracy, a grid density test has been performed for the fin-and-tube heat transfer surface (without winglet). At first, a coarse grid system with the corresponding cell number of about 600,000 has been adopted, and then the grid system is subsequently refined to a finer grid corresponding to the cell number of about 1,200,000. However, the solution becomes grid-independent beyond the grid size of 1,072,000. Thus for the present study, the grid system corresponding to a cell number of 1,072,000 has been adopted for grid generation.

The present numerical model and method are compared with the existing experimental results of Joardar and Jacobi [5] in Fig. 2. The variation of present CFD results are in close agreement with the experimental results [5], and the deviation lies within 0.3–10%. Thus the numerical model and adopted method are reliable for predicting the performance of the fin-and-tube heat transfer surface.

3.2 Performance Analysis of Plain Fin-and-Tube Heat Transfer Surface

The heat transfer performance of the fin-and-tube heat transfer surface is primarily decided by the exposed surface area of the tubes. The circular tubes offer a larger

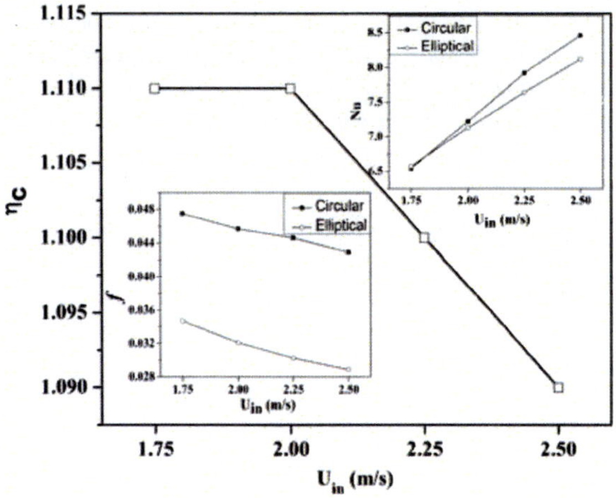

Fig. 3 Variation of Nusselt number, friction factor, and enhancement factor as a function of the frontal velocity of air for circular and elliptical tube geometries

frontal area to the incoming air; however, its wake region contains a thermally isolated zone where the mixing is poor. To overcome this situation, an elliptical tube geometry has been proposed in the present study, and its thermo-fluid performance has been compared to that of the circular tube geometry. The elliptical tube geometry has been chosen such that its surface area remains the same as that of the circular tube.

Figure 3 shows the variation of Nusselt number [Eq. (8)] as a function of frontal velocity. It can be seen from the plot that the elliptical tube geometry offers a relatively lower Nusselt number [Eq. (8)] compared to the circular tube geometry over the entire range of frontal velocity considered. The circular tube having a larger frontal area offers a greater surface for heat transfer compared to the elliptical tube geometry and thus provides greater heat transfer performance. Figure 3 also shows the variation of friction factor [Eq. (10)] as a function of frontal velocity. It can be seen from the plot the friction factor is significantly lower for the elliptical tube geometry compared with the circular one. The larger frontal area of the circular tube covers a greater amount of fluid and thus results in greater form drag. To summarize the overall performance of the heat exchanger, an enhancement factor [Eq. (11)] has also been presented in Fig. 3. The elliptical tube geometry by virtue of its significantly lower pressure drop penalty, offers an enhancement of 9–11% over the circular tube geometry in terms of the enhancement factor.

3.3 Performance Analysis of Winglet Supported Fin-and-Tube Heat Transfer Surface

In the previous section, it was observed that the elliptical tube geometry is more beneficial in terms of lower pressure drop. However, its heat transfer performance is slightly lower than the circular tube geometry. In this section, rectangular winglet type vortex generator is employed over the elliptical tube geometry to enhance its heat transfer capability. The winglets are arranged in a "common flow up" orientation to minimize the thermally isolated wake region. The winglets are placed over various upstream and downstream locations of the central tubes at a nominal attack angle of 10°. The spanwise center–center distance is maintained at 8.6 mm and the streamwise center–center distance considered are −2, 0, and +2 mm. The negative sign indicates a position upstream with respect to the tube center, and the positive sign indicates a downstream placed winglet.

Figure 4 shows the variation in Nusselt number [Eq. (8)] of plain circular tube model and winglet supported elliptical tube models. The plot clearly shows that the winglet supported fin-and-tube heat transfer surface for all the considered streamwise winglet positions performed superior to the plain circular tube model. The winglet pair guides the flow effectively towards the tube wake region and minimizes the poor heat transfer zone. In addition, the rectangular winglets also form a narrow converging passage with the tube for the incoming flow, which then impinges directly at the surface of the adjacent tube and enhances the local heat transfer performance. It can also be observed from Fig. 4 when the winglets are placed upstream with respect to the tube center it provides slightly better heat transfer enhancement.

The temperature contours in Fig. 5 shows that the upstream placed winglet causes a greater temperature rise in the wake region than that of the downstream placed winglet, suggesting attainment of greater thermal mixing in the former case. The friction factor [Eq. (10)] plotted in Fig. 4 also suggests that the modified fin-and-tube heat transfer surface enhances the heat transfer performance significantly without additional pressure drop penalty. In fact, the proposed winglet supported models suffer relatively lesser pressure drop penalty compared to the corresponding plain circular tube geometry. With the upstream placement of tubes, the angular position of the winglet with the elliptical tube and forces the flow to pass through a very narrow constricted passage. When the winglet is advanced downstream, the constricted zone opens up a little, and the flow sufferers relatively lower pressure drop. In order to summarize the overall performance of the winglet supported elliptical tube heat

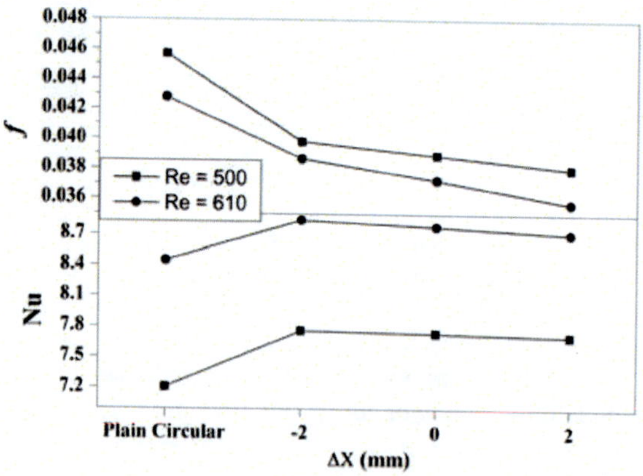

Fig. 4 Variation of Nusselt number as a function of streamwise winglet position

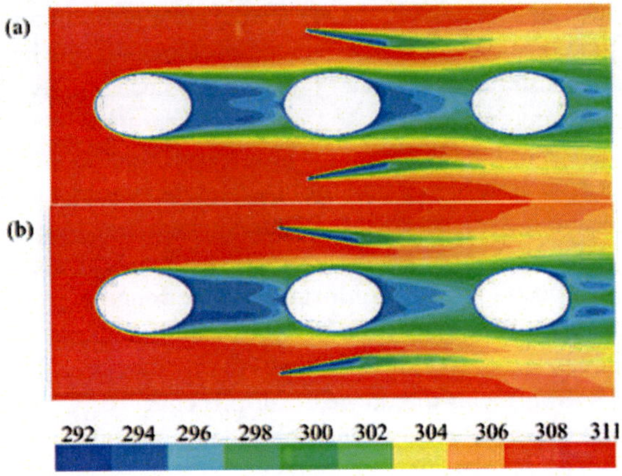

Fig. 5 Temperature contours (unit: K) over the fin surface: **a** $\Delta X = +2$ mm, and **b** $\Delta X = -2$ mm

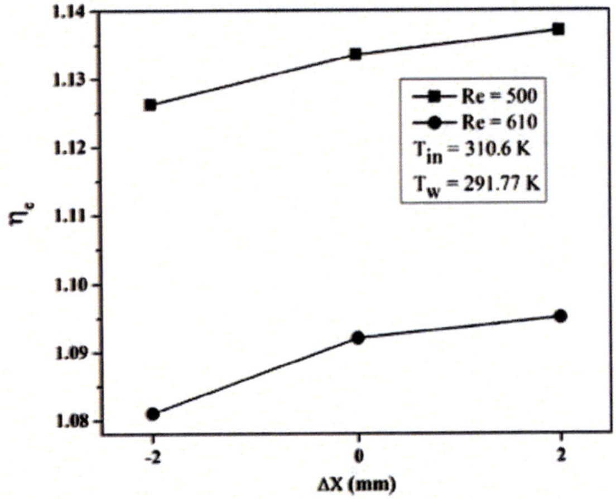

Fig. 6 Variation of enhancement factor as a function of streamwise winglet position

transfer surface, an enhancement factor [Eq. (11)] has been presented in Fig. 6. The downstream placed winglets provide nearly the same heat transfer enhancement but with significantly lesser pressure drop. Thus, the winglets should be placed more towards the downstream side of the tube to achieve greater thermo-fluid performance.

4 Conclusions

The present study has been aimed at comparing the thermo-fluid performance of the plain and winglet supported fin-and-tube heat transfer surfaces having circular and elliptical tube geometries. The fin-and-tube heat transfer surface with a circular tube by virtue of its larger frontal area effects more heat transfer than the elliptical tube type heat transfer surface. However, the larger frontal area of the circular tube also incurs a greater pressure drop penalty. By virtue of its significantly lower pressure drop penalty and moderate heat transfer performance, the overall performance of the elliptical tube heat transfer surface in terms of enhancement factor is significantly higher than the circular tube type fin-and-tube heat transfer surface. The heat transfer performance of the elliptical tube type heat transfer surface gets significantly enhanced when supported with rectangular winglets. The winglets placed at different streamwise locations were found to outperform the plain circular tube geometry both in terms of heat transfer and pressure drop penalty. The winglets streamwise variation does not have a significant effect on heat transfer performance. However, with the downstream placement of winglets, the pressure drop penalty reduces significantly. Thus, the enhancement factor is maximum for the downstream placed winglet supported model.

References

1. Chen Y, Fiebig M, Mitra NK (2000) Heat transfer enhancement of finned oval tubes with staggered punched longitudinal vortex generators. Int J Heat Mass Transf 43(3):417–435
2. Biswas G, Mitra NK, Fiebig M (1994) Heat transfer enhancement in fin-tube heat exchangers by winglet type vortex generators. Int J Heat Mass Transf 37(2):283–291
3. Torii K, Kwak KM, Nishino K (2002) Heat transfer enhancement accompanying pressure-loss reduction with winglet-type vortex generators for fin-tube heat exchangers. Int J Heat Mass Transf 45(18):3795–3801
4. Kwak KM, Torii K, Nishino K (2005) Simultaneous heat transfer enhancement and pressure loss reduction for finned-tube bundles with the first or two transverse rows of built-in winglets. Exp Thermal Fluid Sci 29(5):625–632
5. Joardar A, Jacobi AM (2008) Heat transfer enhancement by winglet-type vortex generator arrays in compact plain-fin-and-tube heat exchangers. Int J Refrig 31(1):87–97
6. Joardar A, Jacobi AM (2007) A numerical study of flow and heat transfer enhancement using an array of delta-winglet vortex generators in a fin-and-tube heat exchanger. J Heat Transf 129(9):1156–1167
7. He YL, Chu P, Tao WQ, Zhang YW, Xie T (2013) Analysis of heat transfer and pressure drop for fin-and-tube heat exchangers with rectangular winglet-type vortex generators. Appl Therm Eng 61(2):770–783
8. Hu WL, Song KW, Guan Y, Chang LM, Liu S, Wang LB (2013) Secondary flow intensity determines Nusselt number on the fin surfaces of circle tube bank fin heat exchanger. Int J Heat Mass Transf 62:620–631
9. Zhou G, Feng Z (2014) Experimental investigations of heat transfer enhancement by plane and curved winglet type vortex generators with punched holes. Int J Therm Sci 78:26–35
10. Li MJ, Zhou WJ, Jhang JF, Fan JF, He YL, Tao WQ (2014) Heat transfer and pressure performance of a plain fin with radiantly arranged winglets around each tube in fin-and-tube heat transfer surface. Int J Heat Mass Transf 70:734–744
11. Arora A, Subbarao PMV, Agarwal RS (2015) Numerical optimization of location of "common flow up" delta winglets for inline aligned finned tube heat exchanger. Appl Therm Eng 82:329–340
12. Sarangi SK, Mishra DP (2017) Effect of winglet location on heat transfer of a fin-and-tube heat exchanger. Appl Therm Eng 116:528–540

13. Naik H, Tiwari S (2018) Effect of winglet location on performance of fin-tube heat exchangers with inline tube arrangement. Int J Heat Mass Transf 125:248–261
14. Sarangi SK, Mishra DP, Mishra P (2020) Parametric investigation of wavy rectangular winglets for heat transfer enhancement in a fin-and-tube heat transfer surface. J Appl Fluid Mech 13(2):615–628
15. Sarangi SK, Mishra DP, Mishra P (2019) Numerical analysis of thermofluid performance of fin-and-tube heat transfer surface using rectangular winglets. J Heat Transf 141(10):101801

Experimental Analysis of a Crossflow Heat Exchanger Using Elliptical Shape Tube for Air Side Heat Recovery

Amruta Sonawane, Neelam S. Gohel, R. S. Jha, and Digvijay Shelar

Nomenclature

Re	Reynolds number
Pr	Prandtl number
Nu	Nusselt number
h	Convective heat transfer coefficient (W/m^{2}K)
Q_{conv}	Heat transfer by convection (Watt)
D_h	Hydraulic diameter (m)
V	Velocity of fluid (m/s)
μ	Viscosity of fluid (N-s/m^2)
ρ	Density of fluid (kg/m^3)
U	Overall heat transfer co-efficient (Watt/m^2)
LMTD	Logarithmic Mean temperature difference
A_s	Surface area of the tube (m^2)

1 Introduction

Crossflow heat exchangers are widely used across the industry for heat recovery. However, recent industry trends and regulations require more energy-efficient devices, and a heat exchanger is an important component to help meet these regulations. In this section, briefly recent research work being done on this topic is described.

A. Sonawane · N. S. Gohel (✉) · D. Shelar
Sinhgad Collge of Engineering, Pune, India
e-mail: gajjar.neelam7@gmail.com

R. S. Jha
Thermax India Ltd, Pune, India

M. Palanisamy et al. (eds.), *Theoretical, Computational, and Experimental Solutions to Thermo-Fluid Systems*, Lecture Notes in Mechanical Engineering,
https://doi.org/10.1007/978-981-33-4165-4_4

Heat transfer in forced convection is affected by numerous parameters such as surface area, the shape of tubes, the orientation of tubes, and flow conditions. Many researchers have explored various aspects affecting these parameters to increase the heat transfer performance. One of the parameters being investigated recently is the shape of tubes such as elliptical shape as compared to traditional circular ones.

Characterization of crossflow cooling of air in an inline elliptical tube heat exchanger, for varying Reynolds number of airside ($1 \times 10^4 < Re < 3.3 \times 10^4$) and for waterside ($1 \times 10^3 < 3.7 \times 10^3$) was considered by Khan et al. [1]. It was observed that increase in heat transfer rate with Reynolds number in power-law fashion. It was also reported that for flat elliptical shape tubes, the heat transfer rate was increased with a higher value of angle of attack. Toolthaisong et al. [2] have also reported a similar observation using elliptical-shaped tubes. It was found that heat transfer performance was noticeably higher as compared to a similar setup using the circular tubes.

Ibrahim [3] investigated thermal performance of elliptical tube bundle (axis ratio 0.25,0.33,0.5,1) experimentally and numerically a flow angles of attack (0–150°) for Reynolds number in the range of 5600–40,000, the result indicated maximum heat transfer coefficient, thermal performance at $\alpha=0°$ and worst performance at $\alpha=90°$ at lowest Reynolds number is achieved. Maximum area of goodness factor at $\alpha=0°$ for all axis ratio followed by $\alpha=30°$ for a given duty is found. Correlation for Nusselt number as a function of axis ratio for different angle of attack.

Shelar [4] has used an inline arrangement for elliptical shaped heat exchanger when water flowed through the tubes and air over them. It was reported that up to 32% enhancement in heat transfer performance could be achieved using elliptical-shaped tubes with 45° angle of attack.

Zukauskas [5] developed general correlation for circular tube bank for inline and staggered arrangement in the Reynolds number in the range of 1 to 2×10^6 and Prandtl number in the range of 0.7 to 10^4.

Chavhan [6] had reported a similar observation for elliptical tube bank at 45° angle of attack when the tube arrangement was staggered.

Ala Hasan [7] have performed experimentation on circual randoval tubes in evaporatively cooled heat exchanger and found that Colburn Factor for oval tube is 89 % of that for the circular tube, average friction factor for the oval tube is 46 % of that of circular tube. Oval tubes shows higher values for the ratio Coulurn factor/friction factor 1.93–1.96 times for the circular tubes.

An egg-shaped tube [8] has a higher favorable pressure gradient at its front end and a lower adverse pressure gradient at its back causes delay boundary layer separation; among the axis ratio 1, 2, 3, 4, 5, the segregation point of the boundary layer moves backward when the axis ratio increases, segregation area reduces and wake disappears at axis ratio 4. A pair of vortexes appear in the wake region for the cases of 1, 2, 3, with the size of the vortex gradually becoming smaller. This suggests its varying characteristics of heat transfer and fluid flow when the air flows around the tube with different axis ratios. The pressure coefficient C_p value is 1 at forward stagnation point; then, it is decreased along the circumference of the tube to reach a minimum value and then increases up to the rear stagnation point. The minimum value of C_p occurs at a position angle smaller than 90°, and as the axis ratio decreases, the angle necessary to provide the minimum C_p decreases. At the axis ratio, four higher pressure gradient in front of the tube and lower pressure gradient at the back of the tube help the flow to inhibit the separation. The maximum value of skin friction coefficient increases with the axis ratio decreases. The highest heat transfer performance achieved at axis ratio 2, and thermal performance decreases as the axis ratio increases due to the thick development of the boundary layer and larger thermal resistance.

Gharbi [9] developed correlation suggested and insight into the characteristics of heat transfer and fluid flow in circular, ellipsoidal and wing-shaped tube bundles arrangements using 2D CFD analysis using ANSYS FLUENT. The result shows higher ($5 \times 10^3 \leq \text{ReDeq} \leq 2.4 \times 10^4$) heat transfer rate, pressure drop in circular tube bank in staggered arrangement compared to wing shape, elliptical tube shape for 0° angle of attack.

Objective of the present work is to understand the behavior of a cross flow heat exchanger with elliptical shaped tubes in inline arrangement with 45° angle of attack. In this work, we would like to report the heat recovery of air which flows through the tubes. Hot water at 80 °C flows over the tubes to maintain a constant temperature boundary condition. Velocity of air was kept in the range of 2–6 m/s at 38 °C.

2 Data Reduction

Thermo-physical properties of fluids mainly depend upon the temperature of the fluid (air).

Curve fitting equations are used to calculate properties of air at average mean temperature.

$$T_{\text{avg}} = \frac{T_{\text{ci}} + T_{\text{co}}}{2} \tag{1}$$

Heat received by air from hot water is given as

$$Q_{\text{air}} = mCp(T_{\text{ci}} - T_{\text{co}}) \tag{2}$$

where m mass flow rate $= \rho\, A_{\text{duct}}\, V_{\text{avg}}$.

The heat transfer rate in the heat exchanger is calculated from the above equation.

The logarithmic mean temperature difference is calculated as

$$\text{LMTD} = \frac{\Delta\theta_1 - \Delta\theta_2}{\ln(\Delta\theta_1/\Delta\theta_2)} \tag{3}$$

$$\Delta\theta_1 = T_{\text{hi}} - T_{\text{co}},\ \Delta\theta_2 = T_{\text{ho}} - T_{\text{ci}}$$

Reynolds Number can be calculated using the following equation

$$\text{Re} = \frac{\rho V D_h}{\mu} \tag{4}$$

$$Q_{\text{air}} = U \times A_s \times \text{LMTD}, A_s = \pi \times D \times L \times \text{Nc} \times \text{Nr} \tag{5}$$

where Nc is No. of Column, Nr No. of row.

Overall heat transfer coefficient is calculated by Eq. (5)

$$\frac{1}{U} = \frac{1}{\text{hi}} + \frac{1}{\text{ho}} \tag{6}$$

Using the above equation inside heat transfer coefficient of air is found

$$\text{Nu}_{\text{i}} = \text{hi} * D_h / k_{\text{air}} \tag{7}$$

where

$$D_h = 4 \times A/P \tag{8}$$

where A is the cross-sectional area of the tube and P is the perimeter of the tube

$$\text{Re} = \nu D_h/\gamma \tag{9}$$

$$\text{Pr} = \frac{\mu Cp}{k} \tag{10}$$

Dittus correlation for circular tube is given by

$$\text{Nu}_{\text{circular tube}} = 0.023\,\text{Re}^{0.8}\,\overset{0.4}{\text{Pr}} \tag{11}$$

3 Experimental Setup

The crossflow heat exchanger consists of 30 tubes in the three-column inline arrangement. The longitudinal and transverse pitch is 52 mm. The tube major axis 48 m and a minor axis 23 mm. These elliptical tubes are arranged at 45° angle of attack in crossflow. Hot water at 1400 lph is supplied with 0.5 hp pump at 80 °C with a constant flow rate over the tube bank, and air at 38 °C passes from the tubes using a blower. Outside of the heat exchanger is insulated with superlon to prevent heat transfer with the environment. High heat carrying capacity water flows over the tube to maintain a constant wall boundary condition over the tube. The temperature of air and water at the inlet and outlet are measured by PT 100 sensor. These sensors are calibrated with ±1.5 °C with the standard before every experiment. The water flow rate is maintained constant at 1400 lph measured with a rotameter. The entire body of the exchanger is made up of SS304 material with a thickness of about 2.5 mm. The heat exchanger is covered with superlon insulation. In this experimental setup, the heat exchanger

Fig. 1 Schematic diagram of crossflow heat exchanger

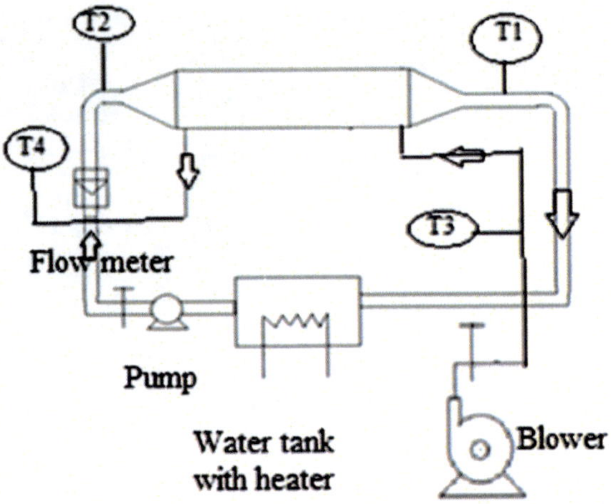

and water tank were insulated as their temperature was quite high compared to the surrounding temperature. Heatx adhesive solution was used to attach insulation on the exterior part of the heat exchanger and water heaters to avoid heat losses. Superlon sheets having 13 mm thickness are used. Elastomeric Nitrile is a 100% closed-cell Nitrile rubber compound with superior insulating qualities. It is flexible microcellular closed-cell elastomeric Nitrile Foam recommended for a temperature range − 40° to 105°. The velocity of air is measured using an anemometer by taking readings at four grid points at the outlet of the heat exchanger. Average air velocity is considered for mass flow rate calculation. Thermo-physical properties mainly depend on the temperature of the fluid. Curve fitting equations are used to calculate properties of air at average mean temperature. Pump having a capacity of 0.5 hp with 2800 lph discharge and head ranges from 7 to 30 m, manufactured by FISHER is used. Rotameter is used as a flow measuring device in the range of 250–2500 lph water flow rate. The blower is used to supply air to inlet duct with mass flow rate 300 m^3/h; pressure head 100 mm water column. Sufficient time was allowed to get the steady-state condition while performing experimentation, which was observed to be about 60 min.

Some of the assumptions made during experimentation are.

i. Conduction and radiation effects are negligible.
ii. A fully developed flow is considered inside the tube (Figs. 1 and 2).

Heat transfer augmentation is studied by varying air velocity in the range of 2–8 m/s (Table 1).

4 Results and Discussions

Figure 3 shows that elliptical-shaped tubes can enhance heat transfer considerably as compared to the one with circular tubes with the same hydraulic diameter. The reason for this enhancement is the shape of the tube and angle of attack, which

Fig. 2 Fabricated laser cut SS304 metal sheet of crossflow heat exchanger

Table 1 Specification of equipment

S. No.	Name of device	Specification	Quantity
1	Centrifugal blower	300 m^3/h	1
2	Water tank	0.75 m × 2 m × 0.40 m	1
4	Centrifugal pump	0.5 hp	1
5	Electric heater	3 KW	4
6	Test section area of crossflow heat exchanger	0.15 × 0.165 × 1 m^3	1
7	Heat exchanger tubes (SS202)	Length = 0.15 m Major axis = 0.048 m Minor axis = 0.023 m	30
8	Temperature measuring instrument	PT-100 RTD	4
9	Anemometer	0–20 m/s	1
10	Rotameter	250–2500 lph	1

generate higher turbulence. Elliptical shape tube has a flat surface that increases the contact time of fluid which increases skin friction drag as compared to circular tube and due to these delayed separations of fluid which enhance mixing hence at 45° angle of attack heat transfer rate from elliptical shape tube bank shows augmentation compared to circular shape tube bank.

Using the experimental observations, a correlation developed for this configuration of a heat exchanger by excel solver. Figure 4 shows the plot using the developed correlation. This also includes the Dittus correlation in the plot for a better comparison. It should be noted that Dittus correlation is valid only for circular shaped tubes. It is clearly seen that correlation found from experimentation for elliptical shape tube performs better in the range of 14,000–22,000 Reynolds number whereas it is comparable at higher Reynolds number.

The ratio of experimental Nusselt number and Nusselt number of the circular tube gives the enhancement ratio. The enhancement ratio of the elliptical tube is between

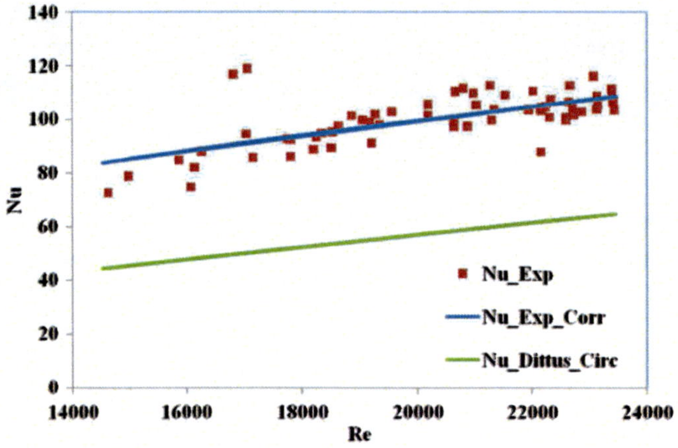

Fig. 3 Plot of experimental Nusselt number for Elliptical Tubes (Nu_Exp), Circular tube Nusselt Number (Nu_ Dittus_circular) against Reynolds Number

Fig. 4 Plot of Nusselt Number correlation based on experimental results (Nu_Exp_Corr), Dittus correlation (Nu_Exp_Dittus) against Reynolds Number

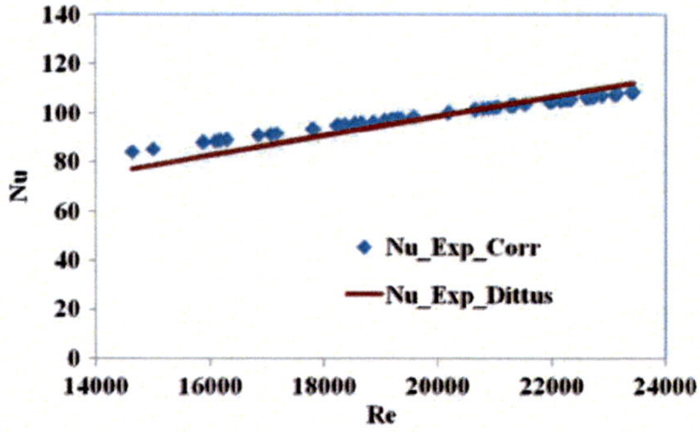

1.67 and 1.9. The enhancement ratio decreases with an increase in the Reynolds number. Figure 5 shows that heat transfer increases as turbulence increases, but after a certain point, it starts to decrease.

5 Conclusions

At the angle of attack 45°, elliptical tubes are more effective in the heat transfer as compared to the circular tubes. Nusselt number increases with the Reynolds number of air. For the same hydraulic diameter, elliptical tubes show higher heat transfer augmentation than circular tubes.

- Thermal hydraulic performance decreases with an increase in turbulence in the range of 14,000–24,000 Re_{Dh}.
- Thermal hydraulic performance of heat exchanger decreases along with pumping power

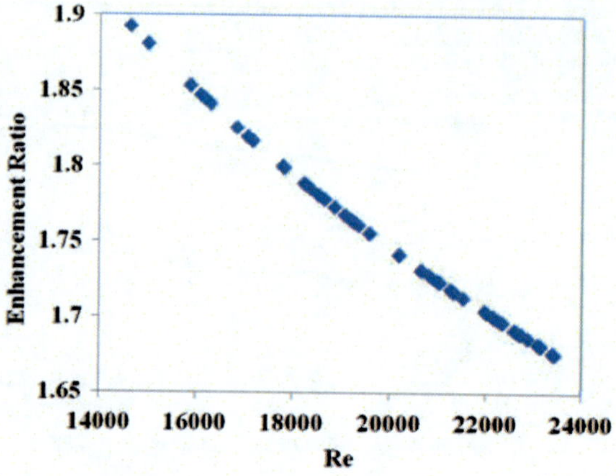

Fig. 5 Enhancement ratio versus Reynolds Number

- Elliptical tubes at 45° angle of attack increase vortex formation, which helps in fluid mixing.
- A general correlation for Nu-Re obtained by curve fitting for 45° configuration $= 0.521\ Re_D{}^{0.542}$ by experimentation and data analysis through excel solver.

Acknowledgements The author would like to thank the Savitribai Phule Pune University for financial support under the Board of College University Development and Thermax India Ltd for continuous support.

References

1. Khan MG, Fartaj A, Ting DS-K (2004) An experimental characterization of crossflow cooling of air via an inline elliptical tube array. Int J Heat Fluid Flow 25:636–648. https://doi.org/10.1016/j.ijmultiphaseflow.2003.09.003
2. Toolthaisong S, Kasayapananda N (2013) Effect of attack angles on air side thermal and pressure drop of the cross flow heat exchangers with staggered tube arrangement. Energy Proc 34:417–429
3. Ibrahim TA (2009) Thermal performance criteria of elliptic tube bundle crossflow. Int J Therm Sci 48:2148–2158
4. Shelar DA (2016) Experimental investigation of elliptical tube bank cross flow heat exchanger with inline arrangement & 45° angle of attack. Int Eng Res J (IERJ) Special Issue pp 725–729, ISSN 2395–1621
5. Zukauskas A (1972) Efficiency parameters for heat transfer in tube banks. Heat Transfer Eng 6:19–25
6. Chavhan SD (2016) Thermal-hydraulic performance of elliptic shape staggered tube cross flow heat exchanger at 45° angle of attack. Int Eng Res J (IERJ) Special Issue pp 725–729, , ISSN 2395-1621
7. Ala Hasan (2004) Performance investigation of plain circular and oval tube evaporatively cooled heat exchangers. Appl Therm Eng 24:777-790
8. Zhang G (2015) Flow and heat transfer characteristics around egg shaped tube. J Hydrodyn pp 76–84
9. Gharbi NE (2015) Numerical optimization of heat exchangers with circular and non-circular shapes, Case Studies in Thermal Engineering 2015

An Experimental Investigation of Tube-in-Tube Type Heat Exchanger to Enhance Heat Transfer Using Titanium Oxide TiO_2-Water Nanofluid

K. Soma Sekhar, P. Vijay Kumar, and B. Kamala Priya

Nomenclature

€	Effectiveness
C_p	Specific heat (J/kg K)
Re	Reynolds number $[\rho UD/\mu]$
Pr	Prandtl number $\left[\mu C_p/k\right]$
Pe	Peclet number $[U/\alpha L]$
Nu	Nusselt number $[hD/k]$
LMTD	Logarithmic Mean Temp Difference
k	Thermal conductivity (W/m K)
D_i	Inner diameter of outer tube (m)
D_0	Outer diameter of the outer tube (m)
L	Length of the heat exchanger (m)
T_i	Inlet temperature of hot fluid (°C)
T_o	Outlet temperature of hot fluid (°C)
U	Overall heat transfer coefficient (W/m^2)
T_{ci}	Inlet temperature of cold fluid (°C)
T_{co}	Outlet temperature of cold fluid (°C)
ϕ	Variable volume concentration of nanofluid
d_i	Inner diameter of inner tube (m)

K. Soma Sekhar · P. Vijay Kumar (✉) · B. Kamala Priya
Department of Mechanical Engineering, Lakireddy Bali Reddy College of Engineeirng (Autonomous) Mylavaram, Krishna-District, AP 521230, India
e-mail: pjoel2013@gmail.com

M. Palanisamy et al. (eds.), *Theoretical, Computational, and Experimental Solutions to Thermo-Fluid Systems*, Lecture Notes in Mechanical Engineering,
https://doi.org/10.1007/978-981-33-4165-4_5

1 Introduction

The technological advancements in nanotechnology and the distinguishing characteristics of nanomaterials have been attracting researchers around the globe for the wide use of nanofluids, which emerged as one of the promising and unique capacity in enhancing the heat transfer rates in heat exchangers used in diverse fields of engineering applications. Nanofluid means suspension of nanoparticles in the base fluid (distilled water). Usually, the base fluids mostly used are distilled water, ethylene glycol, etc. However, colloidal solutions are prepared by adding solid particles to the conventional fluids. These nanoparticles dictate the types of nanofluids and the distinction between them based on the metal from which these are made. The nanofluids can cause is a significant enhancement in heat transfer rate due to high thermal conductivity, viscosity, and specific heat. The application of nanofluids can be found in heat exchangers, cooling of electronic equipment, and chemical process plants. In some cases, magnetic fields are also present, which adds to the complexity of the problems. To address those issues usually, mathematical models are developed and analytical solutions will be used to analyze the problems and parametric studies are carried out to simplify the system for getting feasible solutions.

All the heat exchangers, cooling processes, heating, and chemical processes are involved with heat transfer processes. The ever-existing problem of poor heat transfer properties of common fluids (such as water, mineral oil, and ethylene glycol) compared to most solids is a primary obstacle to the effectiveness of heat transfer processes. However, clogging in the process may be found in the tube when the fluid with large particles is employed. With awareness of this problem, the nanometer particle in the fluid (as called nanofluid is an attractive solution that provides not only enhanced thermal conductivity but also long-term stability and low-pressure drop [1].

Titanium dioxide (TiO_2) is one of the promising materials for heat transfer enhancement purposes due to its excellent chemical and physical stability. In addition, TiO_2 particles are cheap and commercially available. TiO_2 nanoparticles suspended in conventional fluids were extensively utilized in various forms of heat exchangers, including circular tube [2, 3], a double tube [4–7]. The addition of nanoparticles to the base fluid could increase the heat transfer rate in the air conditioner in a shell and tube heat exchanger heat transfer rates under turbulent flow is investigated based on Peclet number, Nusselt number variations and were enhanced by using nanofulid [8]. The convective heat transfer coefficient, inlet temperature, etc., are studied. Based on the results, enhancement of the effectiveness of silver-water nanofluids was observed [9]. Pritam et al. investigated the stability of TiO_2-water nanofluid various surfactants. It was found that acetic acid, sodium dodecylsulfate improved the stability of the suspensions [10]. Murshed et al. investigated the stability of nanofluids with different surfactants such as cetyltrimethylammonium bromide, oleic acid surfactant results have shown that the stability of suspension improved considerably [11]. In a shell and tube heat exchanger the experiments have been carried out to find the Peclet number variations they have investigated with TiO_2/water, Al_2O_3/water nanofluids.

The results yielded that the Peclet number is enhanced with TiO_2/water nanofluid, which was more compared to Al_2O_3/water nanofluid case [12]. Godson et al. investigated the effectiveness of shell and tube heat exchanger by using silver/water Nanofluid. The experimental results indicate an enhancement of effectiveness with the increased particle concentration in the base fluid.

The present work deals with the experimental investigations on the characteristics of heat transfer in a tube in a tube heat exchanger with distilled water as cold fluid in the first stage and TiO_2-water nanofluid as cold fluid in the second stage with variable volume concentrations 0.05, 0.07, 0.1%. The variations in Nusselt number, heat transfer coefficient, and Peclet number are observed under turbulent flow condition. Finally, based on the experimental findings, it was noticed that TiO_2-water nanofluid provides a better heat transfer rate at 0.1% volume concentration.

2 Fabrication of Tube in Tube Heat exchanger-Photographic View

2.1 Specifications of Heat Exchanger (Tube-in-Tube)

1. Material of Inner Tube Copper
   a. Inner diameter of inner tube $d_i = 9.5$ mm.
   b. Outer diameter of inner tube $d_o = 12.5$ mm.
   2. Outer Tube Material Galvanized Iron
   a. Inner diameter outer tube $D_i = 28.5$ mm.
   b. Outer diameter outer tube $D_o = 32.5$ mm.
   3. Length of Heat Exchanger $L = 1500$ mm.
   4. Specific Heat of Water $Cp_w = 4178$ J/kg K.

A tube in the tube heat exchanger as shown in Fig. 1 consists of an electric heater to heat water(act as a hot fluid) to the required temperature, which flows thru the outer tube. Coldwater is used as the cooling fluid, and it can be admitted at either end, enabling the same heat exchanger to run in a counter flow/parallel flow mode. This is achieved by operating different valves provided in the pipeline, the outer tube is adequately insulated to minimize the heat lost to the surroundings. Thermocouples are used to measure the inlet, outlet temperatures of the hot and cold fluids. The flow rates of the cold and hot fluids are measured using a measuring flask or jar. Submersible water pumps of 15 and 25 W capacity are used to pump the hot and cold fluids, thus creating the required pumping pressure for a steady flow of cold and hot fluids through the heat exchanger.

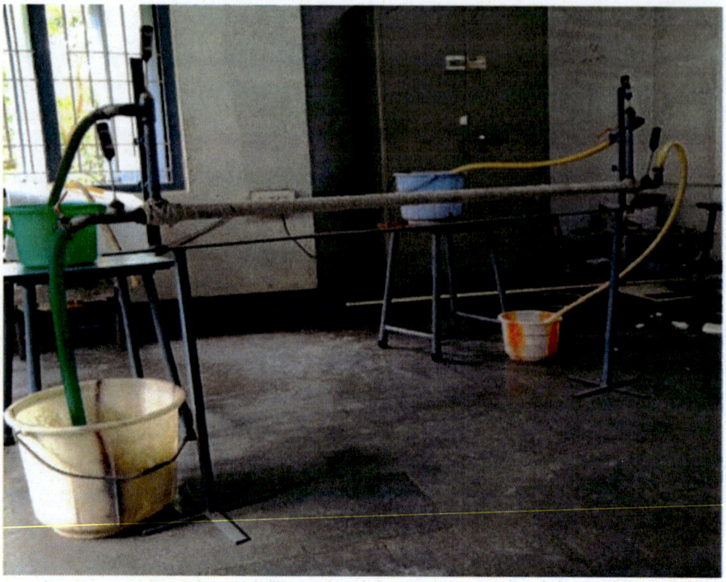

Fig. 1 Experimental setup-photographic view (tube-in-tube heat exchanger)

2.2 *Methodology*

A Tube-in-Tube type heat exchanger with copper tube (inner tube) and GI (outer tube) are considered, and the outer surface of the tube is insulated with the rope, which prevents convective heat losses. Thermocouples are placed at the inlet and outlet of cold and hot fluids to measure the inlet and outlet temperatures accurately after reaching steady-state condition. Firstly, experimentation is conducted using hot water which passes in the outer tube and distilled water(coolant) which passes in the inner tube at fixed temperatures (80 °C, 70 °C, 60 °C, and 50 °C) for the variable mass flow rates 0.025 kg/s, 0.05 kg/s, 0.1 kg/s. The characteristics of heat transfer, that is, effectiveness, Nusselt number, Peclet number, and coefficient of heat transfer were calculated.

Secondly, the magnetic stirring and bath sonicator equipment were used for effective nanoparticles mixing with the distilled water (base fluid), while the magnetic stirring method was used TiO_2-water nanofluids were stirred for four hours and then transferred into bath sonicator with sonication time up to five hours for each concentration of nanofluid and TiO_2-water nanofluids were prepared in three different concentrations (0.05, 0.07, 0.1%).

Thirdly, acetic acid was used as a surfactant, which was added up to 2% for each concentration of nanofluid at the time of preparation to increase the stability of nanofluid for a longer time. Experimentation done for using TiO_2-water nanofluid as a coolant using one concentration at a time and different heat transfer parameters were calculated through the data recorded. The graphs are drawn for comparing the performance of tube in the tube heat exchanger for base fluid and TiO_2-water nanofluids based on the variations of effectiveness, Nusselt number, Peclet number, and coefficient of heat transfer.

3 Results and Discussions

The results and discussion presented in this section for the two cases taken into consideration, namely.

First stage: cold fluid—distilled water (Inner tube), hot, hot water (outer tube),

Second stage: cold fluid—TiO_2-water Nanofluid (Inner tube), hot fluid—hot water (outer tube).

Keeping the other data, such as temperature of hot fluid and mass flow rates constant.

Figure 2 depicts the comparison between effectiveness and Peclet number of distilled water (cold fluid)at different mass flow rates. It is observed that a linear trend of the points located on the graph plotted between the temperature and effectiveness. As Peclet number increased, effectiveness also increased with n effectiveness value of 0.21 being the highest at 0.1 kg/s. A steady increase in effectiveness is identified with increasing mass flow rates.

Figure 3 indicates the variation between effectiveness and Peclet number for different volume concentrations of TiO_2–water nanofluid. It can be clearly seen that as temperature increases, effectiveness also increases, but the linear trend is not seen

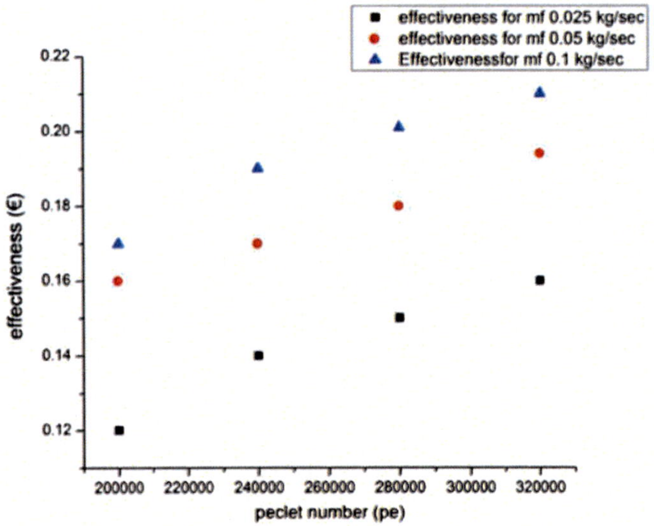

Fig. 2 Effectiveness versus Peclet number (Distilled water as a cold fluid)

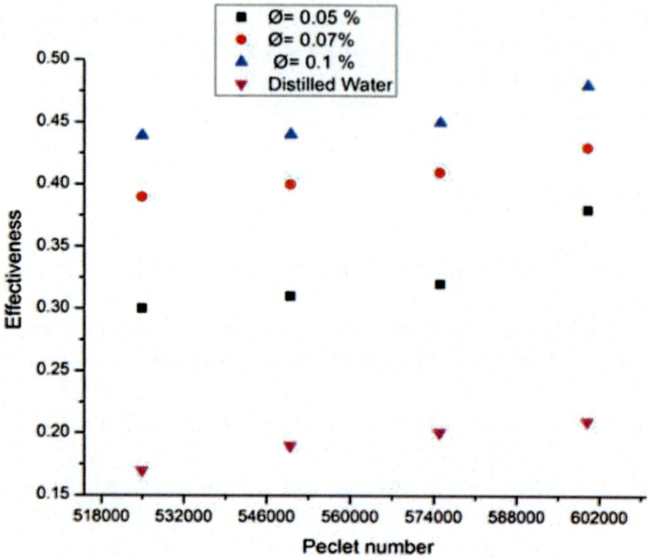

Fig. 3 Peclet number versus effectiveness (mass flow rate 0.1 kg/s)

for TiO_2–water nanofluid with 0.07% concentration due to some losses. There was no much variation observed in effectiveness with the change in Peclet number. However, compared to the effectiveness of water, effectiveness with TiO_2–water nanofluid has reached 0.49, which is very high. The effectiveness is highest at 60,000 Peclet number as the difference is more at that temperature.

The plot is drawn between LMTD and Peclet number in comparison between the coefficient of heat transfer (outer surface area) and temperature, for each concentration are observed that the points closely appear on the plot with not much variation to one another but maintaining the linearity for all the volume concentrations of the TiO_2–water nanofluid in Fig. 4. The coefficient of heat transfer increases steadily with temperature for which no irregular trend is observed. The coefficient of heat transfer is less compared to the coefficient of heat transfer (inside area) as the outer surface is insulated that prevent the convection losses.

Figure 5 is plotted between Nusselt number, and Peclet number wherein the trend followed by the points is linear and directly proportional, that is, as temperature increases, the heat transfer coefficient also increases. All the plotted points on the graph are closely associated with each other, and the coefficient of heat transfer is slightly more because copper is the material for the inner tube, which enhances the conductivity of the cold fluid, that is, TiO_2–water nanofluid.

From Fig. 6 it is observed that for the volume concentration of $\emptyset = 0.1\%$, the overall heat transfer coefficient has significantly improved for all the values of Peclet number compared to other volume concentrations. This is mainly due to the performance of high thermal conductivity of TiO_2–water nanofluid and higher volume

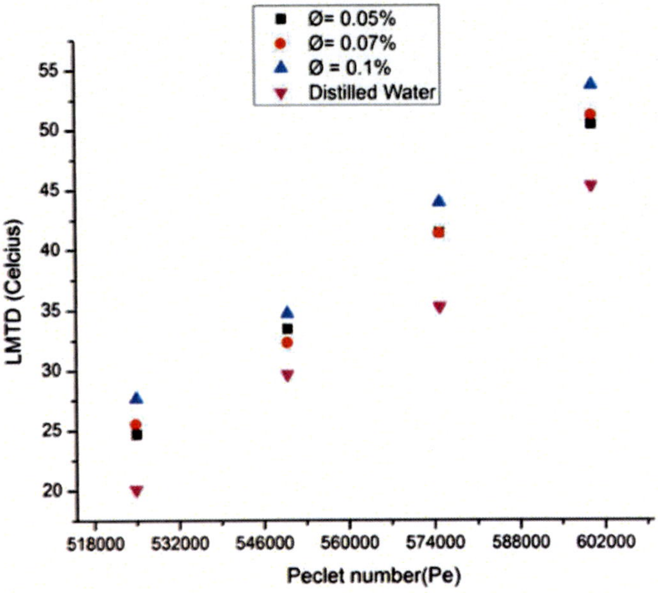

Fig. 4 LMTD versus Peclet number (mass flow rate 0.1 kg/s)

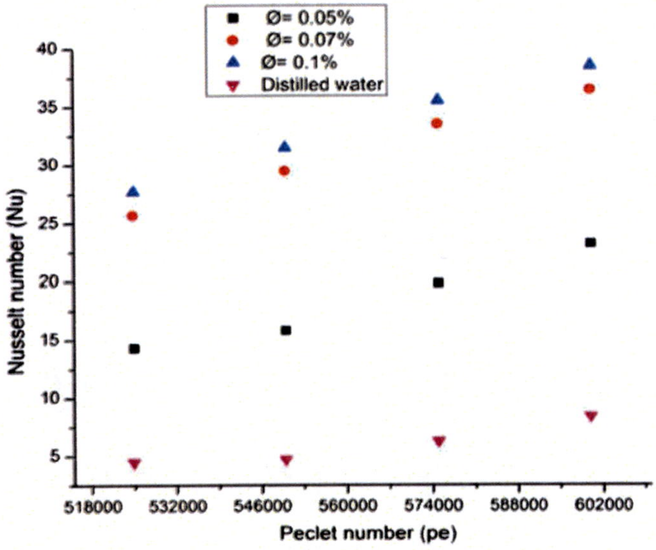

Fig. 5 Variation of Nusselt number with Peclet number for mass flow rate 0.1 kg/s

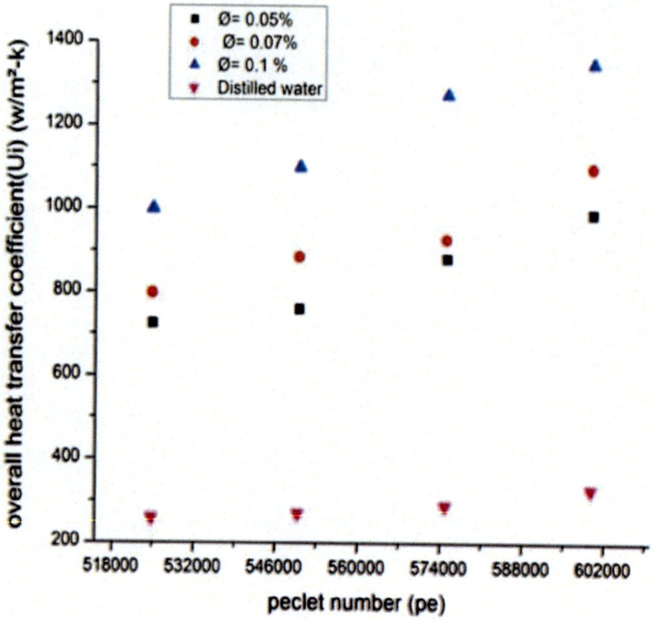

Fig. 6 Variation of overall heat transfer coefficient with Peclet number for a mass flow rate 0.1 kg/s

concentrations. The points located in the graph have shown a steady increase in the coefficient of heat transfer (overall) with an increase in the Peclet number. The highest coefficient of heat transfer (overall) registered was 1350 w/m^2-k at Pe = 600,000.

4 Conclusions

The performance of a heat exchanger (Tube-in-Tube type) with TiO_2–water nanofluid at variable volume concentrations of $\emptyset = 0.05\%$, $\emptyset = 0.07\%$, and $\emptyset = 0.1\%$ has been experimentally tested at different hot water and cold water temperatures and mass flow rates.

The following conclusions were drawn based on the observations made through the experiments.

(i) The characteristics of heat transfer such as Nusselt number, Effectiveness, LMTD are enhanced with the addition of nanoparticles (TiO_2) to the distilled water (base fluid), due to the increased thermal conductivity of the TiO_2–water nanofluid.

(ii) The performance of heat exchanger (Tube-in-Tube type) was enhanced with the increased volume concentration of TiO_2–water nanofluid. The temperature at the outlet of the cold fluid and the effectiveness of the heat exchanger raised with the percentage volume concentration of TiO_2–water nanofluid increased.

(iii) Also the coefficient of heat transfer (overall) is enhanced with the concentration of TiO_2–water nanofluid increases compared to the base fluid. However, the temperature at the outlet of the hot fluid decreases when the TiO_2 nanofluid concentration increases.

(iv) Based on the results observed that the variation of the Nusselt number of the nanofluids is highly influenced by the variable percentage volume concentration of TiO_2 nanoparticles and Peclet number.

(v) The better performance of the tube in the tube heat exchanger was observed at $\emptyset = 0.1\%$ concentration of TiO_2–water nanofluid with a mass flow rate of $m_f = 0.1$ kg/s.

References

1. Pathipakka G, Sivashanmugam P (2010) Heat transfer behaviour of nanofluids in a uniformly heated circular tube fitted with helical inserts in laminar flow. Superlattices Microstruct 47:349–360
2. Pak BC, Cho YI (1998) Hydrodynamic and heat transfer study of dispersed fluids with submicron metallic oxide particles. Exp Heat Transf 11:151–170
3. Sajadi AR, Kazemi MH (2011) Investigation of turbulent convective heat transfer and pressure drop of TiO_2/water nanofluid in circular tube. Int Commun Heat Mass Transf 38:1474–1478
4. Kayhani MH, Soltanzadeh H, Heyhat MM, Nazari M, Kowsary F (2012) Experimental study of convective heat transfer and pressure drop of TiO_2/water nanofluid a double tube. Int Commun Heat Mass Transf 39:456–462
5. Duangthongsuk W, Wongwises S (2010) An experimental study on the heat transfer performance and pressure drop of TiO_2–water nanofluids flowing under a turbulent flow regime. Int J Heat Mass Transf 53:334–344
6. Abbasian Arani AA, Amani J (2013) Experimental investigation of diameter effect on heat transfer performance and pressure drop of TiO_2–water nanofluid. Exp Therm Fluid Sci 44:520–533
7. Chandraprabu V et al (2014) Heat transfer enhancement characteristics of Al_2O_3/Water and CuO/Water nanofluids in a tube in tube condenser fitted with an air conditioning system-an experimental comparison. J Therm Sci Eng Appl 6:1–5
8. Farajollahi B et al (2010) Heat transfer of nanofluids in a shell and tube heat exchanger. Int J Heat Mass Transf 53:12–17
9. Pritam Kumar D et al (2016) Synthesis and characterization of TiO_2–water nanofluids with different surfactants. Int Commun Heat Mass Transf 75:341–348
10. Ebrahimi M et al (2014) Experimental investigation of force convection heat transfer in a car radiator filled with SiO_2-water nanofluid. IJE Trans B: Appl 27:333–340
11. Murshed SMS et al (2005) Enhanced thermal conductivity of TiO_2—water based nanofluids. Int J Therm Sci 44:367–373
12. Godson L et al (2014) Heat transfer characteristics of silver/water nanofluids in a shell and tube heat exchanger. Archives Civil Mech Eng 14:489

Earth Air Tube Heat Exchanger—A Parametric Study

Saif Nawaz Ahmad and Om Prakash

Nomenclature

A	Surface area
C	Specific heat capacity
d	Depth of buried pipe
D	Diameter
EATHE	Earth air tube heat exchanger
f	Friction factor
h	Convective heat transfer coefficient of air
K	Thermal conductivity
l	Length
m	Mass flow rate
Nu	Nusselt number
Pr	Prandtl number
PVC	Polyvinyl chloride
r	Radius of pipe
Re	Reynolds number
R	Thermal resistance
T	Temperature
U	Overall heat transfer coefficient

S. N. Ahmad (✉) · O. Prakash
Department of Mechanical Engineering, National Institute of Technology Patna, Ashok Raj Path, Patna, Bihar 800005, India
e-mail: saif.nawaz.ahmad@gmail.com

O. Prakash
e-mail: om.prakash@nitp.ac.in

M. Palanisamy et al. (eds.), *Theoretical, Computational, and Experimental Solutions to Thermo-Fluid Systems*, Lecture Notes in Mechanical Engineering,
https://doi.org/10.1007/978-981-33-4165-4_6

Greek Letters

ρ	Density
ν	Kinematic viscosity
μ	Dynamic viscosity
ε	Effectiveness

Subscript

a	Air
c	Conduction
i	Inner
p	Pipe
s	Soil
o	Outer

1 Introduction

Space conditioning is a remarkable need, and it's increasing day by day. Most of the conventional space conditioning system works after consuming electrical energy that is to be produced by burning fossil fuels (conventional energy) in a power plant. Since the burning of fossil fuels in power plants emits harmful gases and causes global warming, climate change, biodiversity loss, water crisis, Economic shortage, etc. The rise in conventional energy demands and its price, as well as various environmental threats, motivate the researchers toward the utilization of renewable energy. Hence the various passive technology and renewable energy sources are a suitable alternative for space conditioning. One of the best alternatives for space conditioning is an earth air tube heat exchanger that works on geothermal energy. Geothermal energy is recognized as renewable energy that is clean, free, and environment friendly. The earth has a relatively constant temperature at a specified depth and is lower than the ambient temperature in summer and vice versa in the winter season. This property of earth is very useful in the installation and implementations of earth air tube heat exchanger (EATHE), which makes earth as a source in winter for heating and sink in summer for cooling. In an EATHE system, the pipes of the designed dimension and materials are covered in the earth at a given fixed depth. The soil releases the thermal energy to air that flows in the pipe in winter season for space heating when the outdoor temperature is lesser than underground temperature and in this way earthwork as a source, while in summer season for space cooling the soil absorbs thermal energy from the air through pipe whenever the outdoor ambient temperature is greater than the underground temperature, and hence earth works as a sink. Many researchers have performed the experimental and analytical analysis of EATHE coupled to building in order to investigate its performance and potential to the conditioned space [3, 7–9, 13]. By observing their results, it is clear that EATHE is well suited for cooling/heating in a given region of the world. Kabashnikov et al. [6] performs the numerical and analytical study and found that the rise and drop in air temperature increases with increasing the pipe length. But after some certain length, the heat transfer did not increase more with increasing the pipe length because increasing its length required more air travel time and heat removal rate [4]. Ralegaonkar et al. [12] developed a methodology for designing the length and ducting details of the geothermal cooling system and compare the results in the view of energy consumption by coupling it with a case study of the building. They found that a geothermal cooling system saves 90% energy as compared to air conditioners for cooling, whereas it saves 100% water while comparing this system with evaporative cooling. Misra et al. [10] carried out an experimental study on the prototype of a model of earth air tube heat exchanger for predicting the cooling potential and effect of materials used for making it. Their experiments on the model record 10–15 °C fall in temperature for cooling, and its performance was not affected by the material used for burying pipe. Yassine et al. [17] investigate the design features for attaining the thermal comfort of a building by incorporating it with earth air tube heat exchanger in order to reduce the energy consumption. Singh [14] evaluated the performance of an EATHE system for space cooling using a mathematical model. A numerical calculation has been done for getting the optimum value of dimensions of the tunnel and flow velocity of air inside the tunnel in order to extract the maximum heat from the room at different values of required load and environmental conditions. Niu et al. [11] investigated the effect of inlet temperature on the attainment of earth air tube heat exchanger by developing a mathematical model. It can be concluded from their results that the higher the inlet air temperature, more is the rate of decrease in air temperature in the EATHE pipe. Tiwari et al. [16] designed an Earth air tube heat exchanger for a given dimension of the room by optimizing the various variables such as length of pipe, buried depth of pipe, pipe diameter, and a number of air changes. Their optimized result shows that the outlet air temperature in the summer season decreased by 5–6 °C. Since this decrement in outlet temperature, the diameter and length were optimized as 0.1 m and 21 m, respectively. Ahmad and Prakash [2] performed experimental exergetic analysis of Ground Source Heat Pump System and concluded that the minimum exergy outlet occurs at Ground heat exchanger. Zhao et al. [18] investigating the performance and various influencing parameters of the earth air tube heat exchanger system by experimenting with its model and check its feasibility. Their results show that the cooling and heating capacities are 21.17 and 21.72 kW, respectively, also the temperature extraction efficiencies, increase with pipe length and decrease with its diameter. The present works estimate the required length of the tube in designing the earth air tube heat exchanger system that is affected by various parameters. To this point, this paper deals with the parametric study that emphasizing the effect of inlet temperature and outlet temperature on the length of the tube for the cooling mode of application. The subsoil thermal saturation in the close vicinity of the buried pipe reduced when this system uses for a longer period of time or continuous operation.

So, it may be used for a short period of time or intermittent operation for getting better thermal performance, but this would be uneconomical. So, for increasing the thermal properties of soil, the subsoil moisture content was increased by inserting a necessary arrangement for water impregnation in the close vicinity of buried pipe in the experimental study of Agrawal et al. [1].

2 Thermodynamic Modeling for Length of EATHE System

The heat transfer from air to the inner surface of the tube is due to the convection mode of heat transfer while the transfer of heat from the inner surface to the outer surface of the tube is taken place by the conduction mode of heat transfer and lastly, the heat transfer from the outer surface of the tube to bounded soil is taken place by conduction mode of heat transfer. So, we say that in this system, the number of processes in heat transfer is three while the number of modes of heat transfer is two. Hence the total thermal resistance will be the sum of a convective resistance and two conductive resistances.

Thermodynamic modeling of this system can be written on the basis of the following assumptions.

- The thermal properties of soil are isotropic
- The modeling is done on the basis of steady-state conditions
- A perfect contact exists between the pipe and bounded soil.
- The depth of the buried pipe is constant throughout the whole length.
- The physical properties of the fluid are constant.

Therefore, the convective resistance due to transfer of heat from flowing air to the inner surface of the pipe may be formulated as

$$R_c = \frac{1}{2\pi r_i l_p h} \tag{1}$$

The conductive thermal resistance due to pipe thickness and heat transfer taken place from the inner surface of the pipe to the outer surface of the pipe may be formulated as,

$$R_{\text{pipe}} = \frac{\ln\left(\frac{r_o}{r_i}\right)}{2\pi K_p l_p} \tag{2}$$

The conductive thermal resistance due to transfer of heat from the outer surface of the pipe to bounded soil is given by Holman and Bhattacharyya [5]

$$R_{\text{soil}} = \frac{\text{Cosh} - 1\left(\frac{d}{r_o}\right)}{2\pi l_p K_s} \tag{3}$$

Thus, total thermal resistance would be

$$R_{\text{total}} = R_c + R_p + R_s \tag{4}$$

$$R_{\text{total}} = \frac{1}{2\pi r_i l_p h} + \frac{\ln\left(\frac{r_o}{r_i}\right)}{2\pi K_p l_p} + \frac{\text{Cosh} - 1\left(\frac{d}{r_o}\right)}{2\pi l_p K_s} \tag{5}$$

$$R_{\text{total}} = \frac{1}{l_p}\left(\frac{1}{2\pi r_i h} + \frac{\ln\left(\frac{r_o}{r_i}\right)}{2\pi K_p} + \frac{\text{Cosh} - 1\left(\frac{d}{r_o}\right)}{2\pi K_s}\right). \tag{6}$$

Let us consider

$$X = \left(\frac{1}{2\pi r_i h} + \frac{\ln\left(\frac{r_o}{r_i}\right)}{2\pi K_p} + \frac{\text{Cosh} - 1\left(\frac{d}{r_o}\right)}{2\pi K_s}\right)$$

Since X has been introduced for simplicity of calculation which gives the result of different dimensions and properties of parameters.

$$R_{\text{total}} = \frac{X}{l_p} \tag{7}$$

In the formula of convective resistance, the term convective heat transfer coefficient (h) may be finding using the Nusselt number and the Nusselt number is given as

$$\text{Nu} = \frac{h \times D_i}{K_a} \tag{8}$$

Since Nusselt number is the function of Reynolds number and Prandtl number, that is,

$$\text{Nu} = f(\text{Re}, \text{Pr}) \tag{9}$$

We have available empirical relation for the calculation of Nusselt number, For Laminar flow, that is where the Reynolds number is less than 2300 (Re < 2300)

$$\text{Nu} = 4.36 \tag{10}$$

For fully developed turbulent flow in the buried pipe where Prandtl number lies between 0.5 and 2000, and Reynolds number lies between 2300 and 5 $\times$ 10^6, The given empirical relation for Nusselt number is

$$\mathrm{Nu} = \frac{\left(\frac{f}{8}\right) \times (\mathrm{Re} - 1000) \times \mathrm{Pr}}{1 + 12.7\left(\frac{f}{8}\right)^{\frac{1}{2}}\left(\mathrm{Pr}^{\frac{2}{3}} - 1\right)} \tag{11}$$

Here f is the friction factor and may be calculated as,

$$f = [0.79 \times \ln(\mathrm{Re}) - 1.64]^{-2} \tag{12}$$

And Re is the Reynolds number and may be calculated as

$$\mathrm{Re} = \frac{\rho \times v \times D_i}{\mu} \tag{13}$$

And Pr is the Prandtl number and may be calculated as

$$\mathrm{Pr} = \frac{\mu \times C_a}{K_a} \tag{14}$$

The effectiveness of the system in the form of temperature may be calculated as

$$\varepsilon = \frac{T_i - T_o}{T_i - T_s} \tag{15}$$

However, the effectiveness of system may also be computed by Zukowski and Topolanska [19]

$$\varepsilon = 1 - e^{-NTU} \tag{16}$$

$$\varepsilon = 1 - e^{\frac{-UA}{m_a C_a}} \tag{17}$$

Since,

$$UA = \frac{1}{R_{\mathrm{total}}} \tag{18}$$

Now, putting the value of total thermal resistance in Eq. (18), we get

$$UA = \frac{1}{R_{\mathrm{total}}} = \frac{l_p}{X} \tag{19}$$

Hence using Eqs. (15), (17) and (19), we get the value of the length of the pipe by putting the respective values of used parameters in those equations.

3 Results and Discussion

In order to calculate the length of the pipe of the EATHE system, which is the function of the inlet and outlet temperature, a thermal model has been developed in Sect. 2 of this paper. The average ground temperature using experimental measurement at a depth of 1 m in a hot season was 240 °C. As reported by Tiwari [15], these buried pipes are at a constant underground temperature equal to the average annual ambient air temperature at a depth of 4 m. Now, we applied all the developed mathematical formulas in a Microsoft Excel sheet for calculating the length of the pipe. In this study, a PVC pipe of thermal conductivity 0.15 W/mK is used with an inner and outer diameter of 0.05 m and 0.1 m, respectively. Here two cases are taken into consideration, and in each of the cases, one parameter is varied while keeping all other parameters constant. Now the calculated result of both the cases using the excel sheet has been compiled in Tables 1 and 2.

Table 1 indicates the variation of length with inlet temperature with an outlet temperature remain unchanged, whereas Table 2 shows the results of calculated length by varying outlet temperature keeping the inlet temperature constant.

In order to explain Fig. 1 that formatted using data from Table 1, which indicates that, keeping the outlet temperature constant at 26 °C as per increasing the inlet temperature the calculated length of pipe increases. As per increasing inlet temperature from 34 to 43 °C (26.4%) the length of pipe has increased from 7.66 to 10.51 m (37.2%) by keeping outlet the temperature constant at 26 °C.

Also illustrating the Fig. 2 that has been plotted using data from Table 2 which indicates that decreasing the outlet temperature (cooling need) the calculated length of pipe going to increase by keeping inlet temperature constant at 36 °C. As per decreasing outlet temperature from 34 to 25 °C (26.47%) the length of pipe has increased from 1.38 to 12.03 m that is increased 7.717 times.

Table 1 Variation of length with an inlet temperature

Outlet temperature = 26 °C	
Inlet temperature (°C)	Calculated length of pipe (m)
34	7.66
35	8.13
36	8.42
37	9.04
38	9.37
39	9.7
40	10.09
41	10.09
42	10.51
43	10.51

Table 2 Variation of length with outlet temperature

Inlet temperature = 36 °C	
Outlet temperature (°C)	Calculated length of pipe (m)
25	12.03
26	8.42
27	8.42
28	6.62
29	5.28
30	4.13
31	3.28
32	2.56
33	1.9
34	1.38

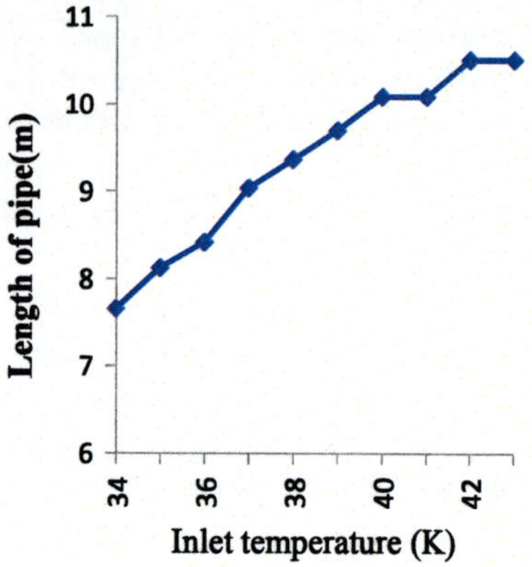

Fig. 1 Variation of length with an inlet temperature

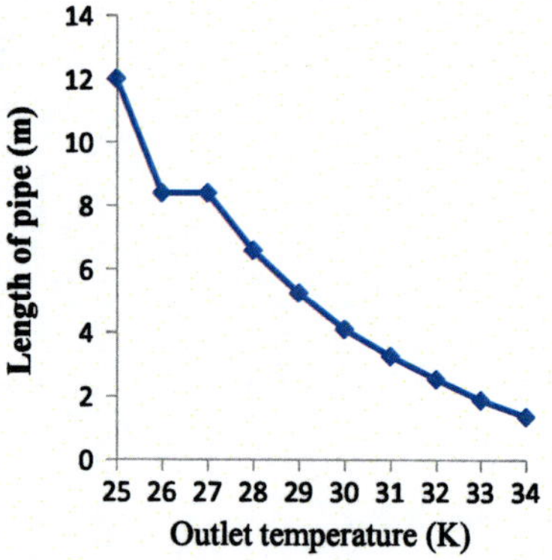

Fig. 2 Variation of length with outlet temperature

4 Conclusions

This paper deals with the parametric study of the earth air tube heat exchanger. In this study, a thermodynamic model for estimating the length of buried pipe has been developed and then inserted in an excel sheet of Microsoft software. The main purpose of this study is to explain the influence of inlet and outlet temperature on the length of the pipe by keeping all remaining parameters constant. From this study, it was concluded that by increasing the inlet temperature of 26.4% with keeping outlet temperature constant calculated length was increased by 37% whereas, by decreasing the outlet temperature of 26.47%, the length was increased by 7.717 times. So as per our need for cooling purposes by keeping the inlet temperature constant more length will be required.

References

1. Agrawal KK, Misra R, Yadav T, Agrawal GD, Jamuwa DK (2018) Experimental study to investigate the effect of water impregnation on thermal performance of earth air tunnel heat exchanger for summer cooling in hot and arid climate. Renew Energy 120:255–265
2. Ahmad SN, Prakash O (2019) Experimental exergy assessment of ground source heat pump system. Adv Energy Res 6(2):161–172
3. Bisoniya TS, Kumar A, Baredar P (2015) Energy metrics of earth–air heat exchanger system for hot and dry climatic conditions of India. Energy Build 86:214–221
4. Derbel HBJ, Kanoun O (2010) Investigation of the ground thermal potential in tunisia focused towards heating and cooling applications. Appl Therm Eng 30:1091–1100
5. Holman JP, Bhattacharyya S (2008) Heat transfer, 10th edn. McGraw-Hill education private limited, New Delhi
6. Kabashnikov VP, Danilevskii LN, Nekrasov V, Vityaz IP (2002) Analytical and numerical investigation of the characteristics of a soil heat exchanger for ventilation systems. Int J Heat Mass Transf 45:2407–2418
7. Khabbaz M, Benhamou B, Limam K, Hollmuller P, Hamdi H, Bennouna (2016) Experimental and numerical study of an earth-to-air heat exchanger for air cooling in a residential building in hot semi-arid climate. Energy Build 125:109–121
8. Mehdid CE, Benchabne A, Roug A, Moummi N, Melhegueg MA, Benabdi MA, Brima A (2018) Thermal design of earth-to-air heat exchanger. Part II a new transient semi-analytical model and experimental validation for estimating air temperature. J Cleaner Prod 198:1536–1544
9. Menhoudj S, Mokhtari AM, Benzaama MH, Maalouf C, Lachi M, Makhlouf M (2018) Study of the energy performance of an Earth air tube heat exchanger for refreshing buildings in Algeria. Energy Build 158:1602–1612
10. Misra AK, Gupta M, Lather M, Garg G (2015) Design and performance evaluation of low-cost earth to air heat exchanger model suitable for small buildings in arid and semi-arid regions. KSCE J Civ Eng 19(4):853–856
11. Niu F, Yu Y, Yu D, Li H (2015) Heat and Mass transfer performance analysis and cooling capacity prediction of earth to air heat exchanger. Appl Energy 137:211–221
12. Ralegaonkar A, Kamath MV, Dakwale VA (2014) Design and development of geothermal cooling system for composite climatic zone in India. J Inst Eng India Ser A 95(3):179–183
13. Ramirez-Davila L, Xaman J, Arce J, Alvarez G, Hernandez-Parez I (2014) Numerical study of earth-to-air heat exchanger for three different climates. Energy Build 76:238–248
14. Singh SP (1994) Optimization of earth air tunnel system for space cooling. Energy Convers Manage 35(8):721–725

15. Tiwari GN (2006) Solar energy fundamentals, design, modeling and applications, 1st edn. Narosa Pub. House, New Delhi, pp 452–487
16. Tiwari GN, Singh V, Joshi P, Shyam Deo A, Prabhakant, Gupta A (2014) Design of an Earth air tube heat exchanger (EAHE) for climatic condition of Chennai, India. Open Environ Sci 8:24–34
17. Yassine B, Ghali K, Ghaddar N, Chehab G, Srour I (2014) Effectiveness of the earth tube heat exchanger system coupled to a space model in achieving thermal comfort in rural areas. Int J Sustain Energ 33(3):567–586
18. Zhao Y, Li R, Ji C, Huan C, Zhang B, Liu L (2019) Parametric study and design of an earth-air heat exchanger using model experiment for memorial heating and cooling. Appl Therm Eng 148:838–845
19. Zukowski M, Topolanska J (2018) Comparison of thermal performance between tube and plate ground air heat exchangers. Renew Energy 115:697–710

Natural Convection in a Square Porous Enclosure with a Pair of Fluid-Pockets

Jayesh Subhash Chordiya and Ram Vinoy Sharma

1 Introduction

Thermal analysis of porous medium has been studied widely by numerous authors over the past few decades. The studies reported can be extensively categorized based on geometry, numerical scheme, flow assumptions, type of model, etc. The research in natural convection within a porous media is now steering more towards finding ways to enhance and/or attenuate the heat transfer rate. For example, the works related to porous media and non-Newtonian fluids like Casson fluids, Nano-fluids, etc., are gaining ground rapidly for enhancement of heat transfer, while attenuation of heat transfer rate is usually encountered in thermal insulation, geophysics, thermal heat storage, packed bed reactors, etc. In the latter case, partitioning the enclosure comes in handy that yields useful and productive results. This paper investigates the effect of employing small-sized fluid pockets within a differentially heated porous enclosure, with a view of reducing the rate of heat transfer and fluid flow across it.

Earlier work by references [1–4] showed how introducing obstructions can reduce the heat transfer in a porous media. Similarly, reference [5] numerically predicted natural convection in a square enclosure with partition; references [6, 7] used Lattice–Boltzmann technique to numerically analyze the effect of partitions attached to its cold walls on heat transfer in an inclined square enclosure; reference [8] studied the influence of internal heat generation in differentially heated and partially divided square cavities; reference [9] investigated the effect on the position of partitions on natural convection flow in square-shaped enclosures; reference [10] reported the effect of multiple internal partitions within a porous enclosure by varying their size

J. S. Chordiya (✉) · R. V. Sharma
National Institute of Technology Jamshedpur, Jamshedpur, India
e-mail: jayesh.subhash@gmail.com

M. Palanisamy et al. (eds.), *Theoretical, Computational, and Experimental Solutions to Thermo-Fluid Systems*, Lecture Notes in Mechanical Engineering,
https://doi.org/10.1007/978-981-33-4165-4_7

and position; references [11, 12] examined the natural convection flow in triangular-shaped porous enclosures and on diagonally divided square enclosure respectively. Recent work by reference [13] reported the effect of a pair of solid partitions within the porous enclosure, while [14, 15] reported the influence of thick-square-wave partition and comparison of various wavy partitions respectively in porous medium.

In all intents and purposes, the majority of the work undertaken for studying the influence of partition or obstructions in a natural convection flow within a porous medium consisted of a solid obstruction. In view of this, the authors are encouraged to examine the effect of a fluid obstruction, its position, and orientation within a differentially heated porous enclosure with a view of obtaining a substantial reduction in heat transfer rate.

2 Mathematical Formulation

A square porous enclosure of side H, has right and left wall at cold and hot temperatures T_c and T_h, respectively. The top and bottom walls are insulated. Two fluid pockets of the same horizontal and vertical dimensions h and l, respectively, are placed at a distance of c_x from the nearest vertical wall and c_y from the nearest horizontal wall, respectively, as shown in Fig. 1.

The flow is assumed to be incompressible, steady, 2D, and laminar. The enclosure is filled with a homogeneous, isotropic, and fluid-saturated porous medium having local thermal equilibrium with the solid matrix. The fluid pockets are assumed to be filled with air. The fluid pockets and porous domain are separated with an infinitesimally thin impermeable diathermal wall. All thermophysical properties are assumed to be constant accept density in the momentum equation (Boussinesq approximation). Viscous and radiation effects are ignored in the energy conservation equation. Since the primary purpose is to access the effect of fluid-pockets within a differentially

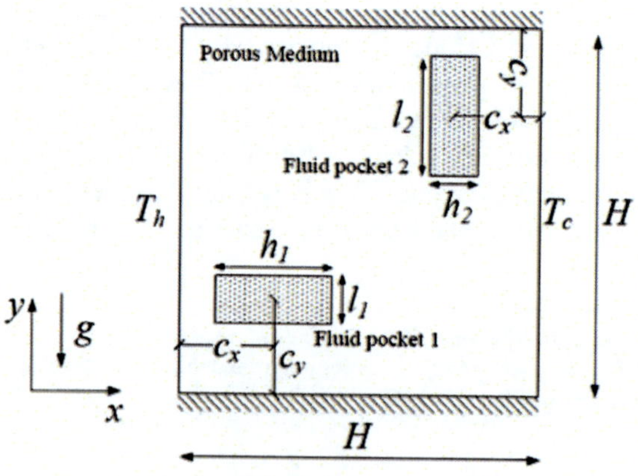

Fig. 1 Schematic diagram and coordinate system

heated porous enclosure, Darcy flow model is used. The dimensionless governing equations for porous domain thus takes the following form.

Porous Region:

$$\frac{\partial^2 \psi_p}{\partial X^2} + \frac{\partial^2 \psi_p}{\partial Y^2} = -\mathrm{RaDa}\frac{\partial \theta_p}{\partial X} \tag{1}$$

$$\frac{\partial \psi_p}{\partial Y} \cdot \frac{\partial \theta_p}{\partial X} - \frac{\partial \psi_p}{\partial X} \cdot \frac{\partial \theta_p}{\partial Y} = \frac{\partial^2 \theta_p}{\partial X^2} + \frac{\partial^2 \theta_p}{\partial Y^2} \tag{2}$$

Fluid Region:

$$\frac{\partial^2 \psi_f}{\partial X^2} + \frac{\partial^2 \psi_f}{\partial Y^2} = \Omega_f \tag{3}$$

$$\frac{\partial \psi_f}{\partial Y} \cdot \frac{\partial \Omega_f}{\partial X} - \frac{\partial \psi_f}{\partial X} \cdot \frac{\partial \Omega_f}{\partial Y} = \Pr\left(\frac{\partial^2 \Omega_f}{\partial X^2} + \frac{\partial^2 \Omega_f}{\partial Y^2}\right) + \mathrm{Ra}\,\Pr\frac{\partial \theta_f}{\partial X} \tag{4}$$

$$\frac{\partial \psi_f}{\partial Y} \cdot \frac{\partial \theta_f}{\partial X} - \frac{\partial \psi_f}{\partial X} \cdot \frac{\partial \theta_f}{\partial Y} = \frac{\partial^2 \theta_f}{\partial X^2} + \frac{\partial^2 \theta_f}{\partial Y^2} \tag{5}$$

Here,

$$X, Y = \frac{x, y}{H}; U, V = \frac{u, v}{\alpha/H}; \theta = \frac{T - T_c}{T_h - T_c}; \mathrm{Ra} = \frac{g\beta(T_h - T_c)H^3}{\alpha v}; \mathrm{Da} = \frac{K}{H^2}; \Pr = \frac{v}{\alpha}$$

Equations (2–5) are subjected to the following boundary conditions,

$$\psi = 0, \theta = 1, \Omega = -\frac{\partial^2 \psi}{\partial X^2} \ldots X = 0, 0 \le Y \le 1$$

$$\psi = 0, \theta = 0, \Omega = -\frac{\partial^2 \psi}{\partial X^2} \ldots X = 1, 0 \le Y \le 1$$

$$\psi = 0, \frac{\partial \theta}{\partial Y} = 0, \Omega = -\frac{\partial^2 \psi}{\partial Y^2} \ldots Y = 0 \text{ and } 1, 0 \le X \le 1 \tag{6}$$

Appropriate conditions at the walls of fluid pockets are set as,

$$\psi = 0, \left.\frac{\partial \theta}{\partial N}\right|_p^- = \left.\frac{\partial \theta}{\partial N}\right|_f^+ \tag{7}$$

Here, N is the normal direction of heat flow, which can be X for vertical walls and Y in case of horizontal walls. Heat transfer is evaluated by calculating local and average Nusselt number, which is the ratio of convection to conduction heat transfer rate.

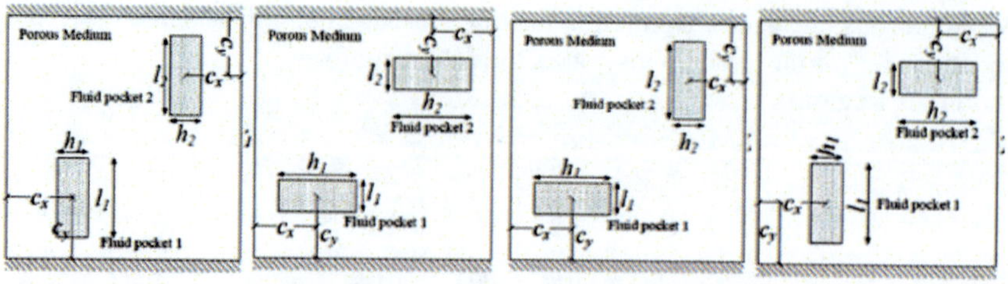

Fig. 2 Type of configurations under study

$$\mathrm{Nu}_L = -\frac{\partial \theta}{\partial X}\bigg|_{X=0,1} \tag{8}$$

$$\mathrm{Nu} = \int_0^1 \mathrm{Nu}_L(Y)\mathrm{d}Y \tag{9}$$

In the current study, four types of configurations are under study, namely VV, HH HV, and VH type. These four types are obtained by changing the dimensions of the fluid-pockets within the porous enclosure, as shown in Fig. 2.

3 Numerical Method

The partial differential equation Eqs. (2–5), corresponding boundary conditions, and appropriate matching condition at fluid-pocket walls are discretized using an overall II order finite difference scheme with central differencing for interior nodes and forward, backward differencing for boundary nodes and nodes at partition. The discretization process converts the differential form into the algebraic form of equations, which are solved using the Successive Accelerated Replacement scheme, whose applicability has been demonstrated by various researchers like [16–19]. The basic approach of this scheme is to approximate a profile that satisfies the boundary conditions for each transport property. Let the error in the equation at (*i*,*j*) of nth iteration be $\tilde{\sigma}_{ij}^n$. The $(n+1)$th approximation to the variable σ is obtained as,

$$\sigma_{ij}^{n+1} = \sigma_{ij}^n - \omega\frac{\tilde{\sigma}_{ij}^n}{\partial\tilde{\sigma}_{ij}^n/\partial\sigma_{ij}^n} \tag{10}$$

Here, i and j vary from 1 to $i_{\max}$ and 1 to $j_{\max}$, respectively. The convergence criteria is,

$$\frac{\sum_{i=2}^{i\max-1}\sum_{j=2}^{j\max-1}\left|\sigma_{ij}^{n+1}-\sigma_{ij}^{n}\right|}{\sum_{i=2}^{i\max-1}\sum_{j=2}^{j\max-1}\left|\sigma_{ij}^{n+1}\right|} < \varepsilon \tag{11}$$

Here, ε is a small tolerance value and ω is the acceleration factor. Nusselt number is calculated locally at every node along the hot and cold wall. These values are further numerically integrated by Simpson's 1/3rd scheme to obtain the average Nusselt number.

4 Results and Discussion

4.1 Validation of Numerical Model

A grid sensitivity test is performed to ensure the solution independent of grid size. A uniform structured square mesh ranging from 60×60 to 100×100 is checked for $Ra = 100$ and 1000. Additionally, the results reported in the literature are also compared with the present solution to validate the code for classical natural convection problem of differentially heated porous enclosure without obstructions. Table 1 shows the test and validation performed for the aforementioned values. It is evident from the table that a grid size of 80×80 is deemed sufficient enough for an accurate solution keeping in mind the computational cost and time. It can be stated from the validations that the current code yields satisfactory results as compared to those reported in the literature, and hence it can be used with greater assurance for the problem addressed in this paper.

In the present study, numerical analysis of natural convection heat transfer and fluid flow is performed for a differentially heated porous enclosure embedded with two fluid pockets. The parameters under consideration are Rayleigh number ($Ra = 10^4$–10^5), Darcy number ($Da = 10^{-2}$), Prandtl number ($Pr = 0.71$), horizontal and vertical fluid-pocket distance ($0.2 \le c_x, c_y \le 0.8$), and dimensions of rectangular fluid-pocket as ($h,l = 0.3, 0.1$) depending upon the case. The value of acceleration

Table 1 Grid sensitivity test and code validation with previously reported results

Ra	Grid size	Nu_h	Nu_c	Iterations	[20]	[21]
100	60 × 60	3.273	3.273	1238	4.2	3.141
	80 × 80	3.146	3.148	1699		
	100 × 100	3.144	3.143	2208		
1000	60 × 60	16.135	16.135	6141	15.8	13.45
	80 × 80	15.63	15.62	8001		
	100 × 100	15.61	15.62	8676		

factor ω is 0.9 to optimize the number of iterations without affecting the accuracy and stability of the solution.

Figure 3 illustrates streamlines and isotherms in all configurations of fluid pockets understudy for $Ra = 10^4$; $Da = 10^{-2}$; $Pr = 0.71$; $h, l = 0.1, 0.3$. The horizontal distance of fluid pockets from the nearest vertical wall, c_x is fixed to 0.2; while the vertical distance, c_y is symmetrically changed to 0.2, 0.5, and 0.8. Thus, the effect of the vertical position of fluid pockets can be manifested from this contour diagram. In case of a porous enclosure without any obstruction or fluid pockets, the value of Nusselt number (*Nu*) and maximum absolute stream function ($|\psi|_{max}$) is about 3.15 and 4.7, respectively. It can be noticed from Fig. 3a, the value of $|\psi|_{\max}$ is reduced to approximately 2.5, 2.9, 2.7, and 2.7 for case I, II, III, and VI, respectively, while *Nu* has dropped to 1.7, 1.9, 1.8, and 1.8, respectively. This decrease is almost around 50%, which may be seen as a substantial drop. This reduction in heat transfer rate is comparable to the centrally located straight vertical diathermal partition inserted within the porous enclosure and can serve as an alternative to it. The reduction in the strength of flow and rate of heat transfer may be attributed to the presence and positioning of fluid-pockets. The isotherms, in the southwest and northeast regions of the enclosure, can be seen to be occurring in a bunch and steep in nature. This indicates a high rate of heat transferring in these particular regions. Case-I yields the maximum reduction because the vertical orientation of the fluid pockets in Case-I is predominantly in these locations. This results in effective obstruction of flow and engaging the flow of heat into the fluid-pockets in the regions of high-temperature gradients.

However, Fig. 3b reveals a further decrease in the values of *Nu* and $|\psi|_{\max}$. This is because, in this case, the fluid-pockets not only affect the high-temperature gradient regions but also interfere with the inner stagnant portion of streamlines. It can be clearly noticed that the size of the inner circular portion of streamlines is relatively Fig. 3b as compared to that in Fig. 3a. Convincingly, in Fig. 3c, since the fluid pockets are positioned away from the high-temperature gradient regions, the drop in *Nu* and $|\psi|_{\max}$ is not very significant. Figure 4 demonstrates the values of *Nu* for all possible positions of fluid pockets within the porous enclosure estimated for all the four configurations. Dark blue colored regions signify the locations where maximum reduction has been obtained, while the red-colored regions indicate the locations with minimum reduction in the values of *Nu*. For almost all the configurations, the region of maximum reduction falls approximately at the same location, that is, centrally positioned in the horizontal direction ($c_x = 0.5$) and quarterly positioned in the vertical direction ($c_y = 0.3$ and 0.7) from the nearest horizontal wall. It can be attributed to the fact that these are the positions where maximum obstruction in the fluid and heat transfer flow is obtained. Similar predictions can be made from Fig. 5a, b. For $c_y = 0.3$, centrally positioned fluid-pocket ($c_x \sim 0.5$) in horizontal direction yields the least value of *Nu* for all types of configurations. Whereas for $c_x = 0.5$, quarterly positioned fluid-pocket ($c_y \sim 0.5$) in vertical direction yields the least value of Nu for all types of configurations. Figure 5c shows the effect of fluid-pockets within a porous enclosure for the optimized location and compared with that of a porous enclosure without fluid pockets. The decrease in *Nu* can be seen to

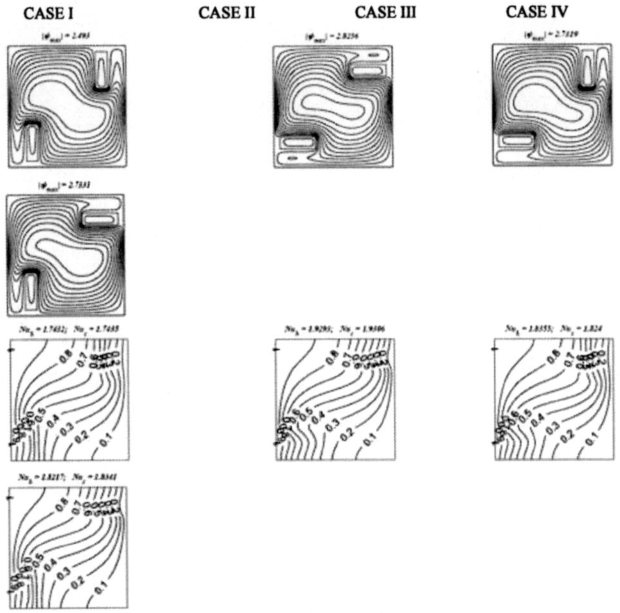

a. $c_x = c_y = 0.2$ ($Ra = 10^4$; $Da = 10^{-2}$; $Pr = 0.71$; h, $l = 0.1, 0.3$)

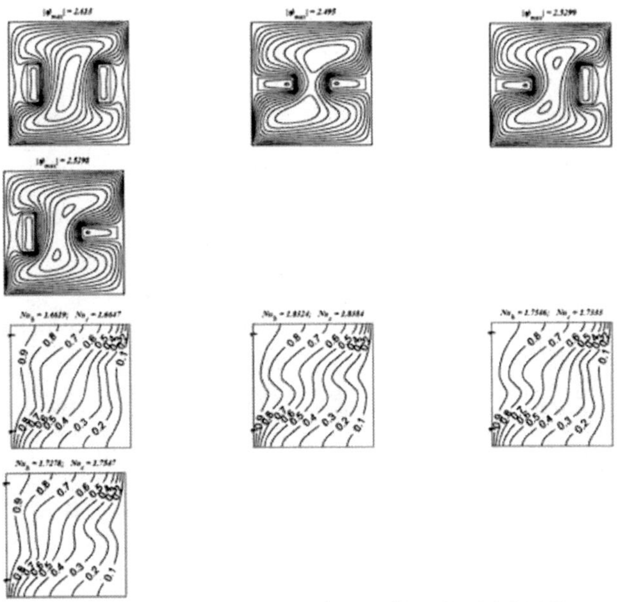

b. $c_x = 0.2$; $c_y = 0.5$ ($Ra = 10^4$; $Da = 10^{-2}$; $Pr = 0.71$; h, $l = 0.1, 0.3$)

Fig. 3 Streamlines and isotherms for all configurations

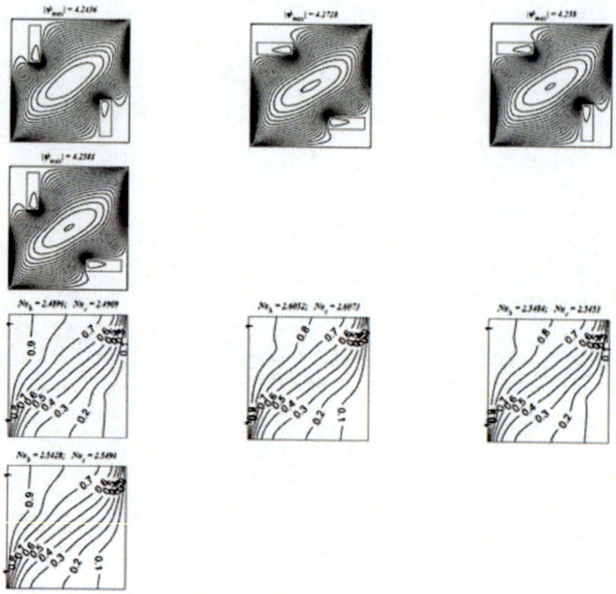

c. $c_x = 0.2$; $c_y = 0.8$ ($Ra = 10^4$; $Da = 10^{-2}$; $Pr = 0.71$; h, $l = 0.1, 0.3$)

Fig. 3 (continued)

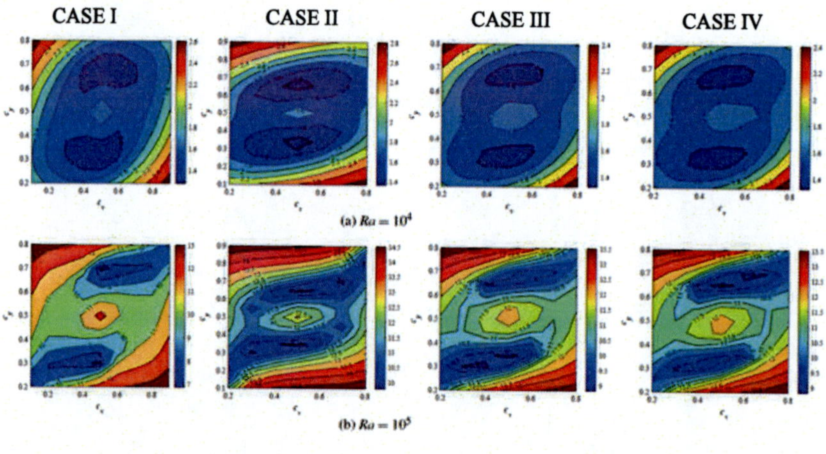

Fig. 4 Values of *Nu* at all combinations of c_x and c_y for all type of fluid-pockets

be approximately 50% for each case, with the highest drop for VV-Type (case-I) of fluid-pocket configuration. This comparison justifies the importance of the current investigation.

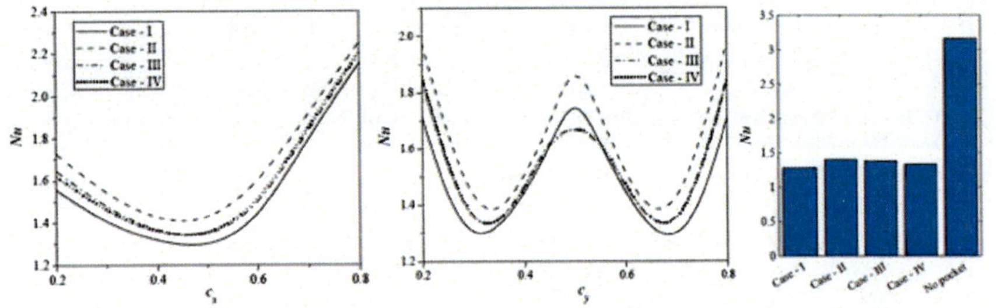

(a) $c_y = 0.3$ (b) $c_x = 0.5$ (c) $c_x = 0.5; c_y = 0.3$

Fig. 5 Comparison of all cases relative to: **a** vertical location and **b** horizontal location **c** Optimized location ($Ra = 10^4$; $Da = 10^{-2}$; $Pr = 0.71$)

5 Conclusions

The present numerical investigation examines the natural convection flow developed within a differentially heated porous enclosure with fluid-pockets. The position and configuration of fluid pockets are estimated for maximum reduction in heat transfer rate. The heat and fluid flow has been observed for four different types of configuration based on dimensions of fluid-pockets using streamlines and isotherms. Heat transfer results have been evaluated by calculating *Nu* for all fluid-pocket locations of each configuration and finally compared it with the *Nu* value of porous enclosure without any fluid-pocket to estimate the degree of reduction. Following conclusions may be drawn from the current numerical investigations:

- The presence of fluid-pockets substantially reduces the rate of heat transfer across a differentially heated porous enclosure.
- The position and dimensions of the fluid pocket play an important role in the degree of reduction in values of *Nu* and $|\psi|_{max}$.
- For two-fluid pockets, centrally located in a horizontal direction and quarterly located in vertical direction yields the maximum reduction in heat transfer rate.
- In addition, the highest drop in the heat transfer rate is obtained when both the fluid pockets are vertically oriented.

References

1. Anderson R, Bejan A (1981) Heat transfer across a vertical impermeable partition embedded in porous medium. Int J Heat Mass Transf 24(7):1237–1245
2. Anderson R, Bejan A (1981) Heat transfer through single and double vertical walls in natural convection: theory and experiment. Int J Heat Mass Transf 24:1611–1620
3. Nansteei MW, Greif R (1981) Natural Confection in Undivided and Partially Divided Rectangular Enclosures. J Heat Transfer 103:623–629
4. Tong TW, Gerner FM (1986) Natural convection in partitioned air-filled rectangular enclosures. Int Commun Heat Mass Transf 13:99–108

5. Kanchan MK, Patankar SV (1990) Numerical prediction of natural convection in square partitioned enclosures. Numer Heat Transf, Part a: Appl: Int J Comput Methodol 17(3):269–285
6. Mezrhab A, Jami M, Abid C, Bouzidi M, Lallemand P (2006) Lattice- Boltzmann modelling of natural convection in an inclined square enclosure with partitions attached to its cold wall. Int J Heat Fluid Flow 27:456–465
7. Mezrhab LB (1999) Radiation-natural convection interactions in partitioned cavities. Int J Numer Methods Heat Fluid Flow 9(2):186–203
8. Oztop H, Bilgen E (2006) Natural convection in differentially heated and partially divided square cavities with internal heat generation. Int J Heat Fluid Flow 27:466–475
9. Selamat S, Hashim I (2014) Effects of position of partitions on natural convection in squares enclosures. AIP Conf Proc 1614:886–890
10. Chordiya JS, Sharma RV (2018a) Numerical study on the effects of multiple internal diathermal obstructions on natural convection in a fluid-saturated porous enclosure. Archive Mech Eng 65(4):553–578
11. Varol Y, Oztop HF, Varol A (2007) Natural convection in porous triangular enclosures with a solid adiabatic fin attached to the horizontal wall. Int Commun Heat Mass Transf 34:19–27
12. Varol Y, Hakan F, Oztop, Pop I (2009) Natural convection in a diagonally divided square cavity filled with a porous medium. Int J Therm Sci 48:1405–1415
13. Chordiya JS, Sharma RV (2018b) Conjugate natural convection in a fluid-saturated porous enclosure with two solid vertical partitions. Heat Transf Asian Res 47(8):1031–1047
14. Chordiya JS, Sharma RV (2020) Conjugate natural convection in porous medium with a thick square-wave partition. ASME. J Thermal Sci Eng Appl 13(1): 011006. https://doi.org/10.1115/1.4046607
15. Chordiya JS, Sharma RV (2020) Numerical analysis on the effect of wavy partitions on natural convection in porous enclosure. ASME. J Heat Transfer 142(9): 092601. https://doi.org/10.1115/1.4047502
16. Chordiya JS, Sharma RV (2018c) Natural convection in inclined rectangular porous enclosure with diathermal partition wall. Carbon Sci Tech 10(1):13–18
17. Chordiya JS, Sharma RV (2019a) Numerical study on effect of corrugated diathermal partition on natural convection in square porous cavity. J Mech Sci Technol 33(5):2481–2491
18. Chordiya JS, Sharma RV (2019b) Study of natural convection in fluid-saturated porous thermal insulations with multiple inclined diathermal partition. Int J Heat Technol 37(1):123–130
19. Chordiya JS, Sharma RV (2019c) Natural convection in fluid-saturated porous enclosure with a pair of vertical diathermal partition. Int J Therm Sci 144:42–49
20. Bejan A (1979) On the boundary layer regime in a vertical enclosure filled with a porous medium. Lett Heat Mass Transf 6(2):93–102
21. Gross R, Bear M, Hickox C (1986) The application of flux-corrected transport (FCT) to high Rayleigh number natural convection in a porous medium. In: Proceedings of the 8th International Heat Transfer Conference, San Francisco, CA

Numerical Analysis of Natural Convection in a Partially Open Square Cavity with Multiple Heat Sources

Gloria Biswal and Aurovinda Mohanty

Nomenclature

- g Acceleration due to gravity m/s^2
- H Height of the enclosure m
- L Length of the enclosure m
- P Dimensionless pressure
- p Pressure Pa
- q Heat flux W/m^2
- U Dimensionless velocity in x direction
- u Velocity in x direction m/s
- V Dimensionless velocity in y direction
- v Velocity in y direction m/s
- X Dimensionless horizontal distance
- x Horizontal distance m
- Y Dimensionless vertical distance
- y Vertical distance m
- T Temperature K

G. Biswal
Former PG Student, Departmant of Mechanical Engineering, VSSUT, Burla, Burla, India

A. Mohanty (✉)
Departmant of Mechanical Engineering, VSSUT, Burla, Burla, Odisha, India
e-mail: amohanty_me@vssut.ac.in

M. Palanisamy et al. (eds.), *Theoretical, Computational, and Experimental Solutions to Thermo-Fluid Systems*, Lecture Notes in Mechanical Engineering,
https://doi.org/10.1007/978-981-33-4165-4_8

Symbols

ρ Density kg/m^3
ν Kinematic viscosity m^2/s
β Thermal expansion coefficient 1/K

1 Introduction

Natural convection heat transfer in the enclosure has been a topic of many experimental and numerical studies. From the technical and practical point of views, it has gained many applications, which consists of solar energy collectors, heating and cooling of houses, energy drying processes, etc. Most of the works that were done earlier consist of a single heat source but work related to a number of heat sources are very less. In electronic systems generally, the chips are embedded in circuit boards, which are the primary sources of heat generation. Along with the increase in electronic miniaturization, the electronic density is also increasing. So an investigation has been carried out to know the effect of the presence of a number of heat sources on the heat transfer rate.

Many research works had been done to analyze the heat transfer inside an enclosure, which has many practical applications in day-to-day life. Some of the earlier experiments and their analysis are given below for a better understanding of the natural convection heat transfer inside an enclosure. Du and Bilgen [1] studied the coupling effect of the conduction in solid wall and convection in fluid flow for the various parameters such as the conductivity ratio, aspect ratio, Rayleigh number, and solid wall thickness. They considered the simple two-dimensional domains with the constant heat flux applied to the solid vertical wall, and the opposite wall is assumed to be insulated. Aydin and Yang [2] numerically investigated the natural connection of air in a vertical square cavity with localized isothermal heating from below and symmetrical cooling from sidewalls. Deng et al. [3] numerically studied a two-dimensional laminar natural convection in a rectangular enclosure with discrete heat sources on walls in the unsteady regime. Ampofo and Karayiannis [4] conducted an experimental study of low-level turbulence natural convection in an air-filled vertical square cavity. Mixed convection heat transfer in a two-dimensional rectangular cavity with constant heat flux from a partially heated bottom wall, while the isothermal sidewalls are moving in the vertical direction, is studied numerically by [5].

Bilgen and Ozotop [6] carried out a numerical study on natural convection heat transfer in a partially opened square cavity which is inclined at some angle. Incropera [7] has discussed the efficient path for heat removal from electronic devices. Mustafa et al. [8] performed an experiment on natural convection in a trapezoidal enclosure with partial heating from below and symmetrical cooling from the sides. Nardini et al. [9] analyzed the influence of the heating source, which is positioned in the lateral walls of the square cavity. Fontana et al. [10] investigated the natural and mixed convection in partially open cavities with internal heat sources using a transient three-dimensional model. Habiba [11] investigated unsteady natural convection laminar flows in a square cavity formed by insulated bottom and top walls, uniformly heated left wall, and cooled right wall.

2 Problem Description

The physical model of the present analysis is shown in Fig. 1, along with the boundary conditions. Analyses have been done in two dimensional systems. It consists of a rectangular enclosure of dimensions, L X H. The side walls are maintained at a constant temperature of T_c. The bottom walls at the point of insertion of heat sources are uniformly given constant heat flux q, and the remaining parts of the bottom wall and the entire portion of the upper walls are adiabatic. In the present study, six different dimensionless spacing between the heat sources, different opening size and opening location are taken for the analysis. The heat source material is taken as aluminum because of its high thermal conductivity.

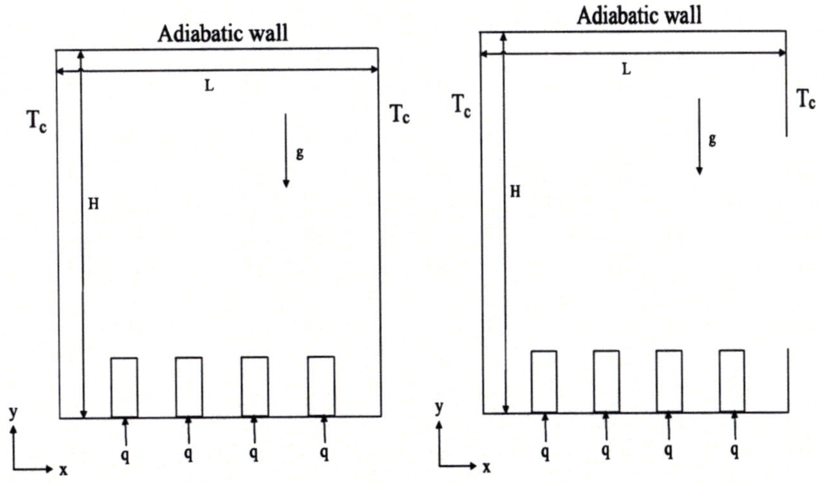

(a) Computational domain without opening (b) computational domain with opening

Fig. 1 Schematic diagram of computational domain

3 Mathematical Modeling

The basic assumptions of this flow model are that the fluid is Newtonian and the flow is steady and laminar. All the thermos-physical properties are assumed to be constant except the density, which is not constant and appears in the momentum equation and signifies the body force term. So to relate the density changes with respect to the change in temperature, Boussinesq approximation is considered. Here viscous dissipation is assumed to be negligible.

The governing equation for steady laminar natural convection can be written as:

Continuity equation:

The continuity equation is

$$\frac{\partial u}{\partial x} + \frac{\partial v}{\partial y} = 0 \tag{1}$$

Momentum equation:

The corresponding x-momentum and y-momentum are given below

$$u\frac{\partial u}{\partial x} + v\frac{\partial u}{\partial y} = -\frac{1}{\rho}\frac{\partial p}{\partial x} + \nu\left(\frac{\partial^2 u}{\partial x^2} + \frac{\partial^2 u}{\partial y^2}\right) \tag{2}$$

$$u\frac{\partial v}{\partial x} + v\frac{\partial v}{\partial y} = -\frac{1}{\rho}\frac{\partial p}{\partial y} + \nu\left(\frac{\partial^2 v}{\partial x^2} + \frac{\partial^2 v}{\partial y^2}\right) + g\beta(T - T_C) \tag{3}$$

Energy equation:

The energy equation is

$$u\frac{\partial T}{\partial x} + v\frac{\partial T}{\partial y} = \alpha\left(\frac{\partial^2 T}{\partial x^2} + \frac{\partial^2 T}{\partial y^2}\right) \tag{4}$$

Boundary condition:

For the existent problem the boundary conditions are defined as follows:

No slip velocity condition:

$$U(X, 0) = U(X, 1) = U(0, Y), U(1, Y) = 0$$

$$V(X, 0) = V(X, 1) = V(0, Y) = V(1, Y) = 0$$

At the top wall: $\frac{\partial \theta}{\partial Y}(X, 1) = 0$

Left wall: $\theta(0, Y) = 0$

Right wall: $\theta(1, Y) = 0$

Bottom wall except at the insertion of heat source: $\frac{\partial \theta}{\partial Y}(X, 0) = 0$

The average heat transfer coefficient: $\overline{h} = \frac{1}{L}\int_0^L h_x dx$

The average Nusselt number can be designed as: $\overline{Nu} = \frac{\overline{h}L}{k}$

4 Numerical Procedure

A set of algebraic equations were obtained by discretizing the governing differential equations and then integrating it over a control volume with the help of finite volume techniques. With the help of a multigrid solver of FLUENT 15, the algebraic equations were solved iteratively by imposing the boundary conditions. For pressure discretization, a body force weighted scheme was used. For the discretization of the convective terms of the momentum and energy equations, second-order upwind (SOU) scheme was used in order to have better accuracy. The diffusion terms in the momentum and energy equations were discretized by using a central difference scheme (CDS). For coupling the velocity and pressure terms in the pressure correction equation, the SIMPLE (Semi Implicit Method for Pressure Linked Equation) algorithm was used. The relative criterion of convergence for energy equation was restricted to 10^{-6}.

For validating the present work, the work of Guo and Sharrif [5] has been considered. Figure 2 shows the comparison of local Nusselt number along with the heat source for a given Reynolds number and Richardson number. The present work is matched with their work. The present computational result agrees well with the computational result as obtained by them, as shown in Fig. 2. Grid refinement work was done in a rectangular cavity with aspect ratio = 1 and Rayleigh number = 10^4 in order to know the effect of grid size on the heat transfer rate. It is observed that after the grid size of 81 × 81, there is very little variation in the average Nusselt number. So grid independency is achieved with 81 × 81 grid size. So all the computations are carried out with this grid size.

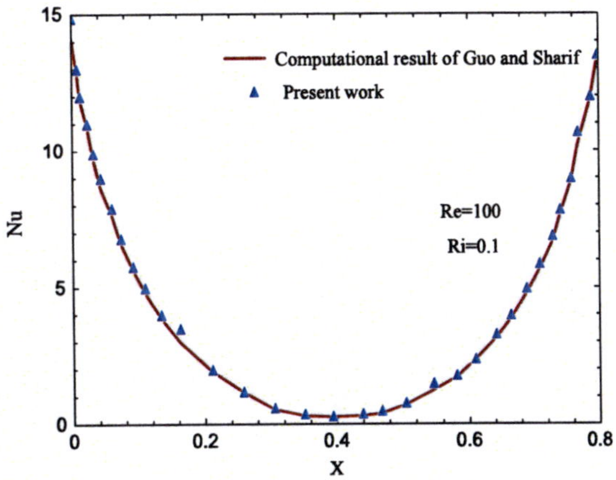

Fig. 2 Variation of local Nusselt number along the surface of heat source

5 Result and Discussion

The results for various input parameters are presented for different Rayleigh numbers.

5.1 *Effect of* s* *on Average Nusselt Number*

The effect of dimensionless spacing (s^*) ratio between the source as a function of average Nusselt number (Nu) and Rayleigh number (Ra) is shown in Fig. 3. The dimensionless spacing ratio is varied from 0.03 to 0.25. The dimensionless spacing ratio (s^*) and the number of heat source are inversely proportional to each other, that is, if the s^* decreases, then the number of heat source will increase and vice versa. It can be observed from Fig. 3 that the Nusselt number increases up to a maximum at around $s^* = 0.2$, and after that, it further decreases for any value of Rayleigh number (Fig. 3).

Figure 4 shows the temperature plume across the heat sources for different dimen-

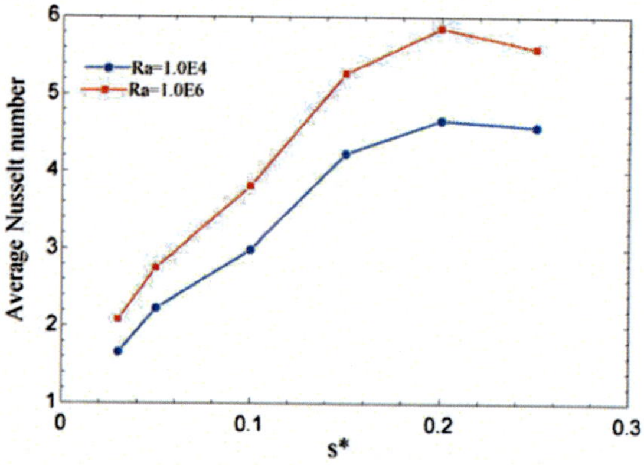

Fig. 3 Variation of average Nusselt number with dimensionless spacing for different Rayleigh number

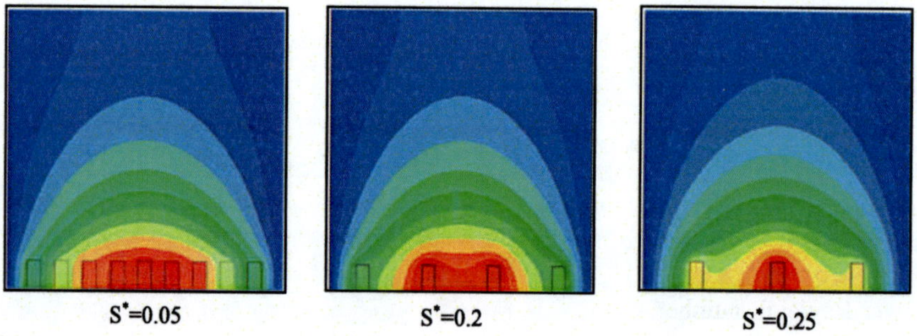

Fig. 4 Temperature plume with varying S*

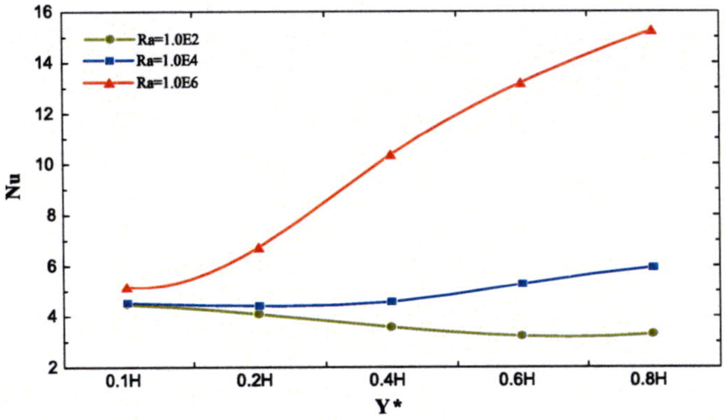

Fig. 5 Variation of average Nusselt number with opening size for different Rayleigh number

sionless spacing ratio. The maximum heat is concentrated at the center of the enclosure because it is surrounded by the heat sources on both sides. As we go away from the center the temperature goes on decreasing because, toward the wall, the heat transfer takes place between the heat source and the cold wall.

5.2 *Effect of the Size of Opening on Average Nusselt Number*

The effect of the size of the opening of the rectangular cavity with an aspect ratio of 1 on the average Nusselt number is shown in Fig. 5. When the Rayleigh number is large, the average Nusselt number increases with the increase of the size of the opening of the cavity because the finiteness of the cavity increases, which helps to remove the heat dissipated from the inserted heat source. This graph shows that the convection strength is a strong function of the Rayleigh number. When the Rayleigh number is low, the heat cannot easily get out of the cavity because the recirculation intensity is less.

6 Conclusion

The simulation has been carried for the range of $Ra = 10^2$ to 10^6. The average Nusselt number becomes maximum when the non-dimensional spacing between heat sources in an enclosure becomes 0.2. The Nusselt number also increases with the increase in Rayleigh number. The average Nusselt number increases with opening size at a lower Rayleigh number. However, at a low Rayleigh number, the Nusselt number does not increase significantly with opening size.

Acknowledgements The authors do hereby acknowledge the Department of Mechanical engineering, VSSUT Burla and TEQIP-III for providing all the infrastructure facility and financial support for carrying out this research work.

References

1. Du ZG, Bilgen E (1992) Coupling of wall conduction with natural convection in a rectangular enclosure. Int J Heat Mass Transf 35(8):1969–1975
2. Aydin O, Yang WJ (2000) Natural convection in enclosures with localized heating from below and symmetrical cooling from sides. Int J Numer Methods Heat Fluid Flow 10(5):518–529
3. Deng Q, Tang G, Li Y (2002) A combined temperature scale for analyzing natural convection in rectangular enclosures with discrete wall heat sources. 45:3437–3446
4. Ampofo F, Karayiannis TG (2003) Experimental benchmark data for turbulent natural convection in an air filled square cavity. Int J Heat Mass Transf 46(19):3551–3572
5. Guo G, Sharif MAR (2004) Mixed convection in rectangular cavities at various aspect ratios with moving isothermal sidewalls and constant flux heat source on the bottom wall. Int J Therm Sci 43(5):465–475
6. Bilgen E, Oztop H (2005) Natural convection heat transfer in partially open inclined square cavities. Int J Heat Mass Transf 48(8):1470–1479
7. Incropera FP (2009) Convection heat transfer in electronic equipment cooling. J Heat Transfer 110(4b):1097
8. Mustafa AW (2011) Natural convection in parabolic enclosure heated from below. Mod Appl Sci 5(3):76–85
9. Nardini G, Paroncini M, Corvaro F (2014) Effect of heat transfer on natural convection in a square cavity with two source pairs. Heat Transf Eng 35(9):875–886
10. Fontana É, Capeletto CA, Da Silva A, Mariani VC (2015) Numerical analysis of mixed convection in partially open cavities heated from below. Int J Heat Mass Transf 81:829–845
11. Habiba F (2017) Natural convection flow in a square cavity with temperature dependent heat generation. IOSR J Math 13(03):10–17

Simulation Studies of Heat Transfer by Natural Convection from an Isothermal Rectangular Cylinder Using ANSYS Fluent

M. Naveen, B. Venkata Sai Raghu Vamsi, M. R. Ch. Sastry, and T. Siva Krishna

Nomenclature

T_S	Inner cylinder surface temperature (°C)
a	Vertical dimension of the inner cylinder
b	Horizontal dimensional of the inner cylinder
a/b	Aspect ratio
g	Gravitational acceleration m s^{-2}
k	Thermal conductivity W m^{-1} K^{-1}
γ	Kinematic viscosity m^2 s^{-1}
β	Thermal expansion coefficient K^{-1}
Pr	Prandtl number
h	Average heat transfer coefficient W m^{-2} K^{-1}

1 Introduction

Convection is a surface phenomenon and is a combination of both conduction and advection, playing a dominant role in the energy transfer in many engineering applications. It is influenced by parameters such as fluid properties, material, geometry, and orientation of the physical structure and the boundary conditions.

At, the surface of the structure, during the energy transfer movement in the fluid, is observed either due to difference in densities (buoyancy effect), termed as natural

M. Naveen · B. Venkata Sai Raghu Vamsi (✉) · M. R. Ch. Sastry · T. Siva Krishna
Gudlavalleru Engineering College, Gudlavalleru, India
e-mail: bvsraghuvamsi@gmail.com

M. Palanisamy et al. (eds.), *Theoretical, Computational, and Experimental Solutions to Thermo-Fluid Systems*, Lecture Notes in Mechanical Engineering,
https://doi.org/10.1007/978-981-33-4165-4_9

convection, or due to the driving force created by external agencies, such as fans, blowers, and pumps, etc., termed as forced convection. An experimental study is performed to study the wall effect on heat transfer from a confined isothermal circular cylinder in natural convection by Marcel Novomestský [1] for maximum Rayleigh number 10^5. Results conclude that the correlation developed between Nusselt number and Rayleigh number with the inclusion of confinement ratio helps to study the natural convection phenomenon better than without confinement proposed by Incropera. Samy et al. [2] developed a computer program to analyze the natural convection from a horizontal square cylinder numerically by varying Rayleigh number from 10^3 to 10^6 under laminar flow conditions. The numerical results obtained are observed as well in agreement with those in the literature.

An elaborated review of the energy transfer due to natural convection from horizontal, and vertical heated cylinders, which included the experimental, numerical, and analytical contributions, are presented by Anu nair et al. [3]. Clemes et al. [4] conducted experiments on long horizontal cylinders to study the heat transfer to air under isothermal conditions with the variation of Rayleigh number from 10^3 to 10^9 for both non-circular and circular sections. Correlations are developed relating to Nusselt number, Rayleigh number, conduction shape factor, and blending coefficient. Ali [5] reported the effect of duct aspect ratio, Grashoff number, and heat flux on the rate of heat transfer, the total and average heat transfer coefficient, and Nusselt number by conducting experiments with a uniform heat flux in natural convection. Correlations are developed using a modified Rayleigh number and compared the results with those existing in the literature. Calcagni et al. [6] analyzed the heat transfer in a square enclosure containing cold vertical walls with the bottom surface having a discrete heat supply, numerically, and experimentally. The effect of the length of the hot surface on the heating surface is studied, with the variation of Rayleigh number.

Jawad et al. [7] studied the free convection heat transfer from an isothermal hemispherical cavity under laminar flow conditions in unlimited space. They had developed correlations for analytical, experimental, and theoretical results independently. Nader et al. [8] reported the heat transfer in a 2-D enclosure, with heating from the bottom surface with a simultaneous cooling from the top subject to various boundary conditions. Chang et al. [9] studied the streamline patterns and temperature distributions in natural convection heat transfer from a concentrically placed square cylinder in a horizontal circular enclosure using the finite element modeling under fixed Prandtl number 0.708, using Galerkin approach. Aspect ratio is varied from 0.2 to 0.4, Rayleigh number from 10^3 to $5x10^5$. Experiments are conducted for aspect ratio 0.4 to validate the results. Smoke visualization technique is adopted to identify the streamlines experimentally.

This work deals with the study of the effect of aspect ratio on the natural convection heat transfer from an isothermal rectangular cylinder. The main objective of the present research is to determine the distribution of parameters such as velocity, static pressure, and temperature.

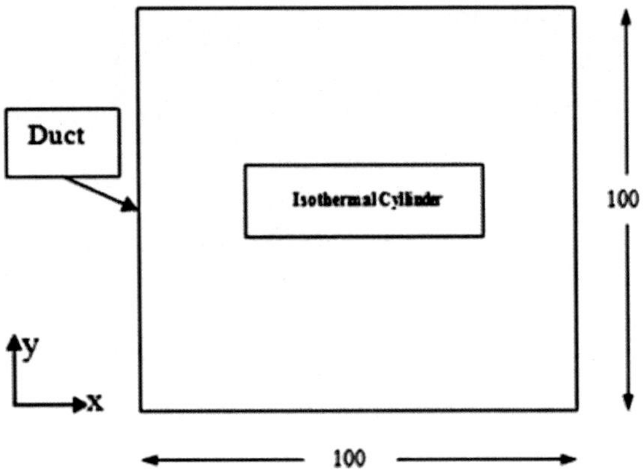

Fig. 1 Geometry (All dimensions are in mm)

2 Problem Definition

In this work, a rectangular cylinder enclosed in a square duct of fixed dimension is considered, as shown in Fig. 1. The duct and cylinder are assumed to be horizontal and concentric to each other, with air-filled between them. The entire inner cylinder is assumed to be having a uniform temperature. Hence the temperature gradient along the length (z-direction) of the cylinder is zero, whereas in the two lateral directions (x and y), the temperature gradient at the surface is non-zero. $\frac{\partial T}{\partial x} \neq 0$, $\frac{\partial T}{\partial y} \neq 0$, $\frac{\partial T}{\partial z} = 0$

Since the inner cylinder is completely enclosed by the outer cylinder, it is fully confined, and the heated air cannot escape. Along the length of the cylinder, the same is observed, and hence, the problem is solved by considering it as a two-dimensional problem.

3 Simulation in ANSYS Fluent

The problem is modeled and analyzed numerically using the CFD tool ANSYS fluent. The duct dimensions are fixed, and the aspect ratio of the inner cylinder is varied. The center to center distance of the duct and inner cylinder is fixed as zero for all the cases.

3.1 Optimality of Mesh

Optimization of the mesh improves the accuracy of the solution. In this work, a grid independence study is performed to optimize the size of the mesh, and the dependence of accuracy in maximum static pressure on the mesh size is presented in Fig. 2.

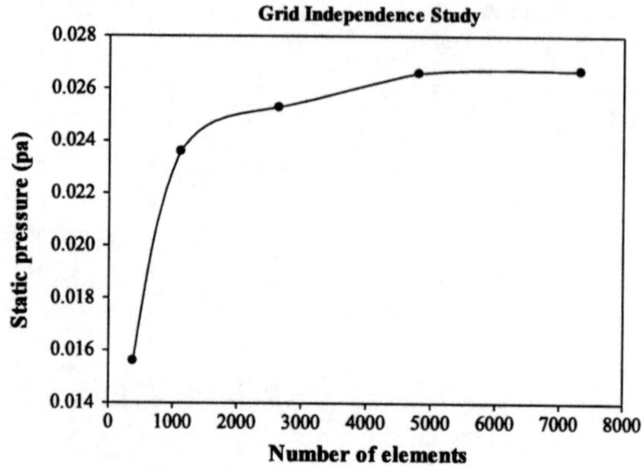

Fig. 2 Grid independence study

It has been observed that an increase in the number of elements beyond 4800 has no effect on the parameters, such as maximum velocity, temperature, and static pressure. Hence, it is taken as an optimum mesh.

3.2 Boundary Conditions

The surface temperature of the inner cylinder is varied from 100 to 200 °C with an increment of 50 °C, keeping the duct temperature constant at 30 °C. The initial temperature of the air is taken as 30 °C. Prandtl number is assumed to be constant as 0.7. Laminar flow conditions are maintained in the duct, and to attain the convergence, the tolerable value of residuals is set as 10^{-6}.

4 Results and Discussion

For the study of natural convection, the following non-dimensional numbers are calculated for all the cases.

Grashoff number,

$$\mathrm{Gr} = \frac{g\beta L_c^3 \Delta T}{\gamma^2}. \tag{1}$$

Rayleigh number,

$$\mathrm{Ra} = \mathrm{Gr} \cdot \mathrm{Pr} \tag{2}$$

For constant Prandtl number 0.7, Nusselt number is calculated using the relation proposed by Samy et al. [2] the

$$\mathrm{Nu} = 0.384 \times (\mathrm{Gr})^{0.252}. \tag{3}$$

Also, average heat transfer coefficient is directly related to the Nusselt number as

$$\mathrm{Nu} = \frac{hL_c}{k} \tag{4}$$

Simulation results of velocity distributions, streamline visualization, static pressure, and temperature distributions for the aspect ratios 0.5, 1.0, and 1.5 for all temperatures are presented below.

4.1 *Velocity Distribution*

4.2 *Stream Line Patterns*

4.3 *Temperature Distribution*

The streamlines indicate that the fluid, around the square and rectangular cylinder and far from it, is entrained toward the cylinder and forced to flow vertically upward above the cylinder. From Figs. 3, 4 and 5, at low aspect ratios of 0.5 and 0.75, one large counter-clockwise vortex and another relatively small clockwise vortex formation at the left and right sides of the inner cylinder has been observed. Because of this, the plume generated is asymmetrical and is shifted toward the right vertical face. With an increase in the aspect ratio, both the vortices attained similar shapes resulting in a symmetrical plume on the top surface. This pattern is observed the same for the aspect ratios 1.0–1.5.

Also, the isotherms in Fig. 6 around the cylinder show a vertical narrow plume. It is cooled by the top and sidewalls of the duct and then again circulated to flow over then hot inner wall surfaces. In addition to the pressure contours, it has been observed that the maximum static pressure is observed at the top surface of the inner cylinder.

Increasing the Rayleigh number speeds up the flow toward the cylinder, and the plume becomes thinner. For the same case, the isotherms shrink around the cylinder, as Ra was increased. Also, due to confinement around the inner cylinder, it has been observed that the hot air is flown away from the inner cylinder and towards the inner surface of the outer duct. The hot air is getting cooled resulting in an increase of density, thereby creating a density gradient across the hot inner surface and duct inner surface. This results in the recirculation of the hot air and the formation of vortices is also observed. At the center of the vortex, the pressure is less and hence

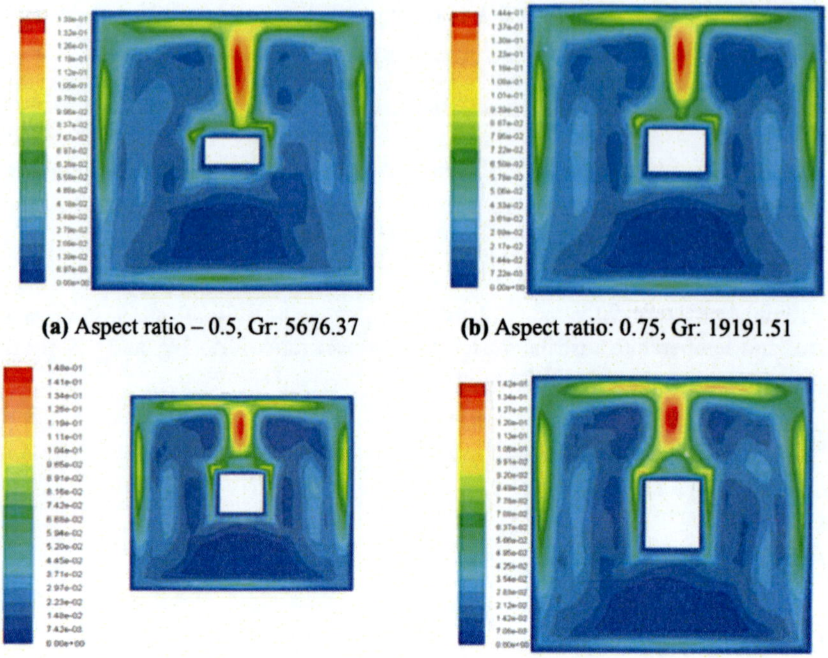

(a) Aspect ratio – 0.5, Gr: 5676.37

(b) Aspect ratio: 0.75, Gr: 19191.51

(c) Aspect ratio: 1.0, Gr: 45490.99

(d) Aspect ratio: 1.25, Gr: 88849.58

Fig. 3 Velocity distributions for temperature, $T_S = 373$ K

it creates the pressure gradient leading to the acceleration of flow around the inner cylinder. These pressure and velocity gradients are increasing with the increase of inner cylinder surface temperature. Also, the number of vortices formed and their size is observed to be increasing with the increase in the surface temperature of the inner cylinder.

A small aspect ratios, the plume flowing in an upward direction is formed away from the center of the top face leading toward the right vertical face of the cylinder. As the aspect ratio increases, the plume formation shifted to the middle of the top surface. This indicates that the aspect ratio has a great influence on the pattern of the plume formed here. An increase in the vertical height of the inner cylinder produces a symmetrical distribution of plume, leading to uniform heat transfer; otherwise, non-uniform heat transfer takes place.

From Fig. 7, the maximum velocity of the plume is observed to be increasing up to an increase in aspect ratio 1.0, and beyond that decreased up to ratio 1.25 and then again increasing beyond that. The same trend is observed for all the surface temperatures, and the maximum velocity is observed for the highest temperature 473 K. This is due to the decrease of viscosity at elevated temperatures, leading to the decrease of viscous force and increasing the fluidity of the air. Also, at a higher temperature, the gravity force also decreases, and the overall effect is an increase

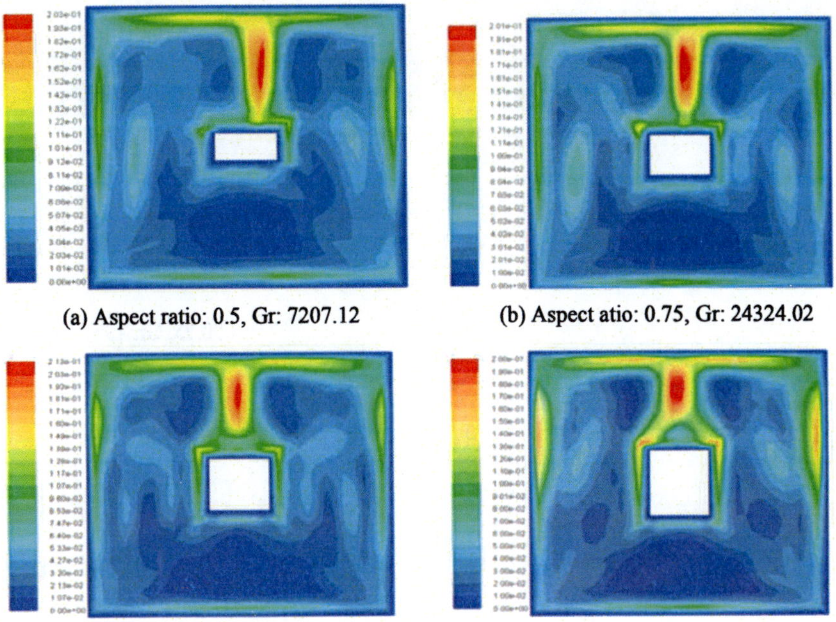

(a) Aspect ratio: 0.5, Gr: 7207.12

(b) Aspect atio: 0.75, Gr: 24324.02

(c) Aspect ratio: 1.0, Gr: 57656.94

(d) Aspect ratio: 1.25, Gr:112611.22

Fig. 4 Velocity distributions for Temperature, T_S= 473 K

in the Grashoff number with an increase in surface temperature, as shown in Fig. 7. The results indicate that the rate of decrease in gravity and viscous force leading to the rapid increase in the Grashoff number depends significantly on the aspect ratio. An increase in aspect ratio resulted in an exponential increase in Grashoff number, as shown in Fig. 8.

It has been observed that the maximum static pressure is increasing with an increase of aspect ratio as well as an increase in surface temperature from Fig. 9.

Figures 10 and 11 show the variation of average Nusselt number (Nu) for natural convection with respect to aspect ratio and Grashoff number, respectively. As shown in the figures, the average Nusselt number increases as the Grashoff number increases. An increase in the aspect ratio also results in the increase of Nusselt number due to the mixing motion of air particles within the confined region. This led to an increase in heat transfer from the inner cylinder to the surrounding air. Hence the vertical dimension of the inner cylinder plays a major role in affecting the rate of heat transfer.

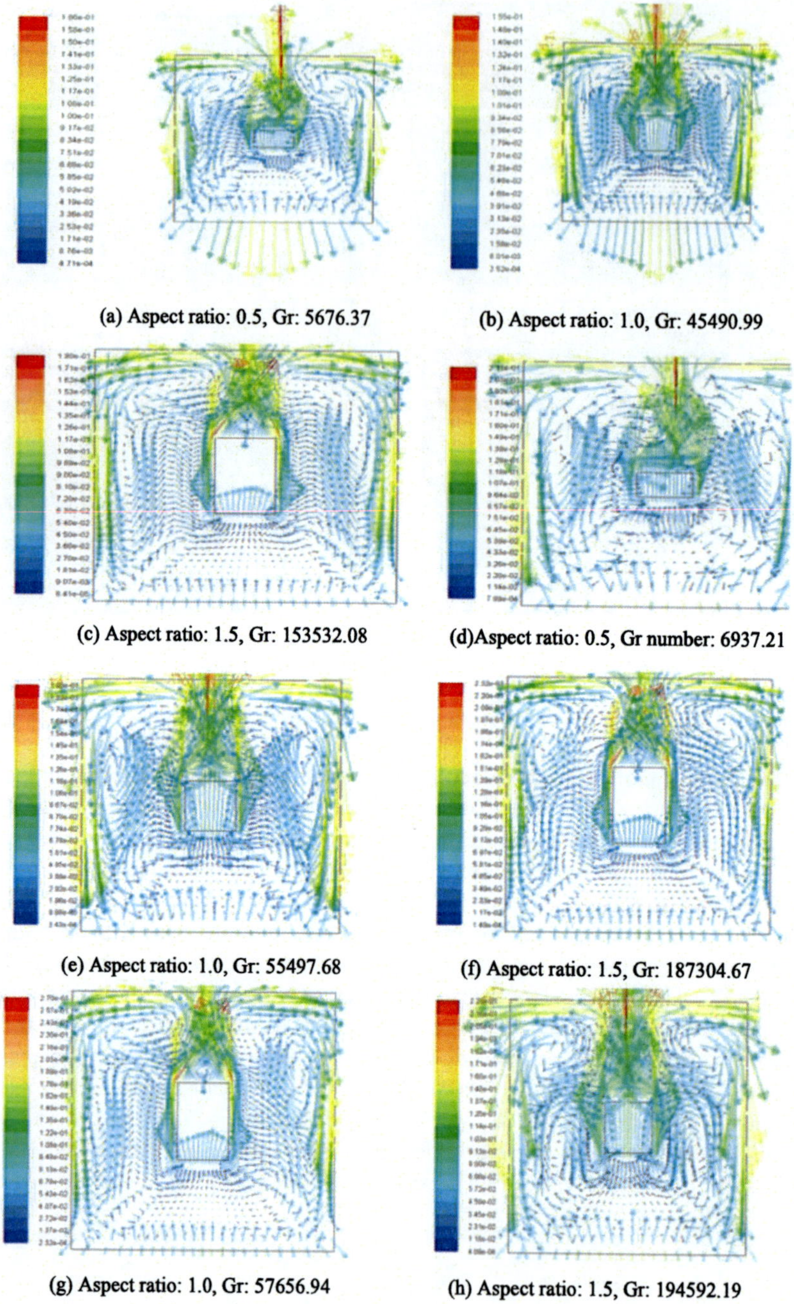

(a) Aspect ratio: 0.5, Gr: 5676.37

(b) Aspect ratio: 1.0, Gr: 45490.99

(c) Aspect ratio: 1.5, Gr: 153532.08

(d)Aspect ratio: 0.5, Gr number: 6937.21

(e) Aspect ratio: 1.0, Gr: 55497.68

(f) Aspect ratio: 1.5, Gr: 187304.67

(g) Aspect ratio: 1.0, Gr: 57656.94

(h) Aspect ratio: 1.5, Gr: 194592.19

Fig. 5 Streamline Distributions

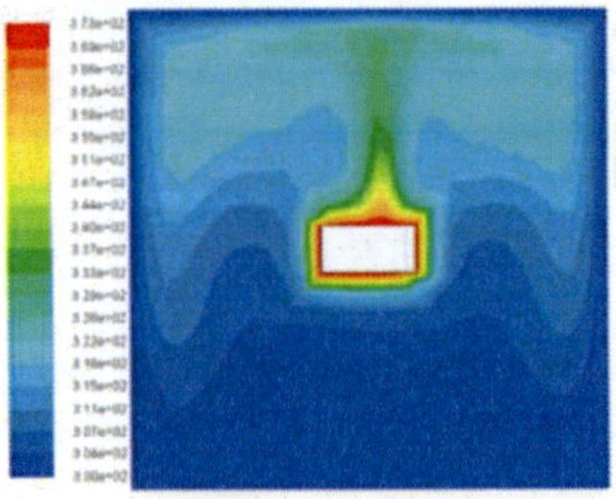

(a) Aspectratio:0.5,Gr:5676.37

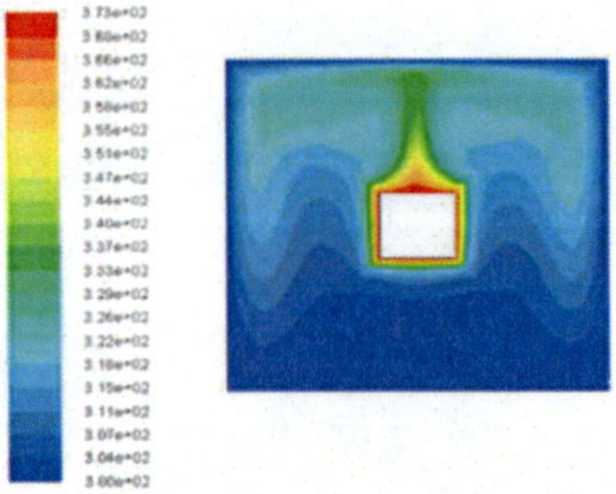

(b) Aspectratio:1.0,Gr:45490.99

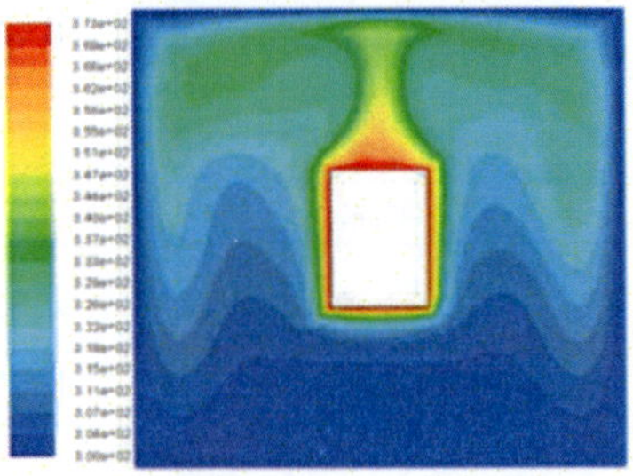

(c) Aspectratio:1.5, Gr: 153532.08

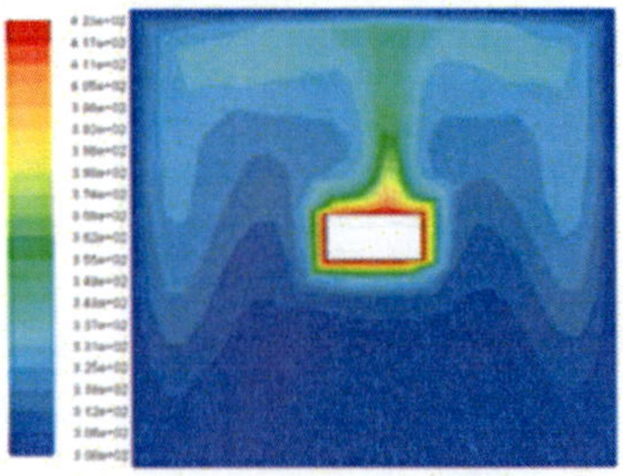

(d) Aspectratio:0.5,Gr:6937.21

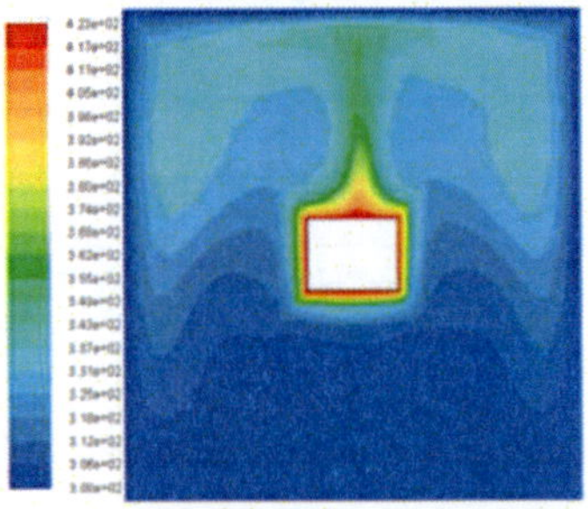

(e) Aspectratio:0.75,Gr:23413.08

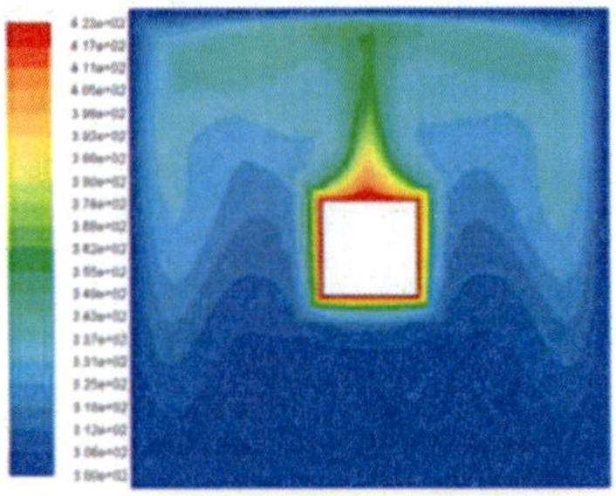

(f) Aspectratio:1.0,Gr:55497.68

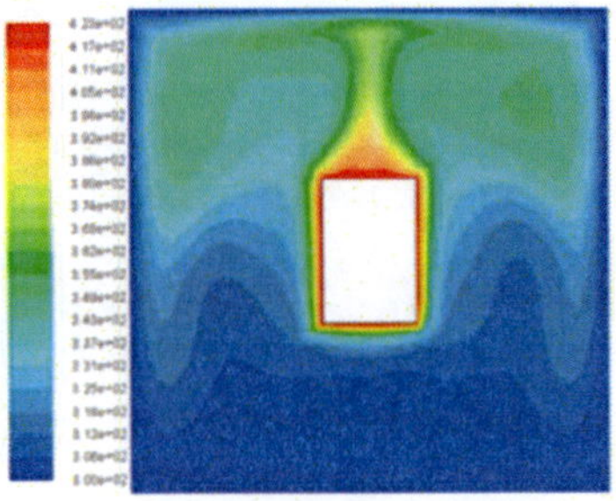

(g) Aspectratio:1.5,Gr:187304.67

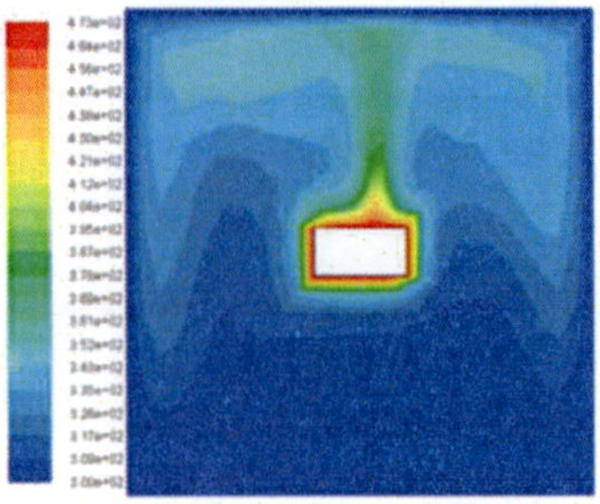

(h) Aspectratio:0.5,Gr:7207.12

Fig. 6 Temperature Distributions

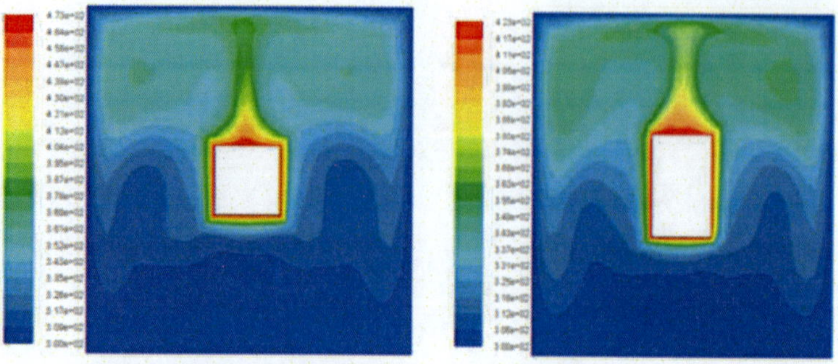

(i) Aspect ratio: 1.0, Gr: 57656.94 (j) Aspect ratio: 1.5, Gr: 194592.19

Fig. 6 (continued)

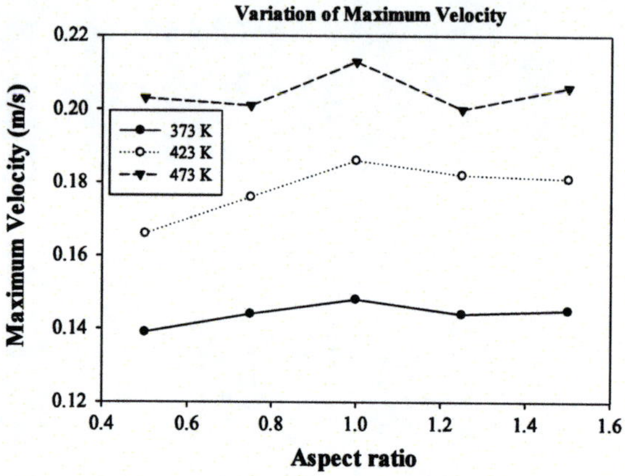

Fig. 7 Effect of aspect ratio on maximum velocity and Grashoff number, respectively

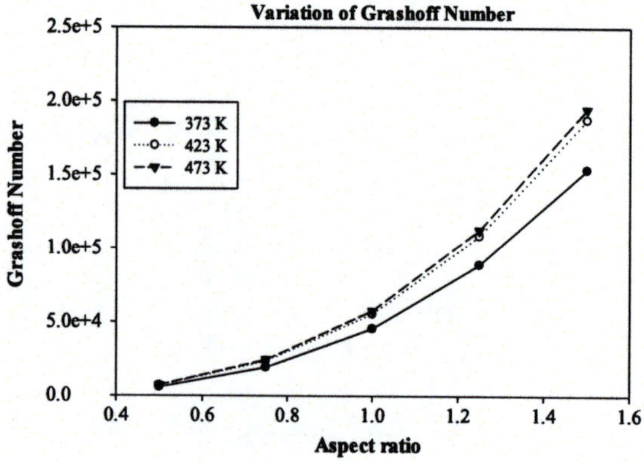

Fig. 8 Effect of aspect ratio on maximum velocity and Grashoff number, respectively

Fig. 9 Effect of Aspect ratio on maximum pressure

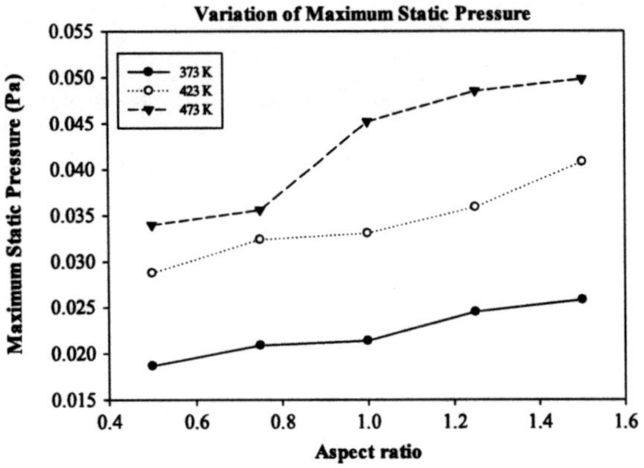

Fig. 10 Effect of Aspect ratio

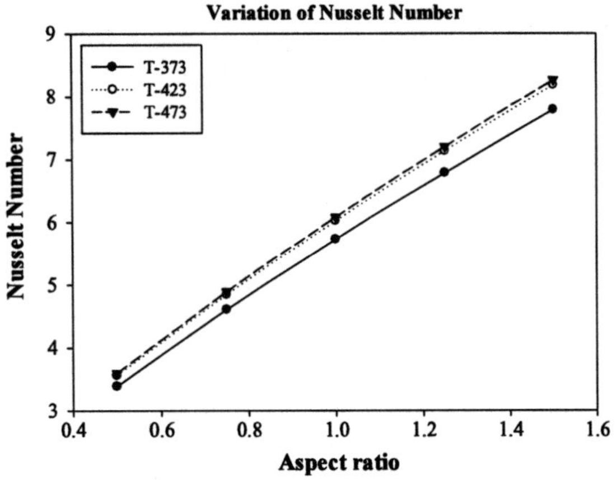

Fig. 11 Grashoff Number versus Nusselt Number

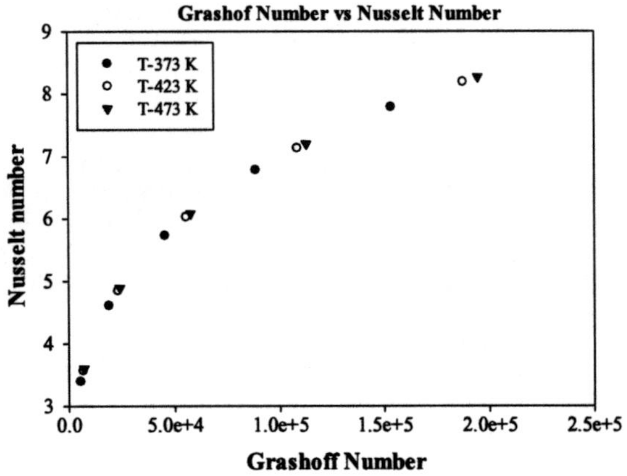

5 Conclusions

Simulation studies are performed to study the effect of aspect ratio on the maximum static pressure, velocity, temperature distributions, and Grashoff number and Nusselt number in natural convection using ANSYS Fluent at different hot surface temperatures. With an increase in aspect ratio, the following conclusions are drawn:

- Static pressure is increasing, with the maximum occurring at 473 K.
- Maximum velocity increases up to aspect ratio 1.0 and then decreases, up to ratio 1.25 and then increasing with maximum occurring at 473 K.
- Grashoff number increases, but the rate of increase is larger at 473 K.
- Nusselt number also increases with maximum values recorded at 473 K.
- The formation of recirculation and vortices is observed, and this trend resulted in an increase in heat transfer.
- The vertical dimension of the inner cylinder has a significant effect on the rate of heat transfer in natural convection.

References

1. Novomestský M, Lenhard R, Siažik J (2018) Natural convection heat transfer around a horizontal circular cylinder near an isothermal vertical wall. EPJ Web Conf 180. https://doi.org/10.1051/epjconf/201818002077
2. Elsherbiny SM, Teamah MA, Moussa AR (2017) Natural convection heat transfer from an isothermal horizontal square cylinder. Alexandria Eng J 56:181–187
3. Anu Nair P, Karuppasamy K, Benziger B, Balakrishnan P (2015) Natural convective heat transfer from horizontal heated plate facing upward in vertical channel—a review. Int J Mech Eng Res 5(1):27–38. ISSN 2249-0019
4. Clemes SB, Hollands KGT, Brunger AP (1994) Natural convection heat transfer from long horizontal isothermal cylinders. J Heat Transf Trans ASME 116
5. Ali ME (2007) Natural convection heat transfer from horizontal rectangular ducts. J Heat Transf Trans ASME 129
6. Calcagni B, Marsili F, Paroncini M (2005) Natural convective heat transfer in square enclosures heated from below. Appl Therm Eng 2522–2531:25
7. Khubeiz JW, Radziemska E, Lewandowski WM (2002) Natural convective heat-transfers from an isothermal horizontal hemispherical cavity. Appl Energy 73:261–275
8. Cheikh NB, Beya BB, Lili T (2007) Influence of thermal boundary conditions on natural convection in a square enclosure partially heated from below. Int Commun Heat Mass Transf 34:369–379
9. Chang KS, Won YH, Cho CH (1983) Patterns of natural convection around a square cylinder placed concentrically in a horizontal circular cylinder. J Heat Transf Trans ASME 105

Forced Convective Wake Dynamics Past a Semicircular Cylinder at Incidence with a Downstream Circular Cylinder in Crossflow

Chitrak Mondal, Sandip Sarkar, and Nirmal Kumar Manna

Nomenclature

C_d	Coefficient of drag
C_l	Coefficient of lift
C_p	Specific heat (J/kg K)
Re	Reynolds number$[\rho U D/\mu]$
Pr	Prandtl number $\left[\mu C_p/k\right]$
St	Strouhal number $[f D/U_\infty]$
Nu	Nusselt number $[h D/k]$
f	Frequency of vortex shedding (Hz)
k	Thermal conductivity (W/m K)
μ	Viscosity of fluid
ρ	Density of fluid (kg/m^3)
θ	Dimensionless temperature $\left(\frac{T-T_\infty}{T-T_w}\right)$
D	Diameter of the circle and the semicircle
u, v	Dimensionless velocity
T_∞	Free stream temperature
T_w	Temperature of the cylinders wall
x, y	Dimensionless coordinate axes
U, V	Velocity in x and y directions
t	Dimensionless time

C. Mondal · S. Sarkar (✉) · N. K. Manna
Department of Mechanical Engineering, Jadavpur University, Kolkata 700032, India
e-mail: sandipsarkar.mech@jadavpuruniversity.in

C. Mondal
e-mail: chitrakmondal800@gmail.com

M. Palanisamy et al. (eds.), *Theoretical, Computational, and Experimental Solutions to Thermo-Fluid Systems*, Lecture Notes in Mechanical Engineering,
https://doi.org/10.1007/978-981-33-4165-4_10

ϕ Angle of incidence
p Dimensionless pressure

1 Introduction

The flow past the bluff body is a topic of great importance for several decades [1–6]. The research persists in this area as it finds several applications from engineering perspectives, starting from aerodynamics to marine and irrigation engineering. Some of the specific areas of applications are: flow through a heat exchanger, flow around chimney stack, offshore structures, and electronic chips, etc. Recently in the field of sports, the analysis of similar flow dynamics is also being used. The basic observation of this kind of flow is that, up to $Re = 5$, a non-separating creeping flow exists. Whereas, in the range of $5 < Re < 40$, a fixed pair of vortices is created. For $40 < Re < 200$, a laminar two-dimensional vortex street appears, and subsequently, a turbulent transition is observed. For $Re > 300$, the development of a turbulent wake takes place. These phenomena are well established experimentally as well as numerically. However, the transition points of the Re are very much dependent on the boundary conditions and the presence of any other object near the target object. Furthermore, the interaction of vortices (between two or more objects in the flow field) gives rise to a more complex flow situation.

In the real world, there can be many cylinders in the vicinity; their arrangement plays an important role in the flow dynamics, and accordingly, the design should be made. In the earlier research works have been considered some arrangements considering two or more bluff-bodies of different shapes and different fluids [1, 2, 6–8]. Bharti et al. [9] carried out numerical simulation for fluid with $0.7 < Pr < 400$ and flow with $10 < Re < 45$. Dhiman et al. [10] performed numerical investigation for the flow field and heat transfer characteristics of a square cylinder placed in cross-stream for $1 < Re < 45$ and $0.7 < Pr < 4000$. The flow over a row of square cylinders have been investigated by Sewatkar et al. [11] with $30 < Re < 140$. They found the effect of the gap ratio on the flow transition and wake dynamics. Sisodia et al. [12] investigated numerically the mixed convective flow past a semicircular cylinder with a square cylinder in tandem with $10 < Re < 45$ and $0 < Ri < 2$, and the incidence angle of the semicircular cylinder as 0–180°. They tried to capture the effect of the orientation of the front cylinder over the flow dynamics.

Though there are many numerical works available in the literature, for a semicircular cylinder and a circular one placed in tandem is yet to appear in the literature. There is an effect of the shape change of the downstream cylinder, so here a region of *Re* is considered where the flow is changing from a stable region to an unstable region. The values of Reynolds number considered are 60,100, and 140, and the angle of incidence of the upstream semicircular cylinder is 0, 45, and 90°. The forced convection is considered, and the *Pr* value of the working fluid is taken as 0.73 (air) with fixed density, thermal conductivity, and heat capacity.

2 Physical Problem Description and Numerical Model

Figure 1 depicts the schematic of the flow domain where the two cylinders are placed one after another in tandem. The upstream one is semicircular, and the downstream is circular in shape. Both cylinders are stationary. The slip boundary, along with the no-heat flux condition, is applied at the top and the bottom walls. The velocity is taken as a uniform in the inlet. The constant pressure boundary is applied at the outlet. The boundary conditions on the cylinder surfaces are taken as no-slip and isothermally heated with temperature T_w ($> T_\infty$, the ambient temperature).

The flow is assumed to be unsteady, incompressible, and laminar. The dimensionless governing equations are:

$$\frac{\partial u}{\partial x} + \frac{\partial v}{\partial y} = 0 \tag{1}$$

$$\frac{\partial u}{\partial t} + u\frac{\partial u}{\partial x} + v\frac{\partial u}{\partial y} = -\frac{\partial p}{\partial x} + \frac{1}{Re}\left(\frac{\partial^2 u}{\partial x^2} + \frac{\partial^2 u}{\partial y^2}\right) \tag{2}$$

$$\frac{\partial v}{\partial t} + u\frac{\partial v}{\partial x} + v\frac{\partial v}{\partial y} = -\frac{\partial p}{\partial y} + \frac{1}{Re}\left(\frac{\partial^2 v}{\partial x^2} + \frac{\partial^2 v}{\partial y^2}\right) \tag{3}$$

$$\frac{\partial}{\partial t} + u\frac{\partial}{\partial x} + v\frac{\partial}{\partial y} = \frac{1}{Re.Pr}\left(\frac{\partial^2}{\partial x^2} + \frac{\partial^2}{\partial y^2}\right) \tag{4}$$

where the non-dimensional parameters $(u, v, x, y, p, \theta, t, Re, Pr)$ are: velocity components of $u = U/U_\infty$ and $v = V/U_\infty$, coordinates of $x = X/D$ and

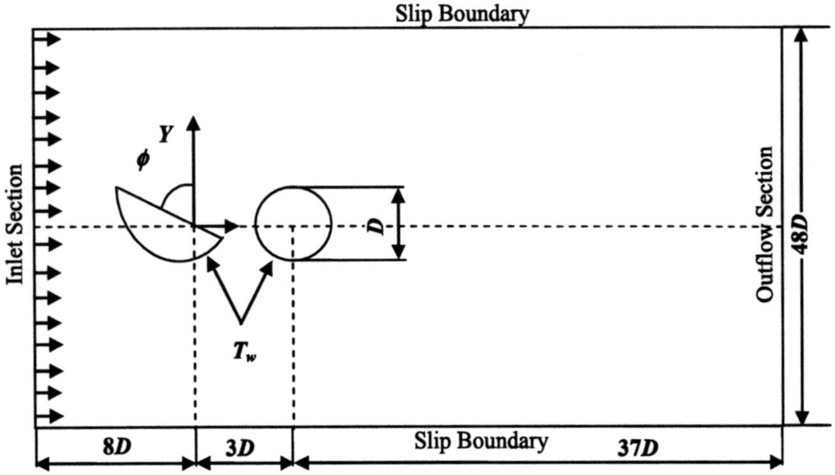

Fig. 1 Schematic of the numerical domain with the boundary conditions

$y = Y/D$, pressure $p = P/\rho U_\infty^2$ and temperature $\theta = (T - T_\infty)/(T_w - T_\infty)$, time $t = \tau U_\infty/D$, Reynolds number $Re = \rho U_\infty D/\mu$ and Prandtl number $Pr = \mu C_p/k$. Here, D is the diameter of the cylinders and the length scale; and U_∞ is a free stream velocity at the inlet and the velocity scale.

The boundary conditions are as follows:

(i) At the inlet, uniform flow velocity along the x-direction with fixed temperature.

$$u = 1, \quad v = 0, \quad \theta = 0 \tag{5}$$

(ii) At the outlet, the flow is being fully developed with a zero-heat flux along the x-direction.

(ii) At the outlet, the flow is being fully developed with a zero-heat flux along the x-direction.

$$\frac{\partial u}{\partial x} = \frac{\partial v}{\partial x} = \frac{\partial \theta}{\partial x} = 0 \tag{6}$$

(iii) At the sidewalls, the upper and lower boundaries are adiabatic and follow slip-condition.

$$\frac{\partial \theta}{\partial y} = 0 = v, \quad u = 1 \tag{7}$$

(iv) The walls of the cylinders have no-slip condition and constant temperature.

$$u = 0, \quad v = 0, \quad \theta = 1 \tag{8}$$

(v) The initial conditions of flow in the domain are written at $t = 0$ as

$$u = 1, \quad v = 0, \quad \theta = 0 \tag{9}$$

To solve the equations of continuity, momentum, and energy, the finite volume method is used. The decoupling of pressure and velocity has been made by employing the SIMPLE (Semi-Implicit Method for Pressure Linked Equations) algorithm. For discretization of the convective terms of momentum as well as the energy equations, a third-order upwind scheme, QUICK is used. The final solutions are achieved when the unsteady solution reaches the dynamic steady-state, and the corresponding curve of the lift coefficient becomes periodic (indicating saturation).

3 Result and Discussion

3.1 *Streamline, Vorticity Contour, and Isotherm*

The flow topology in the computational domain is described with the contours of streamlines, isotherms, and vorticity. The streamlines and the vorticity contours describe the flow field, and the isotherms give an indication of temperature distribution. Here, the incidence angle (ϕ) of the semicircular cylinder is changed to configure different flow situations. The different values of *Re* indicate the different velocity of fluid flow. The overall observation is that as the incidence angle (ϕ) increases, the transition of the *Re*-value becomes larger and larger. As seen clearly from the plots of streamlines indicated in Fig. 2, for the zero-incidence, the flow separates, and the shedding of vortices starts at *Re* 60 and continues as *Re* increase; the wake behind the circular cylinder changes in shape, and the wavelength increases with the different values of increased *Re*. For the incidence of 45° at *Re* 60, the shedding is not present; however, at *Re* 100 the vortex shedding is observed. Therefore, the critical *Re* lays in-between 60 and 100. Furthermore, as *Re* increases the wake length decreases. For the incidence of 90°, the vortex shedding is almost missing up to $Re = 100$. For this incidence angle of the upstream semicircular cylinder, the critical *Re* lies in between $Re = 100$ and $Re = 140$.

Figure 3 presents the vorticity structures for the afore-said streamlines and gives a visualization of the rotationality of the flow field. In the figure, the dashed-lines indicate the negative vorticity and the solid-lines reflect the positive vorticity. Thus, the upper portion of the vorticity is negative in magnitude, and the lower portion

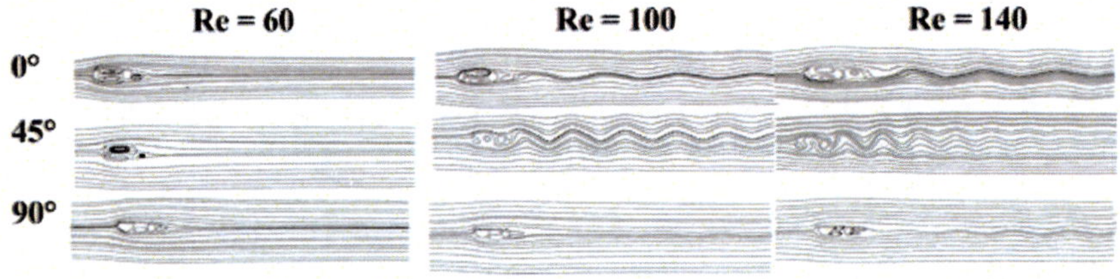

Fig. 2 Instantaneous contours of streamlines

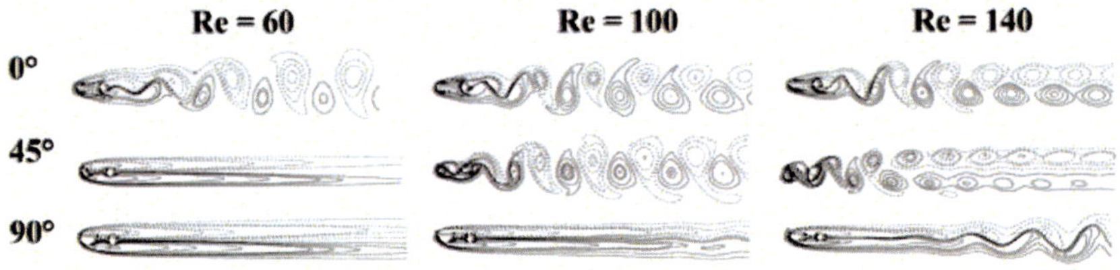

Fig. 3 Instantaneous contours of vorticity

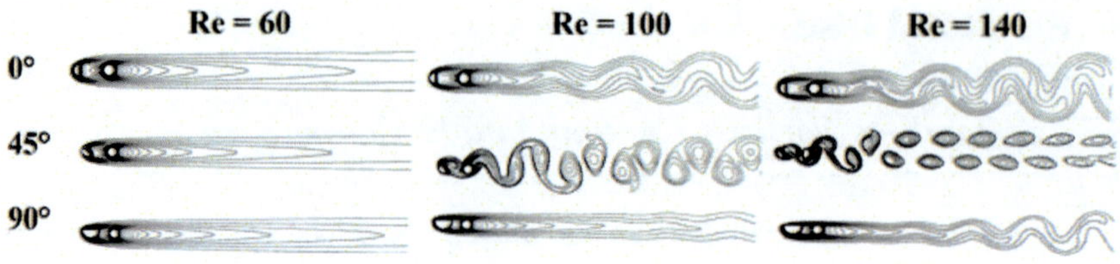

Fig. 4 Instantaneous contours of isotherms

is positive, where the anticlockwise rotation is taken to be positive. These plots of vorticity contours also indicate which flow passes through the critical *Re* value for the vortex shedding.

Finally, the plots of isotherm contours are shown in Fig. 4 for the present case. The presence of vortex shedding influences the distribution of the isotherms. The region of the wake motion is indicated by the wavy isotherms.

3.2 *Force Coefficients, Nusselt Number and Strouhal Number*

Due to the presence of the viscosity, there is a shear force acting parallel to the boundaries of both the cylinders. There also acts a normal pressure force. Combining these two forces yields the net resultant force acting on the bodies. When this net force is resolved along the flow direction and in the perpendicular to that, these components are drag and lift force. The drag force tries to pull the body backward, while the lift force gives a lateral force on the cylinders. Here in the unsteady flow as the dynamic steady state is reached, the periodic values of the force coefficients occur. In the work, the values are presented in Figs. 5, 6, 7 and 8, are the integral value averaged over a cycle.

Figure 5a shows the variation in the drag coefficient (C_d) of the semicircular cylinder with varying incidence angle (ϕ) for various values of *Re*. For $Re = 60$, the drag coefficient increases gradually with the angle; however, for other values of *Re*, the drag coefficient increases up to 45° and then decreases. The drag coefficient is dependent on the *Re* and the relative position of the front cylinder. This is because with increasing *Re*, the flow velocity increases, and as a consequence, the friction drag enhances. On the other hand, the increase in incidence angle restricts the flow rate of the streaming fluid; this eventually results in a separation delay and therefore corresponding enhancement in the form drag. The overall effect culminates in augmenting the drag coefficient. The change in the position of the separation point causes a decrease in the wake width compared to the base condition of the incidence angle of 0°. This causes small suction in the downstream of the semicircular cylinder that reduces the drag coefficient. Figure 5b describes C_d versus incidence angle ϕ for different *Re* values. The drag reduces with increasing angle of incidence for Re

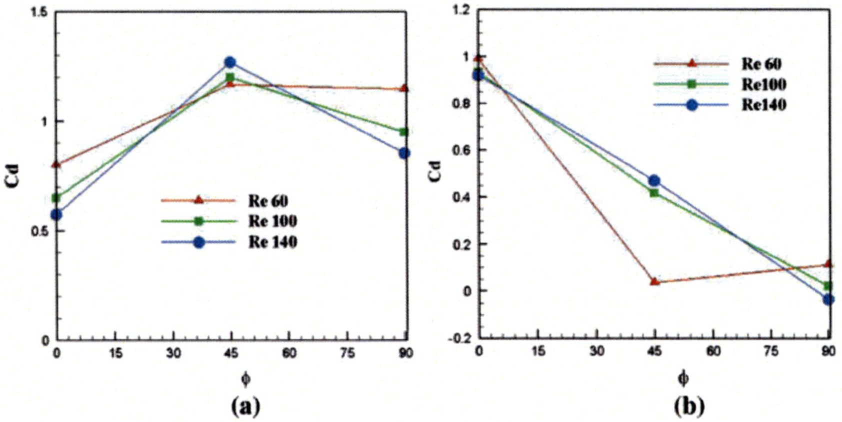

Fig. 5 Drag coefficient (C_d) against ϕ for **a** semicircular cylinder and **b** circular cylinder

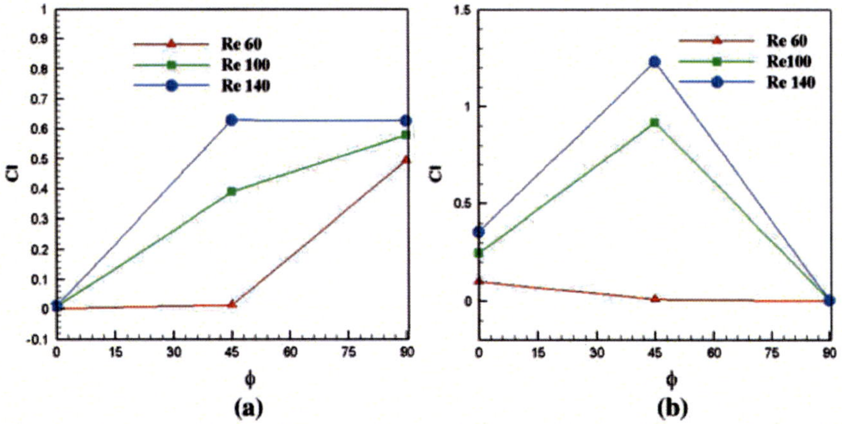

Fig. 6 Lift coefficient (C_l) against ϕ for **a** semicircular cylinder and **b** circular cylinder

100 and 140; while, for $Re = 60$, the C_d reaches a minimum value near 45°, and then C_d increases and achieves almost the same value at other *Re* values.

In Fig. 6a, the lift coefficient variation is shown with the incidence angle ϕ at various *Re*. As *Re* increases, the lift coefficient increases. The difference between the coefficients is found maximum at $\phi = 45°$. However, as the angle of incidence increases the difference decreases. From the streamlines, it is evident that behind the semicircular cylinder, a pair of anti-symmetric vortices is created with the increase in *Re* value. Due to this anti-symmetry lift coefficients increasing with *Re* value for the same incidence angle. In Fig. 6b, the lift coefficient for the circular cylinder is described with respect to the angle of incidence. The maximum lift coefficient is achieved at $\phi = 45°$; then, the lift coefficient decreases to zero value. For the

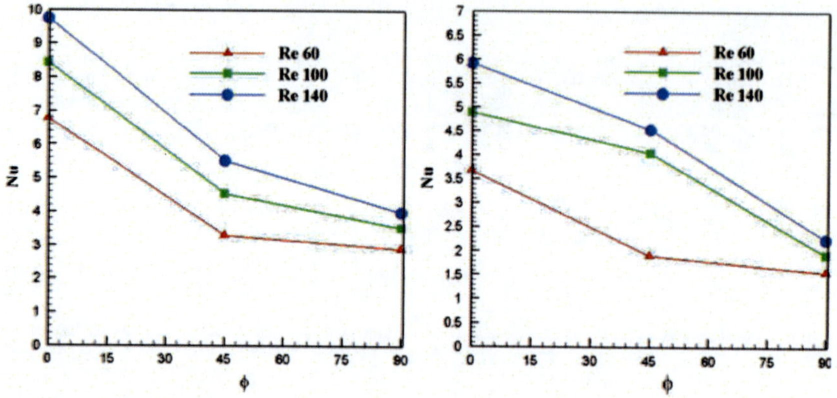

Fig. 7 The average Nu against ϕ for **a** semicircular cylinder and **b** circular cylinder

Fig. 8 The variation of Strouhal number (St) against incidence angle (ϕ)

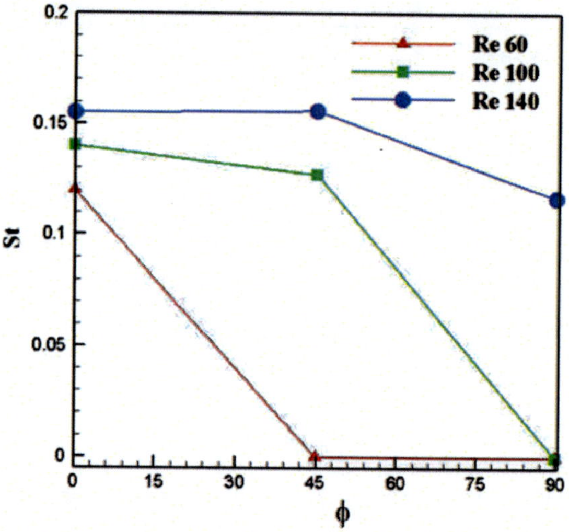

incidence angle of 90°, the lift coefficients are found to be almost zero. It is because the corresponding Strouhal number (St) value is very near to zero. Thus, the state of flow becomes almost steady.

For the characterization of heat transfer, Nusselt number (Nu) is taken as the parameter of consideration. Here, the average Nusselt number (Nu) is computed as the time, and the surface-area averaged Nu. Figure 7a describes the average Nusselt number with respect to the angle of incidence. The value of Nu is low in the case of low values of Re. In any Re case, the Nu value decreases with respect to the angle. The decrease in Nu is due to less recirculation regime in the vicinity of the semicircular cylinder, which causes the thickening of the thermal boundary layer. Figure 7b shows the variation of Nu at the circular cylinder with respect to the incidence angle of the semicircle. In this case also, *Nu* decreases with the increase in incidence angle, and *Nu* value is low at low *Re*. The circular cylinder shows a similar kind of heat transfer characteristics such as the semicircular cylinder due to the presence of vortices. For all the situations, the heat transfer from the semicircular cylinder is greater than that of the circular cylinder. This may be attributed to the flow separation and intense mixing of fluid behind the semicircular cylinder.

Figure 8 describes the variation of Strouhal number (*St*) with respect to the incidence angle (ϕ) of the upstream semicircular cylinder. As incidence angle increases, the Strouhal number decreases, and at the angle of 90°, the value of *St* becomes 0 for all the *Re* values except 140. It signifies that the flow does not cross the critical limit. At a small incidence angle, the transition readily occurs; however, as the angle increases, the transition is delayed.

4 Conclusions

The work investigates a case of crossflow over bluff bodies consisting of hot semcircular and circular cylinders in tandem. Both fluid dynamics and heat transfer characteristics are studied by varying the incidence angle of the upstream semicircular cylinder under forced convection. The analysis is carried out for different Reynolds numbers and incidence angles using streamlines, isotherms, vorticity, average Nusselt number, and Strouhal number. The dynamics of wake generation at different incidence angles and Reynolds number are captured. The values of critical Reynolds number for flow transition are identified for the chosen configuration of semicircular and circular cylinders in tandem. The critical Reynolds number is found increasing with the increase in the incidence angle. The heat transfer from the upstream semicircular cylinder is found more than the downstream circular cylinder irrespective of the values of Reynolds numbers and incidence angles. However, as *Re* increases, *Nu* decreases. The flow transition is also examined using the Strouhal number, which indicates that the increase in incidence angle at lower *Re* leads to transition delays. As *Re* increases, the flow separation takes place.

References

1. Davis RW, Moore EF (1982) A numerical study of vortex shedding from rectangles. J Fluid Mech 116:475–506
2. Monkenwitz PA (1988) A note on vortex shedding from axisymmetric bluff bodies. J Fluid Mech 192:561–575
3. Williamson CHK (1996) Three-dimensional vortex dynamics in bluff body wakes. Exp Thermal Fluid Sci 12:150–168
4. Zdravkovich MM (1997) Flow around circular cylinders, fundamentals, vol 1. Oxford University Press, New York
5. Zdravkovich MM (2003) Flow around circular cylinders, applications, vol 2. Oxford University Press, New York

6. Yang W, Stremler MA (2019) Critical spacing of stationary tandem circular cylinder at Re = 100. J Fluid Struct 89:49–60
7. Chatterjee D, Biswas G (2011) The effect of Reynolds and Prandtl numbers on flow and heat transfer across tandem square cylinders in the steady flow regime. Numer Heat Transf A 59:421–437
8. Chatterjee D, Mondal B (2013) Mixed convection heat transfer from tandem square cylinders for various gaps to size ratios. Numer Heat Transfer, Part A 63:101–119
9. Bharti RP, Chhabra RP, Eswaran VA (2007) A numerical study of forced convective heat transfer from an unconfined circular cylinder. Heat Mass Transfer 43:639–648
10. Dhiman AK, Chhabra RP, Eswaran V (2005) Flow and heat transfer across a confined square cylinder in the steady flow regime: effect of Peclet number. Int J Heat Mass Transf 48:4598–4614
11. Sewatkar CM, Sharma A, Agarwal A (2009) On the effect of Reynolds number for flow around a row of square cylinders. Phys Fluids 21:083602
12. Sisodia SS, Sarkar S, Saha SK (2017) Fluid flow and mixed convective heat transfer around a semicircular cylinder at incidence with a tandem downstream square cylinder in cross flow. Int J Thermal Sci 121:13–29

Electro-thermal Convection in a Rectangular Enclosure

Nirmalendu Biswas, Aparesh Datta, and Nirmal Kumar Manna

Nomenclature

C	Dimensionless injection parameter
E	Dimensionless electric field
G	Gravitational acceleration m/s^2
Gr	Grashof number
H	Cavity height m
L	Cavity length m
M	Mobility parameter
Nu	Average Nusselt number
p	Pressure Pa
P	Dimensionless pressure
Pr	Prandtl number
Q	Dimensionless charge density
R	Electrical Reynolds number
Ra	Rayleigh number
T	Temperature K
$\boldsymbol{T}$	Electrical parameter

N. Biswas
Department of Power Engineering, Jadavpur University, Salt Lake, Kolkata 700106, India
e-mail: nirmalendubiswas@yahoo.co.in

A. Datta (✉)
Department of Mechanical Engineering, NIT Durgapur, Durgapur, West Bengal 713209, India
e-mail: adatta96@gmail.com

N. K. Manna
Department of Mechanical Engineering, Jadavpur University, kolkata, West Bengal 700032, India
e-mail: nirmalkmannaju@gmail.com

M. Palanisamy et al. (eds.), *Theoretical, Computational, and Experimental Solutions to Thermo-Fluid Systems*, Lecture Notes in Mechanical Engineering,
https://doi.org/10.1007/978-981-33-4165-4_11

u, v Dimensional velocities m/s
U, V Dimensionless velocities
x, y Cartesian co-ordinates m
X, Y Dimensionless co-ordinates

Greek Symbols

α Thermal diffusivity m^2/s
β Thermal expansion coefficient K^{-1}
ψ Stream function
θ Non-dimensional temperature
ν Kinematic viscosity m^2/s
ρ Fluid density kg/m^3
κ Electrical conductivity $\mu S/cm$
Φ Dimensionless electric potential
ε Permittivity

Subscripts

$c.\ h$ Cooling heating

1 Introduction

For the past several decades, the fastest growth in various processing industries as well as emerging technologies, to topic electrohydrodynamics (EHD) finds widespread application. EHD essentially indicates the induced secondary flows across a layer of dielectric fluid—when it is subjected to an external electric field. Such an EHD effect improves thermal convection and resulting heat transfer enhancement. This phenomenon is essentially attractive for improving the thermo-convective heat transfer of fluid (which is less conducting) passing through a constricted passage, where the conventional passive cooling technique is neither effective nor easy [1]. Heat transfer augments by means electrical energy input has unique advantages—as no involvement of moving mechanical parts. Such a mechanism is very attractive for rapid and smart control of heat and mass transfer, lowering noise as well as power consumption [2]. A detailed account of the review on the above subject is well documented in ref. [3, 4].

When electrical charge injection occurs at one electrode, it is called unipolar injection. Traoré et al. [5] numerically studied the combined effects of the electrical field and applied temperature gradient to the horizontal layer of dielectric fluid. This results in complex thermo-physical phenomena in the fluid flow. Using the same numerical solver, Koulova et al. [6] studied the electrical effect on heat transfer enhancement in a 2D cavity and found that heat transfer rate is substantially improved by the Coulomb force, whereas Nusselt number is almost independent of the Rayleigh number. Electrohydrodynamic effects on three-dimensional natural convection of dielectric liquid studied by Luo et al. [7]. They found that at a lower Rayleigh number, electrical effect is more prominent for enhancing heat transfer. Lacroix et al. [8] performed a non-linear stability analysis of electroconvection through a layer of dielectric liquid under the influence of unipolar injection. On the other hand, electrothermoconvection (ETHD) through a dielectric liquid confined between two parallel plates is investigated by Wu and Traoré [9] under unipolar injection. In other class of work, electrothermoconvection patterns, as well as heat transfer augmentation unipolar injection, has been investigated as in ref. [10–12]. Recently, Lu et al. [13] investigated electrothermoconvective flow and associated heat transfer through the dielectric fluid filled in a square cavity. It is found that the electric field enhances the convective heat transfer for higher Prandtl number fluid and relatively at lower Rayleigh number.

From the foregoing literature survey, it is noted that electrothermoconvection (ETHD) is not investigated adequately. The objective of the present work is to explore the underneath physics of electrothermoconvection in a differentially heated square cavity filled with a dielectric liquid. The effects of various pertinent parameters on the thermal behavior are assessed thoroughly and explained in great detail. Such a problem is a fundamental configuration of electrothermoconvection problems.

2 Problem Description and Mathematical Modeling Aspect

The graphical diagram of the ETHD problem geometry is basically a two-dimensional square cavity ($L = H$), and it is presented in Fig. 1. The vertical walls are made of metal, which acts as electrodes under constant electric potential and subjected to a fixed temperature. The left electrode is at a higher temperature (T_h), and it releases electric charges (at potential V_0), whereas the right electrode is at a lower temperature (T_c) and collects the electric charges (at potential V_1). The horizontal walls of the enclosure are insulated. The confined space of the cavity is filled with a dielectric liquid. For understanding and analyzing the evolved flow physics, it is assumed that the fluid flow is laminar, incompressible, steady, and Newtonian flow within the validity of Boussinesq approximation. For simplifying the analysis induced magnetic field, the charge diffusion process, viscous dissipation work, Hall effect, and Joule heating are neglected [5]. Evolved governing equations include conservation of mass-momentum equation including electrical effect, energy equation, Poisson equation (Gauss theorem) for electric potential V, equation for the ion charge transport, and the co-relation for the electric field from the electric potential.

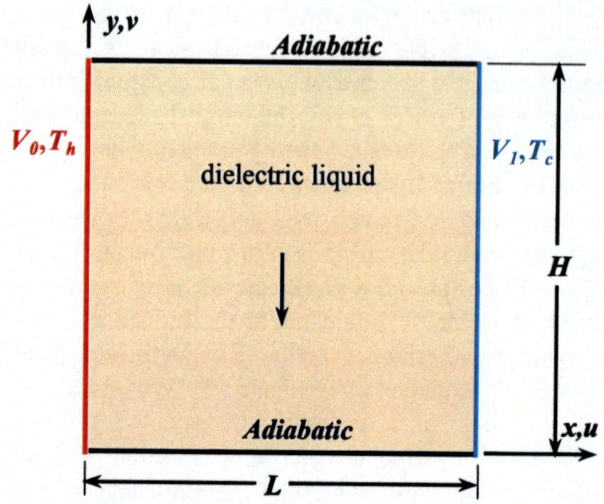

Fig. 1 Physical description of the problem geometry, including boundary conditions

Considering the above assumptions, the governing equations for the ETHD system in the dimensionless form can be expressed as

$$\nabla \cdot u = 0 \tag{1}$$

$$(u \cdot \nabla)u = -\nabla P + \nabla^2 u + RTC_q\mathbf{E} + \frac{Ra}{\Pr}\theta e \tag{2}$$

$$u \cdot \nabla\theta = \frac{1}{\Pr}\nabla^2\theta \tag{3}$$

$$\nabla \cdot \{Q(u + R\mathrm{E})\} = 0 \tag{4}$$

$$\nabla^2\Phi = -C_q \tag{5}$$

$$\mathbf{E} = -\nabla\Phi \tag{6}$$

Above, dimensionless equations are obtained by utilizing the following variables

$$\begin{gathered}(X, Y) = (x, y)/H; \quad (U, V) = (u, v)L/\alpha; \quad \theta = (T - T_c)/(T_h - T_c); \\ P = (p - p_a)L^2/\rho\alpha^2; \quad \Phi = V/(V_0 - V_1); \quad \mathbf{E} = EL/(V_0 - V_1); \; Q = q/q_0; \\ \int \mathrm{Ra} = g\beta(T_\mathrm{h} - T_\mathrm{c})L^3 \Pr/\nu^2; \quad \Pr = \nu/\alpha; \quad T = \frac{\varepsilon_0\Delta V}{\beta_0\nu\alpha_0}; \\ C = \frac{q_0L^2}{\varepsilon_0\Delta V}; \quad M = \frac{1}{\alpha_0}\left(\frac{\varepsilon_0}{\rho_0}\right)^{0.5}; \quad R = \frac{\mathbf{T}}{M^2};\end{gathered} \tag{7}$$

where the dimensionless parameters: X and Y are Cartesian co-ordinates; U and V are velocity components; P and θ are pressure and temperature; $\boldsymbol{T}$ is an electric parameter; C is the injection strength; M is the mobility parameter; R is the electrical Reynolds number; Pr is the Prandtl number.

The boundary conditions are specified as

$$\begin{aligned}
&X = 0 : U = V = 0, \quad \theta = 1, \quad \Phi = 1, \quad Q = 1 \\
&X = 1 : U = V = 0, \quad \theta = 0, \quad \Phi = 0, \quad \partial Q/\partial X = 0 \\
&Y = 0, \quad U = V = 0, \quad \partial\theta/\partial Y = 0, \quad \partial Q/\partial Y = 0, \quad \partial\Phi/\partial Y = 0 \\
&Y = 1, \quad U = V = 0, \quad \partial\theta/\partial Y = 0, \quad \partial Q/\partial Y = 0, \quad \partial\Phi/\partial Y = 0
\end{aligned} \tag{8}$$

The contours of streamlines (ψ), isotherms (θ) are used for the visualization plot; whereas for average Nusselt number (Nu) is utilized for presenting global parameters. The Nu (at the active wall) is computed using the solved temperature field as

$$\mathrm{Nu} = \int_0^1 \left(-\frac{\partial\theta}{\partial X}\bigg|_{X=0,1} \right) \mathrm{d}Y \tag{9}$$

3 Numerical Aspect

For the computation of the problem of ETHD, the set of the coupled partial differential Eqs. (1)–(6) along with initial and boundary conditions as in Eq. (8) are solved numerically in an iterative way using a developed in-house CFD code based on the finite volume approach of discretization and SIMPLE algorithm [14]. For the diffusion and advection terms, the second-order differencing scheme is implemented. When the maximum residuals and the mass-defect is $< 10^{-7}$ and 10^{-9}, the converged solutions are taken. The same code has already been used for simulating different flow situations with the appropriate validations, as presented in many works earlier on mixed [15, 16] as well as natural [17–19] convection. Of course, before conducting the extensive simulation, the grid independency test is carried out, and finally, mesh having 160×160 non-uniform grids is used as the best grid size. Corresponding grid independence test results are not incorporated here for brevity.

4 Results and Discussion

To assess electrothermohydrodynamics (ETHD) behavior of dielectric liquid in a confined cavity, the thermal performance of the non-ETHD system is analyzed first. For exploring the underneath physics, the extensive numerical simulation is conducted for different parametric conditions. The results are represented using contours of streamlines (ψ), isotherms (θ), and average Nusselt number (Nu). The range of pertinent parameters studied: Rayleigh number (Ra $= 10^3$, 10^4, 10^5 and 10^6), and electric parameter ($\boldsymbol{T} = 0$–800), keeping fixed value of injection parameter $C = 10$, mobility parameters $M = 10$, Prandtl number Pr $= 1$.

4.1 Effect of Rayleigh Number (Ra)

For easy understanding of the ETHD effect on the flow structure, temperature distribution, and the heat transfer characteristic at different Rayleigh numbers, the case of non-ETHD is simulated first (as a base case) and illustrated in Fig. 2 using contours of streamlines (first row), isotherms (second row), for Ra $= 10^3$, 10^4 and 10^6. When Rayleigh number is substantially low (Ra $= 10^3$), as expected from differential heating of the cavity (heating at the left wall and cooling at the right wall), the hot fluid (adjacent to the left heated wall) rises upward along the left wall of the enclosure and descends along the cold sidewall. Results, the formation of single circulation cell—rotating in CW direction (as in Fig. 2a). Thus the working fluid carries the heat energy from the heated wall to the cooled wall of the cavity. Corresponding isotherm lines distributed from left to right wall of the enclosure. So it clearly shows thermo-convective phenomena in the enclosure. At this lower Ra value, fluid circulation

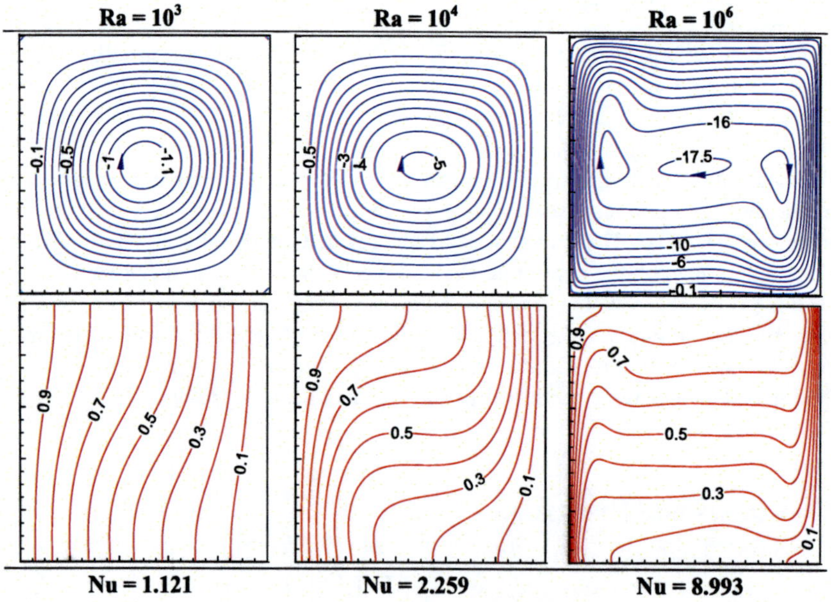

Fig. 2 Effect of varying Rayleigh number (Ra) on streamlines and isotherms without EHD

velocity is weaker; thus the heat transport mechanism from the heated source is mainly dominated by the conduction mode.

As Ra increases from 10^3 to 10^4 and then 10^6, the relative dominance of convection over conduction gradually appears, which is clearly observed from the streamline contour value. Of course, at Ra $= 10^6$, the strength of circulating vortex increases considerably. Instead of one core, there appear multiple circulation cells around mid-horizontal plane and surroundings of which a single circulating cell rotates in CW direction. Corresponding isotherm lines (connected in between the top and bottom adiabatic walls) are clustered near the active side walls (indicating a sharp temperature variation) and stretched almost horizontally inside the cavity.

Now, the combination of imposed temperature gradient and electrical field with higher strength unipolar injection on the fluid-structure, temperature distribution, and associated heat transfer characteristic at different Ra values are illustrated in Fig. 3 (with ETHD) in terms of contours of streamlines (first row), isotherms (second row), for Ra $= 10^3$, 10^4 and 10^6 keeping $\boldsymbol{T} = 400$, $\boldsymbol{M} = 10$, $\boldsymbol{C} = 10$ fixed. Compared to earlier non-ETHD cases, marked changes in streamlines, as well as isotherm patterns, are observed. Thus, it is clear that, with ETHD, circulating fluid velocity is rather higher (by noting streamline value) and substantial changes in the temperature distribution, and this should result in improved heat transfer rate. Resulting increment in average Nusselt number (Nu) estimated at the cold wall (Eq. 9)—as mentioned below, the contour plot. It is found that, for fixed value of $\boldsymbol{T} = 400$, $\boldsymbol{M} = 10$, $\boldsymbol{C} = 10$ heat

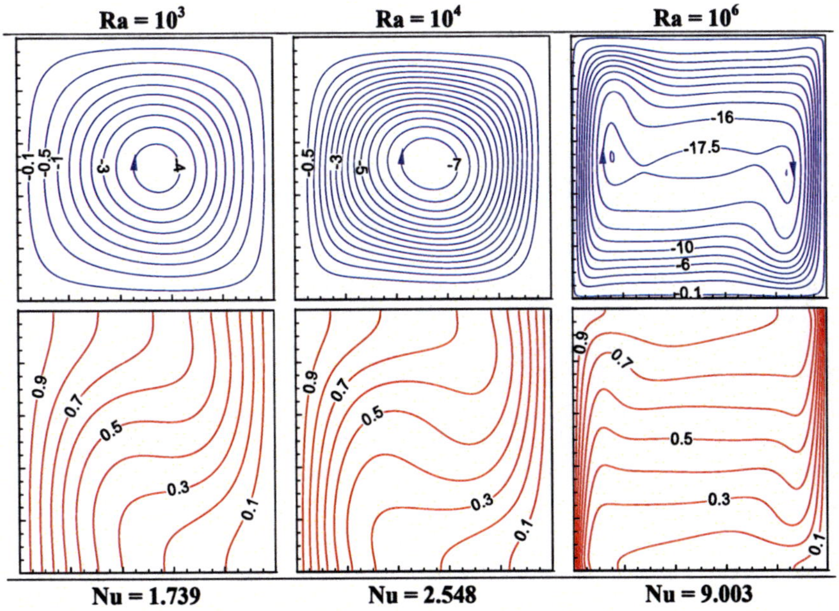

Fig. 3 Effect of varying Rayleigh number (Ra) on streamlines and isotherms with EHD at $\boldsymbol{M} = 10$, $\boldsymbol{T} = 400$

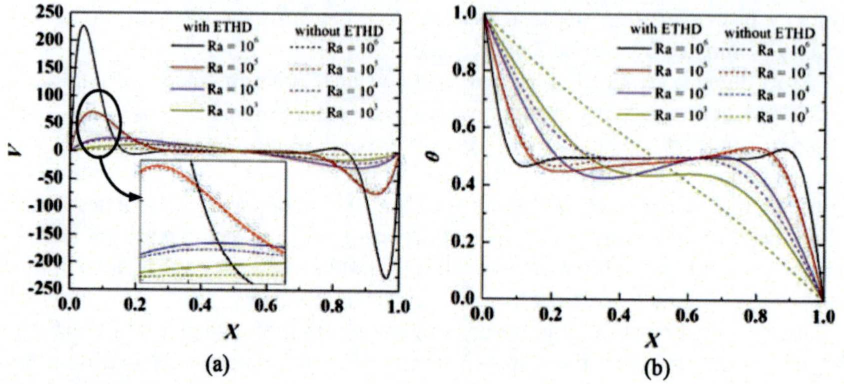

Fig. 4 Variation of mid-plane **a** vertical velocity (V) and **b** temperature (θ) with and without ETHD at $M = 10$, $T = 400$

transfer enhancement is in the range of 55.2% (for Ra $= 10^3$), 12.81% (for Ra $= 10^4$), 1.18% (for Ra $= 10^5$) and 0.11% (for Ra $= 10^6$), respectively.

For further clarity, Fig. 4a shows the variation of vertical velocity (V) component (for different Ra) with varying X at the mid horizontal plane (at $Y = 0.5$) with and without ETHD. It is obvious that with ETHD magnitude of vertical velocity increases, and this increase is more at lower Ra $\leq 10^5$. As the magnitude of the velocity is more, it boosts thermal convection; as a result, more amount of thermal energy is transferred from the heated wall to the cooled wall. Thus average Nusselt number increases with ETHD compared to the non-ETHD case, which is clearly reflected in the Nu value (as indicated below the isotherm contours).

Also, the dimensionless mid-plane temperature (θ) with X for different Ra is presented in Fig. 4b for both ETHD (firm line) and non-ETHD (dashed line) cases. It shows that the magnitude of temperature is significantly lower with the ETHD effect compared to the non-ETHD effect. It is clear that the ETHD enhances the convection process in an enclosure, resulting in better heat transfer from the cavity and lesser temperature rise. Of course, as the Rayleigh number increases beyond Ra $= 10^5$, the heat transfer improvement becomes less significant. Such findings are also reported by Yan et al. [11]. However, it is to be noted that at lower Ra value the buoyancy force is comparatively weaker than that of the ETHD force. Resulting, the significant influence of ETHD on thermal performance as well as enhanced heat transfer. Of course, at higher Ra, the resultant force of buoyancy and ETHD reduced substantially. This results in lowering the improvement in heat transfer with ETHD compared to non-ETHD. So it is pertinent that imposed electric field or electro thermo-convection is only beneficial for enhancing natural convection at lower Rayleigh number (Ra $\leq 10^5$).

4.2 Effect of Electric Parameter T

The impact of the varying electric parameter (***T***) on the fluid flow structure, temperature distribution, and associated heat transfer characteristics are demonstrated using streamlines (first row), isotherms (second row), and average Nu, respectively, in Fig. 5 for a fixed value of Ra = 10^5, $M = 10$. At a lower value of $T = 100$, the fluid flow pattern indicates the presence of multiple vortexes around the mid horizontal plane of the cavity and inside a large outer circulation cell (rotating in CW direction), whereas multiple vortexes disappear with increasing ***T***. Also, the location of the circulating vortex shifts upward. Interestingly, the circulating vortex strength increases (as indicated by the magnitude of streamlines) with increasing ***T***. This fact clearly indicates the enhanced thermal convection, which is obvious from the increasing Nusselt number. Corresponding isotherms are distributed between two active side walls flowing distorted path.

Further to the above, the average Nusselt number (Nu) with increasing electric parameter (***T***) is plotted in Fig. 6 varying the Rayleigh numbers. From the figure, it is clearly observed that Nu increases markedly with increasing ***T*** for Ra $\leq 10^4$. However, the increment of Nu with ***T*** is insignificant for Ra = 10^5 and 10^6.

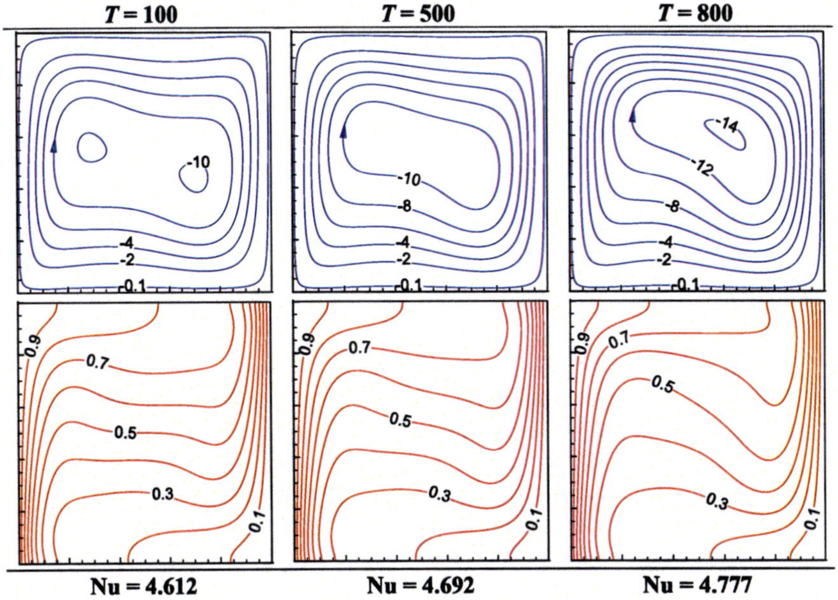

Fig. 5 Effect of electric parameter ***T*** on streamlines and isotherms at $M = 10$, Ra = 10^5

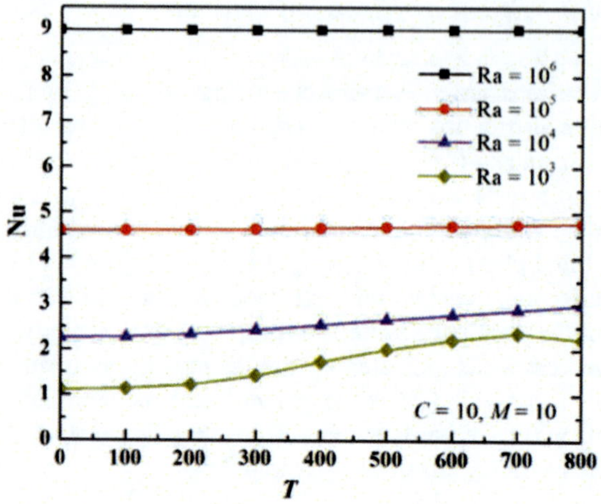

Fig. 6 Variation of average Nusselt number (Nu) with electric parameter T at $M = 10$, $Ra = 10^5$

5 Conclusions

In this work, electrothermal convection phenomena of a dielectric fluid-filled in a 2D square enclosure heated differentially under the actions of electric fields are investigated numerically. The flow structures and temperature field are characterized by the streamlines, isotherms, respectively, and the heat transfer indicated using Nusselt number under various parametric conditions. With the variation of range of pertinent flow parameters such as Rayleigh number, and electric parameter, for the fixed value of injection parameter, mobility parameters, following major observations come out:

(a) It is found that the ETHD effect improves overall thermal performance substantially over the non-ETHD case. For the fixed value of T, M and C the heat transfer enhancement with ETHD is in the range of 55.2% (for $Ra = 10^3$), 12.81% (for $Ra = 10^4$), 1.18% (for $Ra = 10^5$) and 0.11% (for $Ra = 10^6$) respectively compared to non-ETHD.
(b) The magnitude of enhanced heat transfer strongly depends on the Rayleigh number. At relatively lower Rayleigh number, the effect of ETHD is significant, whereas at higher Rayleigh number ETHD effect is insignificant.
(c) For the same injection parameter and mobility parameter, an increase in electric parameter enhances heat transfer.

As an application, such electrothermoconvection is beneficial for enhanced heat transfer, and buoyancy convection is weak.

References

1. Allen PHG, Karayiannis TG (1995) Electrohydrodynamic enhancement of heat transfer and fluid flow. Heat Recovery Sys CHP 15(15):389–423
2. Seyed-Yagoobi J, Bryan JE (1999) Enhancement of heat transfer and mass transport in single-phase and two-phase flows with electrohydrodynamics. Adv Heat Transfer 33:95–186
3. Mutabazi I, Yoshikawa HN, Fogaing MT, Travnikov V, Crumeyrolle O, Futterer B, Egbers C (2016) Thermo-electro-hydrodynamic convection under microgravity: a review. Fluid Dyn Res 48:061413-1-17
4. Martynenko A, Kudra T (2016) Electrically-induced transport phenomena in EHD drying—a review. Trends Food Sci Technol 54:63–73
5. Traoré P, Pérez AT, Koulova D, Romat H (2010) Numerical modelling of finite-amplitude electro-thermo-convection in a dielectric liquid layer subjected to both unipolar injection and temperature gradient. J Fluid Mech 658:279–293
6. Koulova D, Traoré P, Romat H, Wu J, Louste C (2013) Numerical simulations of electro-thermo-convection and heat transfer in 2D cavity. J Electrost 71:341–344
7. Luo K, Li TF, Wu J, Yi HL, Tan HP (2018) Mesoscopic simulation of electrohydrodynamic effects on laminar natural convection of a dielectric liquid in a cubic cavity. Phys Fluids 30:103601-1-11
8. Lacroix JC, Atten P, Hopfinger EJ (1975) Electro-convection in a dielectric liquid layer subjected to unipolar injection. J Fluid Mech 69:539–563
9. Wu J, Traoré P (2015) A finite-volume method for electro-thermoconvective phenomena in a plane layer of dielectric liquid. Numer Heat Transfer Part A 68:471–500
10. Wang BF, Sheu WH (2016) Numerical investigation of Electrohydrodynamic instability and bifurcation in a dielectric liquid subjected to unipolar injection. Comp Fluids 136:1–10
11. Yan YY, Zhang HB, Hull JB (2004) Numerical modeling of electrohydrodynamic (EHD) effect on natural convection in an enclosure. Numer. Heat Transfer A 46(5):453–471
12. Dantchi K, Philippe T, Hubert R (2013) Numerical study of the heat transfer and electro-thermo-convective flow patterns in dielectric liquid layer subjected to unipolar injection. J Electrost 71:970–975
13. Lu Z, Liu G, Wang B (2019) Flow structure and heat transfer of electro-thermo-convection in a dielectric liquid layer. Phys Fluids 31:064103-1-9
14. Patankar SV (1980) Numerical heat transfer and fluid flow. Taylor and Francis
15. Biswas N, Mahapatra PS, Manna NK (2015) Mixed convection heat transfer in a grooved channel with injection. Numer. Heat Transfer A 68:663–685
16. Biswas N, Manna NK (2019) Enhanced convective heat transfer in lid-driven porous cavity with aspiration. Int. J. Heat Mass Transfer 114:430–452
17. Biswas N, Mahapatra PS, Manna NK, Roy PC (2015) Influence of heater aspect ratio on natural convection in a rectangular enclosure. Heat Transfer Eng 37(2):1–15
18. Biswas N, Mahapatra PS, Manna NK (2016) Buoyancy-driven fluid and energy flow in protruded heater enclosure. Meccanica 51:2159–2184
19. Manna NK, Biswas N, Mahapatra PS Convective heat transfer enhancement: effect of multi-frequency heating. Int J Numer Methods Heat Fluid Flow 0961–5539

Numerical Investigation of Aspect Ratio Effect on Biomimetic Heat Sink Model with Two Inlet-Outlet Pairs, for Cooling Rectangular Shaped Electronic Circuits

K. Kandassamy and B. Prabu

1 Introduction

Multicore rectangular-shaped processors require similarly shaped heat sinks to dissipate high heat fluxes for their reliable operation. The straight microchannels arrays offer high heat transfer rates in the developing region and lower values as the hydraulic boundary layer develops [1, 2]. Straight microchannel heat sinks have decreasing temperature gradient in the flow direction due to saturation of the fluid thermal capacity, with higher heat transfer rates in the entrance and reduced values near the exit. One way to increase heat transfer is to introduce interruptions in the boundary layer and redevelopment of flow in the flow direction. Interruptions in boundary layer formation are achieved by the usage of porous substrates, pin fins, wavy and corrugated microchannels, dimples and grooved surfaces, ribs, and cavities [3]. A problem in parallel microchannel arrays is the maldistribution of fluids due to the variation of hydraulic resistances in individual microchannels due to their locational positions between the inlet and outlet manifolds. This maldistribution causes local hot spots and accompanying thermal stresses. This maldistribution is rectified by variously shaped headers or manifold designs, multiple inlet-outlet ports, the relative location of inlet and outlet in the flow field, compartmentalization of portions of heat sinks with separate inlet-outlets for each compartment, and variation of channel aspect ratios in microchannel arrays [4–7]. A way to strike a balance between heat transfer and pressure drop is to mimic biological flow distribution systems as is proposed in the heat

K. Kandassamy (✉)
Department of Mechanical Engineering, Annamalai University, Chidambaram, India
e-mail: kandassamy007@gmail.com

B. Prabu
Department of Mechanical Engineering, Pondicherry Engineering College, Puducherry, India

M. Palanisamy et al. (eds.), *Theoretical, Computational, and Experimental Solutions to Thermo-Fluid Systems*, Lecture Notes in Mechanical Engineering,
https://doi.org/10.1007/978-981-33-4165-4_12

sink models. Split-flow arrangements [8–10] are recommended for reduction in flow length, pressure drop reduction (ΔP), and heat transfer coefficient (h) enhancement. Most of the bio-inspired flow fields encountered in literature have found application in the field of Proton Exchange Membrane fuel cells [11, 12]. Lung and leaf-inspired models have shorter path lengths, uniform species, and velocity distribution for a large number of parallel and higher aspect ratio channels than regular microchannel arrays with longer pathways. Tree type branching with reduced hydraulic diameter (D_h) passages and branching at different angles have been found in literature surveys, known as constructal networks, and these have many levels of diameter reduction with associated manufacturing difficulties [12, 13]. Parallel micro-channel networks with biomimetic inlet and outlet manifolds [9] are found to have more surface area (A), lower ΔP, higher heat evacuation performance, and temperature (T) uniformity for a given pumping power (W_p). They are also easy to manufacture compared to constructal networks. The angle of the branching channels with the axis of the main supply and collecting manifolds is maintained at 90° as the lower angle branching channels (ch) are not able to achieve uniform flow in the flow field. The supply and collecting channel are similar, and their angles are determined by trial and error to minimize the base (b) surface temperature. From the survey of the literature, it is obvious that the application of bio-inspired rectangular flow fields is not experimented or analyzed for application in the field of high heat flux (q) heat sinking, although square-shaped heat sinking has been pursued recently [8, 9]. Hence this numerical study is conducted to identify an aspect ratio for the promising flow field pattern in rectangular shaped models. The performance is compared against literature, and applicability of the new bio-inspired flow fields is gauged by their ability to decrease ΔP across the flow field, uniformity of surface temperature (UST), R_{hy}, and Thermal resistance (R_{th}) at constant pumping power of 2.275 W for comparison with benchmark literature [1].

2 Boundary Conditions and Methodology

The assumptions and boundary conditions are as follows: Laminar, incompressible, steady, and single-phase flow. Buoyancy force due to temperature changes are considered by the inclusion of volumetric source term in the momentum equation. All external surfaces outside the heat sink are adiabatic except the bottom plate, where a constant heat flux of 7.9 MW/m^2 is applied Heat input to the fluid is the sum of convection from the base plate and lateral fin surfaces. The ΔP for the flow fields is maintained between 0.187 and 0.285 MPa. The Re in this analysis does not undergo transition. The analyzed model's aspect ratios are fixed at 6.4, 8.6, 12.9, and 14 by variation of channel height. The inlet flow field temperature is maintained at 283 K for the analyzed models. The non-slip boundary condition is considered for fluid flow, and radiation heat transfer is neglected. The general schematic and geometric parameters for the proposed models are given in Fig. 1. For the biomimetic heat sinks, the basic package dimensions are fixed as per [8, 9]. The channel width and

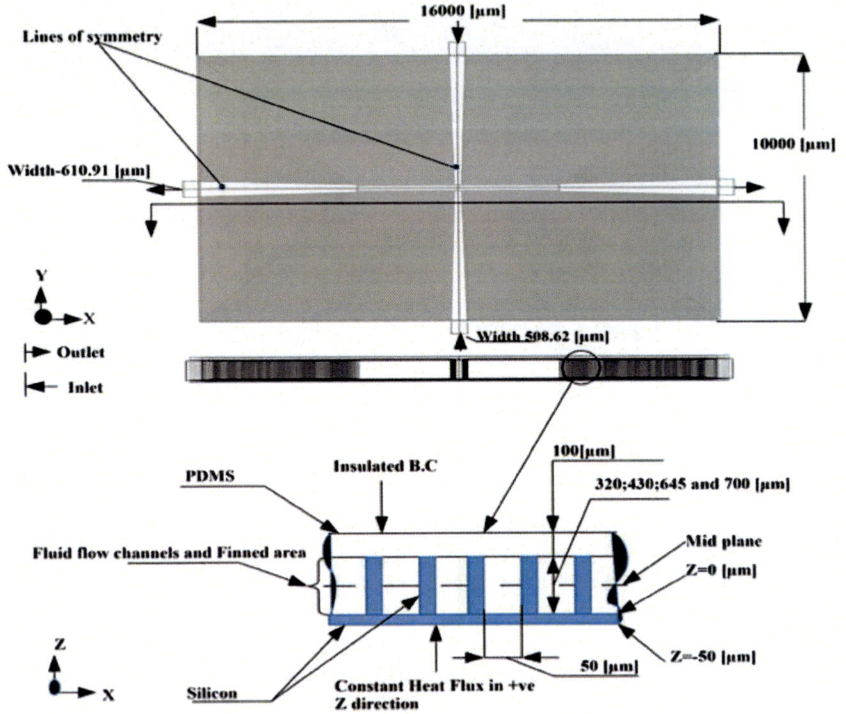

Fig. 1 Basic dimensions of the model taken for study

the wall thickness of the individual microchannel is fixed at 50 μm. The aspect ratio is varied by variation in channel height. The inlet and outlet manifold dimensions are fixed by trial and error for achieving a lower, maximum base temperature. The models are analyzed at different α, constant fin pitch and inlet pressure conditions satisfying the constant pumping power criteria. This is done to study the effect of α on R_{th} and R_{hy}. All the material properties used in the proposed models are as suggested in [14]. Liquid water, Crystalline Silicon, and Poly-dimethyl-siloxane (PDMS) top cover insulation are the materials used in the models.

3 Governing Equations and Data Reduction

The following are the governing equations applicable to the proposed models investigated, as suggested in [15]. The continuity equation is given by Eq. (1) and momentum is given by Eq. (2).

$$\rho\nabla \cdot (u) = 0 \tag{1}$$

$$\rho(u.\nabla)u = \nabla \cdot \left[-pI + \mu\left(\nabla u + (\nabla u)^T\right)\right] + F \tag{2}$$

$$F = -g * \Delta\rho \tag{3}$$

where F is the body force term given by Eq. (3), and $\Delta\rho$-represents the change in density due to thermal heating. The pin fin heat sinks are oriented towards the positive Z direction. The fluid and solid energy are computed using Eqs. (4–5). Where Q is the heat sinking value per unit volume of the heat sink, W/m^3. The conduction Eq. (5) is applicable to solid and fluid regions. The overall Nu and inlet *Re* are calculated using Eqs. (6–7) based on the inlet $D_{h\text{-in}}$. Where h_{avg} is the overall heat transfer coefficient calculated based on the inlet to outlet temperature change of the coolant fluid. The channel *Re* is calculated by evaluating the mid-plane averaged velocity and kinematic viscosity obtained from the simulation results at mid-channel height (in the Z-axis direction) and $D_{h\text{-ch}}$. The R_{th} and R_{hy} for the models are calculated using Eqs. (8–11). In Eq. (8), T_{max} is the maximum base plate temperature near the outlet and T_{min} is the inlet fluid temperature. The overall heat transfer coefficient is calculated using Eq. (12). Where Q_{hy} is the volumetric flow rate, n is the number of inlets, A_{in} is the inlet area, q_b is the heat flux applied to the base plate. The total heat flux magnitude is calculated using Eq. (13). The uniformity of the base surface temperature and pumping power is calculated using Eq. (14–15). A numerical performance index PF with units 1/K is defined similar to [16] as the ratio of heat-sinking to the pumping work. The temperature difference is included to factor in the effect of its non-uniformity in Eq. (16). The Poiseuille number (Po) for the various flow-fields is calculated using the correlation developed by Shah and London, Eq. (17). The channel friction factor-fr is evaluated using Eq. (18).

$$\rho C_P u \cdot \nabla T + \nabla \cdot q = Q \tag{4}$$

$$q = -k\nabla T \tag{5}$$

$$\text{Nu} = \frac{h_{avg} D_{h-\text{in}}}{k_f} \tag{6}$$

$$\text{Re}_{\text{in}} = \frac{\rho u D_{h-\text{ch}}}{\mu} \tag{7}$$

$$R_{\text{th}} = \frac{T_{\text{max}} - T_{\text{min}}}{Q_{\text{th}}} \tag{8}$$

where

$$Q_{\text{th}} = q_b * A_b \tag{9}$$

$$R_{\mathrm{hy}} = \frac{P}{Q_{\mathrm{hy}}} \tag{10}$$

where

$$Q_{\mathrm{hy}} = n * u_{\mathrm{in}} * A_{\mathrm{in}} \tag{11}$$

$$h_{\mathrm{avg}} = \frac{q_b}{T_{\mathrm{max}} - T_{\mathrm{min}}} \tag{12}$$

$$Q_{\text{base and fin services}} = Q_{\mathrm{convection}} + Q_{\mathrm{conduction}} \tag{13}$$

$$\mathrm{UST\%} = \frac{T_{\mathrm{max}} - T_{\mathrm{min}}}{T_{\mathrm{avg}}} * 100 \tag{14}$$

$$w_p = u_{\mathrm{in}} * A_{\mathrm{in}} * \Delta p_{\mathrm{avg}} \tag{15}$$

$$\mathrm{PF} = \frac{Q_{\mathrm{th}}}{\left(T_{b},\mathrm{max} - T_{b},\mathrm{min}\right) * W_p} \tag{16}$$

$$\mathrm{Po} = 24\left(1 - \frac{1.3553}{\alpha} + \frac{1.9467}{\alpha^2} - \frac{1.7012}{\alpha^3} + \frac{0.9564}{\alpha^4} - \frac{0.2537}{\alpha^5}\right) \tag{17}$$

$$\mathrm{fr} = \frac{\mathrm{Po}}{\mathrm{Re}_{\mathrm{ch}}} \tag{18}$$

4 Numerical Analysis

The computational domain is a quadrant section of the three-dimensional model, selected for optimizing computational resources. An unstructured mesh with elemental composition, 75% Tetrahedral, and 25% prismatic elements, having an average element quality (skewness) of 0.6, is used. A multigrid, segregated, iterative solver utilizing GMRES (Generalized minimal residual method) is used for running the simulation. The base temperature parameters are evaluated at $Z = 0$ as it is the first plane of contact of the fluid with the heat flux, and the flow field parameters are evaluated at mid plane, where u_{max} is achieved for a fully developed flow. The grid independence test is done for all the models by increasing the grid density until the change in the maximum temperature of the base plate- $(T_i\_T_{i-1})/T_i$ % is less than 1%.

Table 1 Validation of the heat sink models by heat balance at constant pumping power

ΔP	α	Q_{in}	$T_{avg\text{-}out\text{-}theo}$	$T_{avg\text{-}out\text{-}act}$	Error
285.55	6.4	1264	322.08	322.84	-0.226
236.56	8.6	1264	315.61	316.78	-0.370
196.15	12.9	1264	310.67	310.77	-0.032
188.2	14	1264	307.94	309.80	-0.603

5 Results and Discussion

5.1 Validation

As the biomimetic models proposed in this work are novel, similar models to be used as reference benchmarks are currently unavailable in the literature. Hence the validation of the simulation is done by energy balance. The average temperature of the outlet fluid calculated using the energy balance Eq. (19), and the results of the actual simulation results are tabulated in Table 1.

$$\rho_f * u_{in} * A_{in} * C_{pf} * \left(T_{avg\text{ - }out\text{ - }theo} - T_{avg\text{ - }in}\right) = Q_{th} \tag{19}$$

where $T_{avg\text{-}in}$-283 K and $T_{avg\text{-}out\text{-}theo}$ is calculated. The C_{pf} value is assumed as 4.18 kJ/kg K in the operating temperature range. The weighted average outlet temperature at the outlet obtained by simulation-$T_{avg\text{-}out\text{-}act}$ is given by Eq. (20), where, *n*-normal vector; *T*-Temperature; *A*-Area, *u*-velocity perpendicular to the outlet plane. The comparison between the calculated value ($T_{avg\text{-}out\text{-}theo}$) and the weighted average outlet temperature ($T_{avg\text{-}out\text{-}act}$) in Table 3 shows negligible error. Numerical validation is done by simulating the experimental results of Bhattacharya et al. [17] for a single microchannel, keeping the same input parameters, as shown in Fig. 3. The average element quality is 0.611 with 76,387 elements consisting of 75% Tetrahedra and 25% prismatic elements. The maximum base temperature 306.392 K is compared to 307.8 K in the reference yielding an error of 0.457%, which is negligible, and hence a similar procedure is followed in simulating the proposed heat sink models.

$$T_{avg\text{ - }out\text{ - }simulation} = \frac{\int_A T u \cdot n dA}{\int_A u \cdot n dA} \tag{20}$$

5.2 Thermal Characteristics

The analyzed models are tested with a heat flux of 7.9 MW/m^2. The maximum base temperature occurs near the outlet for the given flow field at all aspect ratios. In the

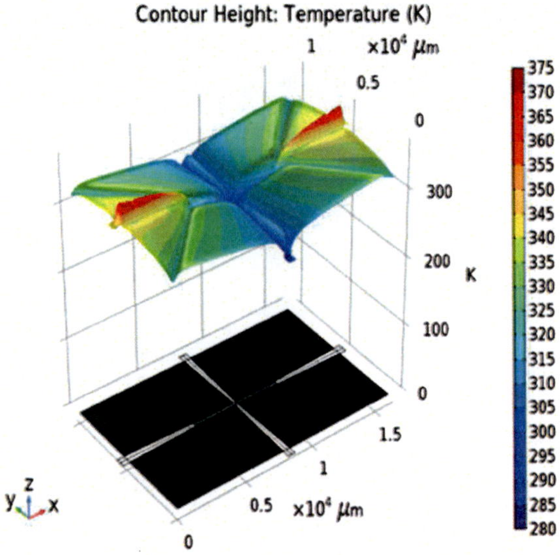

Fig. 2 Temperature contours in K, at the base plane Z = 0 μm for aspect ratio 12.9

models shown in Fig. 2, the temperature rise reaches the maximum value at the outlet manifolds due to adverse temperature gradient. The base temperature is lower in the inlet manifold due to the prevention of boundary layer formation as the fluid cross-section is gradually reduced as it is supplied to microchannel arrays and the 90° bend in the microchannels creates mixing in the fluid due to change in direction, inducing high heat transfer rates. Another area of high heat transfer is the starting point of the outlet manifold where two streams converge, creating mixing of fluid as is shown in Fig. 3. The UST value is roughly 23.42% on an average basis for all the models. The UST value of the current models is higher than [1] by 9.45% as the inlet temperature is nearly 10 K lower than [1]. For the same geometric parameters, an increase in inlet pressure and α gives nearly constant UST at constant pumping power conditions. The lowest UST obtained in this paper is 22.99%, as shown in Table 2. The total local heat flux at inlet and outlet regions of the microchannels is three times higher than the base heat flux $0.88*10^7$ W/m^2 as shown in Fig. 3. The variation in heat flux is minimal in individual microchannels due to the nearly uniform flow in individual microchannels except in the 90° bend regions. The R_{th} is the basis for comparing heat transfer performance of heat sinks, as shown in Table 3. An increase in α and inlet pressure decreases R_{th}. Due to constant pumping power conditions applied at the inlet to the flow fields, the R_{th} remains constant at 0.070 K/W for all aspect ratios. The R_{th} value of the proposed model is lower than [1] by 22.05%. Similar values are found in [18], for comparison an R_{th} value of 0.054 K/W is achieved at an aspect ratio of 15 (channel width 50 μm) and W_p-7.5 W [19]. Generally, for models with lower R_{th}, the PF decreases with an increase in α, indicating that higher W_p is necessary. The models in this analysis showed higher values of PF-7.175 on an

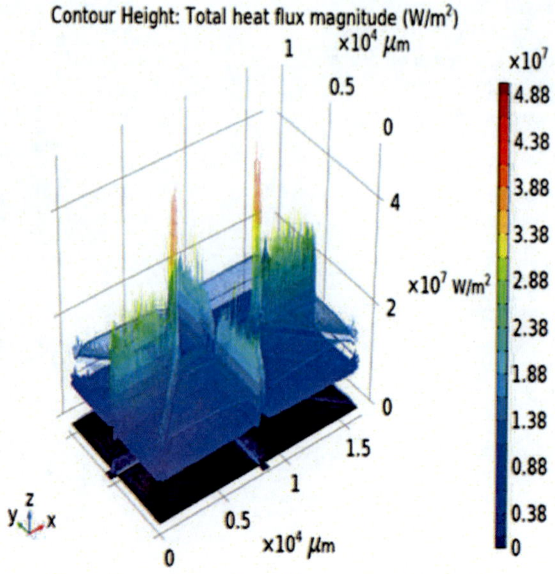

Fig. 3 Total heat flux contours in at the base plane for aspect ratio 12.9

Table 2 UST values for the proposed flow fields at constant pumping power

Model	α	ΔP	T_{max}	T_{min}	T_{avg}	UST
[1]	6.4	206.84	367	296	331.5	21.4
Bio-inspired heat sink	6.4	285.55	371.77	294.16	334.50	23.20
	8.6	236.56	369.45	293.34	331.02	22.99
	12.9	196.15	372.01	294.03	326.67	23.87
	14	187.67	372.15	294.24	329.67	23.63

average basis compared to [1], yielding a 46.72% increase in performance. The inlet averaged $Nu_{in\text{-}avg}$ increases with α as the heat transfer area increases.

5.3 Hydraulic Characteristics

The ΔP decreases for the analyzed models as the aspect ratio increases at constant pumping power, even as higher aspect ratio models require higher pumping power to circulate fluid at the same velocity [4]. The models with aspect ratios 12.9 and 14 have reduced R_{hy} than [1]. The 12.9 aspect ratio model is selected as optimum due to the lower aspect ratio and the inlet pressure is just below the reference value of 206.84 kPa, as shown in Table 3. W_p is the product of volume flow rate and applied ΔP. With branching, ΔP increases due to secondary flows and increase in surface area, but shorter path lengths negate this effect of higher branching in flows. Higher

Table 3 Results of R_{th} and R_{hy} for the proposed flow fields studied at constant pumping power

Model	α	$D_{h\text{-}ch}$	ΔP	R_{th}	R_{hy} *10^{10}	Re_{ch}	fr	PF	$Nu_{in\text{-}avg}$
Bio-inspired heat sink	6.4	86.48	290.82	0.070	3.721	458.87	0.0434	7.15	54.66
	8.6	89.58	242.17	0.068	2.575	384.29	0.0541	7.30	66.58
	12.9	92.80	196.15	0.070	1.691	304.73	0.0713	7.12	78.92
	14	93.33	187.67	0.070	1.548	280.88	0.0779	7.13	81.62
[1]	6.4	98.6	206.84	0.0898	1.88	730	0.030	4.89	16.88

α for the same pitch is a reason for an increase in $W_{p,}$ but the branching lowers the hydraulic resistances; hence lower pressure drops for bio-inspired models, as shown in Table 3. Also, the channel Re decreases with an increase in aspect ratio due to the maintenance of constant pumping power condition at the inlet; this is the reason for the lowering of pressure drop at the inlet. Even the lower flow rate in individual channels is sufficient for heat sinking as the increase in surface area, and higher fluid flow due to the marginal increase in channel hydraulic diameter is sufficient to accomplish the task. The convergent inlets act as a nozzle, increasing the velocity, while the divergent outlet acts as a diffuser, improving pressure recovery at the exit to the flow field, as is evidenced by the pressure contours of the heat sink models. The pressure at the central regions of the heat sinks is higher than at the inlet due to the stagnation pressure rise and reduction in fluid velocities as it comes into contact with the heat transfer surfaces, as shown in Fig. 4. The velocity reduction in the inlet manifolds, and the velocity increase in the outlet manifold, follows a parabolic profile as is seen in the velocity contour graphs in Fig. 5. This gradual reduction and increase in velocities ensure nearly constant velocities in the branching microchannel arrays. As is seen in Table 3, the channel Re of the proposed models is lower than

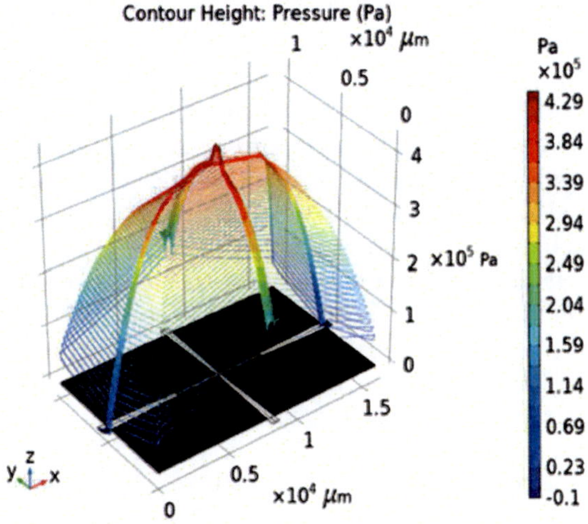

Fig. 4 Pressure distribution in Pa, for the heat sinks at mid-channel height for aspect ratio 12.9

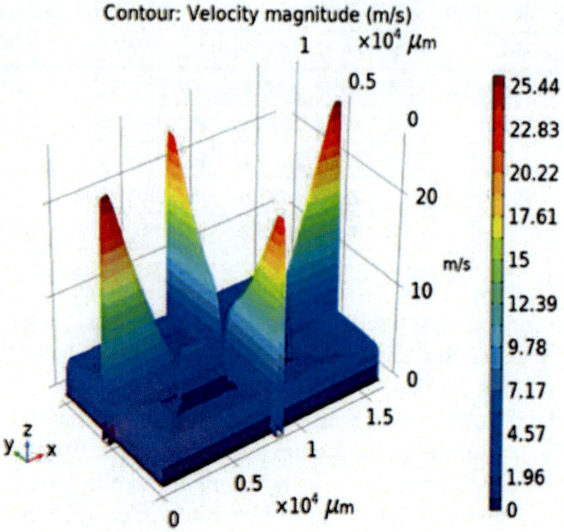

Fig. 5 Velocity distribution for the heat sink models at mid-channel height for aspect ratio 12.9

[1]. This gives scope for increasing the Re in the microchannels without leaving the laminar flow regime and consequentially increases heat transfer. The friction factor decreases with an increase in Re_{ch}, as given in Table 3.

6 Conclusion

The following conclusions are delivered based on the numerical analysis carried out in the present work. The proposed flow fields show a 9.45% higher chip surface temperature variation (UST) than the benchmark conventional flow field of Tuckerman et al., due to higher heat flux dissipation and lower coolant inlet temperature. At constant pumping power condition, R_{th} value of 0.070 K/W for the proposed model is 44.37% lower than the benchmark. Proposed flow fields with aspect ratios 12.9 and 14 have lower ΔP's for a similar pumping power of 2.275 W. At constant pumping power, the proposed flow fields with aspect ratio 12.9 and 14 show 10.05 and 17.66% lower R_{hy}, respectively, when compared to the benchmark, due to flow uniformity and higher fluid flow for a given ΔP. Aspect ratio (α) has an insignificant impact on the thermal characteristics of microchannel heat sinks at constant pumping power inlet condition. In contrast, the hydraulic characteristics improve with aspect ratio. Higher inlet pressure reduces the maximum base temperature and UST (%) value. A higher aspect ratio increases Nu due to the higher heat transfer area. Biomimetic coolant fluid distribution of the proposed heat sink reduces R_{hy} and R_{th} considerably, making this a basis for future contribution in the field of heat sink development.

References

1. Tuckerman DB, Pease RFW (1981) High-Performance Heat Sinking for VLSI. IEEE Electron Device Lett EDL 2:126–129
2. Lee H, Agonafer DD, Won Y et al (2016) Thermal modeling of extreme heat flux microchannel coolers for GaN-on-SiC semiconductor devices. J Electron Packag Trans ASME 138
3. Dewan A, Srivastava P (2015) A review of heat transfer enhancement through flow disruption in a microchannel. 24:203–214
4. Ma DD, Xia GD, Wang J et al (2017) An experimental study on hydrothermal performance of microchannel heat sinks with 4-ports and offset zigzag channels. Energy Convers Manag 152:157–165
5. Manikanda Kumaran R, Kumaraguruparan G, Sornakumar T (2013) Experimental and numerical studies of header design and inlet/outlet configurations on flow mal-distribution in parallel micro-channels. Appl Therm Eng 58:205–216
6. Tang S, Zhao Y, Diao Y, Quan Z (2018) Effects of various inlet/outlet positions and header forms on flow distribution and thermal performance in microchannel heat sink. Microsyst Technol 24:2485–2497
7. Singh V, Kumar H, Sehgal SS, Kukreja R (2019) Effect of plenum shape on thermohydraulic performance of microchannel Heat Sink. J Inst Eng Ser C
8. Kandassamy K, Prabu B (2019) Numerical investigation of bio-inspired pin fin heat sink models for square shaped electronic circuits. J Appl Sci Eng 22:119–133
9. Kandassamy K, Prabu B (2020) Numerical analysis of microchannel based bio-inspired heat sinks with multiple inlet-outlet pairs for cooling square shaped circuits. J Comp Appl Res in Mech Eng (JCARME)
10. Xu S, Hu G, Qin J, Yang Y (2012) A numerical study of fluid flow and heat transfer in different microchannel heat sinks for electronic chip cooling. J Mech Sci Technol 26:1257–1263
11. Roshandel R, Arbabi F, Moghaddam GK (2012) Simulation of an innovative flow-field design based on a bio inspired pattern for PEM fuel cells. Renew Energy 41:86–95
12. Arvay A, French J, Wang JC et al (2013) Nature inspired flow field designs for proton exchange membrane fuel cell. Int J Hydrogen Energy 38:3717–3726
13. Farzaneh M, Salimpour MR, Tavakoli MR (2016) Design of bifurcating microchannels with/without loops for cooling of square-shaped electronic components. Appl Therm Eng 108:581–595
14. Kandassamy K, Prabu B (2020) Numerical modeling of a bio-inspired heat sink with two inlet–outlet pairs for removing 10 MW/m heat flux. Mater Today Proc 2–7. https://doi.org/10.1016/j.matpr.2020.09.697
15. COMSOL (2014) Introduction to COMSOL Multiphysics 5.3. Manual 168
16. Lorenzini-Gutierrez D, Kandlikar SG (2014) Variable fin density flow channels for effective cooling and mitigation of temperature non-uniformity in three-dimensional integrated circuits
17. Bhattacharya P, Samanta AN, Chakraborty S (2009) Numerical study of conjugate heat transfer in rectangular microchannel heat sink with Al_2O_3/H_2O nanofluid. Heat Mass Transf 45:1323–1333
18. Li J, Peterson GP (2007) 3-Dimensional numerical optimization of silicon-based high performance parallel microchannel heat sink with liquid flow. Int J Heat Mass Transf 50:2895–2904
19. Hu G, Xu S Optimization design of microchannel heat sink based on SQP method and numerical simulation. In: International conference on applied superconductivity and electromagnetic devices, pp 89–92

Effect of Filling Ratio on Performance of Two Loop Pulsating Heat Pipe

Est Dev Patel and Subrata Kumar

1 Introduction

The emerging techniques of micro-fabrication in electronic devices require a single or multi-phase flow to achieve heat dissipation or spreading [1]. This transport of heat energy leads to better control and efficiency. Among the multi-phase cooling systems, heat pipes are capable of handling the heat dissipated from the chips and other electronic components. The pulsating/oscillating (PHP/OHP) heat pipe has a thermally driven motion of working fluid to encourage the passive heat transfer for heat management. Oscillating slug-plug and annular flow dominate the flow pattern in a closed-loop wickless system. The liquid slugs entrapped between the vapor plugs travels in oscillatory fashion by the pumping action of bubbles generated in the evaporator. The temperature gradient between the evaporator and condenser creates temporal and spatial pressure disturbances due to simultaneous evaporation and condensation. The sustained flow of working fluid amounts for sensible along with latent heat transfer in the PHPs.

Smaller and faster is the better trend of electronics by 2020 will lead the development in the heat flux density 190 W/cm^2. In the purview of cooling techniques, Mudawar [1] investigated and developed the capabilities of dissipating the high heat fluxes in electronics. He focuses on natural, forced convection, phase change cooling techniques, and recommends that the CHF plays a vital role in the design of phase change cooling. The modernization and improvement of electronic devices day by day lead to explore the enhanced techniques of heat dissipation, which are reliable, cost-effective, high heat dissipation potential, and compact. The CHF is a defining parameter that can be used to predict the performance of any heat dissipating device.

E. D. Patel (✉) · S. Kumar
Indian Institute of Technology Patna, Patna, Bihar 801103, India
e-mail: dev.pme17@iitp.ac.in

M. Palanisamy et al. (eds.), *Theoretical, Computational, and Experimental Solutions to Thermo-Fluid Systems*, Lecture Notes in Mechanical Engineering,
https://doi.org/10.1007/978-981-33-4165-4_13

The cooling method is selected based on the estimation of parameters, such as heat, space, passive or active cooling, etc. Khandekar et al. [2] investigated a single loop pulsating heat pipe of a total length of 190 mm made of glass and copper with ethanol as a working fluid. They realized that the performance would depend on the flow pattern of the working fluid inside the tube. A heat pipe stops working when a complete stopover in the motion happens. The phase change of working fluid and energy exchange follows the p–h diagram, which has been explained about the working of the pulsating heat pipe. A continuous rise in the evaporator temperature can be seen if the net motion does not happen. A review on reliability by Song and Wang [3] for the power electronic systems, focused on reliability assessment, reliability improvement of an existing system, and reliability-oriented design solution. They discussed metrics for the evaluation of power electronic systems comprised of reliability, failure rate, mean time to failure, mean time to repair, and availability. The methods to improve all the metrics are thermal management, fault diagnosis, and improvement in design. Rao et al. [4] conducted a unit cell experiment using a glass tube with working fluid FC-72 and a single bubble to understand the hydrodynamics of the working fluid. The tube sized bubble oscillated between the condenser and evaporator section. The role of evaporation and condensation heat transfer from the thin film discussed, and vapor temperature varies much because of it. The liquid slug performs sluggishly to the temperature variations than the vapor plug.

The vast testing of multi-loop closed pulsating heat pipe of copper tube Mameli et al. [5] found the effects of input heat flux, filling ratio, inclination angle using the working fluid FC-72. The internal diameter of the tube kept lesser, which offers a high surface tension force. The horizontal setup fails to start, but the vertical one stops only at the critical heat flux. In a study of single turn pulsating heat pipe conducted by Saha et al. [6] it was found that the diameter of 4 mm fails to give a stable motion to the working fluid. So lesser diameter recommended by Bond number relation is suitable to perform the pulsating heat pipe experiments. The continuous rise in temperature does not happen like other heat pipes in PHPs Cui et al. [7, 8]. The stagnant heat transfer occurs before the start of the oscillation motion on a small scale. The five loop setup made from 2 mm internal diameter copper tubes contains water, methanol, ethanol, and acetone 20–80 filling ratio as a working fluid. They noticed that a critical start-up heat flux and optimum filling ratio exist for the maximum thermal performance.

In one study, they conducted experiments with the binary mixture to find that the low filling ratio is performing well, and the only benefit is that this delay the dry-out phenomena.

The applications of the heat pipe are extensive in electronic devices, mechanical system, space, etc. The power electronics is one which utilizes this idea much than the others. Kearney et al. [9] conducted a study of the open-loop pulsating heat pipe of 44 channel for two independent dielectric working fluids NovecTM 649 and NovecTM 774; the earlier was performing better. The working fluid with lower pressure operation and low global warming potential are desired for heat pipe applications. They used the pulsating heat pipe as a heat spreading device and identified a range of optimum operating conditions. The heat transfer performance of two single loop heat pipes connected parallel was found lesser than the single two-loop heat pipe by Kim et al. [10]. They used ethanol and water for a fixed filling ratio and different heat inputs. In this setup, a copper tube of 2.2 mm with working fluid ethanol performs better because of high-pressure change for a small temperature variation.

The non-condensable gasses offer some resistance in heat transfer, so the rate of evaporation and condensation are affected. The evacuation pressure role was verified by Sun et al. [11] using working fluid water and HFE-7000 in a copper tube heat pipe of 8 parallel channel of 2.4 and 1.6 mm diameter. The solubility of non-condensable gasses in water depreciates the start-up performance than the HFE-7000. The working fluid characteristics affect the performance of the heat pipe [12]. The properties such as surface tension, pressure change for a given temperature change, latent heat, and presence of non-condensable gasses are some essential points to be considered. The water is readily available having high latent heat is useful, but at the same time, high surface tension makes it vulnerable to start a heat pipe functioning.

The present study deals with a two-loop pulsating heat pipe of 2 mm internal diameter. There are very few experiments carried out at low filling ratio, low heat input at the evaporator in a two-loop pulsating heat pipe. The distilled water as a working fluid and air cooling at the condenser are rare to find in literature. The single loop pulsating heat pipe does not perform while multi-turn does quite well because of uniform pressure distribution. Therefore, a two-loop system would be a great choice between the single and multi-turn to make a compact and reliable heat pipe for low heat dissipating devices.

2 Test Setup and Measurement

The test setup of the pulsating heat pipe, as shown in Fig. 1 includes instrumentation and a model of the pulsating heat pipe installed on a rotatable mount. The PHP has 1 mm wall thickness, 2 mm internal diameter copper tube evaporator, and condenser sections height of 25 and 35 mm, respectively. The quartz glass tube of 202 mm made an adiabatic section. The setup is filled with filling ratios of 5, 30, 40, and 70% with pure degassed water after evacuation to 650 mmHg by a vacuum pump. The bend radius of the copper tube was 10 mm so that it should not collapse to make the U connector to complete the loop. Joule heating to heat the evaporator is achieved through a 32 SWG nichrome wire enfolded around the copper tube and insulation over it. A DC source of accuracy ± 0.5% + 2, KEYSIGHT make model N5751A supplies electrical power to the heater. National Instrument (NI) make NI cDAQ-9189 acquires the signal from temperature and pressure sensors by modules NI9213 and NI9205, respectively. The temperature and pressure transducer read by control software LabVIEW on a computer. Honeywell makes TJE Precision Absolute Pressure Transducer of accuracy 0.1% with internal amplification measures the absolute pressure in this setup while working and degassing.

The Tarson make rocker 300 model 110 mmHg absolute vacuum pump evacuates the pre-filled setup. The temperature of the pulsating heat pipe evaporator, condenser,

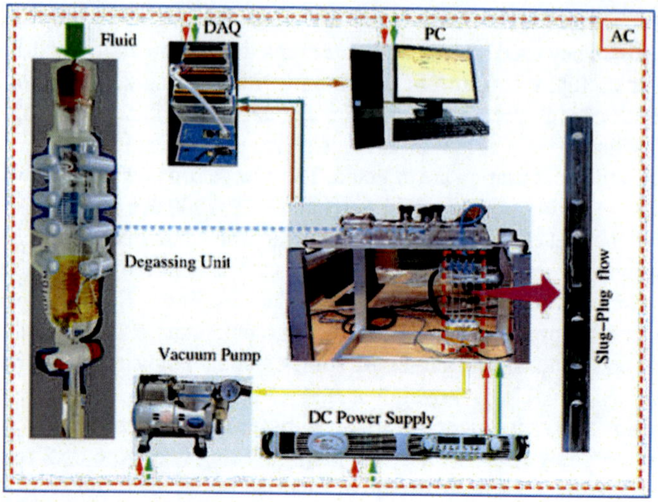

Fig. 1 Test facility of pulsating heat pipe

adiabatic section and cooling air to judge the thermal performance are recorded with calibrated TEMPSENS make T-type thermocouples of 0.5 mm diameter. The proper thermocouple contact with the surface of the heat pipe to avoid the contact thermal resistance is made with a thermal paste. The Swagelok make needle valves, adapters, tubes assembled to make the filling and vacuum port. The condenser section kept in a duct that has a facility that air is axially forced via HICOOL make compact fan model 12A230HSAC axial fan of 120 × 120 mm^2 with an airflow of 95 cfm. The degassed Millipore water from a custom degassing facility is filled with a syringe pump post evacuation of PHP. The presence of air in the system acts as non-condensable gas that interrupts the evaporation and condensation process of working fluid.

3 Performance Calculation

The overall thermal resistance and thermal conductivity estimate the performance [7], [8], and [11] performance of pulsating heat pipe. The heat supplied to the evaporator section is found by Joule heating of the nichrome resistance heater. The heat loss to the ambient doesn't affect the order of thermal resistance and due to heavy glass wool insulation, we have ignored it.

$$Q = V \times I \tag{1}$$

The Fourier's Law of heat conduction provides effective thermal conductivity, using average evaporator T_e and condenser T_c temperature.

$$k_{\text{eff}} = \frac{Q \times L_{\text{eff}}}{A \times (T_e - T_c)} \tag{2}$$

Here the effective length of the pulsating heat pipe L_{eff} is the distance between the evaporator and condenser section.

$$L_{\text{eff}} = \frac{1}{2}(L_e + L_c) + L_a \tag{3}$$

where A is the cross-section area of the pulsating heat pipe. The lengths of the evaporator, condenser, and adiabatic section are L_e, L_c, and L_a, respectively.

$$A = \frac{n \times \pi \times d^2}{4} \tag{4}$$

where n is the number, and d is the internal or hydraulic diameter of channels. The thermal resistance is obtained from electrical analogy,

$$\mathrm{R_{th}} = \frac{T_e - T_c}{Q} \tag{5}$$

4 Results and Discussion

The temporal temperature variations are the main parameters that we required to evaluate the performance of the pulsating heat pipe. The effective thermal conductivity and thermal resistance are evaluated. The effects of different parameters are summarized in the sections below.

4.1 Start-up of the Pulsating Heat Pipe

The working fluid inside the tubes distributes itself in the form of plug-slug due to its surface tension. At ambient condition, the vapor inside the tube is in a saturated state. The plug-slug is unevenly distributed in the parallel channels creating different frictional forces for the directional motion when the working fluid receives heat. The formation of tube size or small bubbles occurs with a rise in temperature. Simultaneous heat dissipation at the condenser section makes the low-pressure zone and vapor volume decrease. Being an isochoric device same vapor generated at the evaporator section makes condition favorable to move the working fluid. This random volume or pressure distribution is responsible for moving the working fluid inside the tube [2]. After the start-up, the temperature of the evaporator falls eventually and remains below the maximum value in oscillating mode while operating. The peak temperature and continuous rise due to heat input increment can be observed in Fig. 2 and Fig. 3. When working with low filling ratios, only chugging motion at the evaporator site happens; no resultant motion of slug-plug train happens, resulting in a continuous rise in temperature. As shown in Fig. 4, at the filling ratio, 70% oscillatory flow occurs once oscillation starts.

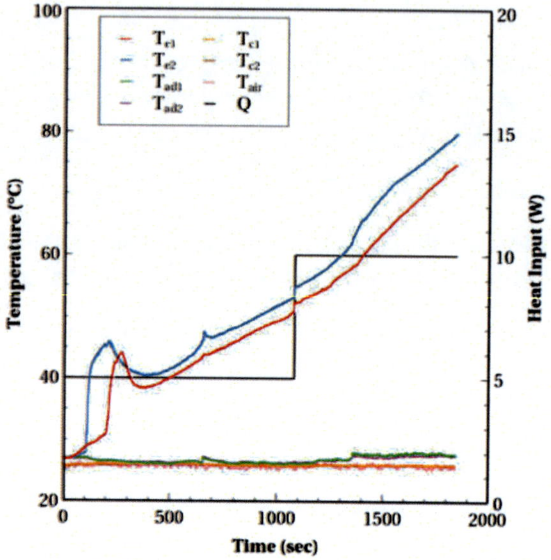

Fig. 2 Temperature variation operating at very low filling ratio 5%

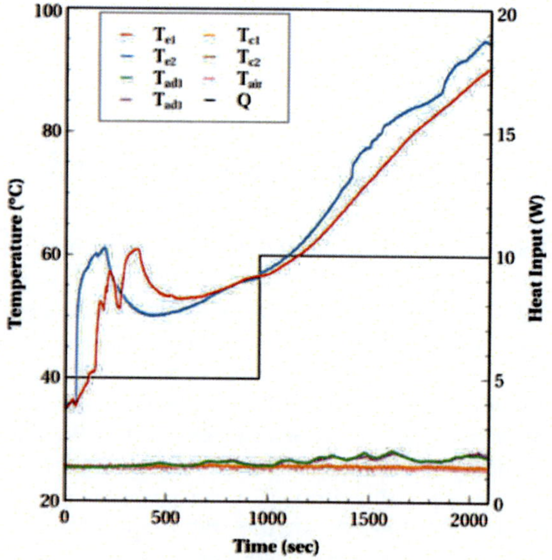

Fig. 3 Temperature variation operating at medium filling ratio 30%

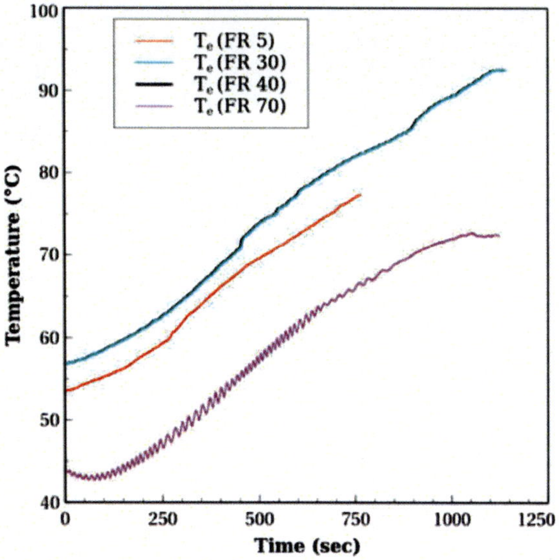

Fig. 4 Average evaporator temperatures at heat input 10 W operating at different filling ratios

4.2 *Effect of the Filling Ratio*

The evacuated tubes of the pulsating heat pipe are filled with the distilled water with a syringe and a one-way valve to a determined filling ratio. The motion of the working fluid is the only reason behind the heat transfer performance of the heat pipe. Proper optimization of parameters needs to be done under which the desired temperature of the electronic equipment can be maintained.

The low filling ratios lead to dry-out phenomena at the evaporator while at high filling ratios performs well, as it has also been discussed by Cui et al. [8]. Due to sudden motion, a fall in the evaporator temperature happens, and it rises when the working fluid is not enough in the evaporator. The working fluid filling ratio 5, 30, and 40% remains deficient, and no oscillatory flow happens, failing the operation, as shown in Fig. 4 at a heat input of 10 W.

4.3 *Effect of the Heat Input*

A threshold value of input heat flux exists at which the working of pulsating heat pipe starts. The boiling of working fluid starts at saturated condition, which initiates small or tube-size bubble formation. The failure of an electronic device might happen if the heat transfer equipment does not remove the heat. So the threshold value of the heat input should be lower than the minimum designed heat dissipated. Kumar et al. [12] also conducted experiments to look after the effect of input heat at the thermal performance of the heat pipe. The effective performance of the pulsating heat pipe

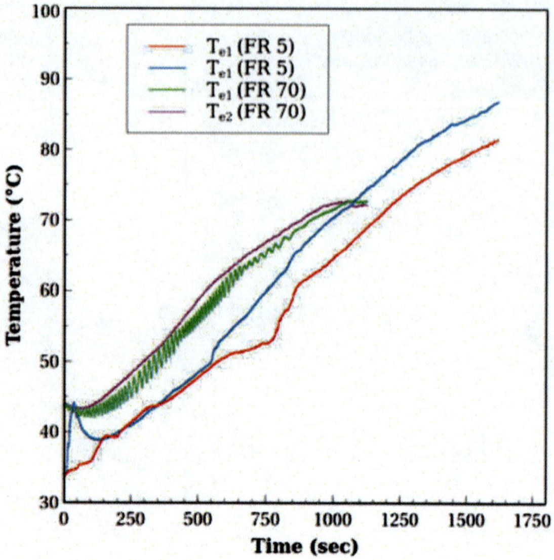

Fig. 5 Temperature variations at the individual turn for heat input 10 W

found to be increasing with the increase in input power because of the more intense motion of the working fluid inside the tube. The desired performance depends on the type of movement, and the circular motion of the working fluid is found best in the literature.

5 Thermal Performance Comparison

The performance in terms of thermal resistance of the present work has been compared, as shown in Fig. 6. Cui et al. [8] conducted experiments with five turns pulsating heat pipe with distilled water with the filling ratios 45 and 70%. Therefore, the performance of the present study was not approaching that level but comparable to their results. The evacuation pressure also affects the rate of heat transfer and the maximum evaporator temperature. In the study conducted by Kumar et al. [12] with two turns pulsating heat pipe, the performance was lower than the present study working also with the distilled water. The repeatability test of the setup shows very similar performance, as shown in Fig. 4 for different evacuation pressures. The filling ratios 30 and 40% perform very identical and follow the same temperature profile.

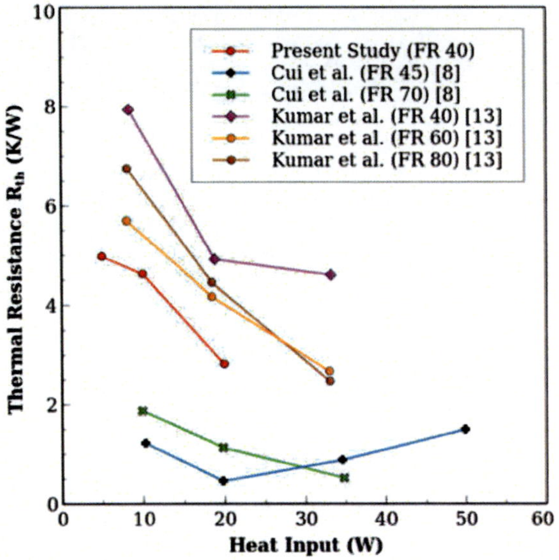

Fig. 6 Thermal performance comparison

6 Conclusion

The experiments conducted to study the pulsating heat pipe for the application like heat dissipation or to spread it over a large area. Low heat input levels evaluate a two-loop semi-transparent pulsating heat pipe performance under different filling ratios. The working fluid in the tube remains saturated. The translating-oscillating motion due to heat addition and rejection makes the pulsating heat pipe an excellent passive cooling device. At low filling ratios (5, 30, and 40%), the heat pipe does not function well, the temperature and pressure piles up in the evaporator, while at 70%, the temperature starts to fall after a threshold value. The heat input of 10 W is enough to start the motion in the vertical orientation. This heat transfer device offers lower thermal resistance has higher heat spreading or dissipating capacity; thus, the more uniform temperature in an electronic device can be maintained.

Acknowledgements This work is supported by Visvesvaraya Ph.D. Scheme of MeitY (Ministry of Electronics and Information Technology), Government of India.

References

1. Mudawar I (2001) Assessment of high-heat-flux thermal management schemes. 24(2):122–141
2. Khandekar S, Groll M (2004) An insight into thermo-hydrodynamic coupling in closed loop pulsating heat pipes. Int J Therm Sci 43(1):13–20. https://doi.org/10.1016/S1290-0729(03)00100-5

3. Song Y, Wang B (2013) Survey on reliability of power electronic systems. IEEE Trans Power Electron 28(1):591–604. https://doi.org/10.1109/TPEL.2012.2192503
4. Rao M, Lefèvre F, Khandekar S, Bonjour J (2013) Understanding transport mechanism of a self-sustained thermally driven oscillating two-phase system in a capillary tube. Int J Heat Mass Transf 65:451–459. https://doi.org/10.1016/j.ijheatmasstransfer.2013.05.067
5. Mameli M, Manno V, Filippeschi S, Marengo M (2014) Thermal instability of a closed loop pulsating heat pipe: combined effect of orientation and filling ratio. Exp Therm Fluid Sci 59:222–229. https://doi.org/10.1016/j.expthermflusci.2014.04.009
6. Saha N, Das PK, Sharma PK (2014) Influence of process variables on the hydrodynamics and performance of a single loop pulsating heat pipe. Int J Heat Mass Transf 74:238–250. https://doi.org/10.1016/j.ijheatmasstransfer.2014.02.067
7. Cui X, Zhu Y, Li Z, Shun S (2014) Combination study of operation characteristics and heat transfer mechanism for pulsating heat pipe. Appl Therm Eng 65(1–2):394–402. https://doi.org/10.1016/j.applthermaleng.2014.01.030
8. Cui X, Qiu Z, Weng J, Li Z (2016) Heat transfer performance of closed loop pulsating heat pipes with methanol-based binary mixtures. Exp Therm Fluid Sci 76:253–263. https://doi.org/10.1016/j.expthermflusci.2016.04.005
9. Kearney DJ, Suleman O, Griffin J, Mavrakis G (2016) Thermal performance of a PCB embedded pulsating heat pipe for power electronics applications. Appl Therm Eng 98:798–809. https://doi.org/10.1016/j.applthermaleng.2015.11.123
10. Kim B, Li L, Kim J, Kim D (2017) A study on thermal performance of parallel connected pulsating heat pipe. Appl Therm Eng 126:1063–1068. https://doi.org/10.1016/j.applthermaleng.2017.05.191
11. Sun CH, Tseng CY, Yang KS, Wu SK, Wang CC (2017) Investigation of the evacuation pressure on the performance of pulsating heat pipe. Int Commun Heat Mass Transf 85(April):23–28. https://doi.org/10.1016/j.icheatmasstransfer.2017.04.005
12. Kumar M, Kant R, Das AK, Das PK (2018) Effect of surface tension variation of the working fluid on the performance of a closed loop pulsating heat pipe. Heat Transf Eng 40(7):1–15. https://doi.org/10.1080/01457632.2018.1436390

Optimization of Process Parameters of Sintered Copper Wick Heat Pipe Using Response Surface Methodology

B. Ch Nookaraju, P. S. V. Kurmarao, and S. Nagasarada

1 Introduction

The field of electronics is the fast-developing science in the present scenario, and its contribution to technology is growing rapidly. Continuous usage of these devices generates high heat. This induces thermal stresses in the electronic circuits, leading to the failure in the components. The generated large heat flux is not removed effectively; it leads to deterioration in the effective functioning of the electronic devices. Also, effective thermal management becomes one of the major serious challenges in many technologies because of constant demands for faster speed and continuous reduction of device dimensions. The heat pipe is a special type of heat exchanging device that transfers a large amount of heat due to the effect of capillary action and phase change heat transfer principle shown in Fig. 1 Verma et al. [1]. It is a simple device with no moving parts that can transfer large quantities of heat over fairly large distances without requiring any power input. Heat pipes can be used extensively in electronics and electrical equipments, energy systems, aerospace and avionics, heat exchangers, gas turbine engines and automotive industries, production tools, ovens and furnaces, manufacturing, transportation systems, and de-icing. Depending on the wick structure, there are different types of heat pipes, such as sintered wick, thermosiphon, and grooved wick heat pipes. Basically, a heat pipe is a sealed slender tube containing a wick structure lined on the inner surface and a small amount of fluid,

B. Ch Nookaraju (✉) · P. S. V. Kurmarao
Gokaraju Rangaraju Institute of Engineering and Technology, Nizampet, Hyderabad, Telangana 500090, India
e-mail: bulipechinanookaraju@gmail.com

S. Nagasarada
Jawaharlal Nehru Technological University, Kukatpally, Hyderabad, Telangana 500085, India

M. Palanisamy et al. (eds.), *Theoretical, Computational, and Experimental Solutions to Thermo-Fluid Systems*, Lecture Notes in Mechanical Engineering,
https://doi.org/10.1007/978-981-33-4165-4_14

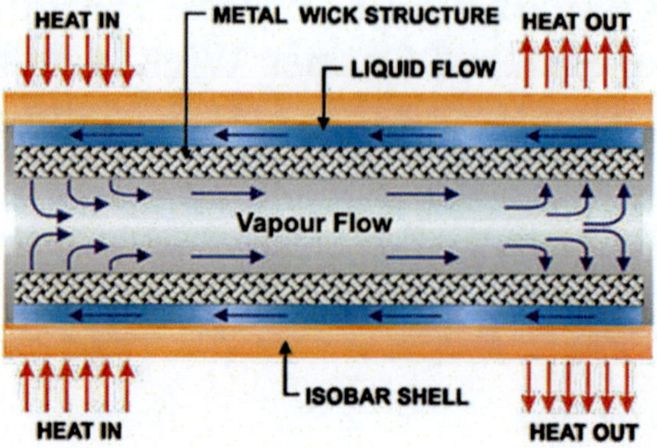

Fig. 1 The principle of operation of the heat pipe

such as methanol at the saturated state. The length of the heat pipe can be divided into three parts viz. evaporator section, adiabatic section, and condenser section.

When the evaporator end of the heat pipe is brought into contact with a hot surface or placed into a hot environment, working fluid absorbs heat and becomes vapor, which causes a rise in vapor pressure. This pressure difference drives the vapor through the core of the heat pipe from the evaporator to the condenser section. At the condenser section, condensation takes place by releasing the latent heat to the coolant, and vapor becomes liquid. The liquid then returns to the evaporator end of the heat pipe through the wick as a result of capillary action, and the cycle is completed, and it repeats. As a result, the heat is absorbed in one end of the heat pipe and is rejected at the other end, with the fluid inside serving as a transport medium for heat.

So far, most of the research work was carried out using only helical grooved heat pipe so sintered wick heat pipe was selected, in the present study which is having very less weight, high performance in gravity-assisted condition and with the minimum drop in radial temperature drop Mozumder*et.al* [2–4]. Till date, very little research work has been done related to the prediction of thermal performance using empirical relationships incorporating process parameters. Moreover, there is no literature available on establishing the relationship between thermal performance and process parameters of the heat pipe. The development of such relationships will also be useful to select optimum parameters that optimize the thermal performance of heat pipe. Hence the present work has been focused on developing a quantitative relationship between the operating parameters of the heat pipe and the response of heat transfer coefficient and on developing empirical relationships to predict the thermal performance of sintered copper wick heat pipe in terms of heat transfer coefficient. Statistical tools such as the design of experiments (DOE), analysis of variance (ANOVA), and response surface methodology (RSM) were used incorporating some of the important heat pipe process parameters, that is, the mass flow rate of coolant (m), the inclination angle of the heat pipe (θ) and heat input (Q) to the heat pipe.

In addition, an elaborate work was also carried out to provide the experimental data on how the Performance of Sintered wick heat pipe varies with varying parameters like Evaporator Fluid (water, methanol), Heat input, Mass flow rate of cooling medium, Orientation of Heat Pipe.

2 Experimental Setup and Procedure

The specifications of the heat pipe are given in Table 1. Heat input is applied at the evaporator section using an electric strip attached to it with proper electrical insulation, and the heater is energized with AC current through a variac. The desired heat input is supplied to the evaporator end of the heat pipe by adjusting the variac (see Fig. 2). The water jacket is used at the condenser end to remove the heat from the pipe. The heat pipe has the ability to transfer the heat through the internal structure. As a result, a sudden rise in wall temperature occurs; this will damage the heat pipe, if the heat produced is not released at the condenser properly. Therefore, before heat is supplied to the evaporator, the cooling water is first recirculated through the condenser jacket. The power input is gradually raised to the desired power level. The temperature of the evaporator and condenser were recorded after the steady-state is reached. Once a steady-state is reached, the input power is turned off, and cooling water is allowed to flow through the condenser to cool the heat pipe and to make it ready for further experiments. Then the power is increased to the next level, and the heat pipe is tested for its performance.

The experimental procedure is repeated for different flow rates, and different inclinations of pipe observations are recorded.

Table 1 Specifications of sintered copper wick heat pipe

S. No.	Properties	Specifications
1	Total Length of heat pipe	565 mm
2	Evaporator length	200 mm
3	Adiabatic length	165 mm
4	Condenser length	200 mm
5	Heat pipe material	Copper
6	Outer diameter of heat pipe	15.88 mm
7	Inner diameter of heat pipe	14 mm
8	Working fluid	Methanol, water
9	Mass flow rate	0.01,0.02,0.03 kg/s
10	Inclination angle	30°,60° and 90°
12	Heat input	100, 150, 200, 250 and 300 W

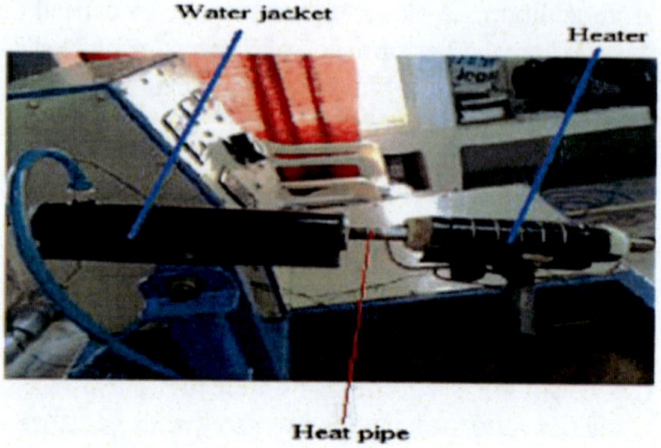

Fig. 2 Experimental set up

Table 2 Heat pipe parameters and their levels

S.No	Factor	Units	Notations	Levels		
				−1	0	1
1	Mass flow rate	kg/s	M	0.01	0.02	0.03
2	Inclination angle	Deg	θ	30	60	90
3	Heat input	W	Q	100	200	300

According to the experimental setup shown in (Fig. 2). Only mass flow rate, angle of inclination, and heat input were considered in this study. The working range was fixed according to the physical constraints. Table 2 shows the working range and levels of input parameters. The mass flow rate, heat input, and angle of heat pipe were varied in the step of 0.01 kg/s, 50 watts, and 30°, respectively.

3 Results and Discussion

The heat transfer coefficient of Sintered wick heat pipe with two different working fluids (DI Water and Methanol) was calculated and compared for all three positions (30°, 60°, 90°). The thermal performance of the heat pipe is quantified in terms of the heat transfer coefficient. The heat transfer coefficient of the heat pipe depends on heat input, the surface area of the evaporator section, and the temperature difference between evaporator and condenser. Overall, the heat transfer coefficient was calculated by using below two Eqs. (1) and (2).

$$h = \frac{Q}{A_s(T_e - T_c)} \quad (1)$$

$$A_s = \pi d_o L_e \quad (2)$$

$$R_{th} = \frac{Q}{(T_e - T_c)} \tag{3}$$

Here h is heat transfer coefficient (W/m^{2}k), as condenser/evaporator surface area (m^2), Te evaporator temperature (°C), Tc is condenser temperature (°C), L_e length of condenser/evaporator and do is the diameter of a heat pipe.

3.1 Heat Transfer Coefficient of Water (H)

From Figs. 3, 4 and 5, the heat transfer coefficient variation with heat input at different flow rates and various inclinations of the heat pipes were explained when Sintered wick heat pipe with DI water as working fluid is used. It is clear that with an increase in heat energy at the evaporator end, working fluid boils inside, and vapor pressure drives the heat at the condenser end by cooling. This is because water boiling capacity

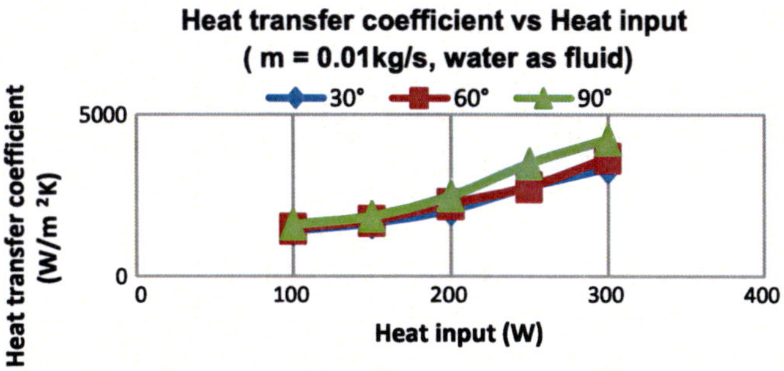

Fig. 3 Variation of heat transfer coefficient with heat input for mass flow rate of water at m = 0.01 kg/s

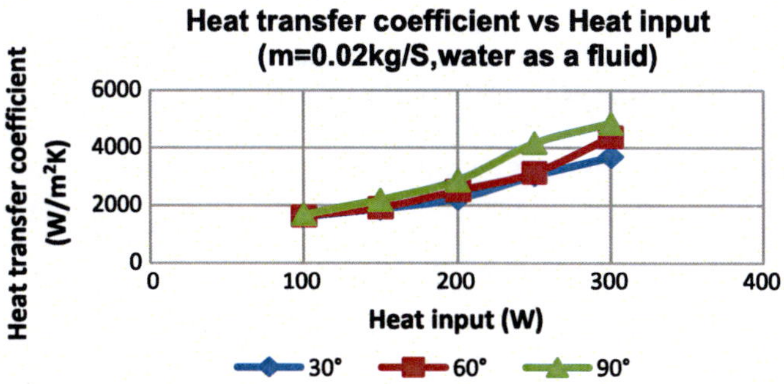

Fig. 4 Variation of heat transfer coefficient with heat input for mass flow rate of water at m = 0.02 kg/s

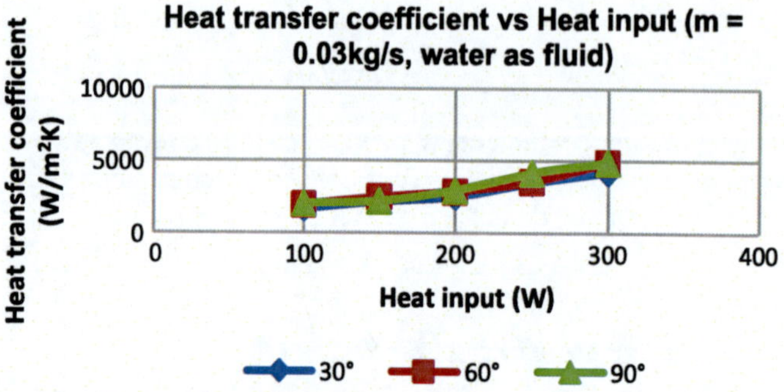

Fig. 5 Variation of heat transfer coefficient with heat input for mass flow rate of water at m = 0.03 kg/s

increases at higher latent heat transfer rates, which enhances the faster cooling rate in the condenser jacket. From the results, it is observed that, as the evaporator section is moving towards the ground, the heat transporting ability of the heat pipe is increasing from 0° to 60° inclinations. The increase in heat transfer coefficient of heat pipe(See in Figs. 3 and 4) is because the condensed liquid returns to the evaporator section easily as the evaporator section moves towards the ground with the assistance of gravitational forces in favor of capillary forces. Whereas from 60° to 90° inclinations of the pipe, there is a decrease in heat transfer coefficient because the gravitational forces will oppose the movement of evaporated fluid from the evaporator section to the condenser section.

For different flow rates, that is, 0.01–0.03 kg/s with an increase in flow rate, more cooling fluid will meet the hot surface of the condenser due to which heat dissipation increases.

3.2 Heat Transfer Coefficient of Methonol (H)

The sintered wick heat pipe with methanol as working fluid, showing the heat transfer coefficient variation with heat input at different flow rates and various inclinations of the heat pipes are shown in Figs. 6, 7 and 8. There is a remarkable increase in heat transfer coefficient from lower heat input (100 W) to higher heat input (300 W) owing to the augmentation of heat transfer rate by the evaporation and condensation process inside the heat pipe at a 50% flow ratio. For the same inputs (300 W), heat transfer coefficients for methanol is 3690 W/m^{2}K, whereas, for water, it is 4250 W/m^{2}K. The performance of heat pipe with methanol is good heat dissipation at lower heat inputs only as compared with water as a working fluid, which is good at higher heat inputs also.

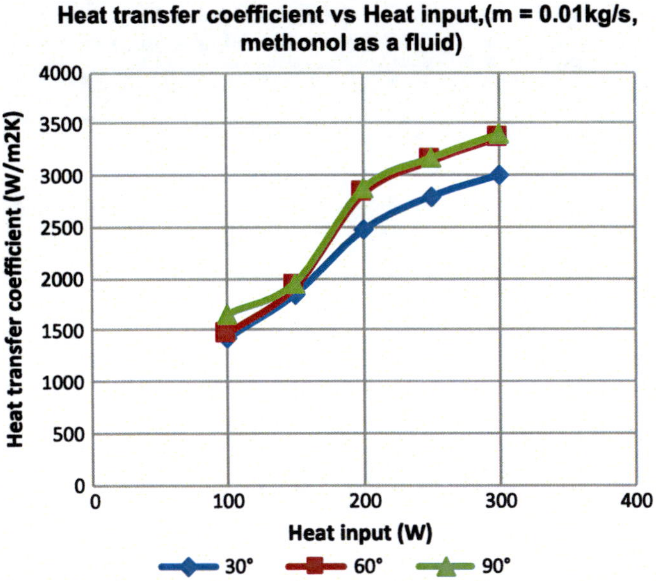

Fig. 6 Variation of heat transfer coefficient with heat input for mass flow rate of methonol at m = 0.01kg/s

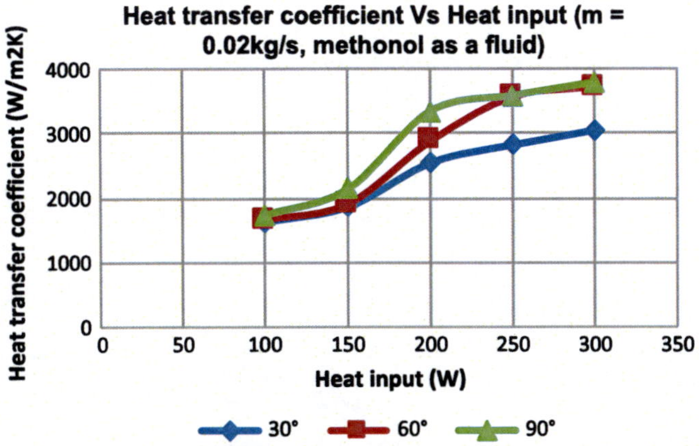

Fig. 7 Variation of heat transfer coefficient with heat input for mass flow rate of methonol at m = 0.02kg/s

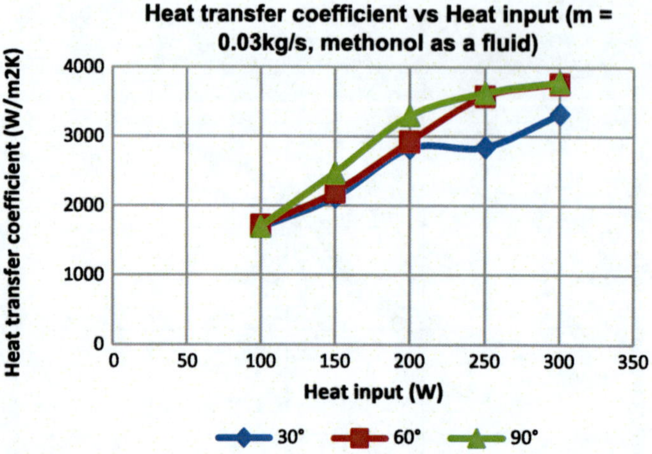

Fig. 8 Variation of heat transfer coefficient with heat input for mass flow rate of methonol at m = 0.03kg/s

3.3 Comparison of Heat Pipe Working Fluids: (Water Versus Methanol)

Experimental heat transfer coefficient of sintered copper wick heat pipe was calculated, and a comparison of variation of heat transfer coefficient of heat pipe at different heat inputs and mass flow rates are shown (See in Fig. 9). It is proved from the results that water has the highest latent heat of evaporation of 4900 W/m^2k at higher heat inputs at 60° heat pipe inclination. But when methanol as a working fluid was

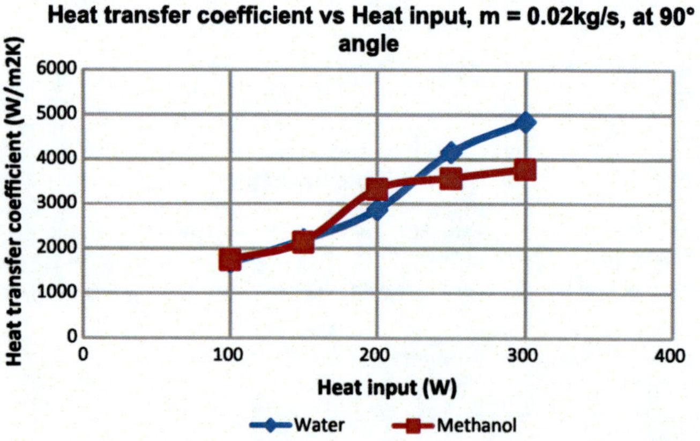

Fig. 9 Comparison of heat transfer coefficient with heat input for mass flow rate of water and methonol at m = 0.02 kg/s at 90° angle inclination

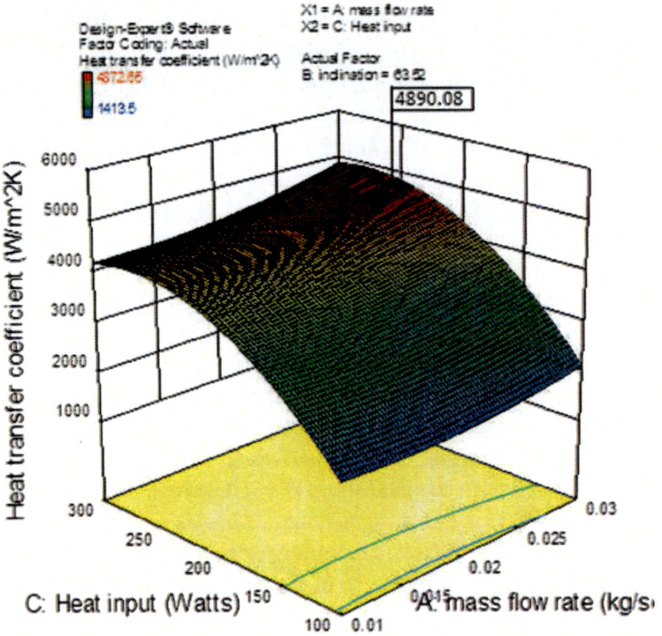

Fig. 10 Variation of heat transfer co-efficient with heat input and inclination for methanol as working fluid

used, the highest latent heat of evaporation was 3600 W/m^2 only for the same heat inputs. Hence it can be concluded that for the same heat inputs, water has better heat absorbing and dissipation capacity than methanol. Water serves as good heat dissipation at higher heat inputs also whereas methanol serves as good heat dissipation at lower heat inputs only.

4 Optimization of Process Parameters by Response Surface Methodology (RSM)

Response surface methodology (RSM) is a collection of mathematical and statistical techniques for empirical model building. By careful design of experiments, we can optimize a response (output variable) which is influenced by several independent variables (input variables). In experiments, there is a series of tests, called runs, in which changes are made in the input variables to identify the reasons for changes in the output response. In the present investigation, RSM was applied for developing empirical relationships in the form of multiple regression equations for the heat transfer coefficient of heat pipes. In applying the response surface methodology, the independent variable was viewed as a surface to which a mathematical model is fitted.

Finally, the application of RSM is aimed to obtain the optimized process parameter to achieve the maximum performance of any process.

Development of empirical relationship is a customary approach to the recently developed process to reveal the effect of each process parameter against the responses, and it is evident from the studies of Senthil Kumar et al. [12], who developed an empirical relationship using Response Surface Methodology (RSM) to predict and optimize the operating parameters. The predominant factors having a significant influence on the thermal performance (in terms of heat transfer coefficient and thermal resistance) are (i) Heat input, (ii) Mass flow rate, (iii) Angle of inclination, (iv) Working fluid inside the heat pipe, (v) Dimensions of heat pipe, and (vi) Heat pipe material. According to the experimental setup, only mass flow rate, angle of inclination, and heat input were considered in this investigation.

The working range was fixed according to the physical constraints. Table 2 shows the working range and levels of input parameters. The mass flow rate, heat input, and angle of heat pipe were varied in the step of 0.01 kg/s, 100 W, and 30°, respectively. As three were factors involved in this investigation and the range is not much wide, three-factor three levels CCD matrix was considered. It consists of 20 sets of conditions; comprising a full replication 3 factors factorial design of 8 points, 6 star points and 6 center points. The upper and lower limits of the parameters were coded as +1 and −1, respectively. The coded values of any intermediate levels can be calculated from Eq. (3)

$$X_i = 2X - \frac{(X_{\max} + X_{\min})}{X_{\max} - X_{\min}} \tag{4}$$

where $X_{\max}$ is upper limit of range, $X_{\min}$ is lower limit of range

By using Response Surface Methodology in Design of Expert software, the optimum values of sintered wick heat pipe with methanol and water as working fluids are worked out. The optimum condition for methanol as working fluid at an inclination of 67.03°, the mass flow rate of cooling water is 0.0202 kg/s at heat input 240.12 W with heat transfer coefficient of 3805.15 W/m^{2}K as shown (See in Fig. 11). Whereas the optimum values of DI Water as working fluid at inclination of 63.52°, mass flow rate of cooling water is 0.029 kg/s at heat input 252.61 W with heat transfer coefficient of 4890.08 W/m^{2}K (Fig. 10).

5 Conclusion

In the present study, the optimization of process parameters using RSM and comparison of two different working fluids, water, and methanol, to maximize the heat transfer coefficient has been done. The following conclusions were drawn. The heat transfer coefficient for DI water is more when compared with the methanol for the same heat inputs. Methanol serves as good heat dissipation at lower heat inputs only, whereas water serves as good heat dissipation at higher heat inputs also. The optimum

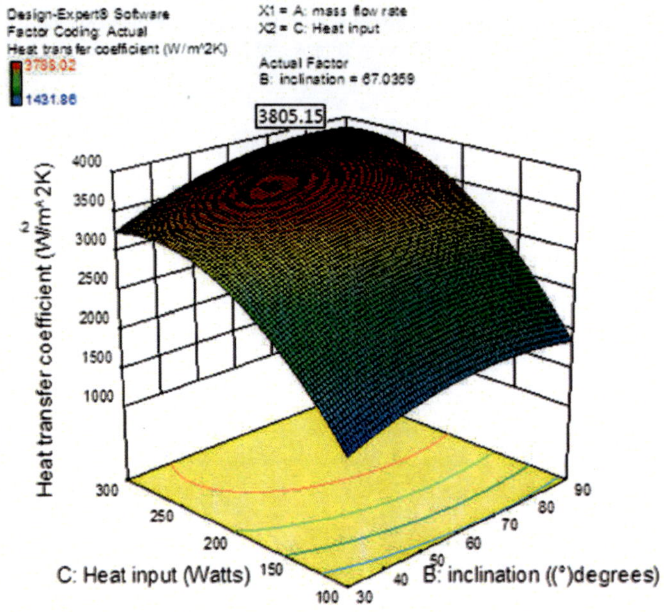

Fig. 11 Variation of heat transfer co-efficient with heat input and inclination water as fluid

condition for methanol as a working fluid at an inclination of 67.03°, at a flow rate of 0.0202 kg/s and heat input of 240.12 W, and the value is 3805.15 W/m^{2}k. Optimum condition for DI water as a working fluid at 63.52° inclination, 0.029 kg/s flow rate and 252.61 W and the value is 4890.08 W/m^{2}k. The heat transfer coefficient was found to increase with the increase in heat input in both cases.

Acknowledgements The work presented in this article was funded by the All India Council for Technical Education (AICTE) under Research Promotion Scheme, File No. 8023/RID/RPS-20/Pvt (II Policy)/2011–12. The authors gratefully express their thanks to them for providing financial assistance to purchase the equipment for this project and all other related expenses.

References

1. Verma B, Yadav VJ, Srivastava KK (2013) Experimental studies on thermal performance of a pulsating heat pipe with methanol/DI water. J Electron Cooling Thermal Cont. Published Online March 2013
2. Mozumder AK, Akon AF, Chowdhury MSH, Banik SC (2010) Performance of heat pipe for different working fluids and fill ratios. J Mech Eng ME 41(2)
3. Vijra N, Sing TP (2015) An experimental study of thermal performance of concentric annular heat pipe. American Int J Res Sci Technol Eng Mathe 176–182. ISSN (Print): 2328–3491, ISSN (Online): 2328–3580

4. Shafii MB, Arabnejad S, Saboohi Y, Jamshidi H (2010) Experimental investigation of pulsating heat pipes and a proposed correlation. Heat Trans Eng 31(10):854–861. https://doi.org/10.1080/01457630903547636

A Numerical Study on the Flapping Dynamics of a Heaving Flexible Foil in a Uniform Flow

Kuntal Patel, K. Supradeepan, and P. S. Gurugubelli

1 Introduction

Flapping dynamics of a flexible foil is an active area of research because of its direct relevance to the development of energy harvesters and bio-inspired locomotion. However, there exists a fundamental difference between energy harvester and flexible-foil propellers in terms of the direction of energy transfer. Energy harvesters draw energy from the fluid flow while flexible-foil propellers impart energy to the fluid via thrust generation [1]. Hence, propellers exhibit reverse Karman Vortex Street. In this section, we will outline some major studies related to the flapping dynamics of a flexible foil. In the first part, flapping dynamics from the perspective of propellers are discussed and then in the context of energy harvesters in the second part.

Flapping dynamics with application to propellers: One of the early theories explaining the swimming of a slender fish was given by Lighthill [2], where swimming movements were restricted to the direction perpendicular to the direction of locomotion. For a slender body to swim at a certain desired speed, it has to generate a bending wave at a speed of 25% higher than that of desired swimming speed [2]. More recently, Quinn et al. [3] proposed scaling laws for the propulsive performance of heaving foils based on an experimental study. It was shown that the increase in heaving frequency leads to an increase in the thrust coefficient. Also, for a given heaving amplitude, the thrust coefficient depends on the Strouhal number. Later, universal scaling laws for the transition to the reverse Karman vortex street (i.e., drag-to-thrust wake transition) were given by Lagopoulos et al. [4]. Different cases of pure pitching, pure heaving, and the combination of pitching and heaving motion were studied using 2D numerical simulations in [4]. They considered a modified

K. Patel · K. Supradeepan · P. S. Gurugubelli (✉)
Birla Institute of Technology and Science-Pilani, Hyderabad Campus, Hyderabad, India
e-mail: pardhasg@hyderabad.bits-pilani.ac.in

M. Palanisamy et al. (eds.), *Theoretical, Computational, and Experimental Solutions to Thermo-Fluid Systems*, Lecture Notes in Mechanical Engineering,
https://doi.org/10.1007/978-981-33-4165-4_15

Strouhal number based on the motion of the chord-line of the foil over a cycle, and it was hypothesized that the onset of reverse Karman Vortex Street takes place when modified Strouhal number approaches unity. The existence of resonant peaks for the thrust power was reported by Alben [5], when plotted against various values of fin rigidity for a given pitching frequency. This resonant peak for the thrust power hits the maximum value when the trailing edge is positioned upwards in a one-quarter-wavelength mode of deflection [5]. Further details on swimming performance and associated vortex dynamics during undulatory and oscillatory motion can be found in [6].

Flapping dynamics with application to energy harvesters: The possibility of harnessing energy through the flow-induced vibrations of a flexible body was demonstrated by Allen and Smits [7] for the first time. Fluid-elastic instability can trigger the self-sustained flapping of a flexible body, and hence capacitive buildup can be achieved by attaching a layer of a piezoelectric patch to a foil undergoing flapping motion. With this motivation, several researchers have explored the flapping dynamic of a flexible foil for different configurations, which mainly includes (1) flexible foil fixed at the leading edge (ordinary flag) and (2) flexible foil fixed at the trailing edge (inverted flag). Flapping dynamics of a flexible foil is governed by mass-ratio (m^*), non-dimensional flexural rigidity (K_B), and Reynolds number (Re); given as

$$m^* = \frac{\rho^s h}{\rho^f L} \quad K_B = \frac{EI}{\rho^f U_0^2 L^3} \quad \mathrm{Re} = \frac{\rho^f U_0^2 L}{\mu^f} \tag{1}$$

where fluid denotes the density of the solid, density of fluid, dynamic viscosity, Young's modulus, and area moment of inertia, respectively. Equation (1) represents the foil thickness of foil, length of foil, and free stream velocity. Connell and Yue [8] presented a detailed numerical study (using a coupled fluid–structure direct simulations) on flapping dynamics of a one-dimensional flag, placed in a two-dimensional flow. Dispersion relation obtained in [8] (from the equation governing flag motion) suggested that a decrease in either mass-ratio or Reynolds number and increase in bending rigidity leads to the stabilizing effect. Further, three different flapping modes were identified in [8], namely fixed-point stability, limit cycle flapping, and chaotic flapping. Detailed review on theoretical, numerical, and experimental studies on flag-flapping can be found in Shelley and Zhang [9]. Inverted flag configuration was studied experientially by Kim et al. [10]. Inverted flag configuration is more promising than an ordinary flag because of the generation of a large amount of strain energy compared to that of ordinary flag configuration [10]. Gurugubelli and Jaiman [11] presented three different stability regimes of an inverted flag: fixed-point stable, deformed steady, and unsteady flapping state. Moreover, unsteady flapping can exist in the form of inverted limit-cycle flapping, deformed flapping, or flipped flapping [11]. They characterized the transition from deformed steady-state to the unsteady flapping state as a consequence of the flow separation at the free leading edge. A recent study from our group (Gurugubelli and Jaiman [12]) on numerical simulations of large-amplitude flapping (LAF) of an inverted foil at Reynolds number Re = 30,000 using variational fluid–solid formulation based large-eddy simulation revealed that LAF of an inverted foil is not driven by classical vortex-induced vibrations rather it originates from the complex interplay of unsteady leading-edge vortex shedding, the inertia of foil and flexibility induced elastic recoil.

The focus of the present numerical work is to study the similarity and differences between the flapping dynamics of a heaving flexible-foil (for different amplitude-frequency combination) and a flexible-foil with a clamped leading-edge. The rest of the paper is organized as follows: In Sect. 2, equations governing fluid and solid dynamics along with the formulation of the present quasi-monolithic fluid–structure solver are discussed. The verification of numerical methodology is presented in Sect. 3. Results obtained from present numerical simulations are discussed in Sect. 4.

2 Governing Equations and Numerical Methodology

Governing equations for fluid and solid dynamics along with the numerical methodology are briefly discussed in this section. The Navier–Stokes equations in ALE reference frame for incompressible flow are given as

$$\nabla \cdot \mathbf{u}^f = 0 \quad \text{on} \quad \Omega^f(t) \tag{2}$$

$$\rho^f \frac{\partial \mathbf{u}^f}{\partial t} + \rho^f (\mathbf{u}^f - \mathbf{w}).\nabla \mathbf{u}^f = \nabla \cdot \boldsymbol{\sigma}^f + \mathbf{f}^f \quad \text{on} \quad \Omega^f(t) \tag{3}$$

where ρ^f, $\mathbf{u}^f$, $\mathbf{w}$, and $\mathbf{f}^f$ are fluid density, fluid velocity, mesh velocity, and volumetric forces acting on fluid, respectively. Further, Cauchy stress tensor $\boldsymbol{\sigma}^f$ can be written as

$$\boldsymbol{\sigma}^f = -p\mathbf{I} + \mu^f \left(\nabla \mathbf{u}^f + \left(\nabla \mathbf{u}^f \right)^T \right) \tag{4}$$

where p is fluid pressure, $\mathbf{I}$ is second-order identity tensor and μ^f represents dynamics viscosity. The equation for the structural dynamics is

$$\rho^s \frac{\partial \mathbf{u}^s}{\partial t} = \nabla \cdot \boldsymbol{\sigma}^s + \mathbf{f}^s \text{ in } \Omega^s \tag{5}$$

where, $\mathbf{u}^s$, $\mathbf{f}^f$, and $\boldsymbol{\sigma}^s$ denote solid velocity, volumetric forces acting on solid and Piola–Kirchhoff stress tensor, respectively. In this work, Saint Venant–Kirchhoff model is used to compute structural stresses. Kinematic and dynamic boundary conditions along the fluid–solid interface Γ are realized by imposing interface velocity and traction continuity conditions; which are given as

$$\mathbf{u}^f \left(\varphi^s(\mathbf{z}, t), t \right) = \mathbf{u}^s(\mathbf{z}, t) \quad \forall \mathbf{z} \in \Gamma \tag{6}$$

$$\int_{\varphi^s(\gamma,t)} \sigma^f(\mathbf{x},t)\cdot \mathbf{n}^f \mathrm{d}\Gamma + \int_{\gamma} \sigma^s(\mathbf{z},t)\cdot \mathbf{n}^s \mathrm{d}\Gamma = 0 \ \forall \gamma \in \Gamma \quad (7)$$

where $\mathbf{n}^f$ and $\mathbf{n}^s$ represent the outward normal vectors to the fluid and solid interface boundaries, respectively. ϕ^s is the position mapping function that maps each structural node in the deformed state at a given time t to its initial position $\mathbf{z}$.

In the present work, Combined Field with Explicit Interface (CFEI) advancing formulation (Liu et al. [13]) is employed. CFEI formulation is implemented in an arbitrary Lagrangian–Eulerian (ALE) reference frame using the Finite Element Method (FEM). This scheme is proved to be stable and second-order accurate in time. The applicability of the solver for problems involving the flapping phenomenon can be found in [14].

3 Verification

Computational set-up for the problem of heaving flexible foil is shown in Fig. 1 with boundary conditions. A flexible foil is placed in a uniform free-stream with a velocity U_0. The leading-edge of the flexible foil is heaved with a certain non-dimensional amplitude A and frequency f. Before we perform a systematic parametric analysis, the solver has been verified and is done by simulating the flapping dynamics of a flexible foil with a fixed leading edge. This verification is done using the mesh obtained from the grid independence test, which consists of 29,549 fluid-nodes, 1071 solid-nodes, 14,657 fluid, and 472 solid second-order triangular elements. Non-dimensional parameters selected for the verification study are $m^* = 0.1$, $K_B = 0.0001$, and Re $= 1000$. Trailing edge displacement obtained from the present numerical simulation is then compared with the numerical result of [8]—shown in Fig. 3. The value of the flapping Strouhal number obtained in the present simulation is

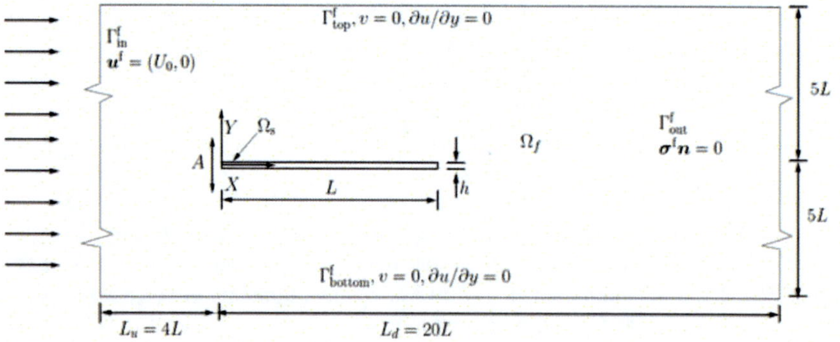

Fig. 1 Computation domain set-up for a heaving flexible foil along with boundary conditions considered for the simulations

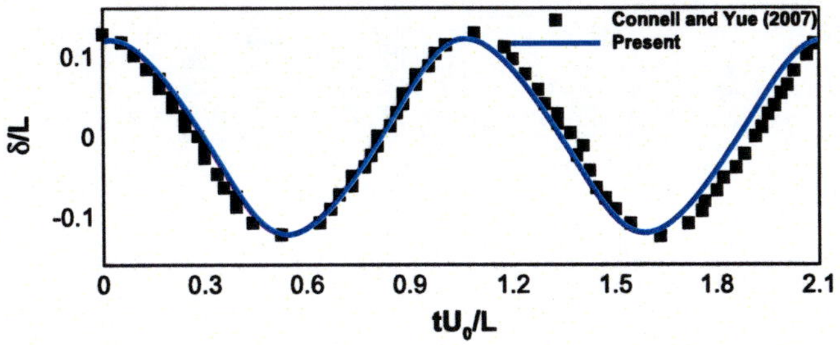

Fig. 2 Comparison of the trailing edge transverse displacement time history with the numerical simulations of [8]

0.2434, while the value of the flapping Strouhal number reported in [8] is 0.2318. The difference between both the values of flapping Strouhal number is within the acceptable limit (Fig. 2).

4 Results and Discussion

Flapping dynamics of a flexible foil—subjected to a heaving motion—is discussed in this section. The heaving motion of a foil is realized by imposing a sinusoidal motion (with a certain amplitude and frequency) at the leading edge (L.E.). Non-dimensional amplitude-frequency combinations corresponding to $(A, f) = \{(0.05, 0.25), (0.05, 0.50), (0.10, 0.25), (0.10, 0.50), (0.15, 0.25), (0.15, 0.50)\}$ are numerically investigated. Fluid–solid dynamics arising from these combinations are explained by comparing it with the flapping dynamics of a flexible foil with a fixed L.E., that is, $A = 0$. In order to quantify the consequences of a heaving motion, transverse displacement and phase portrait for the trailing edge (T.E.) along with the fast Fourier transform are investigated. Further, the vortex structures are presented to understand the similarity and differences in the dynamics as compared to its fixed L.E. counterpart. The parametric simulations have been carried out for $m^* = 0.1$, $K_B = 0.0004$, and $Re = 1000$ (for all amplitude-frequency combinations).

Figure 3 summarizes L.E. and T.E displacement time histories in the transverse direction. Figure 3a presents the time history for the case of fixed L.E., wherein the T.E. exhibits regularized sinusoidal motion. However, this T.E. displacement response gets altered for the non-zero value of heaving amplitude and frequency. In the case of non-zero leading heaving amplitudes, T.E. displacement response exhibits multiple frequencies, except for the amplitude-frequency combination of $(A, f) = (0.15, 0.50)$ (see Fig. 3g). The qualitative trend of T.E. displacement for $(A, f) = (0.15, 0.50)$ and fixed leading edge case is identical. In order to separate the combinations of frequency and amplitude associated with the trailing edge plots in Fig. 3, fast Fourier transform (FFT) is performed. The outcome of FFT for all the simulations is shown in Fig. 4.

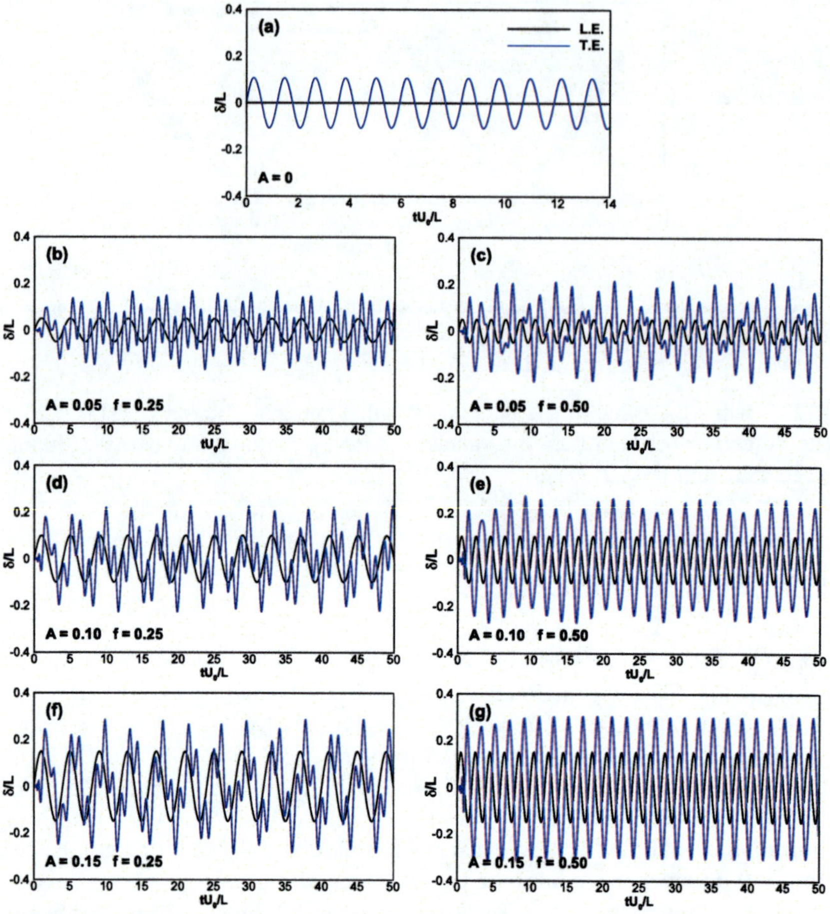

Fig. 3 Time history of the trailing and leading edge cross-stream displacements for $A = \{0, 0.05, 0.1, 0.15\}$ and $f = \{0.25, 0.5\}$

With the exception of FFT plots corresponding to the cases fixed L.E. and $(A, f) = (0.15, 0.50)$ cases, all other simulation cases have exhibit two distinct transverse trailing displacement frequencies. Even though both fixed L.E. and $(A, f) = (0.15, 0.50)$ cases exhibit a single peak frequency, the peak frequency in the case of the fixed L.E. corresponds to passive flapping motion. On the other hand, the peak frequency for $(A, f) = (0.15, 0.50)$ is equal to the heaving frequency of the L.E. For the cases where the trailing edge exhibit multiple frequencies, the first peak, that is, 0.5 in FFT plots for $(A, f) = (0.05, 0.25)$, $(0.10, 0.25)$, and $(0.15, 0.25)$ represents the bending wave generated by the heaving motion, while the second peak corresponds to the passive flapping frequency due to the fluid-elastic instability. Due to the existence of two bending waves, the flapping amplitude strongly depends on the synchronization,

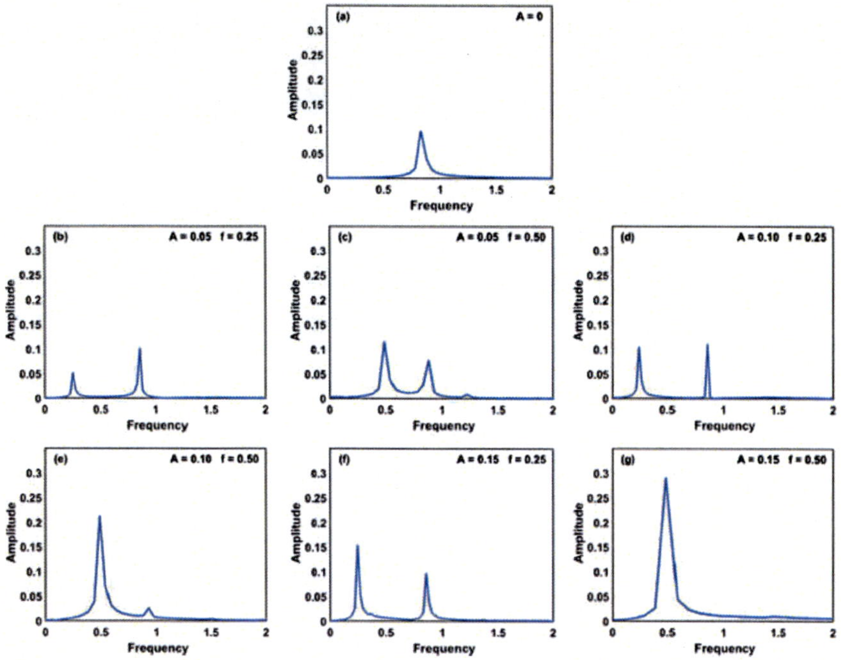

Fig. 4 Fast Fourier transform of signal plots for a transverse displacement of the trailing edge displacements shown in Fig. 3

that is, constructive or destructive interaction between these two waves and magnitude of energy these waves carrying with them. As a result of this, the amplitude of the primary peak (corresponding to the heaving frequency) increases while the amplitude of the secondary peak (corresponding to the frequency of bending wave generated from the fluid–solid interaction) decreases as heaving amplitude is increased. This can be clearly observed by looking at the FFT plots corresponding to $(A, f) = (0.10, 0.50)$ and $(A, f) = (0.05, 0.50)$.

After discussing the transverse displacement of T.E. and their FFT, we can now introduce phase portraits for T.E. position. Phase portraits containing the time history of T.E position are shown in Fig. 5, for all the cases. As depicted in Fig. 5a, for a fixed leading age case, T.E. traverses on a fixed cyclic path once the dynamic equilibrium is achieved. Likewise, for case $(A, f) = (0.15, 0.50)$ also, T.E. travels on a certain fixed cyclic trajectory. The shape of the phase portrait for both the cases is similar, except it is enlarged in the case of $(A, f) = (0.15, 0.50)$ because of the higher flapping amplitude. For all the cases with heaving frequency $f = 0.25$, the maximum distance traveled by T.E. in both X and Y direction increases with an increase in heaving amplitude (because of the occasional superposition of existing bending waves). This results in the enlargement of the phase portrait. For higher heaving frequency, $f =$

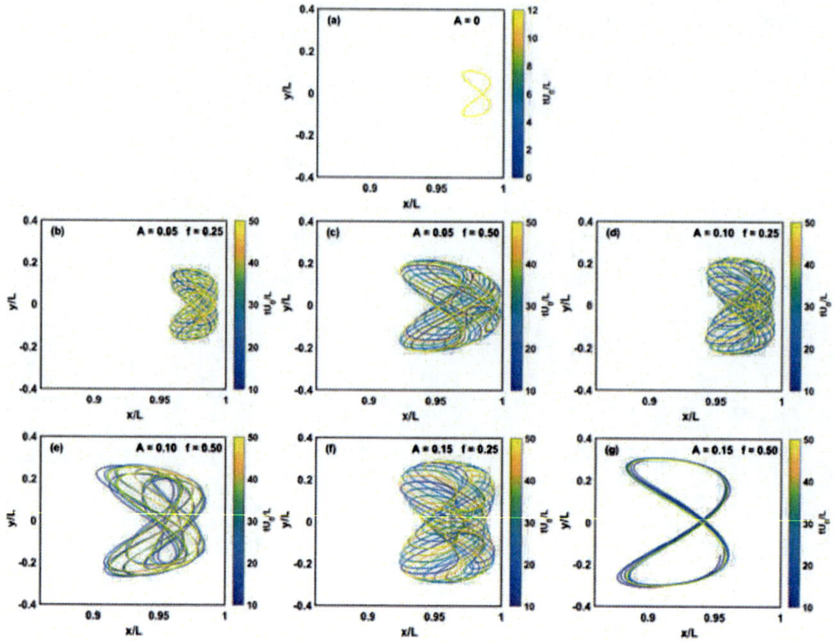

Fig. 5 Phase portraits illustrating traversal of the trailing edge during the flapping motion of a heaving foil

0.5, T.E. trajectories start collapsing with the increase in heaving amplitude and ultimately results in a figure-of-eight trajectory (see Fig. 5g).

Finally, the foil response kinematics (which implicitly represents the fluid dynamics since foil kinematics is a result of feedback forces from fluid) and fluid flow features are illustrated in Fig. 6. For all the cases with frequency $f = 0.25$, vortex street contains the signature of the heaving motion of the foil, which leads to the undulation of the vortex street. However, it should be noted that vortex shedding frequency for these cases is similar to that of a fixed L.E. case (see Fig. 6). This is the indication of the fact that the second peak from the FFT plots of these cases (Fig. 4b, d, f) is responsible for the sinusoidal pattern of the vortex street while the primary peak is responsible for the vortex shedding. In other words, the bending wave generated from the heaving motion does not have any influence on the mechanism leading to the detachment of the vortex; otherwise, it would have resulted in a different vortex shedding frequency along with the change in a vortex structure. For high-frequency cases ($f = 0.5$), the interaction between the bending waves alters the vortex modes (compared to what is observed for fixed L.E. case). Cases with $A = 0.05$ and 0.10 at $f = 0.50$ show chaotic vortex street which may be attributed to the existence of two equally dominate bending waves. A periodic vortex wake is observed for all the cases where there exists a single dominant bending wave. Note

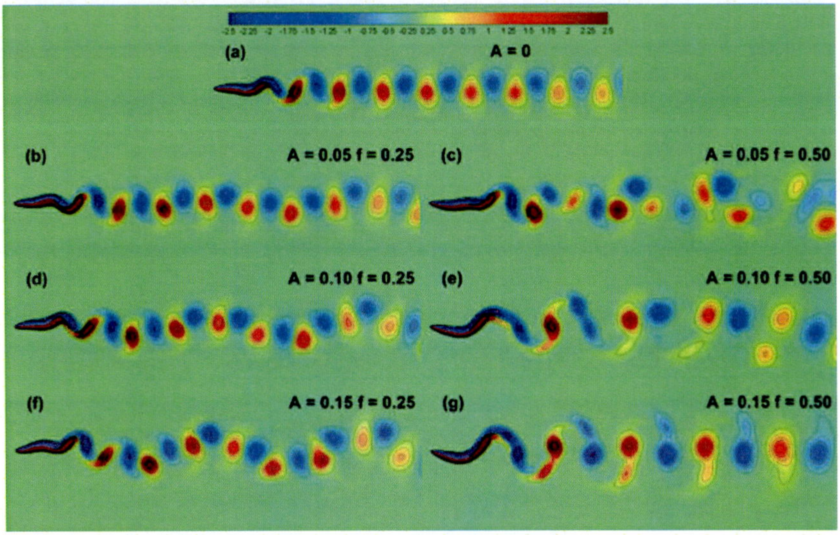

Fig. 6 Effect of heaving motion on the vortex modes generated

that similarity in heaving Strouhal number does not ensure the similarity in flapping dynamics even if they have the identical value of m^*, K_B, and Re (see Fig. 6c, d).

5 Conclusions

Numerical simulations corresponding to non-dimensional amplitude-frequency combinations of (A, f) = {(0.05,0.25), (0.05,0.50), (0.10,0.25), (0.10,0.50), (0.15,0.25), (0.15,0.50)}, and for $m^* = 0.1$, $K_B = 0.0004$, Re = 1000, have been carried out to study the influence of heaving motion on the flapping dynamics of a flexible foil. For the cases with $f = 0.25$, fast Fourier transform of the cross-stream position of the trailing edge showed the existence of two different bending waves, one originated from the heaving motion at the L.E., and another generated from the fluid-elastic instability at the T.E. Further, the vortex shedding pattern in these cases is similar to that of observed in the absence of heaving motion, but with the periodic undulation of the vortex street in the transverse direction because of the heaving motion.

References

1. Xiao Q, Zhu Q (2014) A review on flow energy harvesters based on flapping foils. J Fluids Struct 46:174–191

2. Lighthill MJ (1960) Note of swimming of slender fish. J Fluid Mech 9:305–317
3. Quinn DB, Lauder GV, Smits AJ (2014) Scaling the propulsive performance of heaving flexible panels. J Fluid Mech 738:250–267
4. Lagopoulos NS, Weymouth GD, Ganpathisubramani B (2019) Universal scaling law for drag-to-thrust wake transition in flapping foils. J Fluid Mech872
5. Alben S (2008) Optimal flexibility of a flapping appendage in an inviscid fluid. J Fluid Mech 614:355–380
6. Smits AJ (2019) Undulatory and oscillatory swimming. J Fluid Mech874
7. Allen JJ, Smits AJ (2001) Energy harvesting EEL. J Fluids Struct 15:629–640
8. Connell BSH, Yue, DKP (2007) Flapping dynamics of a flag in a uniform stream. J Fluid Mech581:33–67
9. Shelley MJ, J Zhang (2011) Flapping and bending bodies interacting with fluid flow. Annu Rev Fluid Mech 43:449–465
10. Kim D, Cosse J, Cerderia CH, Gharib M (2013) Flapping dynamics of an inverted flag. J Fluid Mech736
11. Gurugubelli PS, Jaiman RK (2015) Self-induced flapping dynamics of a flexible inverted foil in a uniform flow. J Fluid Mech 781:657–694
12. Gurugubelli PS, Jaiman RK (2019) Large amplitude flapping of an inverted elastic foil in uniform flow with spanwise periodicity. J Fluids Struct 90:139–163
13. Liu J, Jaiman RK, Gurugubelli PS (2013) A stable second-order scheme for fluid-structure interaction with strong added-mass effect. J Comput Phys 270:687–710
14. Bourlet TF, Gurugubelli PS, Jaiman RK (2015) The boundary layer development and traveling wave mechanisms during flapping of a flexible foil. J Fluids Struct 54:784–801

Effect of Heat Shields on Performance of Rolls Integrated with Internal Heaters and Its Simulation

Nalla Shivaprasad and U. S. Jyothi

1 Introduction

Engineering steels have varied to concede for the required properties from ENIA (230M07)—EN41 (905M39), that is, EN1A, EN3B, EN8, EN16, EN19, EN24, EN32, EN36, EN40B, EN41. Rolls of packing machine are manufactured with En8 steel, which has better properties compared to mild steel. Surface processing plays a significant role in having reduced thermal conductivity in order to have low heat transfer from the surface of the rolls, in turn, low heat dissipation.

By flame and induction hardening process, EN8 steel is heat-treated to provide good surface hardness and moderate wear resistance and with the knowledge of slashing interplay between deformation and grain microstructure size. It is Probable to anticipate the behavior of steel when subjected to assorted working conditions [1, 2]. It was studied that the surface boronizing via sandblasting will decrease thermal conductivity because of reduced grain size and increased the volume fraction of equilibrium grain boundaries [3]. In contrast to metals, usually, Ceramics are more resistant to oxidation, corrosion, and wear, as well as being better thermal insulators; these are widely used for thermal barrier coatings like YSZ, PSZ, etc. [4]. It was ascertained that liquid Nanoceramic coatings will create infinitesimal cellular ceramic microspheres, which decreases the thermal conductivity is around 0.001–0.003 W/m K because of the formation of vacuum-hollow balls on high temperature melted ceramic [5]. In the current context, research is targeted to use simulation tools on many aspects such as varying injection pressure, timing, combustion chamber configuration for different fuels in IC engines for temperature variation, and heat flux analysis in heat transfer applications [6].

N. Shivaprasad (✉) · U. S. Jyothi
Mechanical Engineering Department, GRIET, Hyderabad, India
e-mail: shivaese2017@gmail.com

M. Palanisamy et al. (eds.), *Theoretical, Computational, and Experimental Solutions to Thermo-Fluid Systems*, Lecture Notes in Mechanical Engineering,
https://doi.org/10.1007/978-981-33-4165-4_16

Heat shields protect structures from extreme temperatures and thermal gradients by two primary mechanisms, such as thermal insulation and radiative cooling. These relatively isolate the underlying structure from high external surface temperature while emitting heat outwards through thermal radiation. For heat shield applications, low thermal conductivity materials are essential, and stainless steel 316Ti can be chosen for analysis based on its composition and properties for the suitability of heat shield [7, 8]. TIG welding technique is compatible with respect to process and also to retain the shield strength for the fastening of the heat shield in the desired shape of the roll [9, 10]. By inserting the stainless steel in high-temperature molten Li_2BeF_4 (FLiBe) salt, it was reported that the corrosion resistance for the stainless steel will be more [11]. Further, it was concluded with experiments on stainless steel to study the effect of oxidation in supercritical water and reported that the stainless steel is less susceptible to oxidation [12]. The researcher's studies elaborated the methods of surface alloying and surface coating to the stainless steel to increase the wear resistance and to reduce the overall thermal conductivity [13, 14].

From the above discussions, heats shields are most conceding and economical to increase the performance of rolls. Stainless steel 316Ti is decided as heat shielding material because of hierarchical properties such as low thermal conductivity, high resistance to corrosion, and pitting at high temperatures than other conventional chromium-nickel austenitic stainless steels. Heat shield, stainless steel 316Ti are fabricated in the form as the superficial surface of rolls and cladded them to rolls as an interference fit manually.

2 Materials and Methodology

2.1 Materials Used

As per the specifications of the packing machine, the EN8 material is used for the fabrication of rolls. In this study, based on the availability and suitable properties, a stainless steel 316Ti heat shield used to shield on the surface of the rolls to reduce heat transfer and power input to the heater. The composition of the rolls (EN8) and stainless steel 316Ti is shown in Table 1.

Table 1 Composition of EN8 and stainless steel 316Ti

Composition (%)	Cr	Mo	Ni	Mn	S	Si	N	C	Ti	Fe	P
Stainless steel 316Ti	17	1.5	12	1.5	0.03	0.75	0.1	0.08	0.7	63.34	–
EN8 Steel	–	–	–	0.8	0.005	0.3		0.4	–	98.3	0.05

2.2 Experimental Setup

The experiments were conducted on the Packing machine of single cutting-edge rolls integrated with internal heater is used as shown in Fig. 1 and specification are listed in Table 2. The heating element is made up of brass and Nichrome wires and the specification of the setup is shown in Table 1 for analysis purpose; the packing material thickness is adapted for experimentation is 80 gauge polypropylene material and for which the required temperature on rolls of packing is 175 °C. Initially by switching on the power to the heater experiments were conducted on rolls without

Fig. 1 Packing machine with roll. (1) Power switch (2) motor switch (3) ammeter reading (4) vertical temperature controller (5) horizontal temperature controller (6) polypropylene cover (7) chute for cover (8) horizontal rolls (9) oil supply (10) crompton 0.5 HP induction motor (11) 30:1 gear box (11) series of spur gears (12) bevel gear to H-rolls

Table 2 Specifications of the packing machine

Parameter	Specification
Motor power	0.5 HP
Gear box	30:1 worm wheel
Heater	220 V
Speed of motor	1500 RPM
Rolls	EN8
Temperature controller	PID

heat shield for evaluating the of power input required to attain the temperature of requirement. Then the same procedure is repeated with the rolls, which are covered with the heat shield of stainless steel 316Ti for the same temperature of 175 °C for the rolls. Heaters of the horizontal and vertical rolls are connected to the temperature controller, which has the control loop feedback mechanism to regulate the surface temperature of the rolls at the cutting edges. Because of the availability of PID controller, once the desired surface temperature is achieved it automatically cuts off the current supply to the heaters. Further, as time passes, heat from the surface of the rolls will get dissipated, and then the temperature controller will come into effect and switch on the current supply to the heaters to maintain the required surface temperature for packing. The required surface temperature depends on the cover micron we used, that is, for higher thickness, we need high surface temperatures at the cutting edge of the rolls. The standard roll with heat shield (stainless steel 316Ti) of 0.7 mm thickness, as shown in Figs. 2 and 3.

Fig. 2 Standard rolls

Fig. 3 Rolls with heat shield

2.3 Simulation

In order to evaluate the effect of heat shields for different material and variation of any operating/design parameters, instead of wasting the material and lead time for fabrication, an attempt is being done using Simulation techniques. Simulation is done for the rolls with heat shield and without heat shield in ANSYS with steady-state thermal analysis with relevant boundary conditions ascertained from the experiments. Domain size is selected for the numerical solution based on the dimension of rolls of the packing machine, which plays a vital role in terms of saving power by reducing heat transfer to the environment with the aid of heat shields. The images of model of rolls with and without heat shields for simulation analysis are shown in Figs. 5 and 6.

Meshing is the tool used in the software design to divide the entire components into a finite number of small elements as required. The magnitude of the split components must be as tiny as necessary, so that the full number of split components must be as big as necessary, making the findings precise. Meshing for the roll is done by making 32,001 nodes and 18,579 elements for the precise evaluation of results and meshing is shown in Fig. 4 Grid Independence is follows as the meshing type is changing from coarse mesh to fine mesh the results are in favour with respect to experimental results. The boundary condition for the rolls without shield is 465 kJ of heat at the slot of diameter 18 mm and for the roll with shield is 400 kJ to attain a steady temperature of 175 °C.

Fig. 4 Meshing of rolls

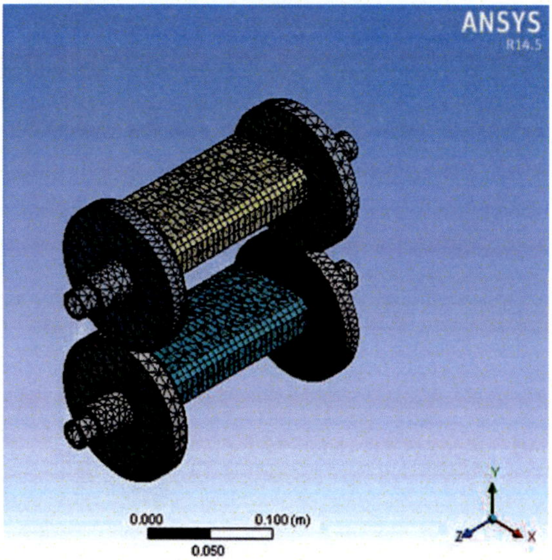

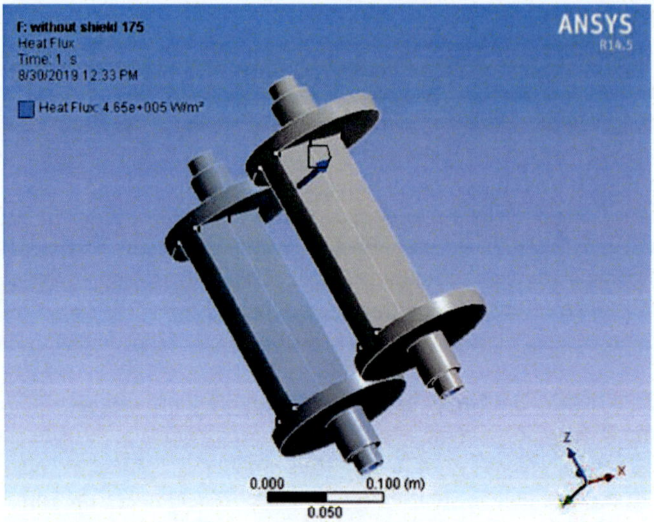

Fig. 5 Model of roll without shield

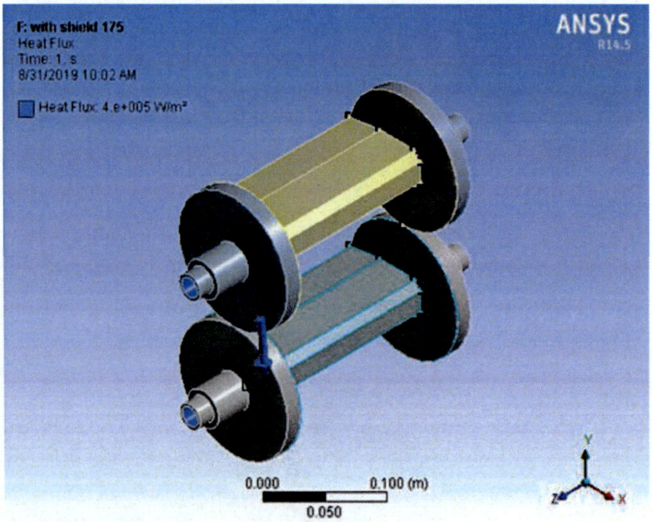

Fig. 6 Model of roll with shield

3 Results and Discussion

3.1 Experimental Results

The variation in temperature for different heat inputs to the heater integrated to Rolls with and without heat shield is shown in Fig. 7. It is observed that there is an increase in temperature with an increase in heat input for two cases of with and without the shield. Due to low thermal conductivity and high reflectivity of heat shield covered over the surface of the roll, the rate of increase in temperature is more in case of shielded roll compared to without shield. The hierarchical properties of 316Ti lead to 14.2% decrease in heat input as compared with standard rolls.

From Fig. 8, with and without a heat shield, the distinction in time to attain the required temperature is presented. The time required to reach the temperature of 175 °C, which is suitable for a packing material thickness of 80 microns is 28% less than without a heat shield roll. The rate of decrease in time is almost uniform till 150 °C, after there is a drastic reduction rate is noticed up to 175 °C. This is because of a decrease in thermal conductivity and heat transfer coefficient between the heat shield and environment at that temperature.

After reaching a steady surface temperature current is cut-off by the temperature controller to the heater; as time passes, heat will be dissipated from the surface of rolls and heat shields. So as to maintain constant surface temperature, T-controller again switches current to the heater. After achieving the desired temperature, let 1750c current to the heaters will be cut-off by the T-controller. Calculation of heat

Fig. 7 Heat input versus temperature

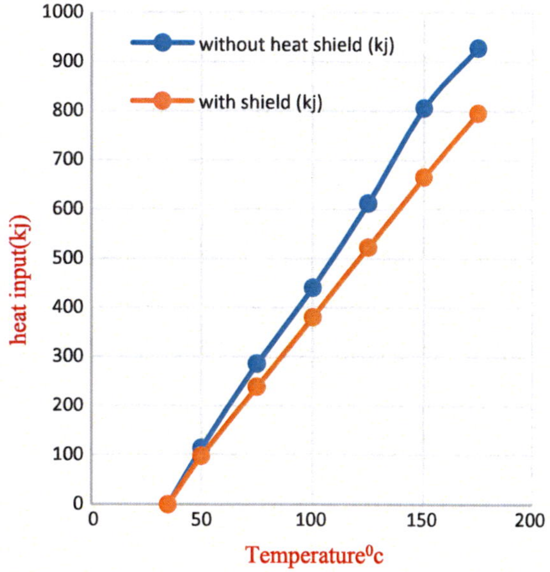

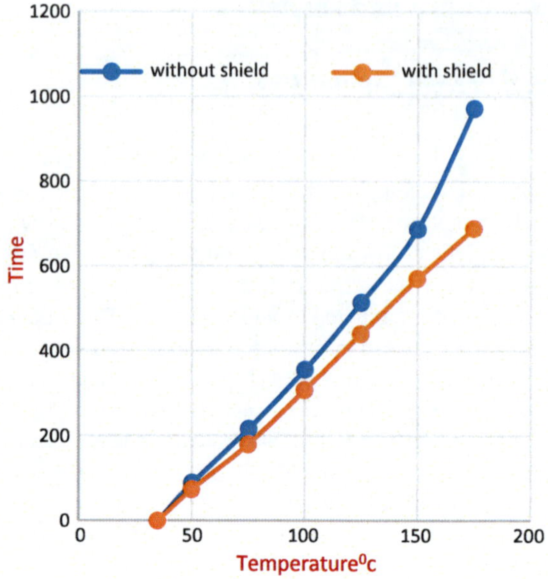

Fig. 8 Time versus temperature

supply for one hour is done to with heat shield and without a heat shield. In an hour, there are current supply 5 amps for 1350 secs for with heat shield and for without heat shield supply 5 amps for 2000 secs. Per hour there is saving 198 W with heat shield after reaching a steady temperature. As per the electricity board, one unit of current is equal to 1 KWH (i.e., 1000 W per hour). It results, in one hour, there is a saving of 198 W. Therefore, there is a saving of 0.19 units of electricity in one hour.

3.2 *Simulation Results*

3.2.1 Heat Flux

Simulation analysis is done for the rolls with and without heat shield rolls using ANSYS with property values of EN8, stainless steel 316Ti, and obtained boundary conditions from the experiments. Heat input is given in the slot where the heater is placed, and heat is distributed in the entire roll according to the geometry. Heat flux distribution is shown in Fig. 9 for without shield and Fig. 10 for with shield. Simulation is done by taking ambient at a constant temperature for both rolls of with and without shield.

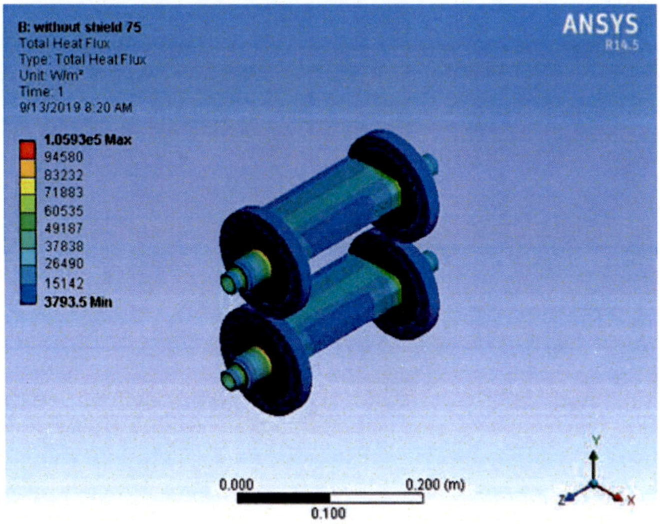

Fig. 9 Heat flux variation without shield

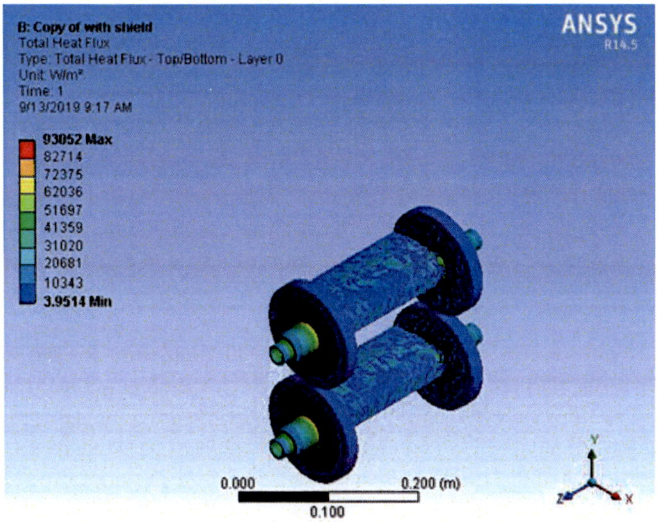

Fig. 10 Heat flux variation with shield

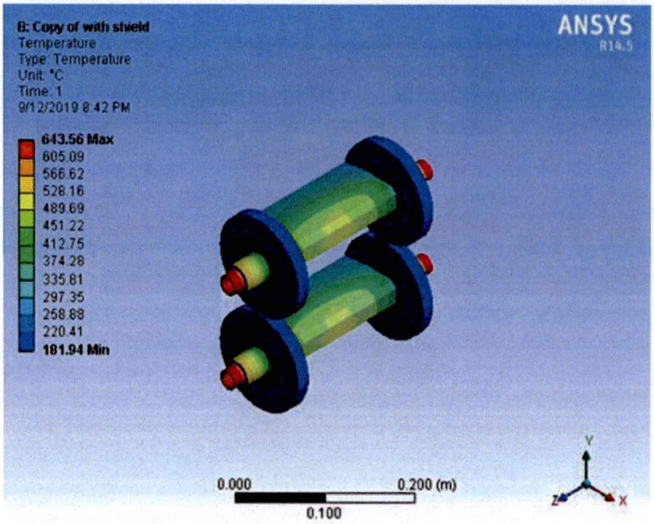

Fig. 11 Temperature simulation with shield

3.2.2 Temperature

Because of heat supply at slot, temperature distribution took place in the entire roll of with and without a heat shield. To reach a minimum temperature of 175 °C, a heat input of 930 kJ of heat is supplied at the slot for the standard roll, and 800 kJ of heat is supplied for the shielded roll. Figure 11 shows the temperature distribution of standard roll in the simulation where the minimum temperature goes up to 181 °C and Fig. 12 shows the temperature distribution of shielded roll in the simulation where the minimum temperature goes up to 183 °C. Because of the ideal condition in simulation, the temperature goes beyond the experimental values.

Figure 13 shows how temperature is varying in simulation for the given heat inputs to the rolls with and without heat shield which are obtained from the experiment. Graph justifies experimental results that, as the temperature is increasing, heat supplied to the shielded roll is decreasing when compared to standard roll. Further, for a case without shielded roll, the slope is streak and more in comparison to shielded roll slope.

4 Conclusions

This paper reports the role of heat shield on the surface of rolls to increase the performance of roll by decreasing heat input and reducing heat dissipation to the environment.

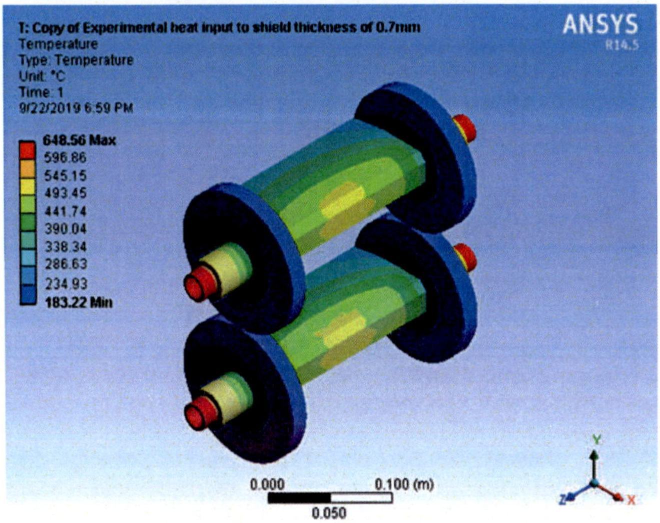

Fig. 12 Temperature simulation without shield

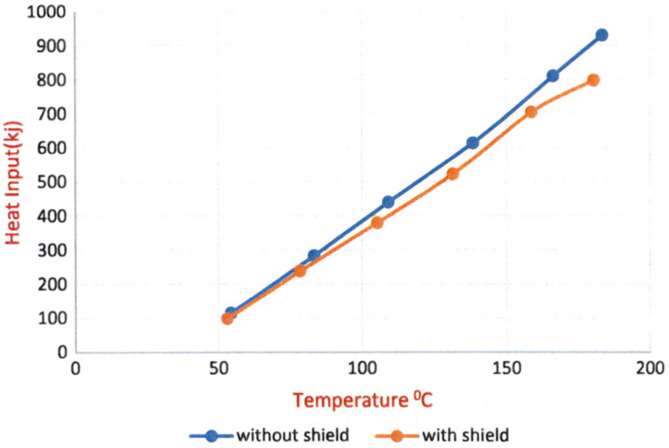

Fig. 13 Heat input versus temperature in simulation

- Shielded rolls are consuming 13.9% less heat and 14.2% less power in contrast to standard rolls.
- After attaining the required temperature for both rolls, shielded rolls are saving 0.19 units of electricity per hour.
- Shielded rolls took 28% less time in comparison to standard rolls to attain the required temperature.

Acknowledgements We would like to express my deep and sincere gratitude to the Department of Mechanical Engineering, Gokaraju Rangaraju Institute of Engineering and Technology (Autonomous), and thankful to **Shiva Snacks and Namkeens Pvt Ltd**. for permitting to utilize the necessary equipment for my experimentation.

References

1. Pashby IR (2003) Surface hardening of steel using a high power diode laser. J Mater Process Technol 139:585–588
2. Zarroug NM (2003) Mild steel (En8) rod tests under combined tension–torsion loading. J Mater Process Technol Ireland 143–144:807–813
3. Balusamy T, Sankara Narayanan TSN, Ravichandran K (2012) Effect of surface mechanical attrition treatment (SMAT) on boronizing of EN8 steel. Surf Coat Technol 213(2012):221–228
4. Cao XQ (2004) Ceramic materials for thermal barrier coatings. J Euro Ceram Soc Germany 24:1–10
5. Bozsaky D (2015) Laboratory tests with liquid nano-ceramic thermal insulation coating. Hung Proc Eng 123:68–75
6. Jyothi US, Vijaya Kumar Reddy K (2020) CFD analysis on particle diameter in hydrogen-enriched diesel engine for different piston bowl geometries. Int J Ambient Energy. https://doi.org/10.1080/01430750.2019.1710565
7. Vaghani M, Vasanwala SA, Desai AK (2014) Stainless steel as a structural material: state of review. J Eng Res Appl 4(3):657–662
8. Vukelic G (2016) Analysis of austenitic stainless steels (AISI 303 and AISI 316Ti) regarding crack driving forces and creep responses. J Brnic Proc IMechE Part L J Mater Des Appl 230(3):699–704
9. Cui Y (2007) Austenite-preferential corrosion attack in 316 austenitic stainless steel weld metals. Carl D. Lundi Mater Des 28:324–328
10. Pradhan R, Krishna Prasad KM, Asif SD, Sai Krishna G, Rama Krishna A, Muruthy DSSK () Experimental investigation and comparative study of MIG and TIG welding on SS202 and SS304 materials. Int J Recent Sci Res 10(04(A)):31678–31683
11. Zheng G, Kelleher B, Cao G, Anderson M, Allen T, Sridharan K (2015) Corrosion of 316 stainless steel in high temperature molten Li2BeF4 (FLiBe) salt. J Nucl Mater 461:143–150
12. Sun M, Wu X, Zhang Z, Han EH (2009) PR China oxidation of 316 stainless steel in supercritical water. Corrosion Sci 51:1069–1072
13. Gupta G (2015) A review Of HVOF thermal spray coating technique on metal plate. Int J Technol Res Eng 3(2):2347–4718
14. Rotshtein VP, Ivanov YF et al (2006) Russia "Surface alloying of stainless steel 316 with copper using pulsed electron-beam melting of film–substrate system. Surf Coat Technol 200:6378–6383

Influence of Geometric and Operating Parameters on the Heat Transfer Enhancement of Synthetic Jet

Aswini Kumar Khuntia, Pandaba Patro, and Sanjoy Ghoshal

1 Introduction

Nowadays, the human dependency on electronic gadgets and components increased drastically, which has resulted in quick development in VLSI technology. This leads to the use of high power, high speed, and high-density microchips. The increase in power levels with the reduction in feature size of the electronics components results in continuously increasing heat fluxes of electronic devices. Mainly, the failures of the electronic components are due to overheating or thermal failure. The heat generated inside the electronics devices during the operation should be effectively dissipated within the specific time for the proper functioning of the device. Therefore a highly efficient and compact cooling solution is essential to keep the consistency of electronic devices, and this need is expected to develop constantly in the near future. A number of cooling methods have been developed in the last few decades for the extraction of heat excess produced in the device. In general, either active or passive cooling can be employed, but active cooling is more suitable as compared to passive cooling in case of limited availability of space. Active cooling using a synthetic jet can be useful to small systems. The use of Synthetic Jet has a lot of advantages over the other methods that are low cost, lightweight, simple structure, and easy installation.

A. K. Khuntia (✉)
NIST, Berhampur, Odisha, India
e-mail: aswinikk@gmail.com

P. Patro
VSSUT Burla, Odisha, India

S. Ghoshal
IIT Dhanbad (ISM Dhanbad), Dhanbad, Jharkhand, India

M. Palanisamy et al. (eds.), *Theoretical, Computational, and Experimental Solutions to Thermo-Fluid Systems*, Lecture Notes in Mechanical Engineering,
https://doi.org/10.1007/978-981-33-4165-4_17

2 Synthetic Jet

A synthetic is produced by the time-periodic motion of an oscillating diaphragm at one end inside a sealed cavity with an orifice opening at the other end. The synthetic jet operating cycle is divided into the suction and ejection cycle. During the suction, the diaphragm is moving outward from the cavity (Fig. 1), drawing in a fluid to fill the cavity, which increases the volume of fluid in the cavity. In the ejection cycle, the diaphragm move towards the orifice; this produces a large amount of force on the fluid in the cavity. Synthetic jet does not require continuous fluid flow as compared with the conventional continuous jet. A synthetic jet has the ability to transfer the linear momentum to the surrounding without the inevitability of a net mass flux, which is the primary advantage of a synthetic jet as compared with the ordinary steady jet.

The performance of a synthetic jet mainly depends upon actuator geometrical constraints: orifice diameter (D_o), orifice depth (h), cavity diameter (D_c), and cavity depth (H); two actuator operating parameters: oscillation amplitude (Δ) and oscillation frequency (f); and also three fluid constraints: temperature (T), density (ρ), and viscosity (μ).

Synthetic jet does not require continuous fluid flow as compared with the conventional continuous jets. Close to the exit plane, the speed decreases continuously, but the local mean velocity is increasing [1]. As synthetic jets are formed totally from

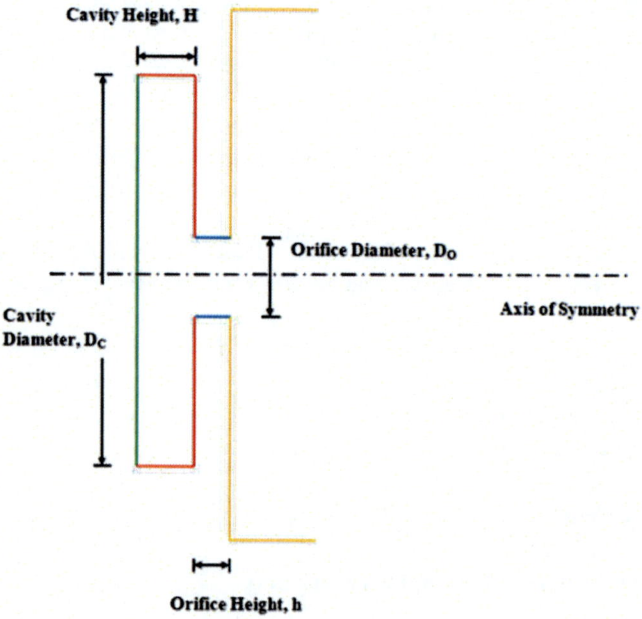

Fig. 1 Schematic of a synthetic jet

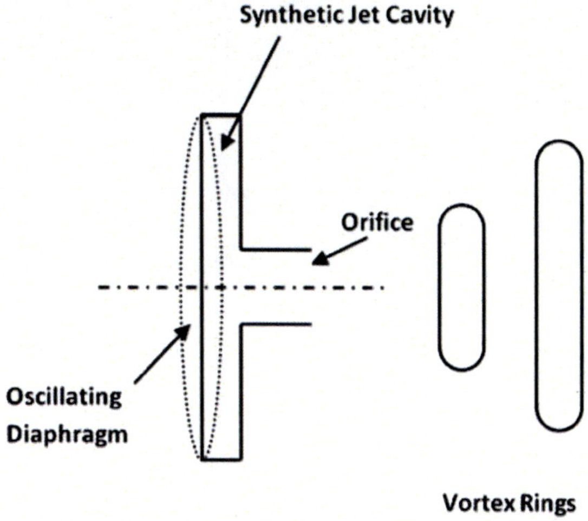

Fig. 2 Synthetic jet

the working fluids of the flow system, they can only transfer momentum with no net mass flow across the flow boundary [2]. The continuous movement of the diaphragm produces a sequence of vortex rings, which make up the body of the synthetic jet and whose presence produces an average velocity along the jet axis away from the orifice. Thus, a synthetic jet is deemed a zero-net mass flux device capable of inducing non-zero momentum along the jet direction [3]. The fluid volume is driven from the cavity in the forward stroke, and the same comes back into the cavity in the succeeding suction stroke. Due to a lot of advantages of a synthetic jet, it has been used in many applications.

A circular synthetic jet actuator, as shown in Fig. 1, consists of three main parts, that is, diaphragm, cavity, and the orifice. The vibrating diaphragm is located at the bottom of the cavity, on the top of which is a round orifice. The diaphragm oscillates up and down with a sine or cosine waveform. Synthetic jets are developed with the periodic movement of the diaphragm inside the cavity, which usually produces an oscillating flow leave into a large volume. Figure 2 shows the oscillatory motion of the diaphragm inside the cavity and flow of fluid through the orifice.

Apart from the heat transfer enhancement application, synthetic jets are used in various fields such as for controlling airflow in aircraft to reduce the noise and enhance the lift, provides thrust for underwater vehicles.

3 Basic Theory and Numerical Procedure

The strength of the synthetic jet mainly depends on the jet velocity, which is derived from the amplitude of vibration of the diaphragm. According to the theory of plate [4], the deformation of the diaphragm is

$$\delta(r,t) = \frac{\Delta}{2}\left[1 - \left(\frac{r^2}{r_c^2}\right) + \left(\frac{2r^2}{r_c^2}\right)\ln\left(\frac{r}{r_c}\right)\right]\sin(2\pi f t) \tag{1}$$

where

- δ Relative deflection of the diaphragm towards the neutral position
- r Radial distance from the center of the diaphragm
- r_c Diaphragm radius
- f Frequency of oscillation of the diaphragm
- Δ Amplitude of vibration

Differentiating Eq. (1) with respect to time, we got the speed of oscillation of the vibrating membrane

$$\frac{\partial\delta(r,t)}{\partial t} = u(r,t)$$
$$= \pi.f.\Delta.\left[1 - \left(\frac{r^2}{r_c^2}\right) + \left(\frac{2.r^2}{r_c^2}\right)\cdot\ln\left(\frac{r}{r_c}\right)\right]\cdot\cos(2.\pi.f.t) \tag{2}$$

The maximum velocity at the exit of the orifice is given by

$$U_{\max} = \frac{\pi}{4}\delta f\left(\frac{D_c}{D_0}\right)^2\cos(2\pi f t) \tag{3}$$

The mean velocity over the complete cycle is given by

$$U_0 = \frac{1}{T}\int_0^{T/2} u_0(t)\mathrm{d}t = \frac{1}{4}\delta f\left(\frac{D_c}{D_0}\right)^2 \tag{4}$$

where, T is the period of oscillation of the membrane (Fig. 3).

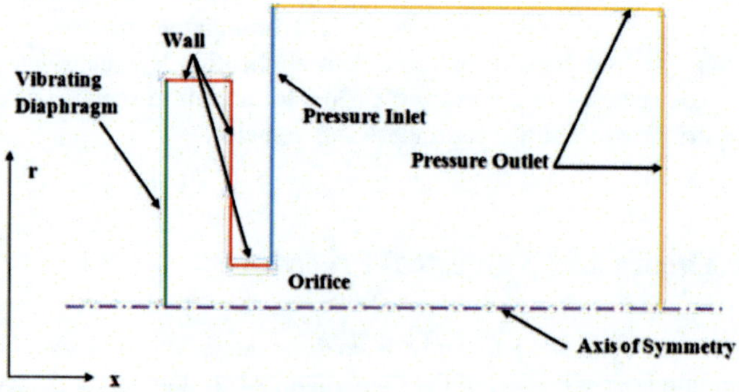

Fig. 3 Computational domain of synthetic jet

The Reynolds number is given by $R_e = \frac{Ud}{\nu}$

The Stokes Number is given by $S = \sqrt{\frac{\omega d^2}{\nu}}$

where U is the maximum instantaneous velocity that is $\max(u)$, ω is the oscillation frequency in radian/sec, d is the diameter of the opening and υ is the kinematic viscosity of the fluid.

The computational domain of a synthetic jet is shown in Fig. 3. Axisymmetric 2D model was used for the numerical simulations. In the present study, the flow characteristics of a synthetic jet are analyzed using the computational fluid dynamics solver Ansys FLUENT. Similar to the experimental set up by Mane et al. [5] and Jain et al. [6], a numerical model was developed to satisfy the jet formation criterion. The dimensions used for the analysis are: orifice diameter $D_0 = 0.00367$ m, orifice length $H_0 = 0.0016$ m, cavity diameter $D_c = 0.03$ m, cavity height $H_c = 0.00955$ m.

The numerical model is validated with the existing experimental data [6]. The number of cells chosen initially was 10,000 cells (Coarse Mesh) then the grid density was increased to two other meshes having 30,000 (Medium Mesh) and 65,000 (Fine Mesh) cells to get the grid-independent solution. From Fig. 4, it was clear that the computational domain with 30,000 cells can be used to get grid-independent solutions.

Figures 5 and 6 show the variation of absolute velocity with respect to t/T for one cycle at two different frequencies 32 and 50 Hz. The x-axis represents the normalized time, and y-axis represents the absolute velocity, and also the results are compared with the previously available experimental as well as numerical results. The present

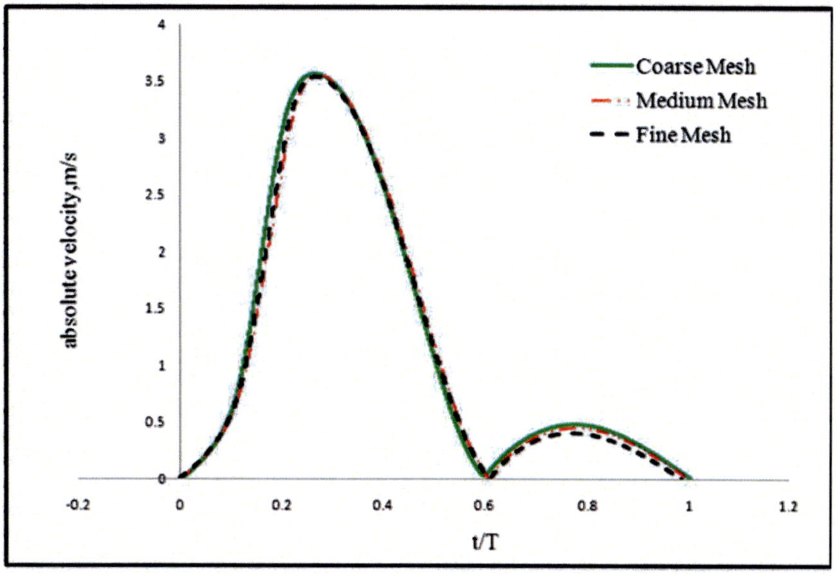

Fig. 4 Variation of absolute velocity with normalized time at 32 Hz frequency

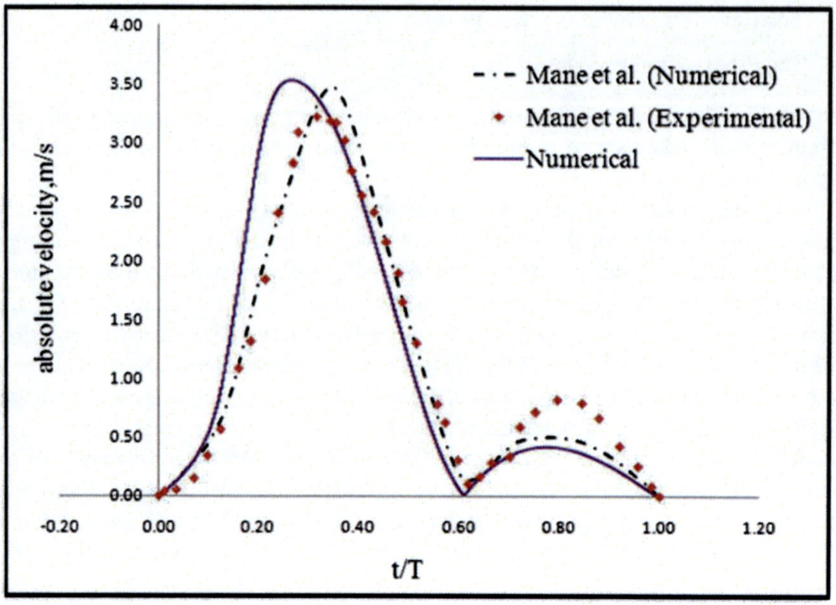

Fig. 5 Variation of absolute velocity with normalized time at 32 Hz frequency

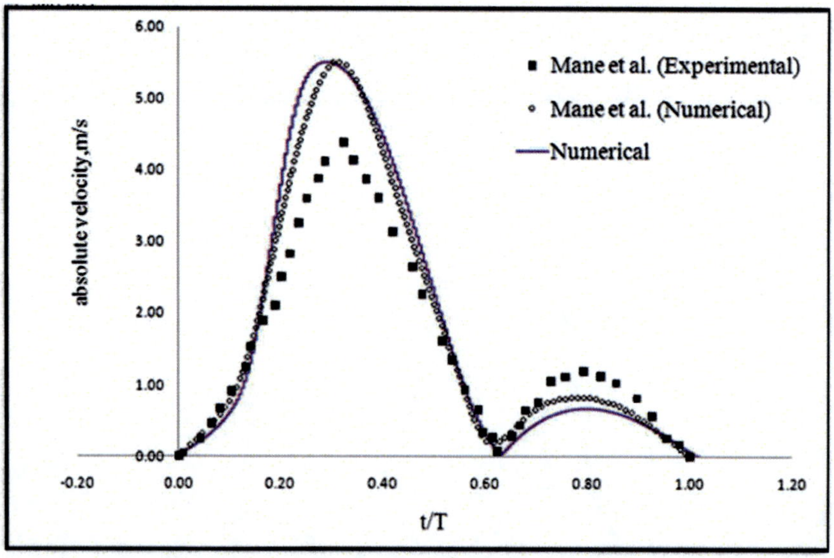

Fig. 6 Variation of absolute velocity with normalized time at 50 Hz frequency

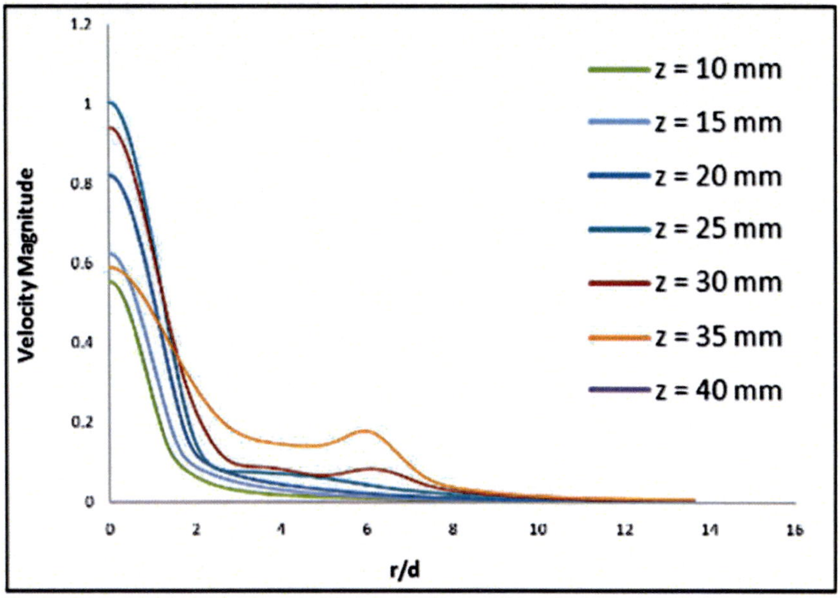

Fig. 7 Radial Variation of velocity at different axial positions at frequency = 100 Hz

simulation results concur very well with experimental data with an error of less than 10%.

The velocity distribution along the radial direction for different axial distance is compared. Figure 7 shows the velocity magnitude at $t/T = 1$. The x-coordinate represents the radial distance from the center line along the length of the heated plate. The result clearly shows that the velocity increases initially with the increase in distance for a particular distance and further increases the velocity decreases.

At first, the surface heat transfer coefficients at a different axial distance from the orifice are compared. Figures 8 and 9 shows the surface heat transfer coefficients initially increases and then decreases with the distance .

Figure 10 describes the comparison between the synthetic jet with the equivalent continuous jet and shows that the synthetic jet gives the better result as compared with the continuous jet.

Figures 11 and 12 shows the prediction of velocity and heat transfer coefficient at different frequency of oscillation of the diaphragm. The result showed that the surface heat transfer coefficient and the velocity magnitude increases with an increase in frequency.

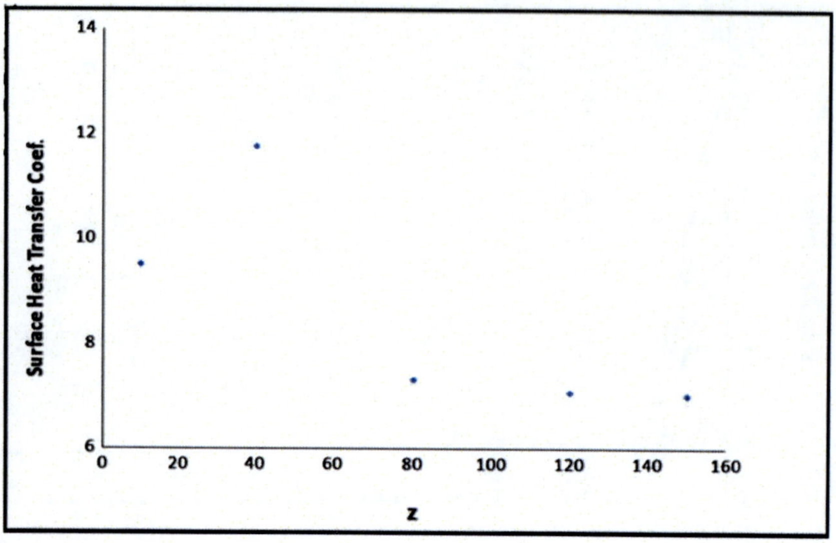

Fig. 8 Variation of surface heat transfer coefficient with axial distance at frequency = 100 Hz

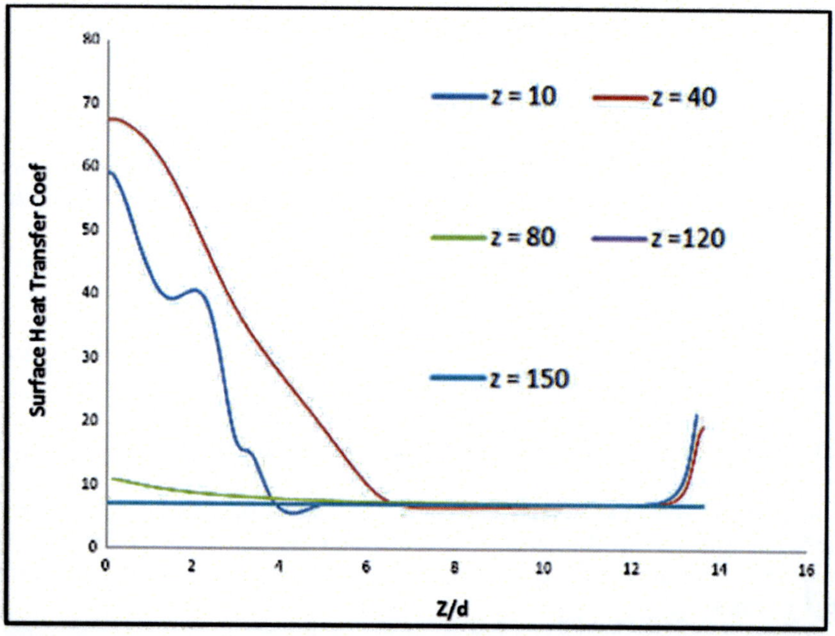

Fig. 9 Variation of surface heat transfer coefficient normalized axial distance at frequency = 100 Hz

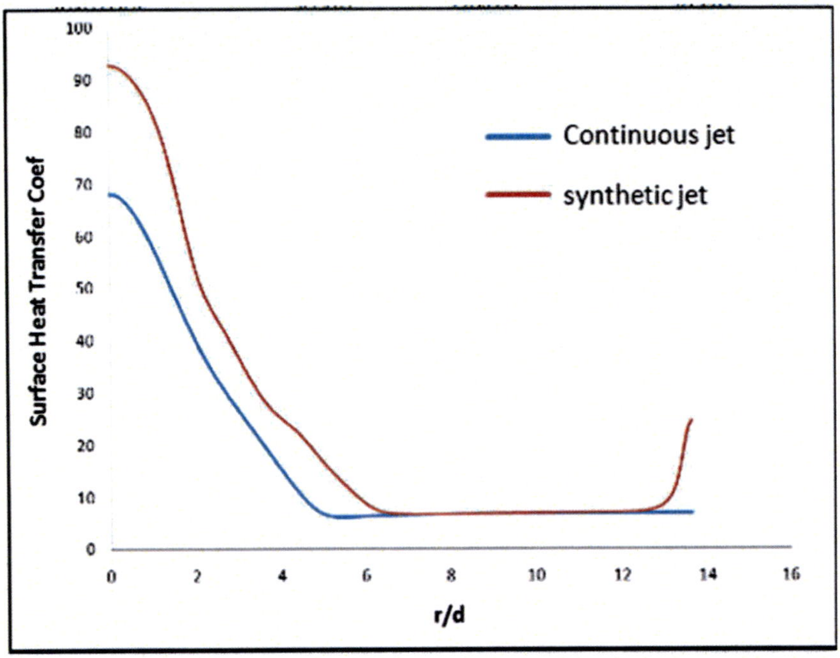

Fig. 10 Variation of surface heat transfer coefficient normalized axial distance at frequency = 100 Hz and $z = 30$ mm

4 Conclusion

Two-dimensional, axisymmetric simulations were performed to study the synthetic jet with a circular orifice at a different frequency of vibration of the diaphragm. A velocity boundary condition was included for a more accurate study by using a user-defined function (UDF). A continuous jet was considered for the comparison with the same boundary condition as that of the synthetic jet. Comparisons were made between a synthetic jet and continuous jet for the prediction of surface heat transfer coefficient. It was found that the computational result agreed well with the previously available experimental as well as numerical results. The synthetic jet gives a better result for heat transfer as compared with that of the continuous jet. Five different models of the synthetic jet were tested by placing the heated plate at a different distance from the orifice, and it was found that the surface heat transfer coefficient gives a good result for a small distance. Finally, the result showed that with the increase in the frequency of vibration, surface heat transfer coefficient, and the velocity magnitude increases. This indicates the strong flow field of the synthetic jet. So, the performance of the synthetic jet is better as compared to the continuous jet. With this advantage, a synthetic jet will be recommended where high exit velocity is required.

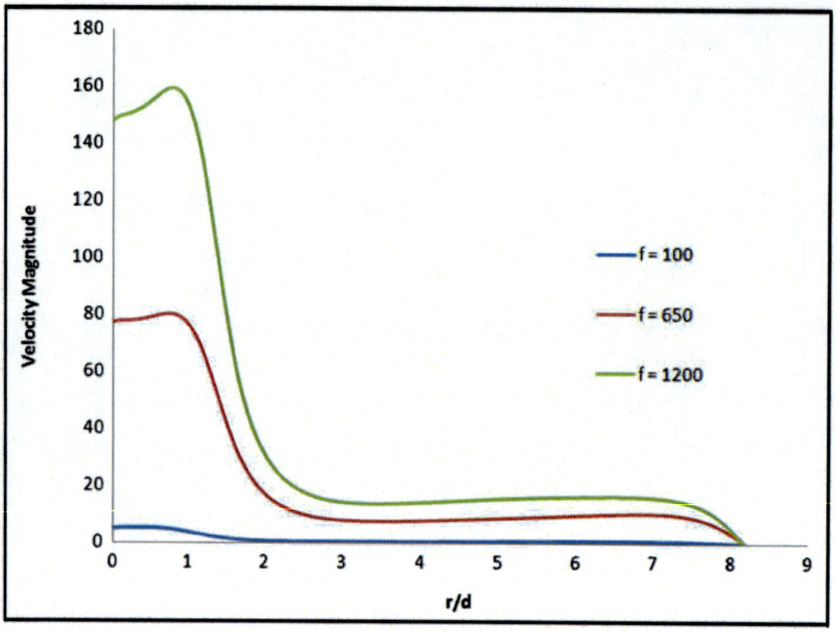

Fig. 11 Variation of velocity magnitude normalized axial distance at $z = 30$ mm

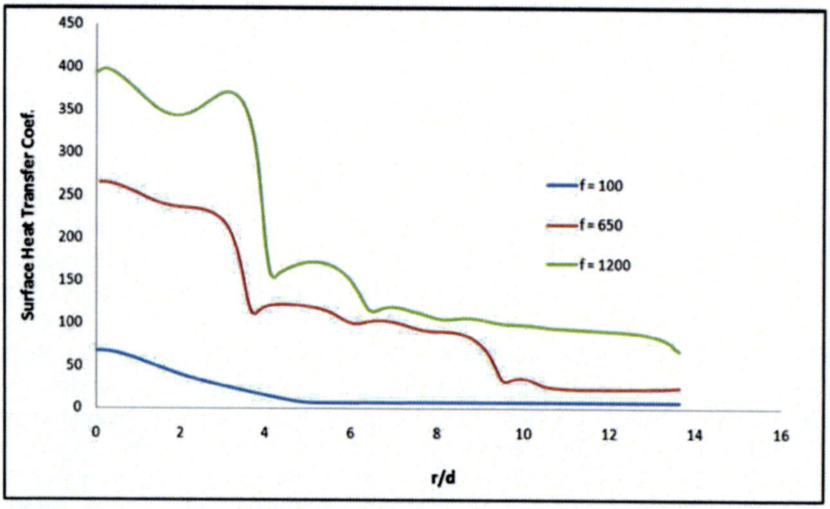

Fig. 12 Variation of surface heat transfer coefficient normalized radial distance at $z = 30$ mm

References

1. Smith BL, Glezer A (1998) The formation and evolution of synthetic jets. Phys Fluids 10:2281
2. Glezer A, Amitay M (2002) Synthetic jets. Annu Rev Fluid Mech 34:503–529
3. Holman R, Cattafesta L, Mittal R, Smith B, Utturkar Y (2005) Formation criterion for synthetic jets. AIAA J 43(10):2110–2116
4. Timoshenko S, Woinowsky-Krieger S (1959) Theory of plates and shells, 2nd edn. McGraw–Hill, London, pp 67–69
5. Mane P, Mossi K, Rostami A, Bryant R, Castro N (2007) Piezoelectric actuators as synthetic jet: cavity dimension effects. J Intell Mater Syst Struct 18:1175
6. Jain Manu, Puranik Bhalchandra, Agrawal Amit (2011) A numerical investigation of effects of cavity and orifice parameters on the characteristics of a synthetic jet flow. Sens Actuat A 165:351–366

A Numerical Investigation of Effect of Shape of Orifice on the Heat Transfer Characteristics of a Synthetic Jet

Aswini Kumar Khuntia, Pandaba Patro, and Sanjoy Ghoshal

1 Introduction

In recent years, the improvements to the cooling of electronics components are highly essential as the life of the electronic device depends on the working temperature. As power levels increase with the reduction in feature size of the electronics components results in constantly increasing heat fluxes of electric components, which increases the temperature of the devices reducing the life period of the device. At the time of working, these devices produce a higher amount of heat that needs to be continuously dissipated to avoid the thermal failure. Therefore, a highly efficient and compact cooling method is essential to keep the consistency of electronic devices. In general, either active or passive cooling can be employed, but active cooling is more suitable as compared to passive cooling. In case of small systems, active cooling using a synthetic jet is more useful as compared to the other techniques. A synthetic jet is produced by the time-periodic motion of an oscillating diaphragm at one end inside a sealed cavity with an orifice opening at other end. The synthetic jet operating cycle is divided into the suction and ejection cycle. During the suction, the diaphragm is moving outward from the cavity drawing in the fluid to fill the cavity, which increases the volume of fluid in the cavity. In the ejection cycle, the diaphragm moves toward the orifice; this produces a large amount of force on the fluid in the cavity. Synthetic jet

A. K. Khuntia (✉)
NIST, Berhampur, Odisha, India
e-mail: aswinikk@gmail.com

P. Patro
VSSUT Burla, Odisha, India

S. Ghoshal
IIT Dhanbad (ISM Dhanbad), Dhanbad, Jharkhand, India

M. Palanisamy et al. (eds.), *Theoretical, Computational, and Experimental Solutions to Thermo-Fluid Systems*, Lecture Notes in Mechanical Engineering,
https://doi.org/10.1007/978-981-33-4165-4_18

does not require continuous fluid flow as compared with the conventional continuous jet. Synthetic jet has the ability to transfer the linear momentum to the surrounding without the inevitability of a net mass flux is the primary advantage of the synthetic jet as compared with the ordinary steady jet.

2 Literature Review

In the recent years, the flow characteristics of synthetic jet have been extensively investigated. The scientific and technical literature documents dealing with synthetic jets are available for the last three decades. Lv et al. [1] numerically investigated the effect of geometric parameters like cavity depth, diameter as well as orifice thickness, diameter, and excitation frequency of the actuator. They demonstrated that the output velocity could be increased if the cavity acoustic resonance frequency coincides with the diaphragm excitation frequency. Liu and Chang [2] showed that a synthetic jet through a diffusion shaped orifice achieved a heat transfer enhancement 30% as compared with the round orifice at an opening angle of 60°, and when the opening angle increased to 90°, heat transfer enhancement was decreased because of increased flow circulation and reduced ejection flow velocity. Zhang and Jun et al. [3] reviewed the development of zero net mass flux jet (ZNMF) and its application in the separation flow control. They studied the vortex ring and slow structure of the ZNMF in the quiescent fluid, as well as the effect of different key parameters on the fluid flow. Cook and Sadri [4] experimentally investigated the use of synthetic jet by placing it just upstream, and the boundary layer was tripped 40° with either side from the separation line. Surface flow visualization demonstrated to measure the effect on the time-averaged separation line due to the synthetic jet around the cylinder at the downstream of the orifice. Tuckerman and Pease [5] investigated liquid cooling for the integrated circuits. They were considered the heat transfer coefficient between the surfaces of the circuit to the coolant, channel width, and high aspect ratio channels that increase the surface area to design new water-cooled system for the electronic circuits. Ziade et al. [6] compared the numerical result with the experimental data and concluded that the synthetic jet nozzle exit property depends upon the internal cavity parameters. They considered three types of cavity geometries, that is, cylindrical, conical and 5th order polynomial contraction for the analysis and found that the cylindrical nozzle transmits more momentum as compared to the others. This was noted that the internal sharp edge plays an important role as the shape changes from the sharper cylindrical cavity changes to the contraction; the momentum flux decreases to 50%. In the cavity, a vortex ring moves downstream due to the prior section stroke, persists the upward motion of the diaphragm in case of all type of cavities. The blockage was more for the conical and contraction cavity, which decreases the momentum flux at the exit of the nozzle. Miro et al. [7] investigated the heat transfer characteristic of a synthetic jet which was enclosed by two isothermal thermal parallel plates with a temperature difference. For this numerical study, the unsteady 3D Navier Stoke equations were solved for a range of Reynolds number from 50 to 1000. It was found out that the heat transfer at the hot wall and the outlet temperature correlated to the Reynolds number.

3 Modeling Parameters of Synthetic Jet

A circular synthetic jet actuator, as shown in Figs. 1 and 2, consists of three main parts, that is the diaphragm, cavity, and orifice. The vibrating diaphragm is located at the left side of the cavity on the right of which is a round orifice. The diaphragm oscillates forward and backward with a sine or cosine waveform. Synthetic jets are

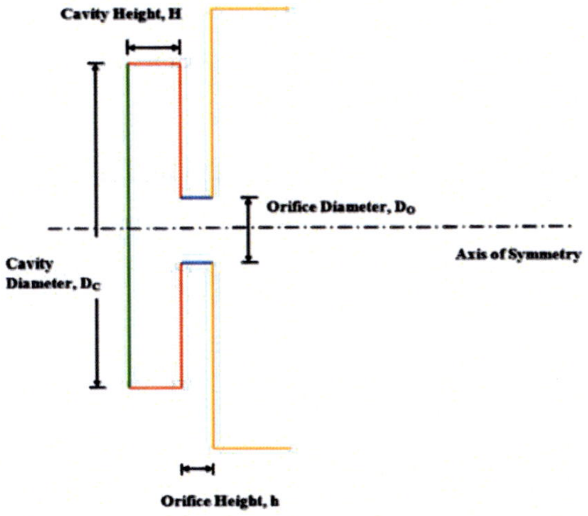

Fig. 1 Schematic of a synthetic jet

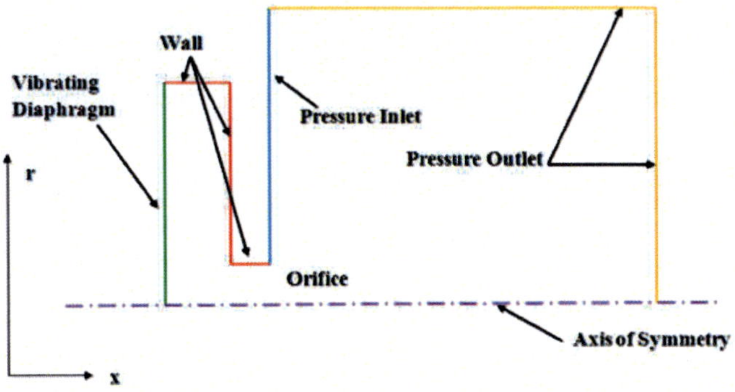

Fig. 2 Computational domain of synthetic jet

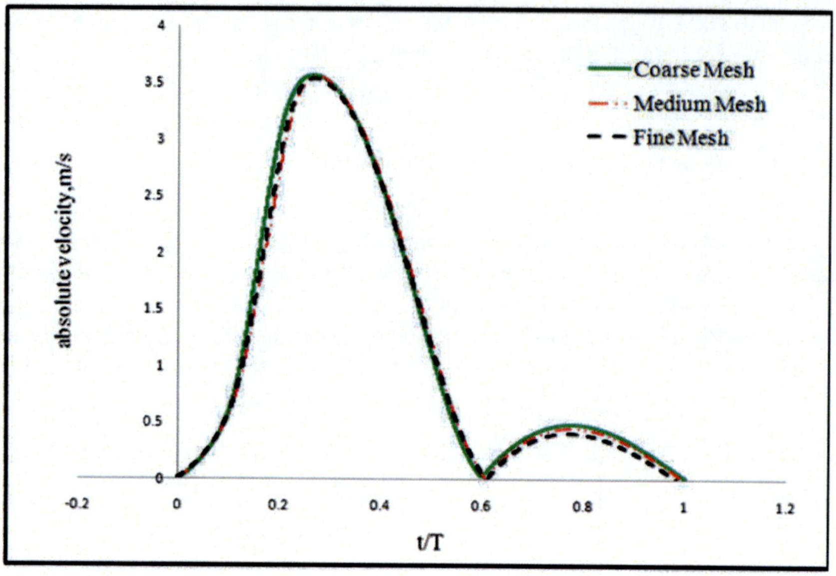

Fig. 3 Grid Independence at 32 Hz frequency

developed with the periodic movement of the diaphragm inside the cavity, which usually produces an oscillating flow leave into a large volume.

The performance of a synthetic jet mainly depends upon actuator geometrical constraints: orifice diameter (D_o), orifice depth (h), cavity diameter (D_c), and cavity depth (H); two actuator operating parameters: oscillation amplitude (Δ) and oscillation frequency (f); and also three fluid constraints: temperature (T), density (ρ) and viscosity (μ).

Synthetic jet can be measured by mainly two dimensionless parameters that are the Reynolds number and the stroke length. The stroke length depends on the diameter of the synthetic jet nozzle. As the velocity profile at the exit of the nozzle is depends on the diameter of the nozzle that is the shape of the nozzle [8]. They were chosen different types of nozzles (sharp-edged with a circular hole, all other were a combination of rounded lips), as well as nozzle plate, sited a little inward to the main nozzle

4 Basic Theory, Numerical Procedure, and Validation

The strength of the synthetic jet mainly depends on the jet velocity, which is derived from the amplitude of vibration of the diaphragm. According to the theory of plate [9], the deformation of the diaphragm is

$$\delta(r,t) = \frac{\Delta}{2}\left[1 - \left(\frac{r^2}{r_c^2}\right) + \left(\frac{2r^2}{r_c^2}\right)\ln\left(\frac{r}{r_c}\right)\right]\sin(2\pi f t) \tag{1}$$

where

- δ Relative deflection of the diaphragm toward the neutral position
- r Radial distance from the center of the diaphragm
- r_c Diaphragm radius
- f Frequency of oscillation of the diaphragm
- Δ Amplitude of vibration

Differentiating Eq. (1) with respect to time, we got the speed of oscillation of the vibrating membrane.

$$\frac{\partial\delta(r,t)}{\partial t} = u(r,t)$$
$$= \pi.f.\Delta.\left[1 - \left(\frac{r^2}{r_c^2}\right) + \left(\frac{2.r^2}{r_c^2}\right).\ln\left(\frac{r}{r_c}\right)\right].\cos(2.\pi.f.t) \tag{2}$$

Differentiating Eq. (1) with respect to time, we got the speed of oscillation of the vibrating membrane. The maximum velocity at the exit of the orifice is given by

$$U\text{max} = \frac{\pi}{4}\delta f\left(\frac{D_c}{D_0}\right)^2\cos(2\pi f t) \tag{3}$$

The mean velocity over the complete cycle is given by

$$U_0 = \frac{1}{T}\int_0^{\frac{T}{2}} u_0(t)\mathrm{d}t = \frac{1}{4}\delta f\left(\frac{D_c}{D_0}\right)^2 \tag{4}$$

The Stroke length L_0 is defined by

$$L_0 = \int_0^{T/2} u_0(t)\mathrm{d}t$$

where, T is the period of oscillation of the membrane, u_0 is the instantaneous velocity.

The Reynolds number is given by $R_e = \frac{Ud}{\nu}$

The Stokes number is given by $S = \sqrt{\frac{\omega d^2}{\nu}}$

where U is the maximum instantaneous velocity, that is $\max(u)$, ω is the oscillation frequency in radian/sec, d is the diameter of the opening, and υ is the kinematic viscosity of the fluid.

The relation between these parameters is as follows: $R_e = \frac{U_o D_n}{\nu}, \frac{L_o}{D_n} = \frac{U_o}{f D_n}$

In this numerical study, the flow and heat transfer characteristics of a synthetic jet impinging on a heated vertical plate that is parallel to the orifice is performed using FLUENT. The synthetic jet was considered here is similar to that of Mane et al. [10] for the model validation. The air was taken as the working fluid with an initial temperature at 300 K and considered incompressible throughout the operation. Other than the heated wall, all other boundaries are adiabatic. The temperature of the heated isothermal wall was taken at 360 K, which is like the surface of an electronics component. As the diaphragm vibrates within the cavity, the jet flows through the orifice in an oscillatory motion in the defined region. The *k*-epsilon RNG turbulence model was preferred for the perfect illustration of the turbulent synthetic jet model. The periodic motion of the diaphragm was simulated using a user-defined function (UDF) along with a dynamic meshing technique. Second-order discretization method was adopted for pressure, momentum, turbulent kinetic energy, turbulent dissipation rate, and energy. For pressure velocity coupling, SIMPLE algorithm was used. The validation of the numerical model was done by considering orifice diameter $D_0 =$ 0.00367 m; orifice length $H_0 = 0.0016$ m, cavity diameter $D_c = 0.03$ m, cavity height $H_c = 0.00955$ m, the amplitude of the diaphragm $A = 0.000365$ m, and for the two vales of frequency of vibration of the diaphragm that is $f = 32$ Hz and $f = 50$ Hz. The simulation results are compared with those of Mane et al. [10] for the effect of various cavity parameters as well as the cavity shapes on the synthetic jet flow.

The grid dependency of the present validation work was tested by studying the variation of the absolute velocity in the solution domain for the different grid sizes considering three different grid size: 10,000 cells (Coarse Mesh), 30,000 cells (Medium Mesh) and 65,000 cells (Fine Mesh) cells to get the best available solution as shown in the Fig. 3. The present simulation results were predicted with the medium mesh. The comparison result shows that the present result is very similar to that of the experimental results with an error of less than 3% as observed in Figs. 4 and 5.

A similar continuous jet was recognized for the comparison with the equivalent synthetic jet. A separate model was prepared by removing the diaphragm and the cavity whereas keeping the rest of the geometry and using the same boundary conditions as like synthetic jet analysis. By matching the Reynolds number of the continuous jet and the synthetic jet, the discharge velocity of the continuous jet was calculated and applied at the orifice inlet.

Figure 6 shows the velocity plots of the continuous jet as well as the synthetic jet at the Reynolds number of 660. From simulation, it is perceived that synthetic jet velocity is high near the centerline area, which can help in more localized cooling. Figure 7 presents the variation of the centreline velocities with Reynolds number for both synthetic jet and continuous jet is plotted versus the non-dimensional axial coordinate x/D_0. In case of a synthetic jet, the velocity begins to sharp decrease from the exit value after $x/D_0 = 5$, while this does not occur after $x/D_0 = 10$. The synthetic jets consistently lie above the continuous jets, indicating that the synthetic jets are consistently better than equivalent continuous jets.

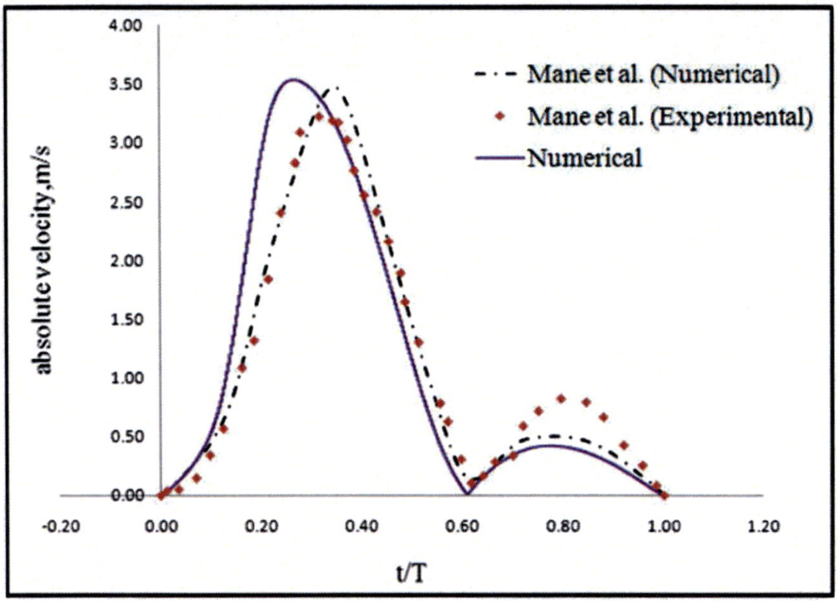

Fig. 4 Validation at 32 Hz frequency

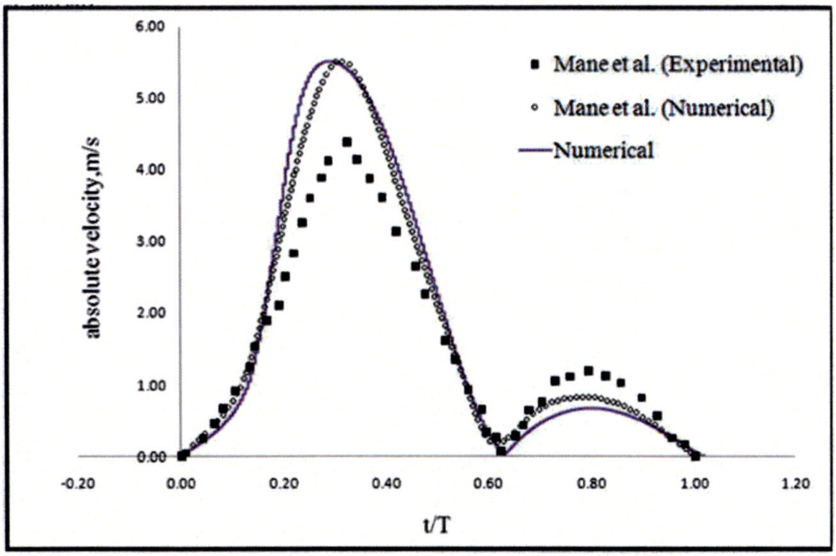

Fig. 5 Validation at 50 Hz frequency

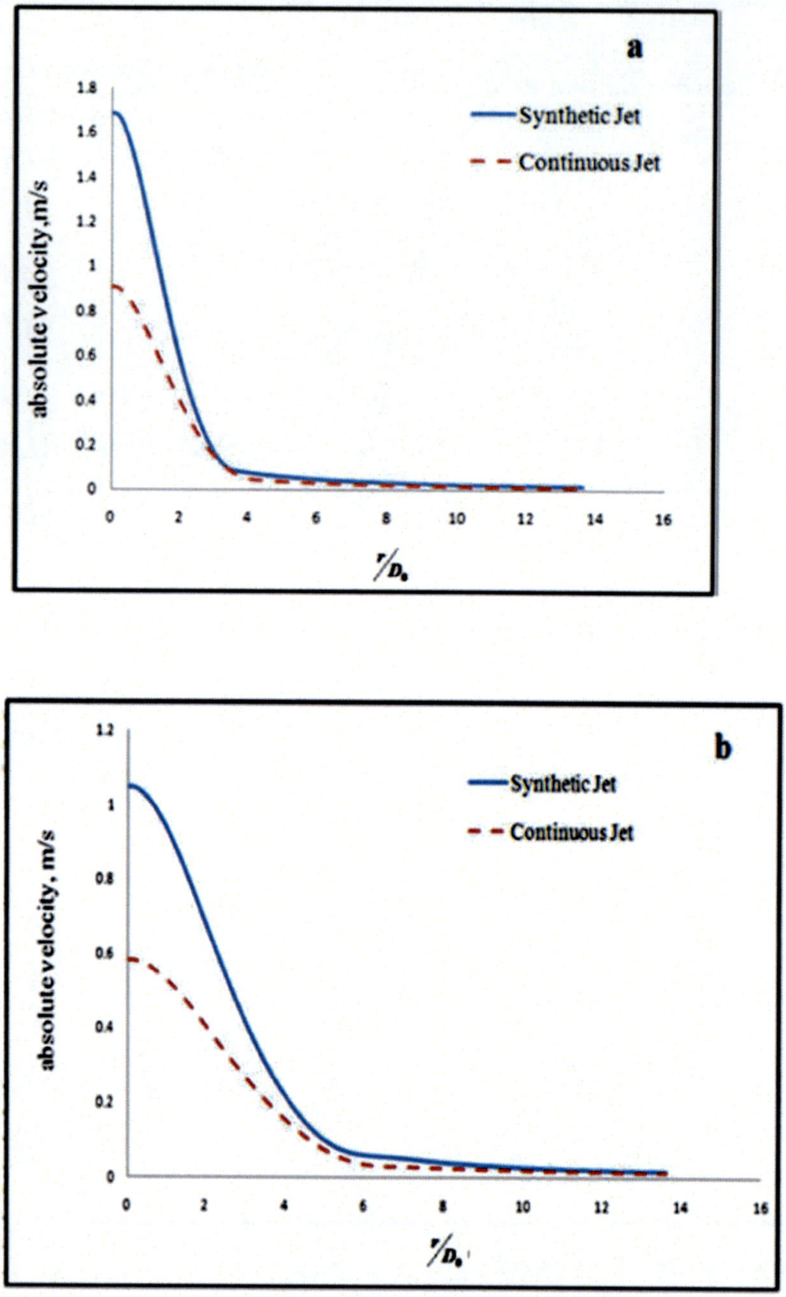

Fig. 6 Velocity versus y/D_0 curve at $R_e = 660$ at an axial distance of **a** 40 mm and **b** 60 mm

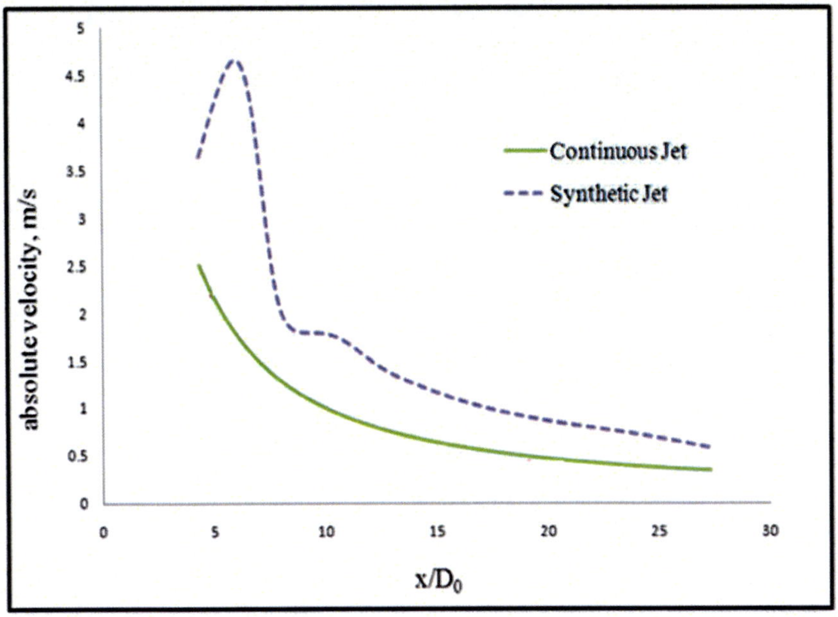

Fig. 7 Centerline velocity variation along the axis at $R_e = 660$

Different nozzle or orifice shape was considered to investigate its effect on the performance of synthetic jets.

Nozzle 1: simple cylindrical with a sharp edge.
Nozzle 2: simple cylindrical round edge.
Nozzle 3: cylindrical and divergent with round edge.
Nozzle 4: Cylindrical with both sides round.
Nozzle 5: Assembled of all types of nozzles but the outer nozzle positioned little outside the downstream (shown in Fig. 8).

Figure 9 signifies the axial velocity as well as the surface heat transfer coefficient with the non-dimensional distance. The highest value of the surface heat transfer coefficient and the axial velocity are attained for the nozzles with cylindrical and divergent shape.

5 Conclusion

In the present research work, numerical simulations were performed for synthetic jets using Ansys Fluent by considering five different shapes of nozzles or orifices. The nozzles are considered from the simple cylindrical with a sharp edge to the addition of a round edge. The next type of nozzle had cylindrical and divergent with

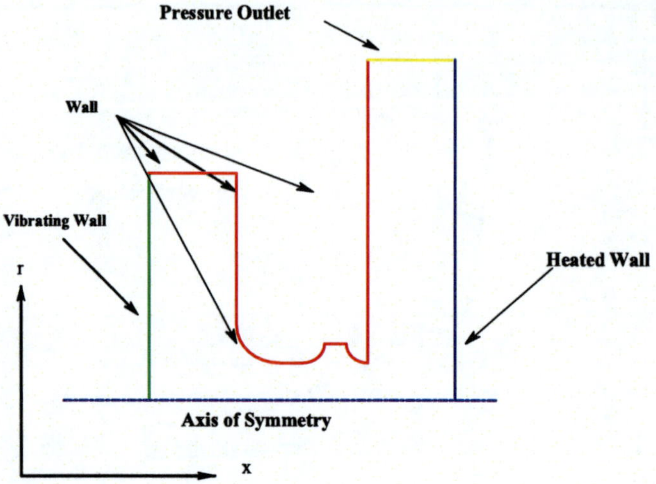

Fig. 8 Computational domain with nozzle -5

a round edge then the both side round. The last nozzle was assembled of all the type of nozzles, but the outer nozzle positioned little outside the downstream. It was found that the jet with cylindrical and divergent nozzles gives a better result for heat transfer as compared to the others. The flow velocity for the nozzles with cylindrical and divergent nozzles is higher than that of the round orifice because of the pressure drop in the ejection. As a consequence, the surface heat transfer is also higher in the case of nozzles with cylindrical and divergent nozzles.

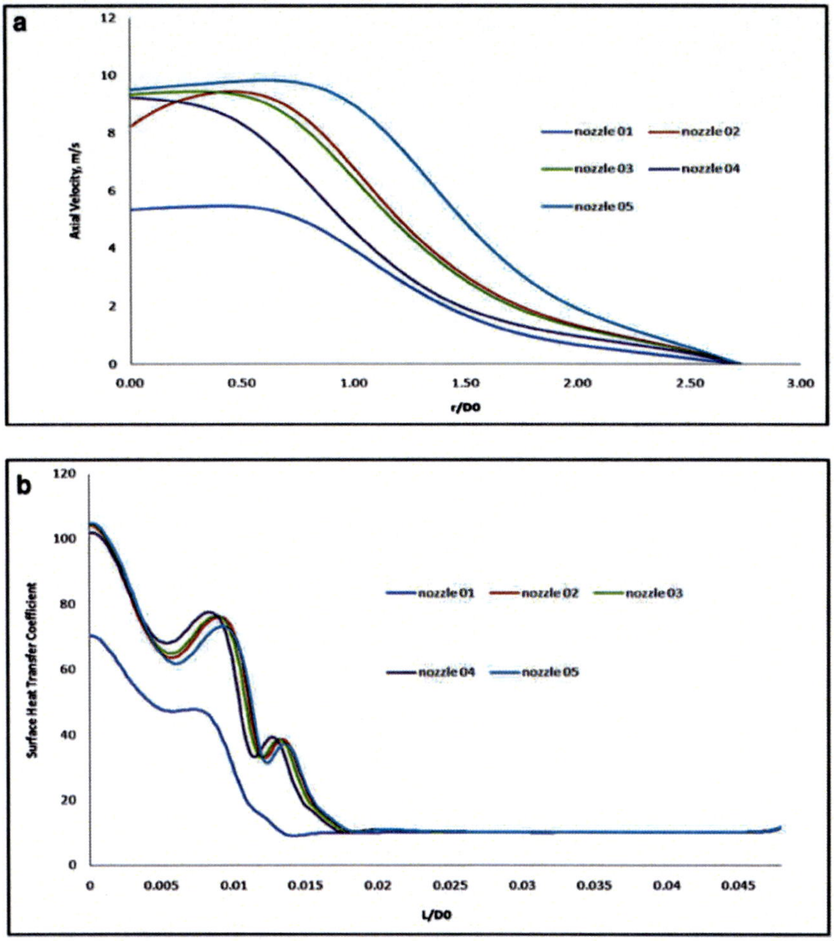

Fig. 9 **a** Axial velocity versus non-dimensional radial distance **b** surface heat transfer coefficient versus length of the heated plate for the different type of nozzles

References

1. Lv Y-W, Zhang J-Z, Y Shan, Tan X-M (2014) Numerical investigation for effects of actuator parameters and excitation frequencies on synthetic jet fluidic characteristics. Sens Actuat A 219:100–111
2. Y-H Liu, Chang T-H, Wang C-C (2016) Heat transfer enhancement of an impinging synthetic jet using diffusion-shaped orifice. Appl Therm Eng 94:178–185
3. Zhang PF, Wang JJ, Feng LH (2008) Review of zero-net-mass-flux jet and its application in separation flow control. Sci Chin Ser E Technol Sci 51(9):1315–1344
4. Crook A, Sadri AM, Wood NJ (1999) The development and implementation of synthetic jets for the control of separated flow. 17th applied aerodynamics conference, AIAA

5. Tuckerman DB, Pease RF (1981) High-performance heat sinking for VLSI. IEEE Electron Dev Lett 2(5):126–9
6. Ziade P et al (2018) A numerical study on the influence of cavity shape on the synthetic jet performance. Int J Heat Fluid Flow 74:187–197
7. Miro A, Soria M, Cajas JC, Rodriguez I (2019) Numerical study of heat transfer from a synthetic impinging jet with a detailed model of the actuator membrane. Int J Therm Sci 136:287–298
8. Kordík J, Trávníček Z (2018) Novel nozzle shapes for synthetic jet actuators intended to enhance jet momentum flux. Actuators
9. Timoshenko S, Woinowsky-Krieger S (1959) Theory of plates and shells. 2nd edn, McGraw–Hill, London, pp 67–69
10. Mane P, Mossi K, Rostami A, Bryant R, Castro N (2007) Piezoelectric actuators as synthetic jet: cavity dimension effects. J Intell Mater Syst Struct 18:1175

Atomization Characteristics of a Porous Injector

M. Jegan and M. Vadivukkarasan

List of symbols

D_l	Diameter of the liquid at the tip of the nozzle (mm)
D_n	Summation of all 1 mm diameter porous hole of the injector (mm)
l_b	Breakup length (mm)
l_i	Jet impact length from the tip of the nozzle (mm)
M	Momentum flux ratio
Re_l	Liquid based Reynolds number
U_a and U_l	Velocity of air and liquid (m/s)
We_e	Aerodynamic gas based Weber number
θ	Spray angle (°)
ν_l	Kinematic viscosity of the liquid (m^2/s)
ρ_a and ρ_l	Density of air and liquid (kg/m^3)
σ	Surface tension (N/m)

1 Introduction

The present study focuses on the preliminary investigation on the atomization characteristics of a newly developed nozzle called air-assisted porous injector. This injector ejects multiple circular liquid jets while airstream is allowed to interact with it radially via multiple holes located at its circumference. Although the analysis was conducted

M. Jegan · M. Vadivukkarasan (✉)
Department of Mechanical Engineering, National Institute of Technology Puducherry, Karaikal, India
e-mail: m.vadivu@nitpy.ac.in

M. Palanisamy et al. (eds.), *Theoretical, Computational, and Experimental Solutions to Thermo-Fluid Systems*, Lecture Notes in Mechanical Engineering,
https://doi.org/10.1007/978-981-33-4165-4_19

using a simple set of experiments and simple design, the quantitative description observed gives us high confidence to apply to a wide range of applications, especially cryogenic engines to achieve better and stable combustion.

The stages of atomization can be classified into (a) a bulk liquid forming into a cylindrical jet or sheet and (b) those cylindrical jet breaking up into several ligaments and drops of different sizes [1]. The former is referred to as primary atomization and the latter as secondary atomization. The different mechanism of interaction of liquid and air with its exchange of momentum determines the dominant instabilities, spray angle, penetration length, distribution, and size of the drops. Depending upon this interaction, the injector or atomizer can be classified into three categories, namely, (i) coaxial atomizer [2], (ii) air-assist atomizer [3] and (iii) airblast atomizer [4]. These are the traditional injectors that rely on the interaction between the liquid and air in the axial direction to enhance the atomization processes. It is well documented that depending upon the values of relative velocities between the liquid and airstream atomizing characteristics falls in one of the above three categories. From the above mentioned existing literature, it is well known that the key ingredient to obtain the smaller drops is to operate an injector at lower liquid velocity and higher air velocity. These smaller drops play a crucial role in the further downstream process in the combustion application for evaporation and reaction. In contrast to these atomizers, a porous injector was introduced for the combustion application to reduce the throttling capability [5–10]. The porous injector relies on the interaction of the airstream in the radial direction.

Typically, coaxial porous injector comprises of two classes. (i) parallel shower-head configuration and (ii) coaxial injector with the porous recessed region. The first class of injector consists of a porous faceplate for the injection of the fuel and a large number of small diameter ports for the injection of the oxidant. These ports are arranged in a non-interacting fashion. The liquid and gas interface undergo interaction immediately after injection. A large number of the small diameter of liquid ports emanates a jet in an oxidant significantly increases the surface area between the fuel and oxidant compared to typical shear coaxial injectors. The basic configuration of the second class of coaxial-porous injector is similar to that of the conventional shear flow coaxial injector. The fuel (liquid) is injected into the combustion chamber in a similar fashion; however, the introduction of oxidant (gas) is different from one another. Due to this, two kinds of forces act on the interface, the shear force and radial force. The atomization is caused due to the transfer of energy by the radial flow of gas to the liquid column as well as the shear force from the axial developed gas. This process would lead to a significant increase in mixing.

In addition, both the classes of porous coaxial injector supplement other advantages such as uniform mixing of spray in the axial as well as radial direction, and the pressure drop can be adjusted by changing the porous size and overcome certain disadvantages that occur in coaxial injectors, in general. However, both the class of coaxial porous injectors were found to be promising candidates for a proper injection system over a conventional coaxial injector. However, these atomizers have received scarce attention from its initial days. Most of the previous studies have been focused on characterizing the combustion process and have not explored the characteristics of atomization in any detail. In this paper, the merits of the above-discussed injectors are exploited and developed a novel design. The design is also based on the theoretical concept by Vadivukkarasan and Panchagnula [11, 12], where this work emphasizes the radial motion destabilization is much more effective than the axial motion. The global parameters or the macroscopic characteristics such as spray angle and breakup length were investigated with the simple experimental setup. To explore the characteristics of the atomizer, the spray imaging techniques have been utilized, the images of near-field spray structure have been acquired. The obtained images have been used to measure the spray angle and breakup length.

2 Experimental Investigation

The primary goal of the current study is to test the newly developed air-assisted porous injector and observe the global parameters such as breakup length and the spray angle. These parameters can be observed using visualization techniques. Henceforth, a test facility is constructed to capture the relevant parameters and it is described below in detail.

2.1 Background for Injector Design

Four kinds of injectors were chosen for this study initially, although this study emphasizes mainly the air-assisted porous injector. All the injectors that have been used for this investigation were designed by using a 3D modeling software called Catia V5 and fabricated through 3D printing.

The four injectors are (a) coaxial injector as depicted in Fig. 1a, where the liquid and gas enter into the injector through the different passage and interacts with each other in the axial direction nearer to the exit of the injector. Secondly, (b) porous injector, similar to the former as shown in Fig. 1b, but the liquid enters in the axial direction, and the air enters in the radial direction via multiple holes and interacts at the exit of the injector. It is to believe that the breakup mechanism and the spray structures are vastly different between the coaxial and porous injector. Thirdly, as shown in Fig. 1c, the parallel showerhead type injector, similar to the coaxial injector, but the liquid is ejected through multiple holes before interacting with the air. This helps to enhance the surface area interaction in the multifold. As previously mentioned, the coaxial injector was explored in detail, while the literature on the porous injector and parallel showerhead type is very scarce. We have explored these two injectors in brief, limited to the operating conditions chosen. With the insights gained from these three injectors with our pilot experiments, we developed a new design as shown in Fig. 1d.

As can be seen that this injector is the combination of a porous and parallel showerhead type injector. In a nutshell, the motivation of a newly designed novel

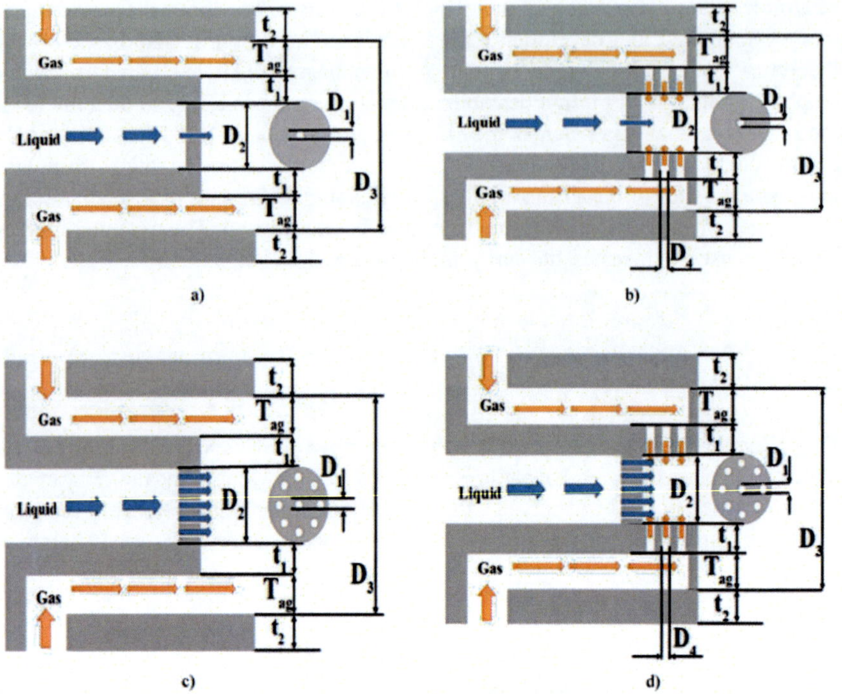

Fig. 1 Cross-sectional view of the injectors. **a** Shear coaxial injector, **b** coaxial porous injector, **c** parallel – showerhead injector and **d** air-assisted porous injector. The blue arrows indicate the pathway of a liquid, whereas the orange arrows indicate the pathway of the gas (air)

air-assisted porous injector arises from the existing design of a porous and parallel showerhead to exploit the advantages of these two. With this, we developed an injector which is the combination of these two, in other words, an atomizer that consists of multiple circular holes for the flow of liquid (parallel showerhead type) along the centerline of the injector, and the air is allowed to interact with liquid in the radial direction (porous injector), again through multiple holes.

The motivation to design such a nozzle is (i) to increase the surface area interaction between the thin circular jet and the air and (ii) to break up those jets via a radial motion of air. It is confirmed that without the presence of air, the atomizer ejects 7 circular jets similar to the single circular jet in the coaxial atomizer. Intuitively speaking, the air which enters in the radial direction in the air-assisted porous injector occupies the space in between the multiple liquid jets and pushes the liquid outwards, thus altering the behavior of spray characteristics. This concept is much different from the jet in a crossflow, wherein a jet in crossflow, the air is allowed to pass from one direction to the other, whereas in the air-assisted porous injector, the air is allowed to strike the liquid jet uniformly in the radial direction. Therefore, it is of our interest to understand the atomization characteristics of such an injector and find its applicability in the rocket engines.

2.2 Experimental Setup

A brief schematic diagram of the experimental setup is shown in Fig. 2, and its geometric details are displayed in Table 1. For the investigation of these four atomizers, the experimental setup was built-up with the injector, air compressed cylinder,

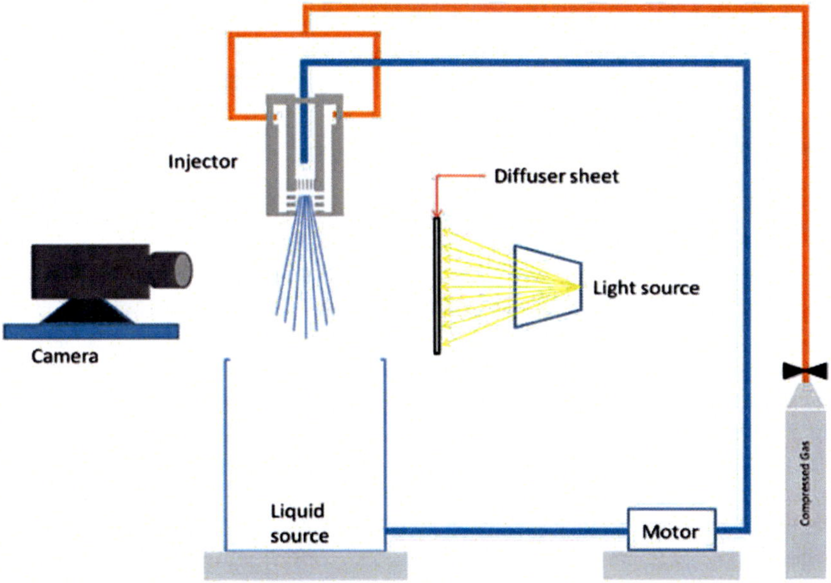

Fig. 2 Schematic diagram of the experimental setup

Table 1 Configuration details of the injector geometry

Parameter	Description	Values (in mm)
D_1	Diameter of the liquid passage(s) at the exit	1
D_2	Initial diameter the liquid passage	5
D_3	Outer diameter of the gas passage	9
D_4	Diameter of the air passage(s) at the exit	1
t_1	Thickness of the primary cylindrical portion	2
T_{ag}	Gas annular gap	2
t_2	Thickness of the secondary cylindrical portion	3

liquid tank, control valves, flow meters, and pressure gauges. The test of investigation at different operating conditions was captured by Nikon D3400 DSLR camera using a shadowgraph photography method by positioning the camera in front of the injector spray and diffuser sheet at the backside followed by LED light source. The measuring instrument had been used to fix the operating range to run the test. Since it is a preliminary study and the objective is to capture only a breakup length and spray angle in a steady spray, the camera which we used is sufficient. However, if one needs to capture the frequency of oscillation of the liquid jet or instabilities associated with the destabilization of liquid jet or the morphology of the breakup mechanism, the high-speed camera that will be operated on the several thousand frames per second is required. We again emphasize the experimental setup chosen in this study is sufficient to capture the two of the global parameters and that will be discussed in the later section

2.3 Operating Conditions

As mentioned earlier, the present work focuses on the two global parameters, spray angle (θ) and breakup length (l_b) of an air-assisted porous injector. The dimensionless quantities that govern these parameters are: (i) the liquid-based Reynolds number, $Re_l = U_l D_l / \nu_l$, where the competition is between the inertial force and viscous force, (ii) the aerodynamic gas-based Weber number, $We_a = \rho_a (U_a\text{-} U_l) 2D_l/\sigma$, where the competition is between the inertial force and the surface tension force, and (iii) momentum flux ratio, $M = \rho_a U_a^2 / \rho_l U_l^2$ represents the competition between air and liquid momentum flux. It is also otherwise referred to as a dynamic pressure ratio. Here, a and l denote air and liquid, respectively, U represents the velocity, ν and σ indicate kinematic viscosity and surface tension, respectively. The density of air and the liquid under the atmospheric conditions are assumed to be 1 and 1000 kg/m^3, respectively. These preliminary experiments were operated for a range of Reynolds number, $Re_l = 1050 - 3150$, Weber number, $We_a = 0$–20, and the Momentum flux ratio, $M = 1$–4. Note that the We_a is chosen less compared to the real operating condition due to the few practical constrains of having a limited size of compressor that can be operated only on the not-so-high air velocity. It should be mentioned to conduct the test on the real operating conditions and it is an expensive and very challenging task. Nevertheless, M appears to be in the range of real operating conditions and it is due to the choice of Re_l.

The preliminary cold flow test of the air assisted porous injector is investigated for a wide range of operating conditions. The preliminary investigation on all the four injectors was performed to understand the atomization performance of each one. Similar to the air-assisted porous injector, the shear-coaxial injector, the showerhead injector and coaxial porous injector were also been tested with the same operating conditions to compare the atomization characteristics. However, the result was presented here is only for newly modified air-assisted injectors owing reason that the other injectors are already tested by a few authors, at least at a preliminary level. The effect of breakup length (l_b) and the spray angle (θ) on the momentum flux ratio (M) for an air-assisted porous injector will be discussed in the next section.

2.4 Air-Assisted Porous Injector Versus Other Injectors

Initially, all the four injectors were operated, and the results were obtained. In a visualization inspection, it was found out that the increase in the airflow rate for a constant liquid flow rate results in accelerating the process alone, and it does not result in altering either the breakup length and spray angle for the coaxial injector and the parallel showerhead injector. This could be also due to the limitation of the operating conditions chosen. However, the porous injector and the air-assisted porous injector showed a significant change in the breakup length and the spray angle for the same operating conditions (even with the same limitation of the flow conditions). This is because of the reason due to the fact that the air interacts with the water in the radial direction that results in vastly different spray characteristics. In these two porous injectors, the air which enters in the radial direction occupies the space in between the single or multiple liquid jets and pushes the liquid outwards, thus altering the behavior of breakup length and spray angle. Since the porous injector (with a single hole at the centerline) can be treated as a special case of the newly developed air-assisted porous injector, we here present only the results of the air-assisted porous injector.

2.5 Spray Imaging

The collage of the near field images of the air-assisted porous injector is captured and has been displayed in Fig. 3. It can be seen that those visualization images clearly show the effect of breakup length and spray angle for a limited range of operating conditions. Although the raw images are presented here, the values of breakup length and the spray angle is calculated from the mean of the images obtained at the different instants for each injector at different operating conditions. The value of these two parameters was obtained from the instantaneous images by using the in-house MATLAB code. Recall that the increase in U_a and the decrease in the U_l increases the value of M according to the definition followed in this study. The value of the Weber number and Reynolds number can be calculated based on the values of U_a and U_l. From these raw images, it can be observed that for the case of $U_a = 0$, the increase in the U_l has no significant effect in either the l_b or θ. On the other hand, for a constant U_l (see the first column of the collage), the increase in the U_a has a significant effect. It can be observed that the increase in U_a decreases the l_b and increases the value of θ. Consider the second column, where the U_l is maintained as constant while U_a is increased. As can be seen, an increase in U_a decreases the l_b and increases the spray angle (θ). But notice that there is no effect from increasing

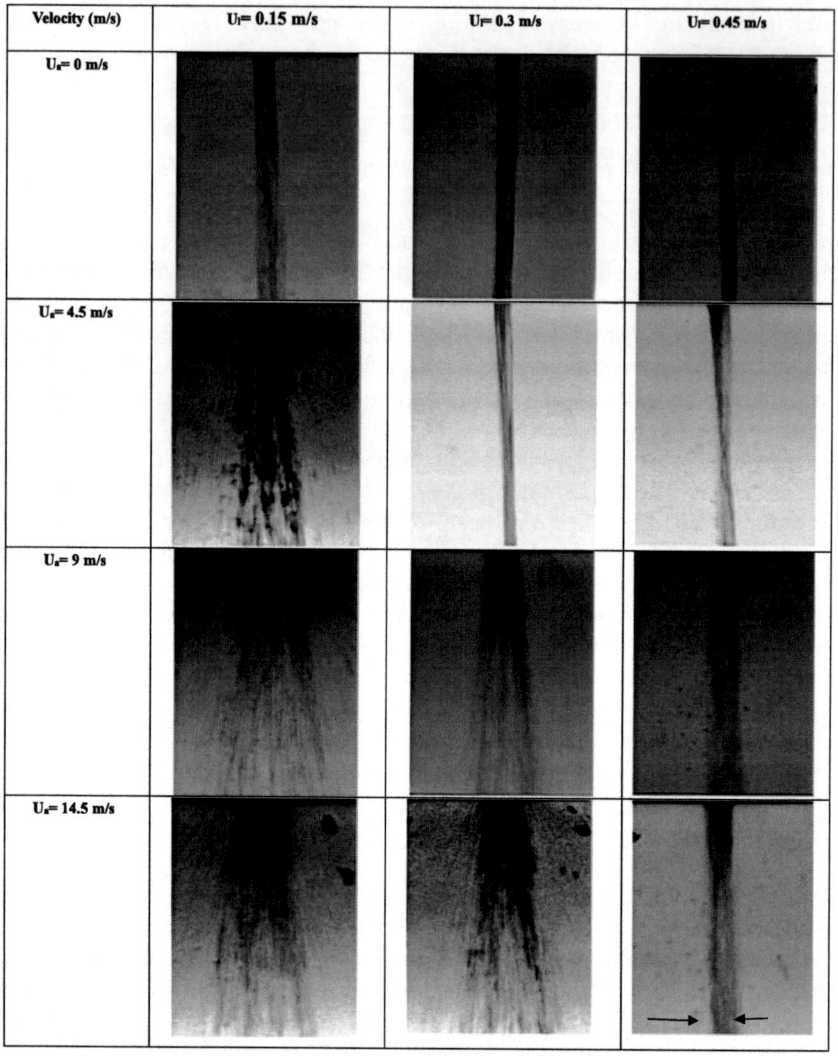

Fig. 3 Instantaneous (representive) images of an air-assisted porous injector at different operating conditions. The arrow mark indicates the cone-angle detection

the U_a from 0 to 4.5 m/s. It is due to the fact that the quantity of air that interacts with air is less than the sufficient to destabilize the liquid jet. Further increase in U_a from 4.5 m/s to higher values, the effect of U_a is pronounced. Likewise, for the case of $U_l = 0.45$ m/s, only $U_a = 9$ and 14.5 m/s show a little effect on the l_b and θ. It is an obvious fact that for a particular amount of liquid, a required amount of air is necessary to show a significant effect.

In other words, $\rho_a U_a^2$ should be always greater than $\rho_l U_l^2$, that is, M should be preferred such that it should be always greater than 1. As a closing remark, as an increase in the value of momentum flux ratio (or the dynamic ratio) between the air and the liquid, it was observed that the penetration length decreases and the spray angle increases. In general, it is due to the fact that the breakup length decreases and the spray angle increases because due to the higher the dynamic pressure difference between the liquid and the air accelerates the process and in turn increases the interaction area between the liquid and air interface, causing the instabilities in both the axial and radial directions This is the artifact of enhancement of the shear via axial and radial motion.

Consequently, this behavior is the indication of yielding the smaller size of droplets along the downstream process as one increases the value of M. As mentioned earlier, all the experiments were performed under atmospheric pressure and temperature. The results could significantly vary if the same experiments were conducted in the elevated pressure and temperatures. The above-discussed inferences can be seen through the plots, which are discussed in the next section.

3 Results and Discussion

In this section, the result is presented in the form of graphs for the breakup length (l_b) and spray angle (θ) with respect to the momentum flux ratio (M) alone, though it can be plotted with respect to varying Reynolds number and Weber number, otherwise. It is well known that the global parameters such as l_b and θ are of practical interest and used to characterize any kind of spray. Remember that the behavior of these two parameters occurring on the course of the primary atomization process causes a significant difference during the further downstream processes such as secondary atomization, evaporation, and reaction.

The experiments were performed for $M = 1$–4 for a limited $We_a = 0$–20. The reason is attributed to the limitation of the experimental conditions, that is, beyond $We_a = 0$–20, the breakup morphology differs, and the liquid jet tends to oscillate with high frequency, thus altering the breakup mechanism. Therefore, we limit our scope of this study to the $We_a = 0$–20 and varying $M = 1$–4.

3.1 Breakup Length

The distance from the nozzle tip to a point in the vertical direction along the spray direction at which the liquid jet passes through the nozzle tip is referred to be breakup length. It is confirmed that the measurement was taken up the distance where the liquid is intact and before the liquid is atomized into droplets. In this subsection, the l_b is captured for a limited range of M. Here, l_b represents the average breakup length obtained by the sequence of images of the steady spray. Also, l_b is l_i/D_n, where l_i

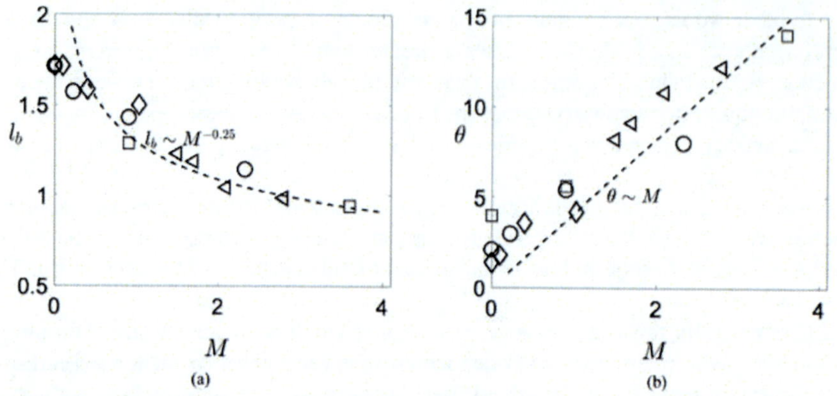

Fig. 4 Effect of **a** breakup length (l_b) and **b** spray angle (θ) for different values of momentum ratio (M). The symbol 'circle' indicates the $U_l = 0.15$ m/s for the different values of U_a. Likewise, the symbols 'square' and 'diamond' denote the $U_l = 0.3$ and 0.45, respectively. These spray patterns for these three flow conditions are also shown in Fig. 3 via shadowgraphic images. As can be seen, $l_b \sim M^{-0.25}$ and $\theta \sim M$

represents jet intact length from the tip of the nozzle to the point before it atomize, and D_n indicates the summation of all the porous hole diameter of the injector. In general, from the previous works on the injector, it is well known that the breakup length will decrease if the momentum flux ratio is increased, and the same is also observed in our study. Figure 4a depicts the behavior of l_b for different values of M. From this plot, and it can be observed that for a particular velocity of the liquid, an increase in the air velocity decreases the breakup length. At the same time, it was also observed that for a particular velocity of the liquid, the value of the velocity of air should be greater such that $M > 1$ for the breakup to occur. In general, as the existing literatures point out, we also observed l_b decreases as M increases and l_b scales as $M^{-0.25}$. Here, the variation in the breakup length with respect to M indicates the corresponding variation in the size of the droplet and its distribution downstream of the injector.

3.2 *Spray Angle*

Following the breakup length, the spray angle is the utmost parameter to be known. The measurement of the angular deviation of the liquid jet ejected out the tip of the injector with respect to the vertical axis is referred to as a spray angle. The spray angle, herein, is referred to as the angle between two straight lines drawn along the jet from the injector exit. The spray angle can be controlled by the operating conditions as well as the internal geometry of the injector. Figure 4b shows the variation of the spray angle for the different values of M. As can be seen, as the M increases, the spray angle also increases. The results are consistent with the existing literatures. Again here, the variation in the spray angle with respect to momentum ratio indicates the corresponding variation in the size of the droplet and its distribution downstream of the injector.

4 Conclusion

A newly developed air-assisted porous injector was characterized by the water and air as a working fluid for a limited range of operating conditions for Reynolds number, $Re_l = 1050–3150$, Weber number, $We_a = 0–20$ and the Momentum flux ratio, $M = 1–4$. The present study focuses on exploring the applicability of the newly developed injector with the help of visualization techniques. The results showed the effect of breakup length and spray angle on the momentum flux ratio is consistent with the existing literature. From the results discussed, it could be concluded that our novel design of air-assisted porous injector will be efficient in atomizing, and it appears to be a promising candidate for the application of the rocket engines where the water and the air are replaced by the aviation fuel and the liquid oxidizer. With these promising results, this study can be extended to high-speed imaging that can further provide much more insights about the quality of the spray characteristics. Further comprehensive and rigorous investigations may unravel the morphology of the primary and secondary atomization process.

Acknowledgements This work was supported by Department of Science and Technology INSPIRE Faculty Award/Batch-12/2017 (IFA17-ENG220).

References

1. Lefebvre AH (1989) Atomization and sprays. In: Hemisphere. CRC Press, New York
2. Lasheras JC, Hopfinger EJ (2000) Liquid jet instability and atomization in a coaxial gas stream. Annu Rev Fluid Mech 32:275–308
3. Marmottant P, Villermaux E (2004) On spray formation. J Fluid Mech 498:73–111
4. Park J, Huh KY, Li X, Renksizbulut M (2004) Experimental investigation on cellular breakup of a planar liquid sheet from an air-blast nozzle. Phys Fluids 16:625–632
5. Kim DH, Shin JH, Lee IC, Koo JY (2012) Atomizing characteristics of coaxial porous injectors. J ILASS 17(1):35–44
6. Suslov DI, Hardi J, Knapp B, Oschwald M (2016) Optical investigation of the LOX-jet disintegration processes at high pressure conditions in a LOX/H2 single coaxial injector combustion chamber. Space Propulsion
7. Kim D, Lee I, Koo J (2014) Macroscopic spray characteristics of a porous injector. J Atomization Sprays 24(1):1–22
8. Lee K, Kim D, Koo J (2016) Disintegration process and performance of a coaxial porous injector. JTherm Sci 25:394–401
9. Zhukov VP, Suslov DI (2015) Wall heat fluxes at different oxidizer to fuel ratios in rocket combustion chamber with porous injector head (EUCASS, 2015), 1–7

10. Lux J, Suslov DI, Haidn O (2008) On porous liquid propellant rocket engine injectors. Aerosp Sci Technol 12:469–477
11. Vadivukkarasan M, Panchagnula MV (2016) Helical modes in combined Rayleigh-Taylor and Kelvin-Helmholtz instability of a cylindrical interface. Int J Spray Comb Dyn 8(4):219–234
12. Vadivukkarasan M, Panchagnula MV (2017) Combined Rayleigh-Taylor and Kelvin-Helmholtz instabilities on an annular liquid sheet. J Fluid Mech 812:152–177

Numerical Simulation of PCM-Based Heat Sink with Plate Fins for Thermal Management of Electronic Components

Anuj Kumar, Rohit Kothari, Pushpanjay K. Singh, M. P. Paulraj, Santosh K. Sahu, and Shailesh I. Kundalwal

1 Introduction

The demand of advanced features, functionality, higher processing speed, and compactness of portable electronic devices leads to higher heat generation, which in turn increases the operating temperature of the electronic devices. Higher processing speed and highly dense electronic circuitry are the two key factors of generating higher heat flux in electronic devices, such as mobile microprocessor, LEDs, computer CPU, and missile control systems. If the rise in temperature of the electronic devices exceeds the allowable temperature, their efficiency and performance get affected or even they become the key reasons for their failure [1, 2]. Conventional heat exchange techniques such as air and water jet cooling system transfers higher heat flux at the expanse of increased power consumption, space, and extra maintenances.

Therefore the development of novel cooling technology is a prime need to maintain the recommended operating temperature of the electronic devices with reduced power consumption, less space, and lower cost. Latent heating thermal energy storage system (LHTESS) using phase change material (PCM) is very prominent to reserve the thermal energy released by electronic circuitry due to its higher energy storage density and ideally isothermal solid–liquid phase transformation. PCM consumes the waste heat released by electronic components during its working in the form of latent heat, gets liquefied and releases this heat to the ambient when the device is in off mode. PCM utilizes latent heat of fusion which is thermal energy stored during the melting process to achieve the required cooling effect. PCM possesses amicable thermophysical and chemical properties namely, high latent heat, high specific heat,

A. Kumar (✉) · R. Kothari · P. K. Singh · M. P. Paulraj · S. K. Sahu · S. I. Kundalwal
Department of Mechanical Engineering, IIT Indore, 453552 Indore, India
e-mail: kumarguptaanuj93@gmail.com

M. Palanisamy et al. (eds.), *Theoretical, Computational, and Experimental Solutions to Thermo-Fluid Systems*, Lecture Notes in Mechanical Engineering,
https://doi.org/10.1007/978-981-33-4165-4_20

high energy density on volumetric basis, good thermal stability, congruent melting, non-toxicity, and non-corrosiveness [3, 4], which makes it an ideal material to employ it in LHTESS. However, the major concern with PCM is its thermal conductivity. PCM possesses very low thermal conductivity which enables slow melting and solidification of PCM. In order to increase its melting rate, various thermal conductivity enhancers (TCEs) [5–7] such as metallic fins, metallic nanoparticle and metallic foams can be integrated.

Various experimental and numerical investigations [8–14] have been carried out in recent years to better understand the heat transfer and fluid flow phenomena during PCM melting and solidification. Arci et al. [8] carried out numerical investigation on melting of paraffin wax integrated with Al_2O_3 at numerous volume fractions in a rectangular enclosure with partially active walls. It was reported that integration of Al_2O_3 in pure PCM improves the thermophysical properties of PCM and orientation of the wall significantly affects the melting rate and stored energy in the PCM. Effect of TCEs on melting and solidification of PCM for solar chimney applications was performed out by Yongcai et al. [9]. They studied four different configurations of TCEs such as horizontal plate fins, vertical fin, honeycomb structure and square cell structure experimentally. It was reported that melting time of PCM reduces by 8, 12, 14.5, and 16% for vertical fin, horizontal fin, honeycomb structure, and squre cell structure, respectively as compared to pure PCM sample. Both experimental and numerical studies were carried out by Dhaidan et al. [10] on melting of PCM/nanoparticle suspension in a square container subjected to constant power input. They concluded that optimal concentration of nanoparticle in pure PCM has enough capacity to improve the traditional thermal energy system. Bondareva et al. [11] simulated 3-D natural convection melting in a cubical cavity subjected to constant temperature from bottom wall. They analyzed the effect of Rayleigh number and dimensionless time for Prandtl and Stefan numbers on streamline, isotherm profile of temperature and velocity as well as mean Nusselt number at the heat source. Experimental investigation of the heat transfer through a vertical stack of rectangular cavities filled with PCM was carried out by Soares et al. [12]. They concluded that conduction mode of heat transfer is dominant during charging of PCM without any thermal stratification. Chenzhen et al. [13] carried out numerical investigation on PCM heat transfer inside the PCM-based rectangular thermal energy system. Heater is placed on sidewalls and RT 42 is taken as PCM. Different orientation of fins was studied numerically. Stritih [14] performed experimental investigation on heat transfer characteristics with finned surface to study the melting and solidification processes of PCM. A correlation between Nusselt number and Rayleigh number was proposed.

It is evident from literature that, significant work has been reported on melting and solidification of PCM with internal fins experimentally. However, limited study is found on numerical investigation of variation of number of fins in heat sink for thermal management of electronic components. Moreover, a closed PCM system using the VOF model to characterize the PCM-air gap problem is scarcely discussed. Therefore in the present study 2-D numerical model for melting of PCM in a finned heat sink filled with PCM is developed. Heat sink and fins are made up of aluminum and paraffin wax is taken as PCM. Numerical simulations are performed to study the effect of number of fins on the melting of PCM.

2 Physical Model

The 3-D model of PCM filled finned heat sink for thermal management system reduced into 2-D geometric model for simplified numerical simulations. The schematics diagram of different heat sinks filled with PCM are presented in Fig. 1a–c. The plate fin made up of aluminum is vertically placed on the aluminum heat sink. The overall dimension of heat sink is taken as constant for each case ($110 \times 110 \times 25$ mm^3). The depth of heat sink cavity is taken as 20 mm. A gap of 2 mm is left unfilled to facilitate the volume expansion of the PCM upon melting. The heater is considered on the bottom surface of the heat sink. A constant input heat flux such as 1.00 kW/m^2, 1.50 kW/m^2 and 2.00 kW/m^2, respectively is supplied from the bottom surface of the heat sink. Top and sidewall of the heat sink is considered to be adiabatic. The detailed dimension of each heat sink design is depicted in Fig. 2. The thickness of

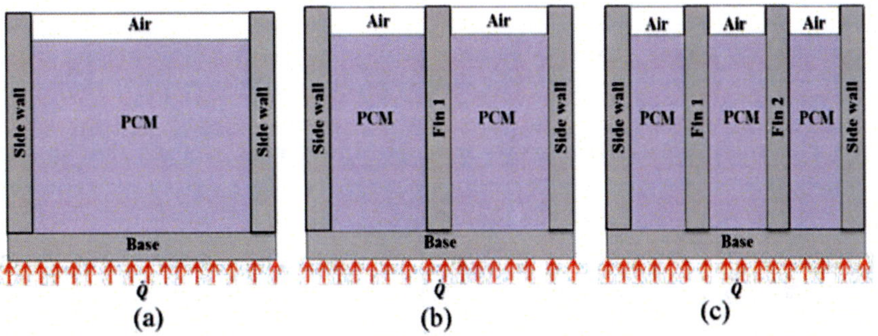

Fig. 1 PCM-based heat sink **a** without fin, **b** one plate fin, and **c** two plate fins

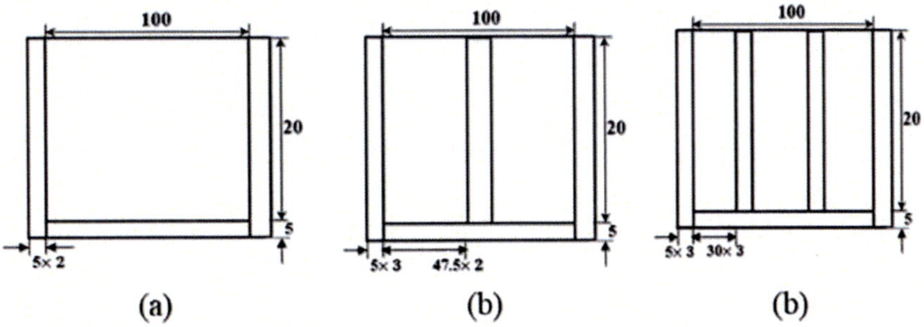

Fig. 2 Dimensions of heat sink design: **a** without fin **b** one plate fin and **c** two plate fins

Table 1 Thermophysical properties of paraffin wax and aluminum

Property	Paraffin wax	Aluminum	Air
Solidus/Liquids Temperature (°C)	58/62	–	–
Density (solid) (kg/m^3)	900	2719	1.225
Thermal conductivity (W/m K)	0.21	202.4	0.024
Dynamic viscosity (Pa s)	0.0127	–	1.8e-5
Specific heat (J/kg.K)	2890	871	1006
Heat of fusion (J/kg)	194,200	–	–
Thermal expansion coefficient (K^{-1})	0.0001	–	–

fins is taken as 5 mm for each heat sink design. A commercially available paraffin wax, from Sigma Aldrich USA is selected as PCM for thermal energy storage. Thermophysical properties of PCM, aluminum heat sink and air are described in Table 1 [15].

3 Assumptions and Governing Equation

Here, enthalpy porosity model [16] approach is employed for modeling the solid-liquid phase change during the PCM melting process. In this approach, the whole computational domain is considered as porous zone and porosity of each cell is associated with melt fraction varying from 0 to 1. The melt fractions are obtained at different time interval. Various assumptions are accounted while developing the numerical model and are illustrated below.

(a) Flow of liquid PCM and air are considered to be laminar, incompressible, and Newtonian.
(b) Unsteady and two-dimensional.
(c) Constant thermophysical properties.
(d) Radiation heat transfer is neglected.
(e) The side wall and top edge of heat sink are considered as adiabatic.

Boussinesq approximation is employed to incorporate the effect of natural convection of PCM melting, which is valid for density variation of buoyancy force:

$$\rho = \frac{\rho_l}{(\beta(T - T_l) + 1)} \tag{1}$$

With the above methodologies and assumptions Ansys Fluent software is utilized to solve the following conservation equation for PCM-filled finned heat sink.

Continuity equation:

$$\frac{\partial \alpha_n}{\partial t} + u_i \frac{\partial \alpha_n}{\partial x_i} = 0 \tag{2}$$

Momentum equation:

$$\frac{\partial(\rho u_i)}{\partial t} + \frac{\partial}{\partial x_j}(\rho u_j u_i) = \mu\frac{\partial^2 u_i}{\partial x_j \partial x_j} - \frac{\partial p}{\partial x_i} + \rho g_i + S_i \quad (3)$$

Energy equation:

$$\frac{\partial}{\partial t}(\rho h) + \frac{\partial}{\partial x_i}(\rho u_i h) = \frac{\partial}{\partial x_i}\left(k\frac{\partial T}{\partial x_i}\right) \quad (4)$$

where, α_n nth fluids fluid's volume fraction in the computational cell, component, x_i is the Cartesian coordinate, u_i is the velocity,ρ, k, μ are the density, thermal conductivity and, dynamic viscosity respectively. S_i is the momentum source term and h is the specific enthalpy which is defined as the sum of sensible and latent heat enthalpy: $h = h_{\mathrm{ref}} + \left(\int_{T_{\mathrm{ref}}}^{T} C_{\mathrm{p}}\mathrm{d}T\right) + \gamma L$. Where, h_{ref} is the reference enthalpy, T_{ref} is the reference temperature, C_{p} is the specific heat, L is the specific enthalpy of melting, and γ is the melt fraction during solid-liquid phase change over a temperature of $T_{\mathrm{s}} < T < T_{\mathrm{l}}$ and is defined by following relations.

$$\text{If}\, T < T_s \gamma = 0, \text{If}\, T > T_L\, \gamma = 1,\ \text{If}\, T_s < T < T_L\, \gamma = \frac{T - T_s}{T_L - T_s} \quad (5)$$

The source term in momentum equation is given by $S_{\mathrm{i}} = A(\emptyset)u_{\mathrm{i}}$. Here, $A(\emptyset)$ is the "porosity function" defined by Brent et al. [17] to make the momentum equation mimic Karman-Kozeny equations for flow in porous media: $A(\emptyset) = C_{\mathrm{mush}}\left(1 - \emptyset^2\right)/\left(\emptyset^3 + \varepsilon\right)$ where C_{mush} is mushy zone constant which reflects the behavior of PCM melting and is set to be 10^5 in this study. ε is the small computational constant and used to avoid division by zero and set to be 0.001 in this study.

4 Numerical Solution and Validation

Conservation equations such as mass, momentum, and energy are solved by employing finite volume method (FVM), using Ansys Fluent software. In order to describe the PCM-air gap with a moving internal interface but without interpenetration of the two-fluid VOF model has been utilized. The SIMPLE algorithm is employed for pressure-velocity coupling and Presto! Scheme is selected for pressure correction equation. The momentum and energy equations are discretized using second-order upwind scheme. The convergence is checked at each time step, with convergence criterion of 10^{-5} for continuity, 10^{-6} for momentum and 10^{-8} for energy equation. The time step and grid size independent test have been carried out, to avoid their effects on numerical accuracy. By careful comparisons, the mesh with 21,923

Fig. 3 Validation of present model with experimental result of Hosseinizadeh et al. [18]

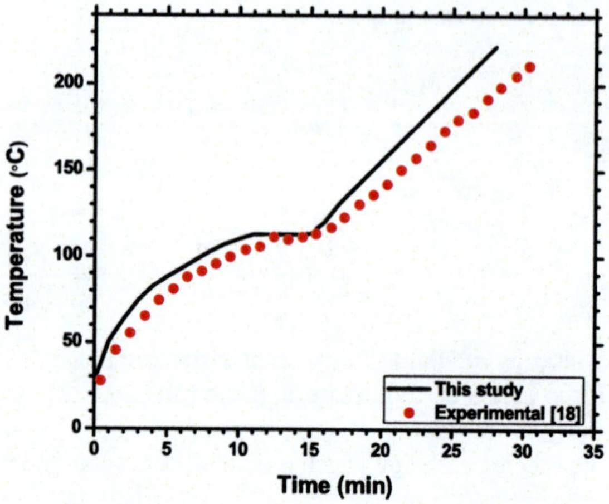

cells and time step of 0.7 s are taken in this study. All the meshes are generated in quadrilateral cells.

To validate the present numerical model, the result of the present study is compared with the experimental results of Hosseinizadeh et al. [18] and presented in Fig. 3. The current validation is based on heat sink with three fins subjected to 45 W of input heat from the bottom of the heat sink. The temperature profile with time is obtained and compared. It is found that maximum differences between experimental and numerical data are lower than 12%. This discrepancy in experimental and numerical results might be due to the variation in PCM properties with temperature in the experimentation, whereas in numerical investigation, PCM properties are assumed to be independent of temperature. In general, the present developed numerical model agrees well with the experimental result and can be devoted for the phase change simulation.

5 Results and Discussion

5.1 *Effect of Heat Flux on Heat Sink Configuration*

Performance of PCM-based finned heat sink is evaluated at various heat flux values such as 1.00 kW/m^2, 1.50 kW/m^2 and 2.00 kW/m^2. Figure 4a–b presents the effect of heat flux on the thermal performance of plain and one fin heat sink filled with PCM. A closer look to Fig. 4a reveals that at 1.0 kW/m^2, the phase change begins after 4130 s, while for heat flux of 1.50 kw/m^2 and 2.00 kW/m^2 value it starts after 2730 s and 1960 s, respectively. It can also be seen that latent heating phase duration decreases with an increase in heat flux values. Similar results were obtained by [18] in their experimental investigation.

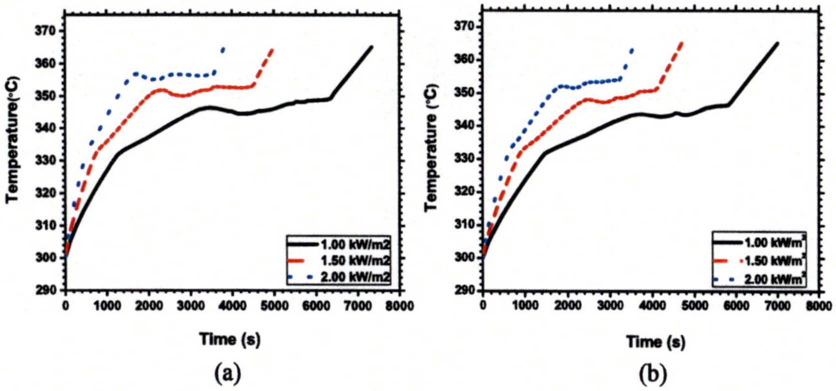

Fig. 4 Effect of heat flux **a** unfinned heat sink **b** Heat sink with one fin

5.2 *Enhancement of PCM Melting with Addition of Fins*

Melt fraction (γ) of PCM during melting process is defined as:

$$\gamma = \iint \frac{S_{[T58\,^{\circ}\mathrm{C}]}}{S_{[\mathrm{Intial}]}} \mathrm{d}x\mathrm{d}y \tag{6}$$

Here, $S_{[T>58\,^{\circ}\mathrm{C}]}$ represents area of the melted PCM if their temperature is more than 58 °C and $S_{[\mathrm{Intial}]}$ represents area of PCM initially in solid state. Figure 5a–b presents the variation of melt fraction of PCM with respect to time at different heat flux values, for heat sink with one and two fins. From Fig. 5, it can be seen that as the heat flux values increase the melting rate of PCM increases. It can be noted that for heat sink with one fin, time taken by the PCM for complete melting are 6300 s at 1.00 kW/m^2 whereas, it takes 4500 s and 3500 s at heat flux of 1.50 kW/m^2 and

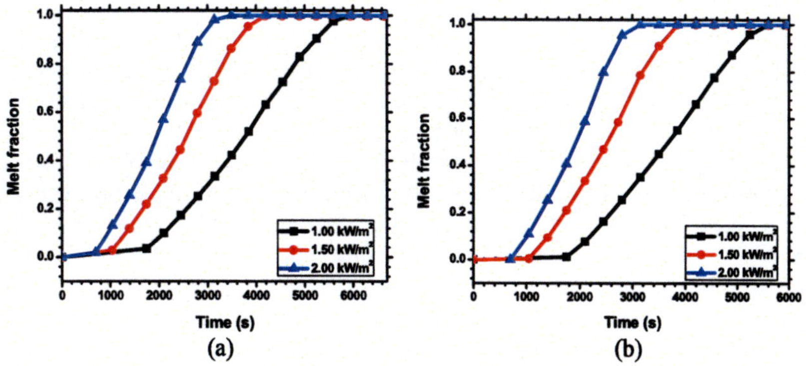

Fig. 5 Variation of melt fraction with time **a** heat sink with one fin **b** heat sink with two fins

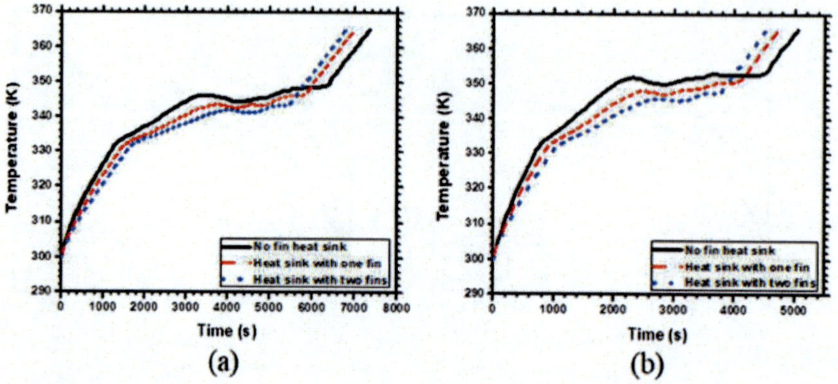

Fig. 6 Time temperature variation of different heat sink design at various heat fluxes **a** 1.00 kw/m^2 **b** 1.50 kW/m^2

2.00 kW/m^2 respectively. The variation of melt fraction with time after initiation of melting is almost linear. Before melting starts the heat transfer mode is mainly conduction whereas when temperature reaches beyond the melting point of PCM, heat transfer mode is convection dominant.

5.3 *Effect of Heat Sink Configuration*

Figures 6a–b present the effect of number of fins on the heat transfer ability of PCM-based heat sink at various input heat flux values (1.00 and 1.5 kW/m^2). Temperature at the center of base of the heat sink is used to represent the transient temperature variation. It is observed that heat sink with two fins performs better than the heat sink with one fin and heat sink with no fin irrespective of the heat flux. This is because uniformity in temperature increases with increasing the number of fins. The increase in temperature uniformity is due to the increase in thermal diffusivity of the TCE (fins). Increasing the number of fins inside the heat sink causes the heat from the base of the heat sink to diffuse more rapidly and thus, preventing the local overheating of base of the heat sink. In addition to this, the melting of PCM starts from several sides in case of finned heat sink instead of only from the bottom for heat sink in case of heat sink without fins. However, increasing the number of fins reduces the amount of PCM inside the heat sink. The effect of which can be seen in post melting process. Temperature at the base of heat sink after post melting of PCM increases sharply for heat sink with two fins compared to the heat sink with one fin and heat sink with no fin. This is because of reduction in amount of PCM due to insertion of fins. For heat sink with no fin amount of PCM is higher, hence its latent heating duration is also higher. Therefore there exists an optimum distribution of PCM and TCE for better thermal management.

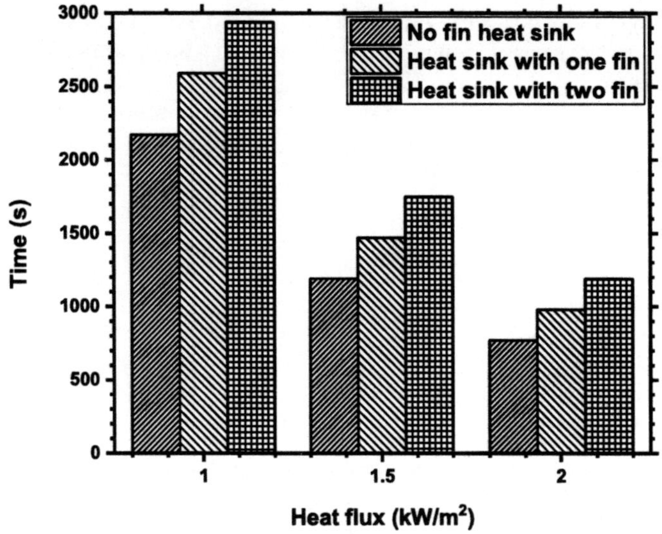

Fig. 7 Time to reach SPT of 65 °C at various heat fluxes

5.4 Enhancement in Operation Time to Reach Set Point Temperature (SPT)

Figure 7 presents the effect of fins in enhancing the operation time to reach a critical set point temperature at various heat flux values. Critical SPT represents the maximum temperature beyond which electronic components are not functioning efficiently. Here, SPT of 65 °C (338 K) is selected. It is evident from Fig. 7 that as the input heat flux increases, the operating time decreases linearly for a specific set point temperature. It can be noted that two finned heat sink takes more time to reach the SPT of 65 °C compared to one fin and unfinned heat sink at various heat fluxes values.

6 Conclusion

In the present study, 2-D numerical model is developed to model the plate finned heat sink filled with PCM. Heat flux is varied from 1.00 to 2.00 kW/m^2. To simulate the melting process of PCM in the heat sink enthalpy-porosity model is employed. The result shows that increase in the value of heat flux increases the corresponding melting rate and base temperature. It can be concluded that increase in the number of fins inside the heat sink enhances the performance of heat sink by decreasing its base temperature. However, this happens only up to the melting of PCM. After the melting of PCM, base temperature of heat sink with no fin is lower than finned heat sink. This is because heat sink with no fin contains more PCM than finned heat sink. Therefore there exist an optimum number of fins and amount of PCM that needs to be investigated.

Acknowledgements The financial support received from DST, Government of India (DST Sanction No.: DST/TMD/MES/2k17/65) is gratefully acknowledged by the authors.

References

1. Viswanath R, Wakharkar V, Watwe A, Lebonheur V (2000) Thermal performance challenges from silicon to systems. Int Tech J Q3. ISSN 1535-766X
2. Kothari R, Mahalkar P, Sahu SK, Kundalwal SI (2018) Experimental investigation on thermal performance of PCM based heat sink for passive cooling of electronic component. In: International conference on nanochannels, microchannels, and minichannels, ICNMM-7732
3. Baby R, Balaji C (2011) Experimental investigation on phase change material based finned heat sinks for electronic equipment cooling. Int J Heat Mass Transf 55(5–6):1642–1649
4. Srikanth R, Nemani P, Balaji C (2015) Multi-objective geometric optimization of a PCM based matrix type composite heat sink. Appl Energy 156:703–714
5. Sharma R, Ganesan P, Tyagi VV (2015) Developments in organic solid–liquid phase change materials and their applications in thermal energy storage. Energy Conver Manage 95:193–228
6. Ashraf MJ, Ali HM, Usman H, Arshad A (2017) Experimental passive cooling: parametric investigation of pin fin geometries and efficient phase change material. Int J Heat Mass Transf 115:251–263
7. Mansin S, Diani A, Doretti L, Hooman K, Rossetto L (2015) Experimental investigation of phase change phenomenon of paraffin waxes embedded in copper foams. Int J Thermal Sci 90:79–89
8. Arici M, Tutuncu E, Kan M, Karabay H (2017) Melting of nanoparticle-enhanced paraffin wax in a rectangular enclosure with partially active walls. Int J Heat Mass Transf 104:07–17
9. Li Y, Liu S, Shukla A (2016) Experimental analysis of use of thermal conductivity enhancers (TCEs) for solar chimney applications with energy storage layer. Energy Build 116:35–44
10. Dhaidan N, Khodadadi JM, Al-Hattab T, Al-Mashat SM (2013) Experimental and numerical investigation of melting of phase change material/nanoparticle suspensions in a square container subjected to a constant heat flux. Int J Heat Mass Transf 56:672–683
11. Bondareva NS, Shermet MA (2017) 3-D natural convection in a cubical cavity with a heat source. Int. J Thermal Sci 115:43–53
12. Soares N, Gaspar AR, Santos P, Costa JJ (2015) Experimental study of heat transfer through a vertical stack of rectangular cavities filled with phase change materials. Appl Energy 142:192–205
13. Ji C, Qin Z, Low Z, Dubey S, Choo FH, Duan F (2018) Non-uniform heat transfer suppression to enhance PCM melting by angled fins. Appl Thermal Eng 129:269–279
14. Stritih U (2004) An experimental study of enhanced heat transfer in rectangular PCM thermal storage. Int J Heat Mass Transf 47(12–13):2841–2847
15. Ganatra Y, Ruiz J, Howarter JA, Marconnet A (2018) Experimental investigation of phase change materials for thermal management of handheld devices. Int J Thermal Sci 129:358–364
16. Voller VR, Prakash C (1987) A fixed grid numerical modeling methodology for convection-diffusion mushy region phase-change problem. Int J Heat Mass Transf 30(8):1709–1719
17. Brent AD, Voller VR, Reid KJ (2007) Enthalpy porosity technique for modeling convection-diffusion phase change: application to the melting of a pure metal. Num Heat Transf 13(3):297–318

18. Hosseinizadeh S, Tan FL, Moosania SM (2011) Experimental and numerical studies on performance of PCM-based heat sink with different configurations of internal fins. Appl Thermal Eng 31(17–18):3827–3838

Mixed Convective Power-Law Fluid Flow and Heat Transfer Characteristics Past a Semi-circular Cylinder Mounted with a Splitter Plate

Souparna Banerjee, Samrat Banik, Chitrak Mondal, Sandip Sarkar, and Nirmal Kumar Manna

1 Introduction

In recent years, considerable research efforts have been devoted in studying the momentum and heat transfer characteristics past bluff bodies of various geometrical configurations, viz. circular, square, semi-circular, etc. The flow of fluids past the bluff bodies represents an idealisation in many industrial applications like heat exchanger systems, cooling towers, agricultural products, oil and gas cooling towers, cooling of electronic components, etc. Despite the fact that two-dimensional and three-dimensional vortical instabilities are of great interest to engineers as well as scientists, a complete understanding of the flow over bluff bodies poses a great challenge. Bluff body wakes are complex as they involve the interaction of three shear layers in the same problem, namely, boundary layer, separating free shear layer and a wake [1]. The knowledge of the flow is also an important aspect in wind and civil engineering. Aerodynamic actions such as vortex shredding has a great impact on

S. Banerjee · S. Banik · C. Mondal · S. Sarkar (✉) · N. K. Manna
Department of Mechanical Engineering, Jadavpur University, Kolkata 700032, India
e-mail: sandipsarkar.mech@jadavpuruniversity.in

S. Banerjee
e-mail: souparnajume@gmail.com

S. Banik
e-mail: samratbanik.stu@gmail.com

C. Mondal
e-mail: chitrakmondal800@gmail.com

N. K. Manna
e-mail: nirmalkmannaju@gmail.com

M. Palanisamy et al. (eds.), *Theoretical, Computational, and Experimental Solutions to Thermo-Fluid Systems*, Lecture Notes in Mechanical Engineering,
https://doi.org/10.1007/978-981-33-4165-4_21

the design of tall buildings [2], or else it may cause structural damage due to periodic surface loading which increases drag, and thereby shortens life.

In the last decade, research in the field of steady flow control has shown that a splitter plate attached to the bluff body, reduces drag and may inhibit vortex shredding [3, 4]. In general, initially two wakes are formed at the rear end of the cylinder. The size of these wakes depends on the Reynolds number (Re), i.e., for low Re, the wakes are small and for higher Re, not only the size of the wakes increases but also the instability. At about $90 < \mathrm{Re} < 200$, oscillatory wakes are formed, and for $250 < \mathrm{Re} < 2 \times 105$, vortex shredding occurs, which is known as the Von-Karman vortex street. The Von-Karman vortex instability for a bluff body is characterised by the vortex shredding frequency (f), and in the non-dimensional form it is written as the Strouhal number (St), where $\mathrm{St} = fD/U\infty$, where $U\infty$ and D, are the freestream velocity and diameter of the cylinder respectively. This causes fluctuating drag and lifts forces to act on the body caused due to periodic pressure fluctuations. Various flow separations devices were used to control vortex shredding, but because of its simplicity, a single splitter plate is used as an effective measure to control vortex shredding past any bluff body [5]. Splitter plates have been termed as wake stabilizers and are used in the downstream of the cylinder. In a circular cylinder the flow separation depends on the Reynolds number (Re), whereas in case of a semi-circular cylinder, the flow separation points are the sharp edges [5, 6]. Fluid particles separated at the sharp edges travel a larger distance downstream to form the separation bubble. Hence, the effectiveness of the splitter plate is different for circular and semi-circular geometries [7].

As far as known to us, there is no prior study on mixed convective flow of power-law fluids over semi-circular cylinder mounted with a splitter plate. The objective of the present study is to extend the understanding of mixed convection and to investigate the flow separation control mechanism using a splitter plate at the rear end of a semi-circular cylinder at low Reynolds number flow (Re = 100). In this study, the effect of splitter on vortex shredding and some of the flow characteristics like drag and lift co-efficient were numerically investigated in a two-dimensional regime. Generally, for a real flow the vortex shredding is formed in three-dimensional, but the two-dimensional model covers the flow pattern and serves the purpose. The above assumption has been used by many researchers in their study of flow past a circular cylinder [8].

2 Grid Generation and Boundary Conditions

The Computational domain for the present work is shown in Fig. 1. Computations were performed for a semi-circular cylinder of diameter D, with which a wake splitter plate of length L equal to the diameter D ($L = D$) and of small thickness (= 0.01 D) is attached. The following boundary conditions were used. The inlet is at a distance 8 D from the centre of the semi-circular cylinder where $U_o = 1$ and $V_o = 0$. The Reynolds number was constant and had a value Re = 100. The thickness of the wake

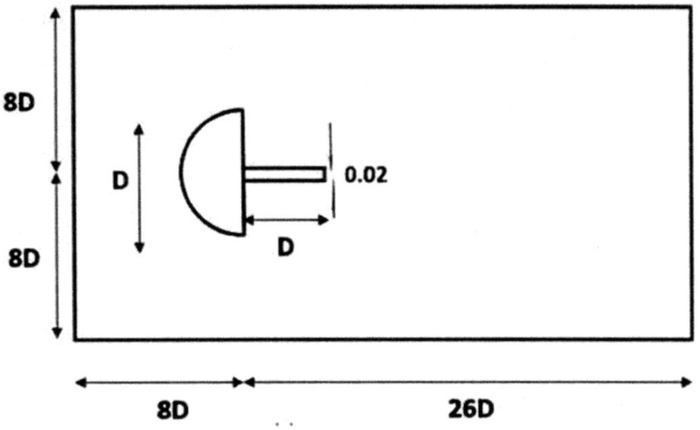

Fig. 1 Computational domain for studying the flow past the semi-circular cylinder with a splitter plate

splitter plate was assumed to be 0.02 D so that the pressure drag of the splitter plate is negligible. In the present study, a non-uniform mesh, with finer around the cylinder and coarse towards the far domain is adopted. A total 240 grid points were used to define the semi-circular cylinder. The fluid enters the channel at a fully developed condition with velocity U_o (= 1) and at a constant temperature $T_o = 300$ K. The walls of the channel are at a constant temperature T_w. The non-Newtonian power model (*m*, *n*) is independent of the temperature.

3 Governing Equations and Numerical Methodology

In this study, the incompressible continuity equation, the Navier-Stokes and energy equation in the Cartesian Co-ordinate system form the governing equations for the flow. The buoyancy-driven flow from the heated surface interacts with the laminar flow to yield mixed convection conditions. The continuity, momentum and thermal energy equations are written in their dimensionless form with Boussinesq approximation:

$$\frac{\partial u}{\partial x} + \frac{\partial v}{\partial y} = 0 \tag{1}$$

$$\frac{\partial u}{\partial t} + u\frac{\partial u}{\partial x} + v\frac{\partial u}{\partial y} = -\frac{\partial p}{\partial x} + \frac{1}{Re}\left(\frac{\partial \tau xx}{\partial x} + \frac{\partial \tau yx}{\partial y}\right) \tag{2}$$

$$\frac{\partial v}{\partial t} + u\frac{\partial v}{\partial x} + v\frac{\partial v}{\partial y} = -\frac{\partial p}{\partial y} + \frac{1}{\mathrm{Re}}\left(\frac{\partial \tau xy}{\partial x} + \frac{\partial \tau yy}{\partial y}\right) + \mathrm{Ri}\,(\theta) \tag{3}$$

$$\frac{\partial \theta}{\partial t} + u\frac{\partial \theta}{\partial x} + v\frac{\partial \theta}{\partial y} = \frac{1}{(\text{Re.Pr})}\left(\frac{\partial^2 \theta}{\partial y^2} + \frac{\partial^2 \theta}{\partial y^2}\right) \quad (4)$$

where, Ri = Gr/Re^2, Gr = g/U_2, is the Richardson number, g is the acceleration due to gravity and Re is the Reynolds number. For Ri = 0, the situation corresponds to the thermofluidic characteristics of forced convection. The length and velocity scales that are used to non-dimensionalise the governing equations are the cylinder diameter (D) and the average velocity (U_o), respectively. Thus, the pressure is non-dimensionalised by diving by ρU_0^2, and the stress components are non-dimensionalised using the factor m (U_o/D)^n. The non-dimensional temperature θ is defined as [(T − T_w)/(T_w − T_o)]. The Reynolds number is given by Re = $\rho U_0^{2-n} D^n/m$ where n is the power-law index and m is the flow consistency index. The Prandtl number is given by Pr = mC/kU_0^{n-1}. Clearly, $n = 1$ shows the Newtonian behaviour, $n < 1$ shows shear thinning behaviour and $n > 1$ shows the shear thickening behaviour.

The domain length is chosen long enough so that the thermal boundary layer is fully developed before the outlet. Finally, the continuity, momentum and energy equations along with the boundary conditions map the flow domain and local and global flow quantities are obtained, e.g.: local and average Nusselt number, drag coefficient, lift coefficient, etc. The following quantities are defined as:

Nusselt number: $Nu_l = hD/k = q_w D/k(T_w - T_b)$

Lift Coefficient: $C_l = F_l/\left(1/2\ \rho U_0^2\right)$

Drag Coefficient: $C_d = F_d/\left(1/2\ \rho U_0^2\right)$

here, h is the convective heat transfer coefficient, q_w is the wall heat flux, T_b is the bulk temperature, F_l is the lift force and F_d is the drag force as defined. The bulk mean temperature is defined as (where C is the specific heat of the fluid):

$$T_b = \frac{\iint \rho UCT\, dA}{\iint \rho UC\, dA}$$

4 Results and Discussions

The time-averaged field is obtained by making an average of the instantaneous velocity and temperature over consecutive vortex shredding cycles. The time-averaged contours of isotherms, vorticity and streamlines are shown below for different ranges of the Richardson number (Ri) and the power-law index (n). The plot of the drag and lift coefficients and the average Nusselt number for the cylinder as well as the splitter plate are also shown in (Fig. 2).

From the above results, it is clear that the region where most of the isotherms cluster around the cylinder or plate is the location of a high temperature. The regions away from it are of gradually decreasing temperature. The time-averaged contours of isotherms show symmetric distributions about the line of symmetry. On the other

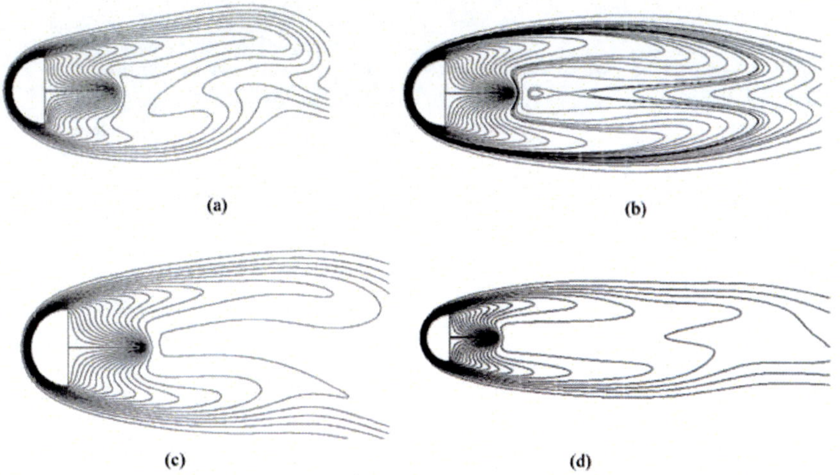

Fig. 2 Time-averaged contours of isotherms **a** Ri = 1; $n = 0.6$ **b** Ri = −1; $n = 0.6$ **c** Ri = −1; $n = 1.6$ **d** Ri = 1; $n = 1.6$

hand, the time-averaged vorticity contours consists of two strands of oppositely signed vortices formed at the rear end of the cylinder. Due to the presence of the splitter plate the separation time for the vortices increase. It is observed that as the value of the power-law index (n) increases, the separation time also increases, and is greatest for $n = 1.6$. For a constant Reynolds number (Re = 100), with the increase in time, the wakes become more unstable and shredding occurs. The instantaneous contours of streamlines do not yield symmetrical distribution. It is observed that the lower half of the cylinder shoulder shows more curvature than that of the upper half. Hence, it can be concluded that the lower half of the cylinder will experience more fluid dynamic forces than the upper half (Figs. 3, 4, 5 and 6).

5 Conclusions

A numerical investigation is carried out to analyse the mixed convective flow and heat transfer characteristics of a power-law fluid past a semi-circular cylinder mounted with a splitter plate. The power-law indices are taken to be 0.6, 1.0 and 1.6, thus covering both the shear thinning and shear thickening purpose. The Richardson number (Ri) is taken to be as −1, 0 and −1. The contours of the isotherms, vorticities and streamlines are shown and have been analysed in detail. From the results, it may be concluded that

- At regions near to the cylinder, there is a cluster of isotherms and hence the temperature is high at such regions. Away from the bluff body, the temperature gradually decreases.

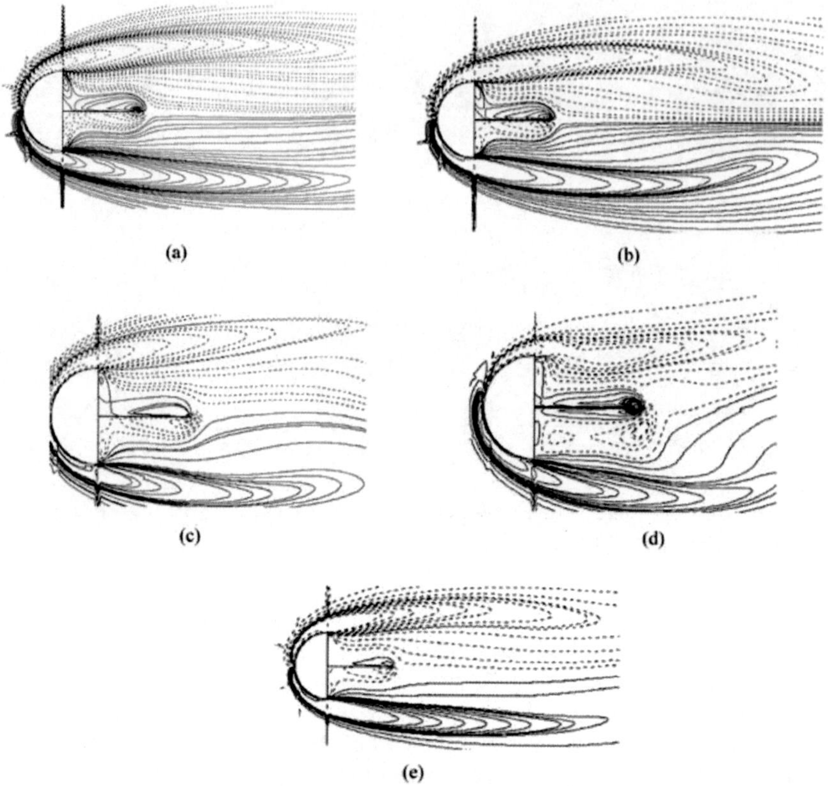

Fig. 3 Time-averaged contours of vorticity **a** Ri = 0; $n = 1.0$ **b** Ri = 0; $n = 0.6$ **c** Ri = −1; $n = 1.6$ **d** Ri = 1; $n = 0.6$ **e** Ri = 1; $n = 1.6$

- The time-averaged isotherms are symmetrical about the channel centerline, which indicates that the temperature distribution on the upper and lower half of the bluff body is uniform.
- Due to the presence of the splitter plate, the separation time for the wakes increases. The separation time increases as the value of the power-law index increases and is highest for $n = 1.6$.
- The streamlines do not appear to be symmetrical, but the curvature is more on the lower half than on the upper half, hence the lower half will experience more hydrodynamic forces.
- From the lift coefficient plot, it is observed that the lift force decreases with an increase in power-law index (n).

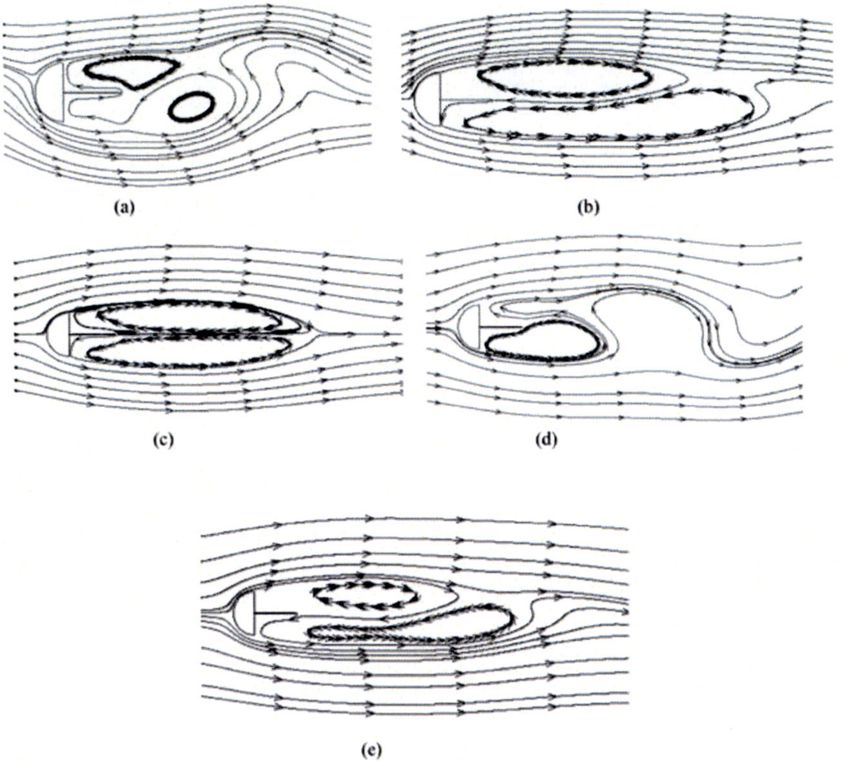

Fig. 4 Instanteneous contours of streamlines after $t = 75$ s **a** Ri = 1; $n = 0.6$ **b** Ri = 0; $n = 1$ **c** Ri = −1; $n = 0.6$ **d** Ri = −1; $n = 1.6$ **e** Ri = 1; $n = 1.6$

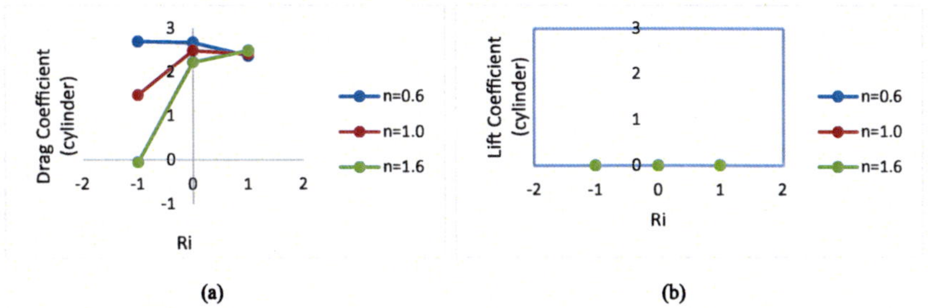

Fig. 5 Plot of Lift and Drag Coefficient with Ri of cylinder for different values of Ri and n

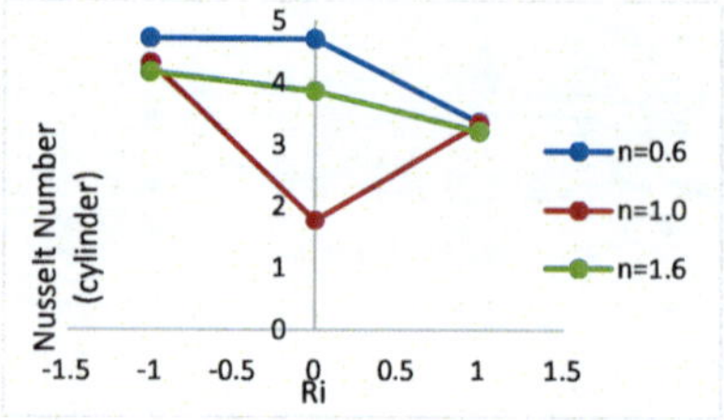

(c)

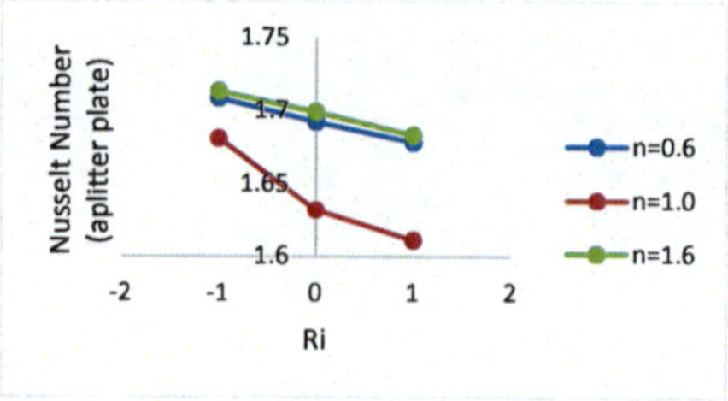

(d)

Fig. 6 Plot of average nusselt number *v/s* Ri of cylinder and splitter plate for different values of Ri and *n*

References

1. Biswas G, Sarkar S (2009) Effect of thermal buoyancy on vortex shedding past a circular cylinder in cross-flow at low Reynolds numbers. Int J Heat Mass Trans 52:1897–1912
2. Bearman PW, Wadcock AJ (1973) The interaction between a pair of circular cylinders normal to a stream. J Fluid Mech 61:499–511
3. Meneghini JR, Saltara F, Siqueira CLR, Ferrari JA (2001) Numerical Simulation of Flow Interference between two circular cylinders in tandem and side by side arrangement. J Fluids Struct 15:327–350
4. Sisodia SS, Sarkar S, Saha SK (2017) Fluid flow and mixed convective heat transfer around a semi-circular cylinder at incidence with a tandem downstream square cylinder in cross flow. Int J Therm Sci 121:13–29
5. Sarkar S, Ganguly S, Dalal A (2014) Analysis of entropy generation during mixed convective heat transfer of nanofluids past a rotating circular cylinder. J Heat Trans 136:062501-1-10
6. Sarkar S, Ganguly S, Biswas G, Saha P (2016) Effect of cylinder rotation during mixed convective flow of nanofluids past a circular cylinder. Comput Fluids 127:47–64
7. Sarkar S, Ganguly S (2012) Biswas, Mixed convective heat transfer of nanofluids past a circular cylinder in cross flow in unsteady regime. Int J Heat Mass Transf 55(17–18):4783–4799
8. Verma N, Dulhani JP, Dalal A, Sarkar S, Ganguly S (2015) Effect of channel confinement on mixed convective flow past an equilateral triangular cylinder. J Heat Transfer 137:121013

Design and Analysis of Cold Box for Helium Liquefier/Refrigerator

Kamlesh Kumar, Sandip Pal, and Murugan Sivalingam

1 Introduction

The Helium Refrigerator/Liquefier (HRL) system comprises of the following sub-Systems:- Helium Cycle Compressors, Oil removal module (ORM), Cold box, Liquid Helium (LHe) Dewar, Valve Box, Purification System, Helium Storage Tanks (Buffer), Programmable Logic Controller (PLC). The losses associated with this system degrade the performance and liquefaction rate. Here, the main loss associated with the system is the pressure loss and the heat in-leak from ambient to the equipment maintained at cryogenic temperature. This interaction can be conduction through solid materials, molecular conduction of gases, and radiation. The pressure loss can be reduced by minimizing the number of bends, elbows, length of the pipes, etc. This leads to developing a compact cold box but makes the system difficult to fabricate. The leakage of heat from the system to ambient can also be minimized by using proper design. Actually, it is not possible to stop leakage of heat from outside to the system but by using an optimized design, losses can be minimized [1]. To visualize the internal part of the cold-box [2], a 3-D model is developed by using Solid Works software.

The optimized placement of different components inside the cold box is planned based on their temperature distribution from 300 K to cryogenic temperature [3].

K. Kumar
Indian Institute of Technology, Patna, India

S. Pal (✉)
Variable Energy Cyclotron Centre, Kolkata, India
e-mail: sandip@vecc.gov.in

Homi Bhabha National Institute, Mumbai, India

M. Sivalingam
National Institute of Technology, Rourkela, India

M. Palanisamy et al. (eds.), *Theoretical, Computational, and Experimental Solutions to Thermo-Fluid Systems*, Lecture Notes in Mechanical Engineering,
https://doi.org/10.1007/978-981-33-4165-4_22

The requirement of refrigeration at 4.5 K for superconducting cyclotron (SCC) of Variable Energy Cyclotron Centre (VECC) is met by any of the two helium liquefiers/refrigerators (Air Liquide, France make). The liquefaction capacity of the two liquefiers are 50 L/h and 85 L/h, respectively, without liquid nitrogen (LN_2) precooling. At the refrigeration mode, the refrigeration capacity of the two liquefiers are 250 watts and 415 watts at 4.5 K, respectively without LN_2 precooling. After the liquefaction, the liquid helium (LHe) is stored in dewars having the capacity of 1000 L. From dewars, LHe is supplied to the SCC through valve box, which is used to switch among the two liquefiers, for cooling of magnet section and cryopanels.

As the 50 L/h capacity helium liquefier is almost 20 years old, there is a need to develop a liquefier of capacity 50 l/h without LN_2 precooling for that cold box design is under process. The new helium liquefier system consists of six heat exchangers, among them E401 and E404 are three stream exchangers, E402, E403, E405 and E406 are two stream exchangers. Their process connections are shown in Fig. 1a and its corresponding temperature distribution in the T-S diagram is shown in Fig. 1b. There are two turboexpanders—warm and cold. The outlet of the cold turboexpander passes through the heat exchanger cold return stream results in decrease in JT valve inlet temperature following the Joule–Thomson effect or Joule–Kelvin effect or

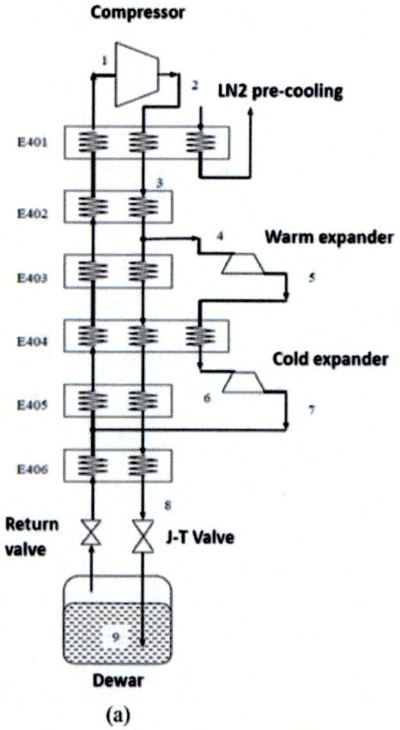

(a)

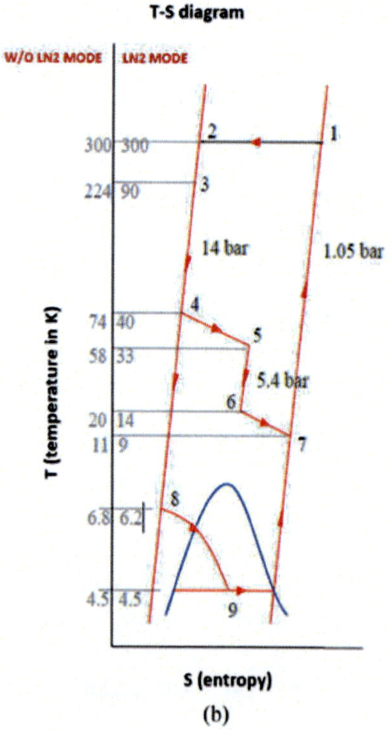

(b)

Fig. 1 **a** Helium liquefier **b** T-S Diagram of Helium liquifier

Kelvin–Joule effect [4], while the valve is insulated from the surroundings. Through this process, a decrease in temperature will be possible in helium, neon and hydrogen, while inlet temperature is lower than the JT inversion temperature. Hence precooling is required for all these three gases in the liquefaction process.

As per overall dimension of the cold box, complete modeling has been done by using solid works software to visualize the components inside the system. The complete layout inside the cold box is shown in Fig. 2. The layout of the inner part of the cold box is the one of the most important things in designing the whole system. The major components used inside the system are six heat exchangers (E-1, E-2/3, E-4, E-5/6), two turbo expanders (hot and cold), J–T valves, liquid nitrogen precooling section, two adsorbers, two filters, and complete piping systems. Besides the main equipment, pressure transducers, temperature measuring devices, and other

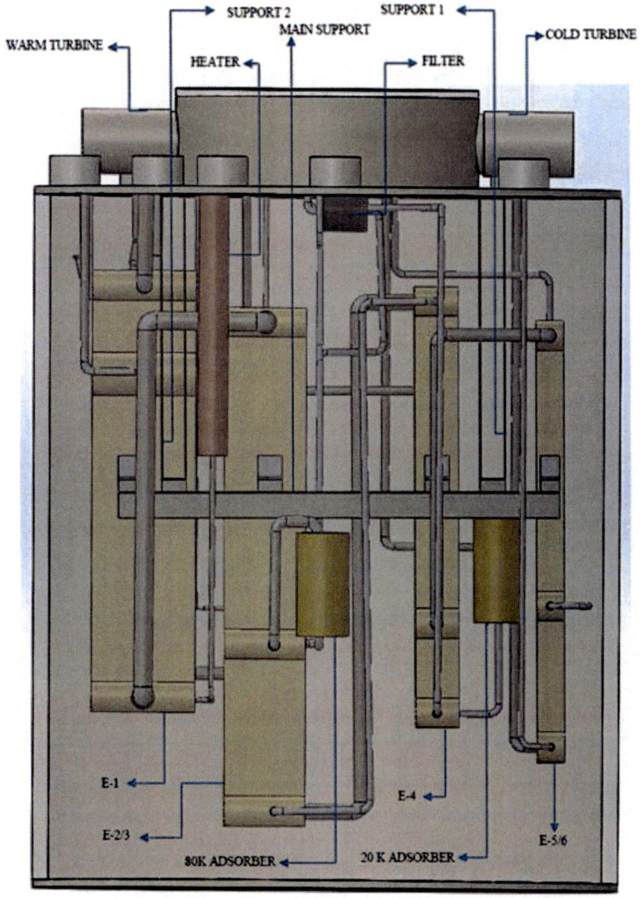

Fig. 2 Proposed cold box of Helium liquefier system

instrumentation are installed. But these components cannot be arranged haphazardly, because wrong positioning reduces the overall efficiency. Therefore, the positioning of different equipments inside the system must be carefully done. In actual practice, in order to reduce the space inside the cold box, heat exchangers E402 and E403 are geometrically attached as E-2/3 and placed vertically E402 on top of E403 and similarly E405 and E406 as E-5/6.

2 Analysis and Discussion

2.1 Analysis of Complete System

Modelling of inner part of the cold box has been done, as shown in Fig. 2, by using SOLID WORKS and conduction heat transfer through the support system has been analysed by using ANSYS software. It is very much important to know about the stress-induced on the support of the system. To know about the conduction effect from the support system, all the piping has been neglected inside the system, and only support has been taken for the analysis as shown in Fig. 3.

Here the meshed view of the system (Fig. 3.) is shown in Fig. 4. Meshing has been done in ANSYS fluent in creating nodes throughout the system. In ANSYS, the calculation is always done at the nodal points. Hence it can be inferred that more the nodal points more will be the computation load, and analysis will be more complicated. But this makes the result better. Assumptions made for this analysis are:

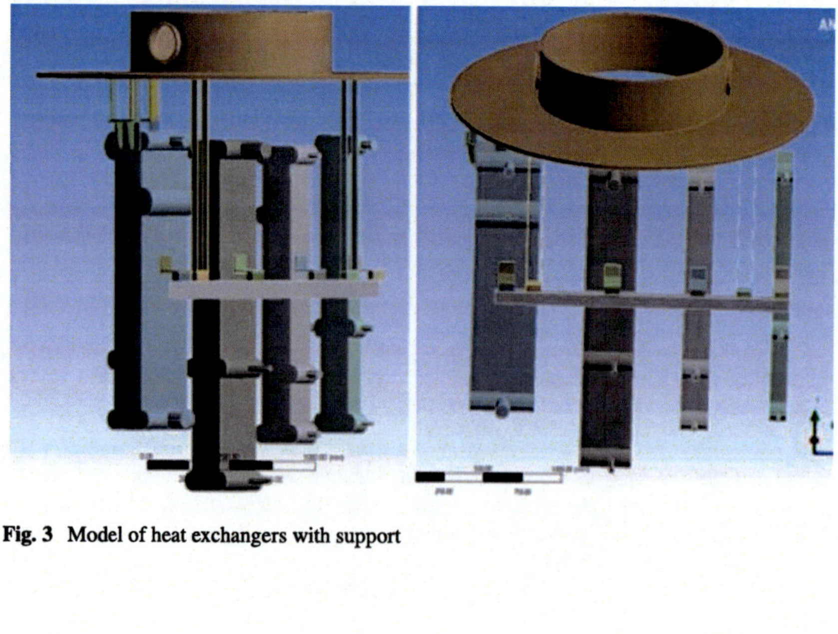

Fig. 3 Model of heat exchangers with support

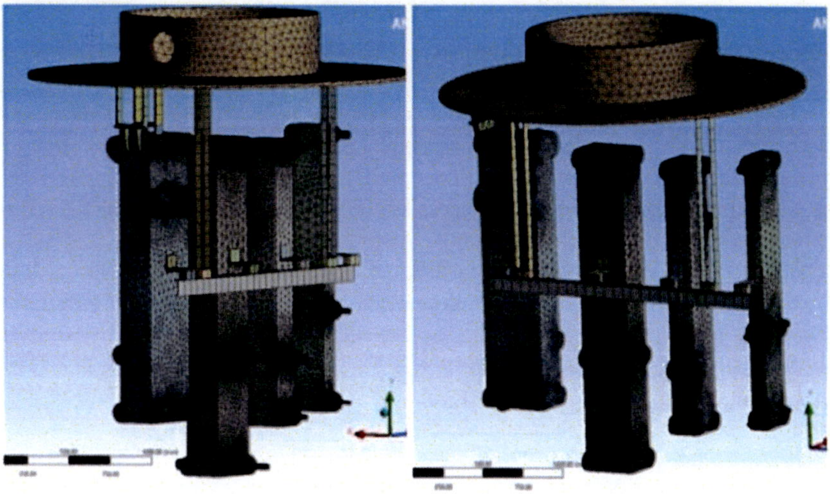

Fig. 4 Meshed view of heat exchangers with support

- The cover plate is taken at a temperature of 300 K (ambient temperature).
- The system neglects the conduction through all the piping systems. The study has been done on the conduction effect inside our system and neglected molecular conduction and radiation effect throughout the system.
- Inlet temperature and outlet temperatures have been fed as input to the heat exchangers, to introduce boundary conditions.

2.2 *Analysis of Heat Exchangers*

Offset strip plate-fin heat exchangers have been used in the cold box, as shown in Fig. 5 [5]. For the simulation in ANSYS (Fluent), it is complicated to analyze only

Fig. 5 The solid portion of the Offset strip plate-fin heat exchanger

the solid portion. Hence for simplification, fluid portion flowing inside offset strip fin with parting sheet has been drawn, as shown in Fig. 6, by using Solid works software. Thermal analysis is too complex due to its complex geometry; hence the smallest part of it is taken for our study, as shown in Fig. 7. Heat exchangers used in cryogenic applications need to have very high effectiveness to preserve the produced refrigerating effect [6]. Normally the heat exchangers used in cryogenic refrigerators and liquefiers should have effectiveness of the order of 0.95 or higher [7]. If the effectiveness of the heat exchangers falls below the design value, there may not be any liquid yield. In this system, the offset strip plate-fin heat exchanger is used for higher effectiveness [8]. An offset path creates turbulence and due to this turbulence, the heat transfer rate will be better.

Fluid 1-Low pressure cold helium gas (fin height-3 mm)
Fluid 2-High pressure warm helium gas (fin height-8.8 mm)

Fig. 6 The fluid portion of offset strip plate-fin heat exchanger with parting sheet

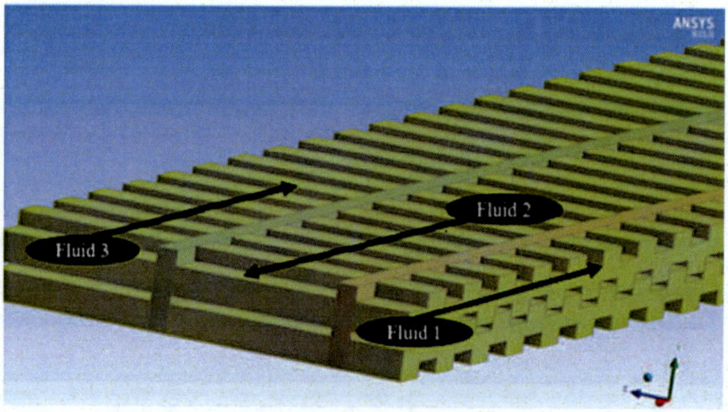

Fig. 7 Part of flowing fluid inside the heat exchanger

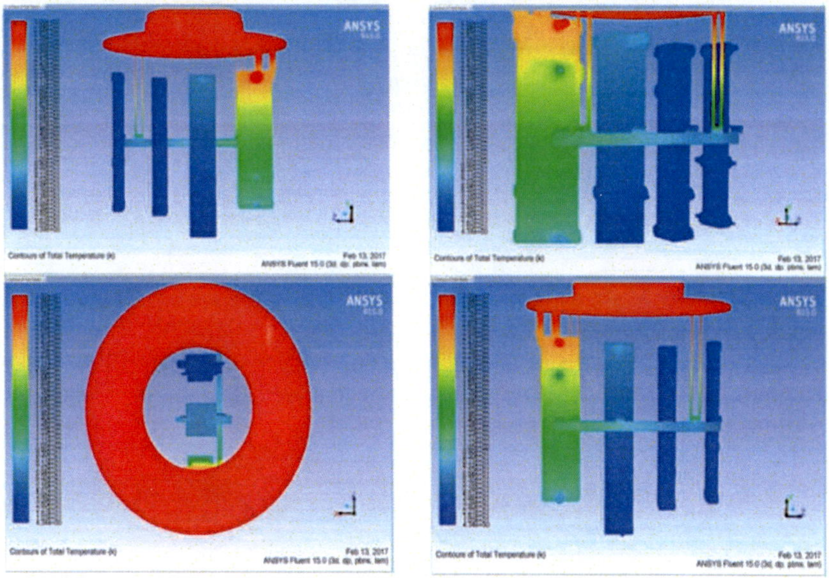

Fig. 8 Temperature contour of the system (Heat exchangers with support)

Fluid 3-Liquid nitrogen (fin height-8.8 mm).

3 Results

3.1 Temperature Distribution in Heat Exchangers

In the model, only conduction effect of solid supports has been considered, neither radiation nor gaseous molecular conduction. Along with this, the piping system is removed from the system. The temperature distribution on all heat exchangers with support system and cover part has been obtained, as shown in Fig. 8.

3.2 Comparison of the Temperature Distribution in All Heat Exchangers and Support System

The temperature distribution curve with the position for all heat exchangers along the length has been obtained, as shown in Fig. 9, and temperature distribution along the length of the support system has been obtained, as shown in Fig. 10. The temperature of the first heat exchanger varies from the minimum temperature of 100 K to the

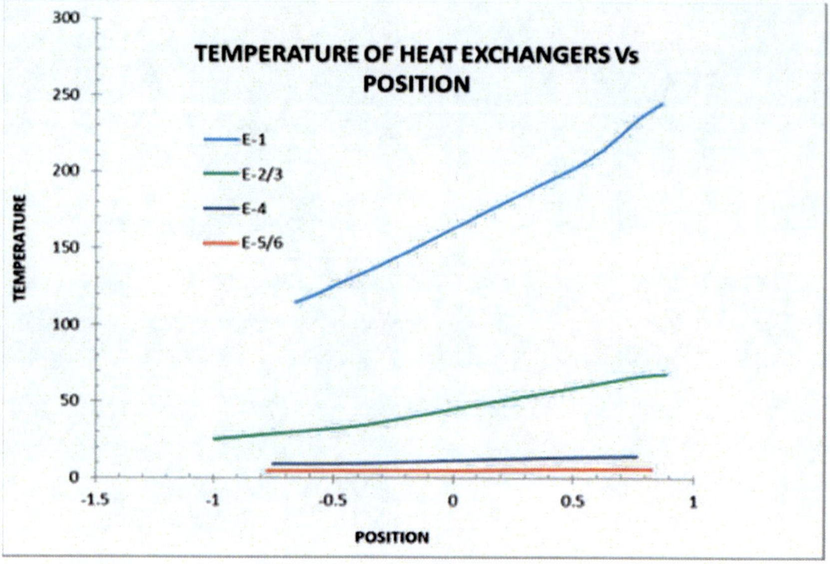

Fig. 9 Variation of temperature (K) with position for all heat exchangers along the length (m)

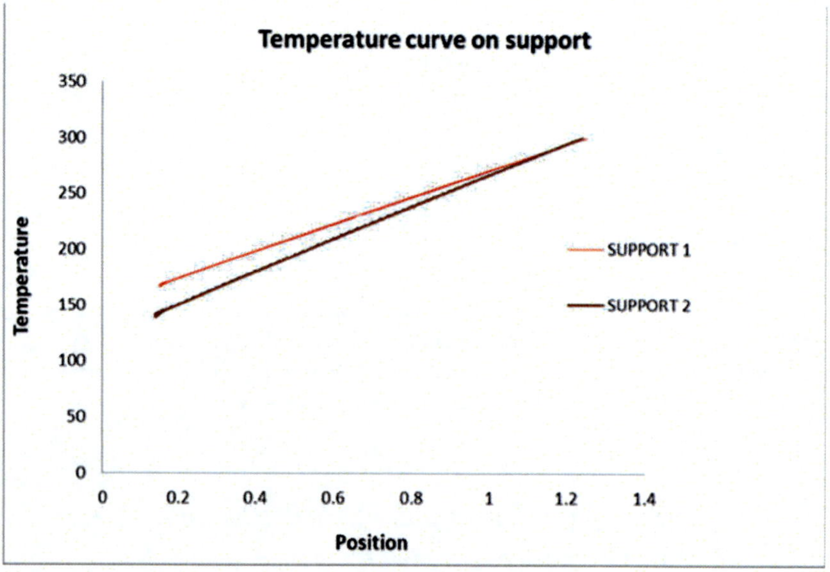

Fig. 10 Variation of temperature (K) on support along position (m)

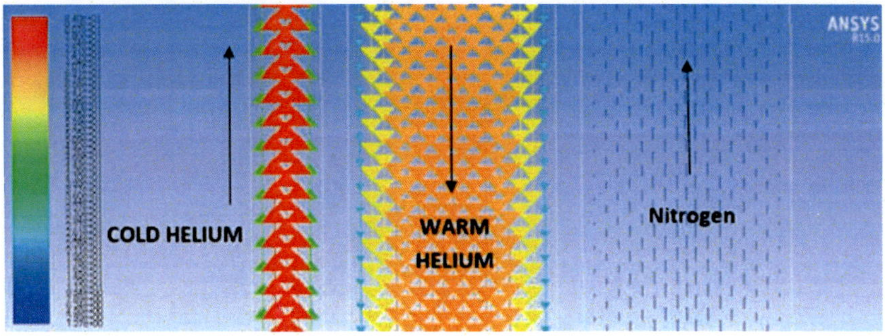

Fig. 11 Temperature distribution of flowing fluid inside heat exchanger

maximum temperature of 240 K. This large temperature gradient is due to the Liquid nitrogen precooling at 77 K. This liquid nitrogen precooling enhances the liquefaction rate. The liquid nitrogen can also be used as a thermal shield to reduce the radiation effect from the 300 K (ambient) environment to the internal system. Hence, it is beneficial to use liquid nitrogen precooling in the system. The important thing is, an 80 K adsorber is attached to trap all the nitrogen, oxygen, and remaining water molecules from this heat exchanger. Hence, a temperature lower than 80 K is needed at this portion. It can be seen that the temperature of E-2/3 varies from 25 to 70 K. The temperature of E-4 varies from 8 to 13 K. Because of this, a 20 K adsorber is attached so that it removes the remaining gases like Argon, Neon, etc. After passing from this 20 K adsorber, some part of the gas passes from the warm turbine for expansion to provide a cooling effect to the high-pressure line. The temperature of E-5/6 varies from the minimum temperature of 5 K to the maximum temperature of 6.5 K approx. After reaching the minimum temperature, fluid passes through the J–T valve for liquefaction.

3.3 *The Fluid Inside the Offset Strip Fin*

A portion of the fluid section has been used for simulation, and variation of temperature and fluid direction inside the heat exchanger has been obtained after simulation, as shown in Fig. 11. From this result, the pattern of fluid flowing inside the strip of the heat exchanger can be obtained.

3.4 *Temperature Distribution in Fluid in the Heat Exchanger*

The temperature distribution for liquid nitrogen with respect to the position has been acquired, and it varies from 77 to 85.9 K as shown in Fig. 12. Because of

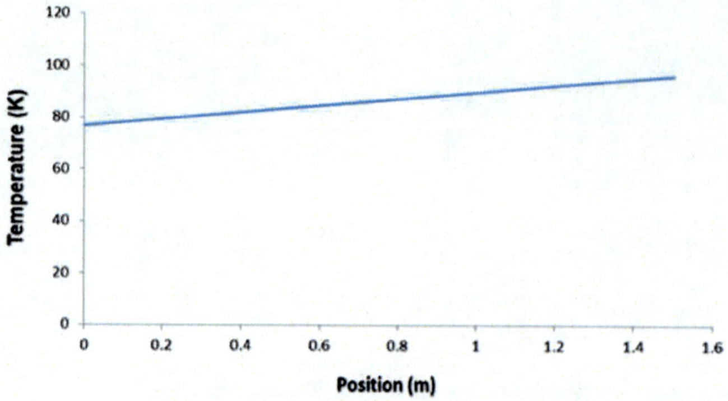

Fig. 12 Temperature distributions inside heat exchanger E401 for LN_2

LN_2 precooling, the performance, as well as the productivity of the system, will be improved. As the cost of the liquid Nitrogen is very less, it is also used as a thermal shield in the cryogenic system frequently.

The temperature distribution of high-pressure liquid (warm helium) with the position is acquired and it varies from 300 to 80 K, as shown in Fig. 13. This output temperature is achieved by using liquid nitrogen bath and passage of low-pressure cold helium in the adjacent channel of the heat exchanger, as shown in Fig. 11.

The temperature distribution for low-pressure cold helium is acquired, and it varies from 100 to 240 K, as shown in Fig. 14. This Helium is using to decrease in temperature of upcoming helium in the next channel. With the help of these results, thermal stress on fin and plates can be easily calculated. With the help of this temperature variation, the thickness of the thermal shield and fin can be improved to achieve higher effectiveness of the system.

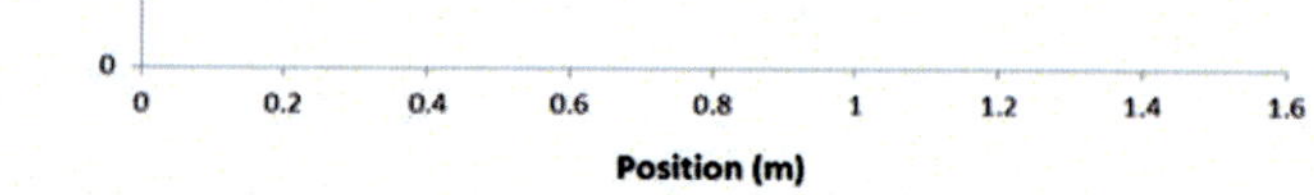

Fig. 13 Temperature distributions for high-pressure warm helium

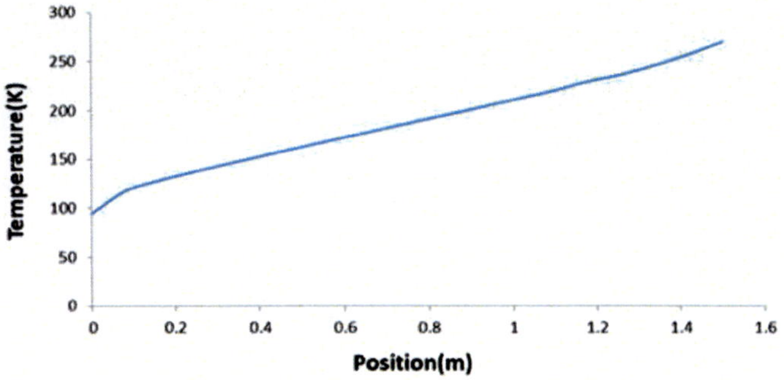

Fig. 14 Temperature distributions for low pressure cold helium

4 Conclusion

- The components should be arranged in such a way that proper temperature gradient can be maintained. That means there must be proper segregation of hotter as well as colder parts.
- The arrangements should be such that the warmer parts would be installed at the top portion and the colder parts at the lower portion inside the cold box. Therefore, colder regions are hanged from the top flange of the cold box.
- The material, having very low thermal conductivity, should be used as a separator in horizontal and vertical support systems. This arrangement will be useful to maintain a higher temperature gradient in horizontal and vertical support systems.
- This 3-D modeling has been done for proper visualization by using Solid-Works software, which shows all the components within the system including heat exchangers, valves, the piping systems with turbines and adsorbers. This will help to segregate all the parts.
- After modeling, the conduction effect has been analyzed. From this result, it can be concluded that most of the heat loss from the support system is through conduction heat transfer.

Acknowledgements The authors sincerely thank the Variable Energy Cyclotron Centre, Kolkata for providing the facilities to carry out the project.

References

1. Yu F, Li Y, Zhu Y (2009) Numerical and experimental investigation on the thermal insulation performance of low temperature cold box. Int Commun Heat Mass Trans 36(9):908–911
2. Brownell LE, Young EH (1959) Process equipment design: vessel design. Wiley, New York
3. Hauviller C (2007) Design rules for vacuum chambers
4. Timmerhaus KD, Thomas MF (2013) Cryogenic process engineering. Springer, New York

5. Patankar SV, Prakash C (1981) An analysis of the effect of plate thickness on laminar flow and heat transfer in interrupted-plate passages. Int J Heat Mass Transf 24(11):1801–1810
6. Gentzlinger RC, Kirk EC (1991) A guideline for design, analysis, and fabrication of vacuum vessels for cryogenic accelerators. In: IEEE particle accelerator conference, vol 4. IEEE
7. Nayyar ML (1992) Piping handbook, vol 1, no 45370771. Mcgraw-hill, New York
8. Geyari C (1976) Design considerations in the use of stainless steel for vacuum and cryogenic equipment. Vacuum 26(7):287–297

Effect of Porosity on the Performance of an GM-Type DIPTR Using CFD

Pankaj Kumar, Ajay Kumar Gupta, Sanjay Kumar Gupta, and R. K. Sahoo

1 Introduction

The first report on pulse tube with a difference of pressure profile wave at lower frequency in early 1960s [1], which is called as pulse tube refrigeration (PTR). After that in 1964 Gifford-McMahon describe the development in PTR. Numerous progresses in such area have been described, whereas GM-type cryocooler is used for higher cooling [2–4]. The important arguments, need of examining a pulse tube is still a question mark or requirement, remains debatable. There is still requirement to understand the phenomena of Pulse tube for the desired phenomenon. Therefore, it is necessary to establish a new methodology using CFD software [5]. So the basics requirement is to establish new research on the basis of earlier literature. From the aforementioned literature review can be decided that 3D problem analysis of PTR is a basic important need that has not been described in detail till now a day. Depending upon the acoustic still, there is an absence of work in this field of thermoacoustic. So, the present work focus on satisfactory effort from the prescience of industrial design. This paper underlines a process for optimizing the value of porosity in GM-type PTR that will afford the maximum probable cooling capacity for a stable compressor. Achieving this goal requires an analysis of the loss mechanisms in the PTR system. For the past many years, various Techniques and approaches have been followed at the

P. Kumar (✉) · S. K. Gupta
Department of Mechanical Engineering, GMR IT, Rajam, AP, India
e-mail: pankajkumar@gmrit.edu.in

A. K. Gupta
Department of Mechanical Engineering, CSIT, Durg, Chhattisgarh, India

R. K. Sahoo
Department of Mechanical Engineering, NIT Rourkela, Rourkela, Odisha, India

M. Palanisamy et al. (eds.), *Theoretical, Computational, and Experimental Solutions to Thermo-Fluid Systems*, Lecture Notes in Mechanical Engineering,
https://doi.org/10.1007/978-981-33-4165-4_23

Cryogenic field to develop a highly efficient pulse tube refrigerator system and also to understand the physical mechanisms that help to find out its better performance [6]. In recent times, numerous literatures explain the potential of CFD in analysing inertance pulse tube refrigerator [7–9]. This proved us to establish robust numerical method to estimate the cooling capacity of a GM pulse tube. However, due to geometrical complexity, it is still a challenge for researchers. In present extensive numerical investigation, the operating and geometrical properties of an GMPTR, taken from literature [10], have been analyzed to attain the lowest cooling temperature at cold heat exchangers. To start with, the excellence in GMPTR has been absorbed and a numerical technique has been used. On different issues in CFD investigations have been reported in detail. However, the reference has been made to achieve steady-state condition with low computational time.

2 CFD

The heat transfer and flow simulation model integrate the fluid dynamic equations of energy in the fluid and porous medium, continuity equation and Navier-Stokes equation [11]. The various regenerator and heat exchangers are modeled as porous medium through corresponding material properties of the heat exchangers and regenerator. It is the responsibility of a regenerator to transfer a particular acoustic power to the cold end from the compressor in such a way that it would contain minimum losses. The governing equations for energy, mass, and momentum [12, 13] corresponding to porous media have been solved concurrently with the gas-phase equations corresponding to an PTR.

2.1 Essential Boundary Conditions with Important Assumptions

From latest literature review, it can be observed that the simulation analysis of GM-type PTR in computational fluid dynamics (CFD) still carry as a hard task for research scholars of relevant area due to computational timing overhead. The present extended part of numerical investigation attempts to resolve such matter in detail. The computational fluid dynamics of the GM pulse tube refrigerator has been investigated in ANSYS/Fluent® platform for a given dimension [14]. ANSYS meshing module is performing finite element discretization. Number of nodes can be reduced by dropping the mid-side node. Fluid flow problem can be solved by boundary layer meshing. A pressure profile of very large amplitude such as 5 bar on the operating pressure of 16 bar is produced in the current situation. So that the effect of the boundary layer (BL) is minimized due to oscillation to reduce the computational time overhead. Firstly, in the numerical simulation analysis, the behavior of temperature in the system has been

Table 1 Dimension and thermal boundary conditions of GM-type PTR

S.N.	Components	Radius (m)	Length (m)	Thermal BCs
a.	Transfer line	0.0053	0.1150	Isothermal
b.	After cooler	0.022	0.020	Isothermal
c.	Regenerator	0.020	0.21	Adiabatic
d.	CHX	0.015	0.18	Adiabatic
e.	Pulse tube	0.015	0.25	Adiabatic
f.	HHX	0.022	0.020	Isothermal
g.	Reservoir	0.10	0.30	Isothermal

constant that assumed to be at 293 K. Table 1 shows the essential thermal boundary conditions on all nine components, equivalent to the executed experiment [15]. The numerical simulation has been taking place using some assumptions given below:

- The computational time overhead is depending upon the number of elements as well as nodes. Furthermore, the motion of piston simulation done by applying dynamic meshing which adds more significantly on the computational overhead.
- The minimization of computational overload for parametric optimization is possible by the help of symmetric section analysis on the simulation.
- In the numerical simulation, ideal gas model is considered by solving different equation like energy, momentum, Navier-Strokes continuity equations with click on $(k - \epsilon)$ turbulent model in fluent to simulate the behavior of flow into the porous material.
- Considered 16 bar operating pressure and 2 Hz frequency of compressor piston arrangement motion in GM-type PTR [15].
- In real-time investigation [15], the after cooler (AFT), regenerator (REGEN), cold heat exchanger (CHX), and hot heat exchanger (HHX) have been ready with the help of fine-wire meshes, i.e., Copper (Cu) wire of 110 mesh size). In numerical simulation, some components out of nine have been created as porous material media applying equivalent viscous resistance (1.085e9 1/m2), inertial resistance (5750 1/ m) and porosity (0.69), following empirical models [16].
- The equations of conduction w.r.t porous material model have been resolved with the help of steel as material away from CHX, by taking Cu.

3 3D-Axisymmetric Analysis of GM-Type Pulse Tube Refrigerator

The stability and reliability of the proposed model, can be justified by taking literature review of a GM-type PTR, has been experimentally examined [17]. The dimension of all the components of GM-type PTR has been tabulated in Table 1. The

Fig. 1 Geometry of the GM-type PTR under analysis; **a** 3D geometry, and **b** sectional geometry

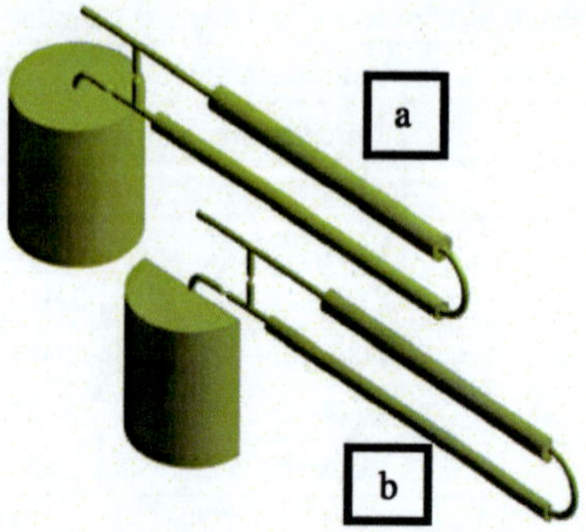

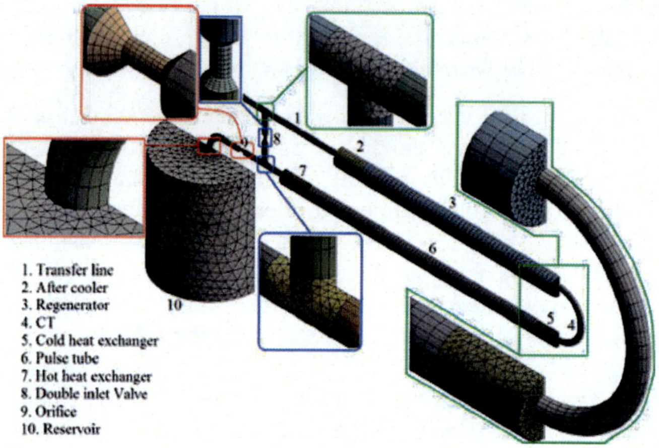

Fig. 2 Meshing of 3D-axisymmetric parts of an GM-type PTR

detailed research has been carried out to determine the complexities i.e. the necessity of preferred meshing, proper node contact between two-part, different boundary conditions (BCs), solver formations, and implementation of User Defined Functions (UDF) used in numerical simulation to find out the desired results. Figure 1 shows the complex geometry of the GM-type pulse tube refrigerator for investigation. To save the computational overhead time, only take axis-symmetric sectional geometry has been simulated during CFD (Fig. 2).

4 Governing Equation

Subsequent paragraphs, however, are indented. All over the system, continuum-based conservation equations can be implemented. The energy, mass and momentum equations explained by fluent (CFD) are as follows [18]:

$$\frac{\partial \rho_f}{\partial t} + \frac{1}{r}\frac{\partial}{\partial r}(r\rho_f V_r) + \frac{\partial}{\partial x}(\rho_f V_x) = 0 \tag{1}$$

$$\frac{\partial}{\partial t}(\rho_f \vec{v}) + \nabla(\rho_f \vec{v}\vec{v}) = -\nabla p + \nabla.[\overline{\overline{\tau}}] \tag{2}$$

$$\frac{\partial}{\partial t}(\rho_f E) + \nabla \cdot [\vec{v}(\rho_f E + p)] = \nabla \cdot [k_f \nabla T + (\overline{\overline{\tau}} \cdot \vec{v})] \tag{3}$$

where

$$E = h - \frac{p}{\rho} + \frac{v^2}{2} \tag{4}$$

One point is very important, i.e., all the properties are valid only when we take helium as a working fluid. All the above equation is relevant except four components, i.e., after cooler, regenerator, HHX, and CHX. Behaviour of these four components such as after cooler, regenerator, HHX, and CHX working as a porous media. During porous media model, resistance of gas flow behavior is solving by some additional source term like momentum to the standard fluid flow equations. An additional pressure drop term is work as source term due to concept of porous media come in picture. Source term can be divided into two different parts: first one is an inertial loss term and other is a viscous loss term. The source term for homogeneous porous media is [5].

$$S_i = -\left(\frac{\mu}{\alpha}\bar{v} + C_2\frac{1}{2}\rho_f|\bar{v}|\bar{v}\right) \tag{5}$$

where, C_2 and α are the inertial resistance factor and the permeability, respectively. Thus, the energy, mass, and momentum equations for the porous components can be expressed as [19]

$$\frac{\partial}{\partial t}(\phi\rho_f) + \frac{1}{r}(\phi r \rho_f v_r) + \frac{\partial}{\partial r}(\phi\rho_f v_x) = 0 \tag{6}$$

$$\frac{\partial}{\partial t}(\phi\rho_f \vec{v}) + \nabla(\phi r \rho_f \vec{v}\vec{v}) = -\phi\nabla p + \nabla.[\phi\overline{\overline{\tau}}] + S_i \tag{7}$$

$$\begin{aligned}&\frac{\partial}{\partial t}(\phi\rho_f E_f + (1-\phi)\rho_s E_s) + \nabla \cdot [\vec{v}(\rho_f E_f + p)] \\ &= \nabla \cdot [\phi k_f + (1-\phi)k_s \nabla T + (\phi\overline{\overline{\tau}} \cdot \vec{v})]\end{aligned} \tag{8}$$

where 'ϕ' is Porosity of the porous components; k_f and k_s is 'Thermal conductivity' of fluid structure and solid structure, respectively. For the solid component fix the axial thermal conductivity up to be 10% of the screen material by the thermal resistance contact between the layer of screen [20]. During the porous media, the following 1-D empirical equations are considered for the pressure gradient [19]:

$$\frac{\partial p}{\partial x} = -f_{\mathrm{F}} \frac{2\rho_{\mathrm{f}} v^2}{d_h} \tag{9}$$

where f_{F} is shows fanning friction factor on steady state flow correlation and expressed are as follows:

$$f_{\mathrm{F}} = \frac{n\phi {d_h}^2}{4l\beta}\frac{33.6}{R_{\mathrm{eh}}} + 0.337\frac{n\phi d_h}{4\beta^2}, \; R_{\mathrm{eh}} = \frac{\rho_{\mathrm{f}} v d_h}{\mu} \tag{10}$$

Putting the value of Eq. (10) into the Eq. (9) after that in the porous media, pressure gradient becomes

$$\frac{\partial p}{\partial x} = -\frac{33.6 n\phi v\mu}{2l\beta} - 0.337\frac{n\phi^2 \rho_{\mathrm{f}} v^2}{2\beta^2} \tag{11}$$

where 'l' is known as mesh distance; 'β' is known as the ratio of screen opening area; 'dh' is known as the hydraulic diameter, 'n' is the number of packed screens/length; 'μ' is known as viscosity. Comparison of Eq. (11) with Eq. (5) tends to

$$D = \frac{1}{\alpha} = \frac{33.6 n\phi}{2l\beta}; \; C_2 = \frac{0.337 n\phi^2}{\beta^2} \tag{12}$$

For a perfect square type mesh screen during which no inclination causes by the weaving of the wires and separation of screen layers are not done, ϕ and β can be calculated with the help of given formula [21]

$$\phi = 1 - \frac{\pi}{4x_t}, \; x_t = \frac{0.0254}{d_{\mathrm{w}} m}, \; \beta = (x_t - 1)^2 / {x_t}^2 \tag{13}$$

where, 'm' is mesh/inch; 'd_{w}' is screen wire diameter (m). Porous medium of regenerator, 200 mesh of 304SS wire mesh size and for another heat exchangers (HE) copper wire mesh of 100 mesh size is used during the numerical simulation.

5 Numerical Optimization

Till now, it has been observed that the numerical simulation needs computational optimization to decrease the computational time overheads. The present research has planned to reduce the computational overheads. In general, the piston simulation can be carried out by implementing the dynamic meshing and motion of a piston [16]. First, a user-defined function (UDF) of pressure has been implemented at inlet (inlet of transfer line) to simulate the piston oscillation, mentioned below [22].

```
# include "udf.h" Define_Profile (unsteady_pressure, thread, position)
{
face_t f;
real t = Current_Time;
begin_f_loop (f, thread)
{
F_PROFILE (f, thread, position) = 101325.0 × (5:0 ×
Sin (2 × 3:1416 × 2000 × t));
}
end_f_loop (f, thread)
}
```

As the number of nodes and elements are increase, the computation overhead time for each iteration is high. It should also be noticed that the GM pulse tube runs generally less than 5 Hz and in the present case it is 2 Hz. So the simulation time in transient analysis is high. In order to reduce the computational time, an initial temperature profile has been applied on regenerator to pulse tube.

In order to remove aforementioned numerical error, next, a sinusoidal temperature profile UDF has been applied on regenerator and pulse tube. From current literature review, it can be noticed that the temperature along the axial direction of the pulse tube and regenerator varies very closely to a quarter wavelength sine wave.

By observing the above work we can say that implementation of pressures UDF to solve the piston and transient model simulation by steady-state solution using sinusoidal profile temperature reduces the computational time. In line with this proposed method, the GM type PT has been explored. From experiment data, it has been detected that the lowest temperature, at CHX in steady state is 64.7 K.

Therefore, the steady-state temperature is assumed to be 95 K at CHX for simulation. The consistent cooling temperature profile curve for 25 s with a time-step of 0.0015 s has been given in Fig. 3, overlapped with experimental result [17]. Figure 3 shows that the minimum temperature at CHX reach is very near to 64.7 K.

For comprehensive information, the contour diagram of pressure, temperature and density configuration have been take out and which is display in Fig. 4. From contour diagram, it has been clearly understood that the temperature variation distribution progressively decreases in regenerator (REGEN) but in pulse tube (PT) it behavior is not that steady. The observed axial temperature over GM pulse tube has been shown in Figs. 5 and 6.

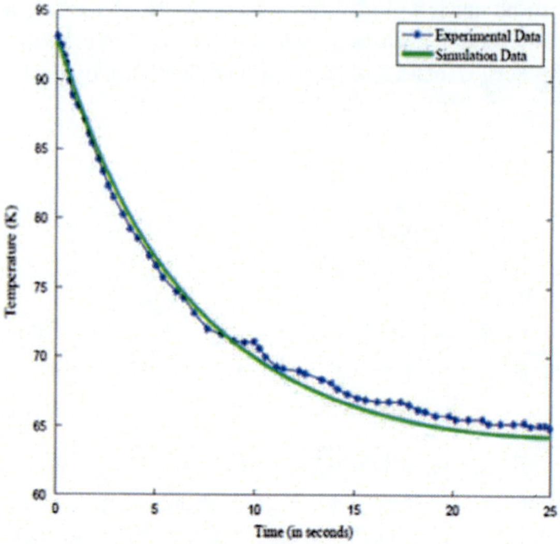

Fig. 3 Profile for cooling curve at CHX, validation with experimental data

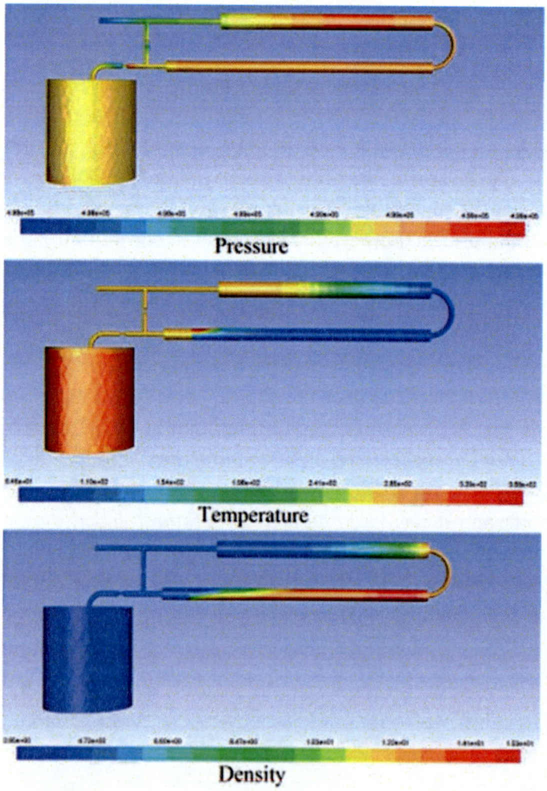

Fig. 4 Pressure, temperature and density contour of the GMPTR

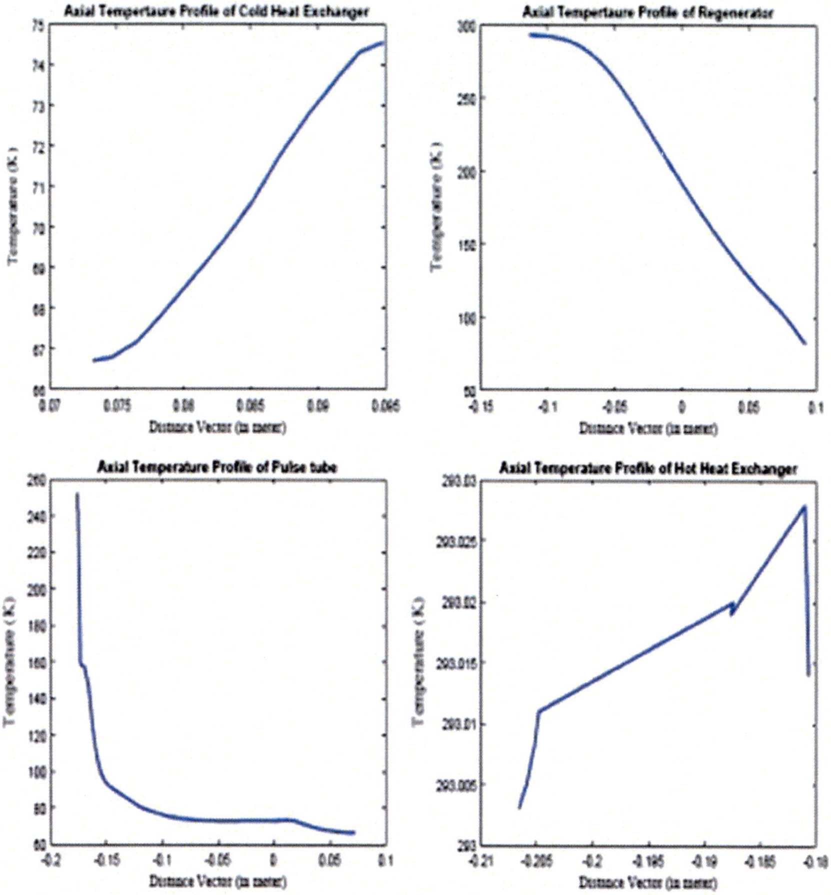

Fig. 5 Temperature profile of steady-state at the axis of the GM-type pulse tube

6 Porosity Effect on GM-Type Diptr

Regenerator is the most precious part of GM-type PTR. Effectiveness of regenerator is one of the most important parameters on which performance of the PTR is depended. The GM pulse tube refrigerator is in general used in a situation where cryogenic temperature is required with high capacity. In other words, such cryocooler is mostly used to produce liquid nitrogen, etc. However, the performance of GM-type PTR depends upon the Porosity. So the major challenge is to estimate the optimal value of porosity which in general used to be done by changing the value of porosity in different number of case studies (Tables 2 and 3).

In this paper, a different type of case study has been conducted to find out the value of porosity through which it will get a better cooling performance at the end of cold

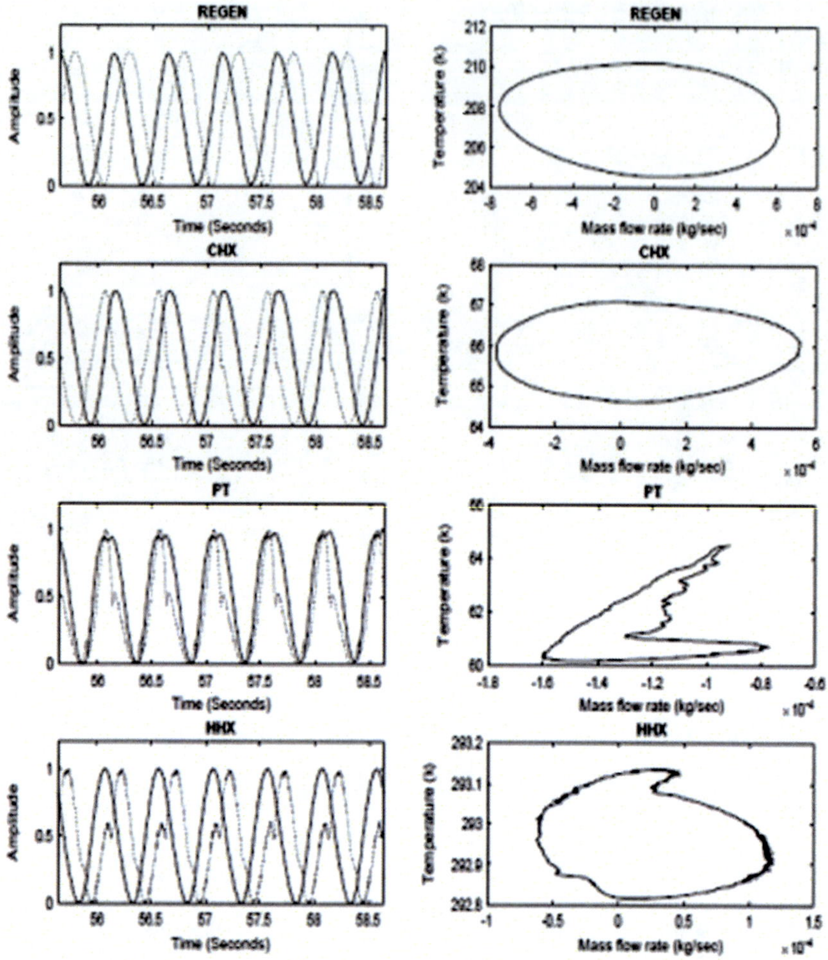

Fig. 6 Profile of Mass flow versus temperature at Mid portion of HHX, PT, CHX and REGEN

Table 2 Variation in porosity [23]

Case-A	Case-B	Case-C	Case-D	Case-E	Case-F
0.4	0.5	0.6	0.7	0.8	0.9

Table 3 Estimated steady-state temperature at CHX corresponding to variation in porosity

Case-A	Case-B	Case-C	Case-D	Case-E	Case-F
56.55 K	59.79 K	45.6 K	40.89 K	66.84 K	69.6 K

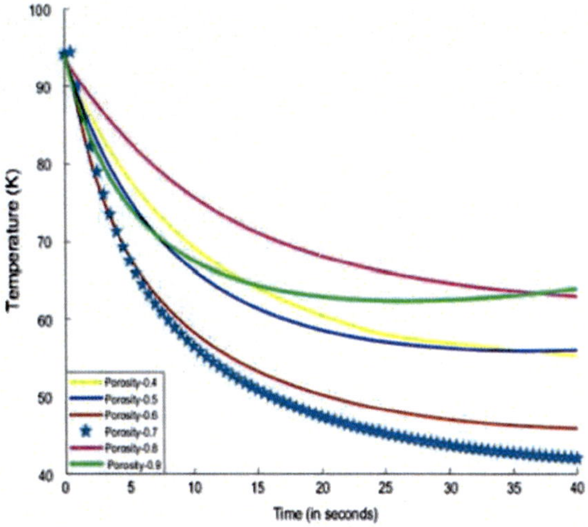

Fig. 7 Cooling curve at CHX over time

end of the pulse tube. From the CFD analysis, it was establish that the 0.6 porosity value gives the optimal result found comparative to different value of porosity which is shown in Fig. 7.

In order to demonstrate the porosity variation in a GM pulse tube, the temperature, pressure, and density of the contour plot have been retrieved and describe in Fig. 8. A detailed investigation has been done by extracting the temperature and mass flow rate versus temperature (m–T) profile at mid-point of regenerator, cold heat exchanger, PT and hot heat exchanger at steady state. The curves have been shown in Fig. 9. The observed phase angle at mid of CHX, REGEN, HHX and PT are 73.02, 131.36, 151.23 and 62.72°, respectively. Comparing with experimental data observation, the phase angle decreased at CHX, PT and increased at HHX, REGEN. Moreover, from m–T curves, it can be inferred that the bounded area has been increased at CHX, REGEN and decreased at HHX, PT.

7 Conclusions

The following results can be encapsulated as mentioned below:

- GM-type pulse tube, which had been experimentally investigated, has been taken for numerical investigation, where the demonstrated numerical result agrees to experimental result adequately.
- The investigations of current numerical simulation to found the robust numerical study of a GM pulse tube, however, applying suggested methodology such as implementing UDFs for eradicating the necessity of simulating piston and initialization of transient simulation from a steady state.

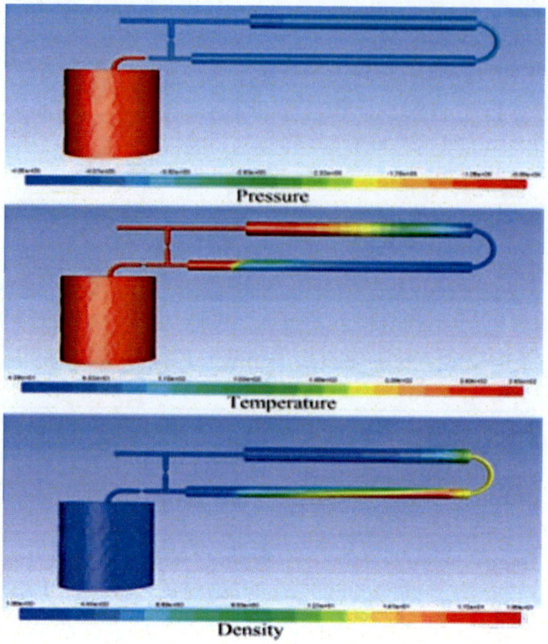

Fig. 8 Pressure, temperature and density contour of the optimized GMPTR

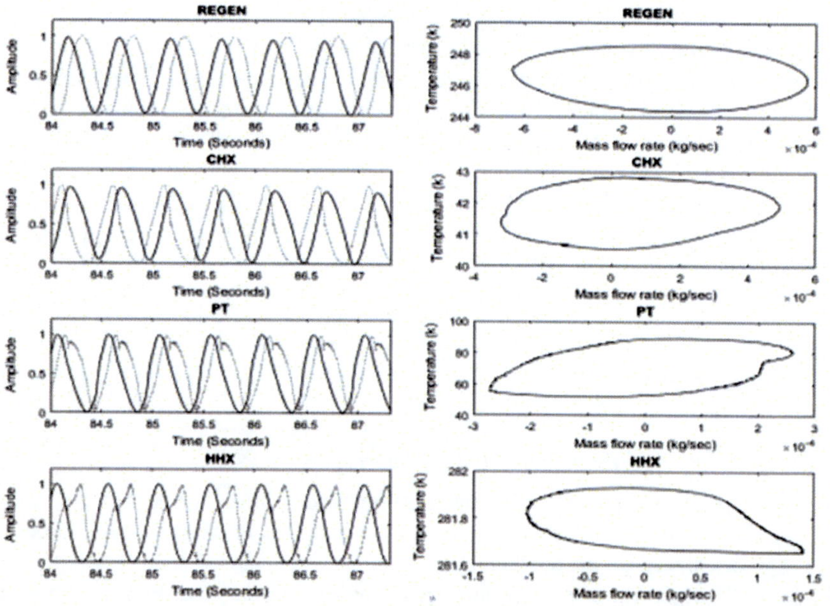

Fig. 9 Mass Flow versus temperature (MF-T) profile at mid-point of REGEN, CHX, PT and HHX

- The proposed method such as introducing a porosity in GM-type pulse tube refrigerator, the efficiency of the present system has been improved significantly. The lowest temperature at CHX has been observed at 45.6 K, which is quite impressive.

References

1. Gifford WE, Longsworth R (1964) Pulse-tube refrigeration. J Eng Indus 86(3):264–268
2. Yuan J, Maguire J (2005) Development of a single stage pulse tube refrigerator with linear compressor. In: Cryocoolers 13. Springer, pp 157–163
3. Fujimoto S, Kang Y, Matsubara Y (2002) Development of a 5 to 20 w at 80 k gm pulse tube cryocooler. In: Cryocoolers 10. Springer, pp 213–220
4. Ravex A, Poncet J, Charles I, Bleuze P (1998) Development of low frequency pulse tube refrigerators. In: Advances in cryogenic engineering. Springer, pp 1957–1964
5. Ansys AF (2009) 12.0 Theory guide, ANSYS. Inc., April
6. Radebaugh R (2000) Development of the pulse tube refrigerator as an efficient and reliable cryocooler. Proc Inst Refrig 96:1999–2000
7. Cha J, Ghiaasiaan S, Desai P, Harvey J, Kirkconnell C (2005) Cfd simulation of multi-dimensional effects in an inertance tube pulse tube refrigerator. In: Cryocoolers 13. Springer, pp 285–292
8. Gustafson S, Flake B, Razani A (2006) Cfd simulation of oscillating flow in an inertance tube and its comparison to other models. In: AIP conference proceedings, vol 823. AIP, pp 1497–1504
9. Luo E, Radebaugh R, Lewis M (2004) Inertance tube models and their experimental verification. In: AIP conference proceedings, vol 710. AIP, pp 1485–1492
10. Matsubara Y, Gao J (1995) Multi-stage pulse tube refrigerator for temperatures below 4 k. In: Cryocoolers 8. Springer, pp 345–352
11. Das MK, Mukherjee PP, Muralidhar K (2018) Equations governing flow and transport in porous media. In: Modeling transport phenomena in porous media with applications. Springer, pp 15–63
12. Wang C, Gu W, Liaw B (1998) Micro-macroscopic coupled modeling of batteries and fuel cells i. model development. J Electrochem Soc 145(10):3407–3417
13. Wang C, Cheng P (1997) Multiphase flow and heat transfer in porous media. In: Advances in heat transfer, vol 30. Elsevier, pp 93–196
14. Kasthurirengan S, Jacob S, Karunanithi R (2000) Development and studies on convection free single stage pulse tube cooler operating at 77 k. Technical report, Final technical report, April 2000, Centre for cryogenic Technology IISC, Bangalore
15. Kasthurirengan S, Jacob S, Karunanithi R, Nadig DS, Behera U (2001) Indigenous development of rotary valve for cryocooler applications. J Instrum Soc India
16. Banjare Y, Sahoo R, Sarangi S (2009) CFD simulation of a Gifford–Mcmahon type pulse tube refrigerator. Int J Therm Sci 48(12):2280–2287
17. Jacob PS (2007) Cryogenic research and technology at IIsc. Technical report, Centre for Cryogenic Technology Indian Institute of Science, Bangalore, India
18. Barrette F, Arsalan R (2004) Modeling pulse tube cryocoolers with CFD. Adv Cryogenics Eng 49:1493–1499
19. Kwanwoo N, Sankwon J (2005) Novel flow analysis of regenerator under oscillating flow with pulsating pressure. Cryogenics 45:368–379
20. Cha JS, Ghiaasiaan SM (2006) Multi-dimensional flow effect in pulse tube refrigerators. Cryogenics 46:658–665
21. Robbert A (1997) A cryogenic regenerative heat exchangers. Plenum Press, New York, pp 48–54

22. Kumar P, Gupta AK, Sahoo RK, Jena DP (2019) Numerical investigation of a 3D inertance pulse tube refrigerator from design prospective. Cryogenics 98:125–38
23. Rout SK, Choudhury BK, Sahoo RK, Sarangi SK (2014) Numerical study and analysis of inertance-type pulse tube refrigerator. WSEAS TRANSACTIONS on HEAT and MASS TRANSFER, pp 1–7

Numerical Analysis of Pulse Tube Cryocooler and Optimization Using Taguchi Method

Ajay Kumar Gupta, Pankaj Kumar, and R. K. Sahoo

1 Introduction

The Coefficient of Performance (COP) of pulse tube cryocooler mostly depends on the dimensional parameter that is the length and diameter of pulse tube and regenerative. By taking all the data, the optimum COP of Pulse tube refrigerator has been found out by making software in MATLAB and then optimized in MINITAB. So the main objective of the project is to evaluate the actual COP (cryogenic temperature) of a pulse tube cryo-cooler through numerical and optimized the parameters effecting the PTR in MATLAB and MINITAB software and the process parameters are cooling capacity, cold end temperature, effectiveness, ineffectiveness, diameter and length of pulse tube and regenerator, mass flow rate in PT, volume of the compressor of the PTR.

The first report was generated on pulse tube with a difference of pressure profile at low frequency wave in early 1960s [1], which is known as pulse tube refrigeration (PTR). In 1964 Gifford–McMahon explain the development in PTR. So many number of progresses in the cryogenics area have been described, whereas different types of cryocooler are used for higher cooling performance [2–4]. The computational overhead time is the one of reason which may stop so many research scholars to

A. K. Gupta (✉)
Department of Mechanical Engineering, Shri Rawatpura Sarkar University, Raipur, India
e-mail: ajaythermal@gmail.com

P. Kumar (✉)
Department of Mechanical Engineering, GMR Institute of Technology, Rajam, India
e-mail: pankajkumar@gmrit.edu.in

R. K. Sahoo
Department of Mechanical Engineering, National Institute of Technology, Rourkela, India
e-mail: rksahoo@nitrkl.ac.in

M. Palanisamy et al. (eds.), *Theoretical, Computational, and Experimental Solutions to Thermo-Fluid Systems*, Lecture Notes in Mechanical Engineering,
https://doi.org/10.1007/978-981-33-4165-4_24

solve in 3D. So the complexity during meshing needs to find out adequately with certain level of meshing quantification. The possibilities in reducing the number of nodes need to be examined [5].

Computational fluid dynamics (CFD) software package has been used by Barrett et al. [6] to model the oscillating fluid flow under a pulse tube refrigerator (PTR). Sources for modelling of pulse tube refrigerator by CFD are done with some different case studies and the results described. Yarbrough et al. [7] have done another computational fluid dynamics modelling related to pressure drop in the regenerator with different wire mesh geometry. They have investigated the three different types of regenerator models by using CFD software. Anjun et al. [8] have investigate the combined both the study like experimental as well as numerical to find out the heat transfer effect of cryogens i.e. (He) gasses with thermo physical properties which is temperature dependent in a miniature tube. They have found that the heat transfer characteristics of cryogenic gas with temperature-dependent thermophysical properties (TDTP) are very different from those in the atmospheric condition with constant thermophysical properties. Cha et al. [9] have modeled the two inertance tube pulse tube refrigerator (ITPTR) systems operating under a various thermal boundary conditions using a computational fluid dynamics (CFD) code.

An artificial neural network (ANN)-based process model is proposed to establish a relation between input parameters and the responses. The model provides an inexpensive and time-saving substitute to study the performance of ITPTR. The model can be used for selecting ideal process states to improve ITPTR performance [10].

Xiao et al. [11] has done the computational fluid dynamic (CFD) simulation of an inter-phasing pulse tube cryocooler (IPPTC). Due to a reduction in turbulence, this IPPTC provides improved performance compared with single OPTR. An approximate numerical simulation analysis is conducted with obtaining approximate numerical solutions for predicting the behaviour of cooling capacity of OPTR while keeping reasonable bounds on errors [12]. Banjare et al. have performed the CFD simulations for OPTR [13] and ITPTR [14] respectively at different frequency by using a dual opposed piston compressor. Roach et al. [15] has performed the theoretical analysis of the behaviour of a typical pulse tube regenerator. de Waele [16] have discussed the dynamic behaviour of the temperature distribution in the regenerator and in the gas near the hot and cold ends of the pulse tube. With some simplifying assumptions done on the analysis, the basic properties of the temperature profile or distribution in the regenerator and in the tube are understood. The research on pulse tube cryocoolers has undertaken in the Los Alamos National Laboratory by Wheatley [17] who used a thermo-acoustic pressure wave generator instead of a mechanical wave generator. Jin et al. [18] has developed the single-stage co-axial pulse tube refrigerator (PTR) driven by a standing wave thermo-acoustic prime mover and a minimum temperature of 117.6 K is achieved in the tube.

2 Methodology

The present work uses formulations from the different papers on PTR and applied for developing codes on MATLAB. Validation of developed code with the references. Optimization of various parameters of pulse tube cryocooler such as cooling capacity, power required, dimension, etc., using MINITAB-based Taguchi method.

The goal of the research is to develop a reliable and scalable design tool for pulse tube refrigerators. To achieve the objective, we are developing a Matlab code following the flow chart and data below (Table 1).

The methods define the correlation factor and response affecting the process called DOE or Design of experiments. It can also find out the cause and effect correlation. In order to optimize the response, it is necessary to manage the output.

DOE abilities offer a technique for instantaneously exploring the properties of multiple variables on a response (output variable). These investigations contain a series of experiments or runs, tests, in which attentive changes are made to the variables in input or factors, and data are collected at each run. It helps to maximize the result (Fig. 1).

Table 1 Design data for PTR models

Components	Parameters
Compressor	Dead volume Swept volume
Regenerator	Length, $Lrg = 0.3$ m Dia., $drg = 0.032$ m Porosity = 0.7 Void volume, $Vdrg = 0.00015$ m^3 Hydraulic diameter, $d_h =$ 0.04 mm
Pulse tube	Length, $L_t = 0.8$ m Dia., $d_t = 0.02$ m
Cold end block	Dead volume, Vdcx = 0.00002 m^3
Hot end block	Dead volume, Vdhx = 0.00002 m^3
Cold end temp	100 k
Hot end temp	300 k
Orifice and DI valve	Diameters = 1 mm for both
Helium gas at 16 bar and 300 K temperature	Viscosity (Dynamic), $\mu = 15.21 \times 10^{-6}$ N s/m^2 $\rho = 2.389$ kg/m^3 $C_p = 5193.0$ J/kg K $R = 2074.6$ J/kg K, $\gamma = 1.67$

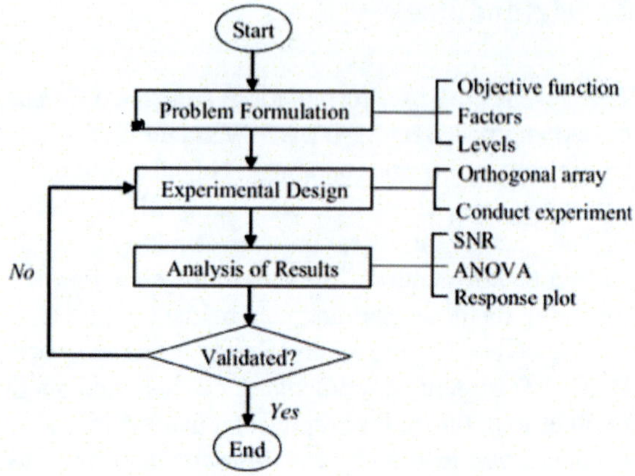

Fig. 1 Flow diagram of the Taguchi method [20]

3 Theoretical Study on Ptr—Adiabatic Gas Behaviour

3.1 Formula Used

1. Change in compressor volume:

$$\frac{dp_{cp}}{dt} = \frac{k}{V_{cp}}[-m_{cp}RT_{cp} - P_{cp}\frac{dV_{cp}}{dt}]$$

2. Pressure variation at pulse tube:

$$\frac{dP_t}{dt} = \frac{R(m_{cp} - m_{dt}) - \frac{T_h}{T_c}(m_0 - m_{dt})R - \frac{V_{ac}}{T_h}\frac{dP_{cp}}{dt}}{[\frac{V_{drg}}{T_{rg}} + \frac{V_{dcx}}{T_c} + \frac{V_t}{kT_c} + \frac{V_{dnx}}{T_c}]}$$

3. Pressure variation at buffer/reservoir:

$$\frac{dP_t}{dt} = \frac{1}{V_t}(-m_0RT_h)$$

4. Mass flow through regeneration:

$$m_{rg} = \frac{\rho\pi D^2{}_{rg}D^2{}_h\varnothing^3}{4\times 150L_{rg}\mu(1-\varnothing)^2}(\Delta P) = C_{rg}(P_{cp} - P_t)$$

where,

$$C_{rg=}\frac{\rho\pi D^2{}_{rg}D^2{}_h\varnothing^3}{4\times 150L_{rg}\mu(1-\varnothing)^2}$$

5. Mass flow through orifice:

$$m_0 = C_d A_0 \sqrt{2\frac{k}{k-1}\frac{P_t}{V_t}\left[\left(\frac{P_r}{P_t}\right)^{\frac{2}{k}} - \left(\frac{P_r}{P_t}\right)^{\frac{k+1}{k}}\right]} \text{ where , } P_t > P_r$$

$$m_0 = -C_d A_0 \sqrt{2\frac{k}{k-1}\frac{P_r}{V_t}\left[\left(\frac{P_t}{P_r}\right)^{\frac{2}{k}} - \left(\frac{P_t}{P_r}\right)^{\frac{k+1}{k}}\right]} \text{where, } P_t < P_r C_d \text{ is constant.}$$

$$\text{putting } V = \frac{RT}{P} \text{ for ideal gas}$$

$$m_0 = -C_d A_0 \sqrt{2\frac{k}{k-1}\frac{P^2{}_t}{RT_h}\left[\left(\frac{P_r}{P_t}\right)^{\frac{2}{k}} - \left(\frac{P_r}{P_t}\right)^{\frac{k+1}{k}}\right]} \text{where , } P_t < P_r$$

6. Mass flow through double inlet valve

$$m_{dt} = -C_{dt} A_{dt} \sqrt{2\frac{k}{k-1}\frac{P^2{}_{cd}}{RT_h}\left[\left(\frac{P_t}{P_{cd}}\right)^{\frac{2}{k}} - \left(\frac{P_t}{P_{cd}}\right)^{\frac{k+1}{k}}\right]} \text{where , } P_{cp} > P_t$$

$$m_{dl} = -C_{dl} A_{dl} \sqrt{2\frac{k}{k-1}\frac{P^2{}_t}{RT_h}\left[\left(\frac{P_{cp}}{P_t}\right)^{\frac{2}{k}} - \left(\frac{P_{cp}}{P_t}\right)^{\frac{k+1}{k}}\right]} \text{where, } P_{cp} < P_t C_{dl} = \text{constant} \ldots$$

3.1.1 Operating Condition of Adiabatic Model

Temperature Cold end = 100 K
Temperature Hot end = 300 K
Frequency = 2 Hz
Average pressure = 10 bar.

3.1.2 Fluid Data for Adiabatic Model

Helium gas at 200 K and 10 bar temperature have the following values:

$\mu = 15.21 * 10^{-6}$ N s/m^2
$R = 2074.6$ J/Kg K
$\rho = 2.389$ kg/m^3
$C_p = 5193.0$ J/Kg K
$K = 1.67$.

Table 2 Design data for adiabatic mode

Components	Parameters
Compressor	Dead volume = V_Q = 0.13 × 1O"^{3} m^3 Swept volume = V = = O.S x 1O"^{3} m^3
Regenerator	Length = L^, = 0.21 m Diameter = d_{r3} = O.Q2rn Porosity = O.T" Hydraulic diameter = dr, = 0.04 mm
Pulse tube	Length = L- = U.25 m Diameter = dt = O.OISm
Cold end block	Dead volume = V_{dCK} = 0.00002 m^2
Hot end block	Dead volume = $V_d$tix = O.QQQQ2m^3
Orifice	Diameter = 1 mm
DI valve	Diameter = 1 mm
Reservoir	Volume = O.Slftre
Average pressure	16 bar
Frequency	2 Hz
Cold end temperature	lOOK
Hot end temperature	300 K
Helium gas At 16 bar and 300 K temperature	Dynamic viscosity = p = 15.21 × 10"= IMs/m£ p = 2.3S9 ikg/rrr c_p = 5193.0 J/kg K R = 207A - &3 J kgi - K *r* = 1.67

Design Data for Adiabatic Model

See Table 2.

4 Result and Discussion

Model prepared in MATLAB gives the dimension parameters which are optimized in MINITAB based on Taguchi Method. Results of this section are obtained using the equations as provided in Sect. 3 (Fig. 3).

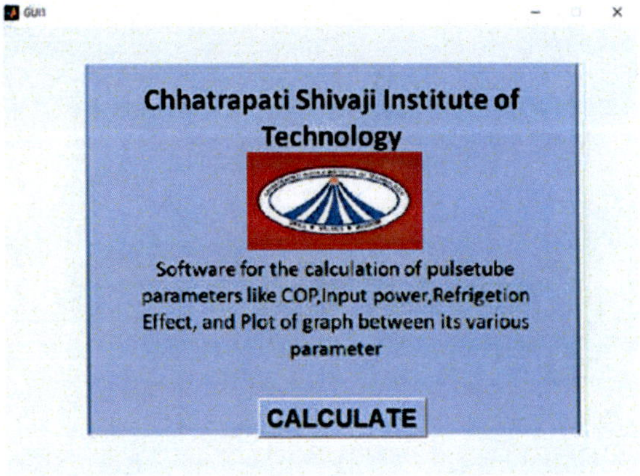

Fig. 2 Cover of software created in MATLAB [19]

4.1 Modelling

4.1.1 Model Prepared in MATLAB

See Fig. 2.

4.1.2 Representation of Figure

Figure 4 represents (A) Pressure at compressor, regenerator and pulse tube versus Time. (B) Mass flow rate through the hot end section, Double inlet valve, orifice versus Time. (C) Mass flow rate at cold end section, Dynamic pressure at pulse tube versus Time (D) Mass flow rate through the regenerator, Double inlet valve, orifice versus Time.

In our numerical experiments, we considered sinusoidal and step-function pressure variations shown in Fig. 5 and Fig. 6 represents the variation between mass flow rate at cold end section and the Dynamic pressure at Pulse Tube Time.

Figure 7 represent the graph between Mass flow rates through the regenerator, orifice, Double Inlet Valve versus Time and Fig. 8 represents the Mass flow rate through the hot end section, Double inlet valve, orifice versus Time (Fig. 9).

4.1.3 Data Interpretation and Analysis of Results

From the developed GUI in Matlab as shown in Fig. 10. The data has collected which is being used for optimization in the Tagughi method using MINITAB as shown in Fig. 10.

From Fig. 11 that is main effect plot for Signal Noise Ratios shows that optimum value of COP (Larger is better, that value is taken) is coming Regenerative length

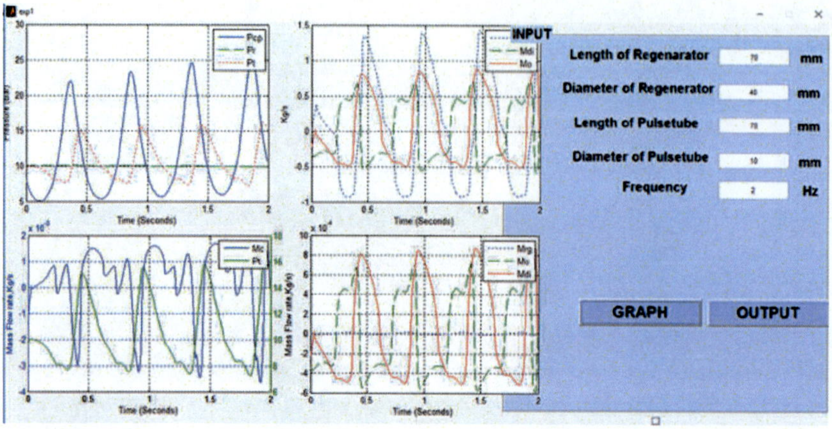

Fig. 3 Graph between pressure versus time [19]

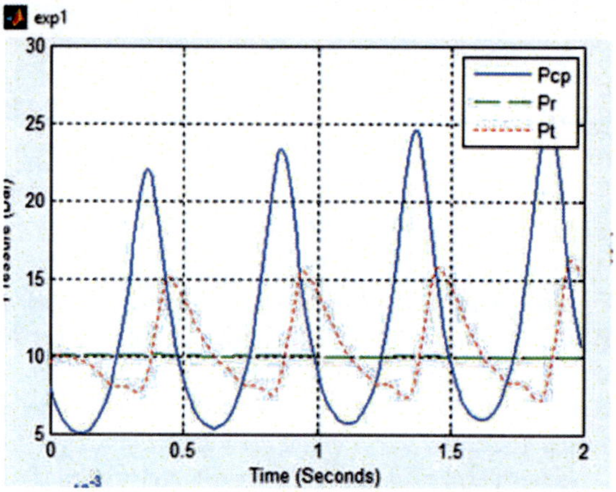

Fig. 4 Pressure at compressor, regenerator and pulse tube versus time [19]

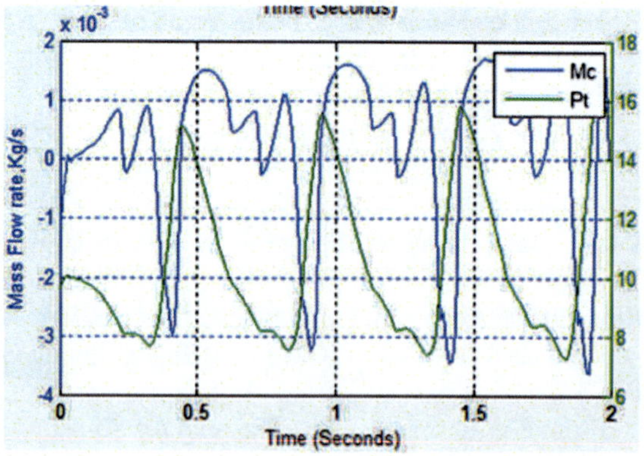

Fig. 5 Mass flow rate (PT) versus time [19]

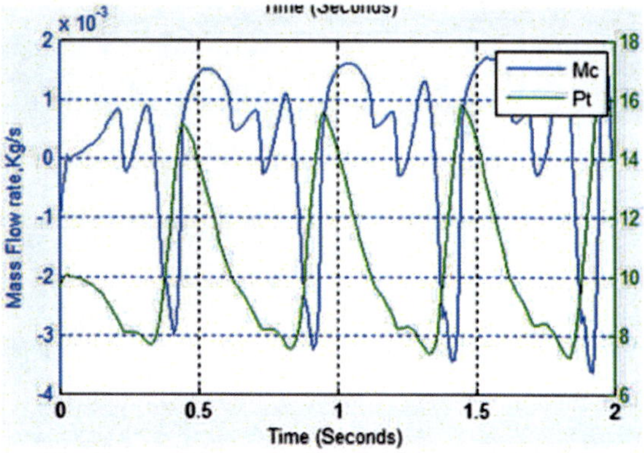

Fig. 6 Mass flow rate (Reg.) versus time [19]

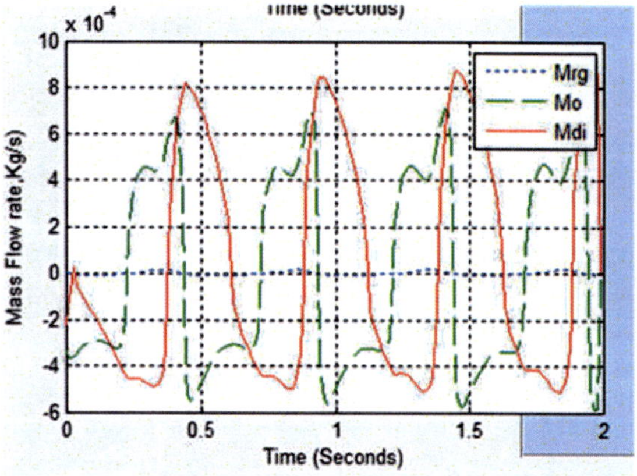

Fig. 7 Mass flow rate of hot end section versus time [19]

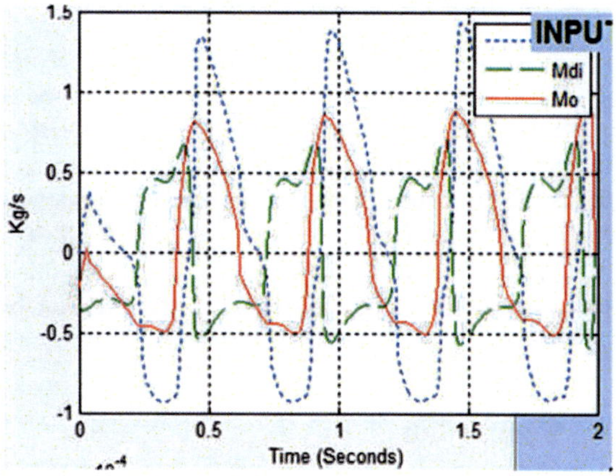

Fig. 8 Calculation for power and COP [19]

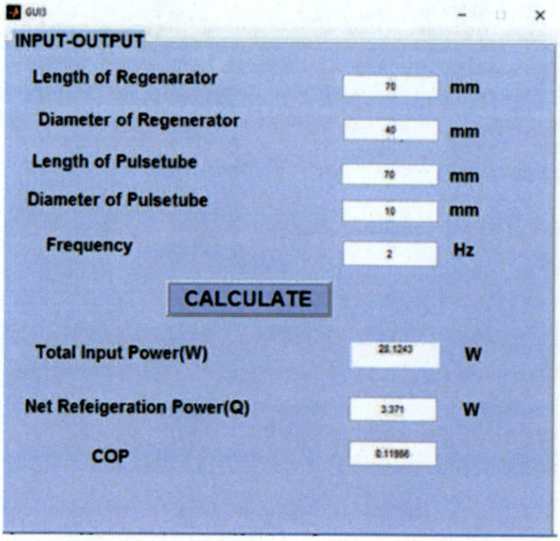

Fig. 9 Parameter optimization done in MINITAB [20]

	C1	C2	C3	C4	C5
	RGL	RGD	PTL	PTD	COP
1	50	25	50	10	0.11082
2	50	25	70	20	0.1108
3	50	40	50	20	0.11125
4	50	40	70	10	0.11125
5	70	25	50	20	0.11413
6	70	25	70	10	0.11413
7	70	40	50	10	0.11986
8	70	40	70	20	0.11986

Fig. 10 Parameter optimization done in MINITAB [20]

(RGL) has the most effect on COP of PTR and the corresponding value of Regenerative diameter (RGD) has less effect than the RGL. The Pulse tube length and diameter have the least effect on COP of PTR. Thus the optimized value of COP is 0.11986 and the corresponding dimension parameters that are RGL, RGD, PTL and PTD are 70, 40, 70 and 20 respectively or 70, 40, 50 and 10. All the dimensions are in mm. The best performances of the pulse tube refrigerator are in these dimension parameters are then verified from the numerical analysis.

5 Conclusion

A one-dimensional model has been developed for simulating the oscillatory gas flow in the tube section of a PTR. In this model, the two factors length and diameter of the

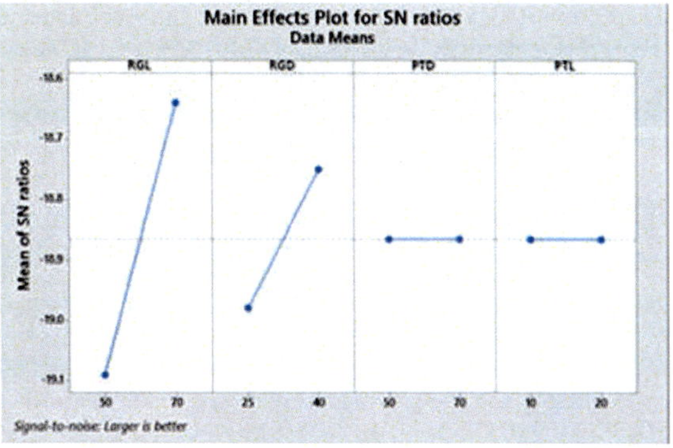

Fig. 11 Main effects plot for SN ratio [20]

Pulse tube and regenerator are considered to be the most detrimental factor which affect the behaviour of PTR. The present method turns out to be more accurate and versatile for calculating the size of pulse tube and regenerator before implementing in the physical model.

Pressure variation of pulse tube has been determined at different orifice opening. Pressure variation at regenerator inlet, pulse tube and reservoir has been shown assuming constant average pressure of the reservoir.

A comparison between theoretical pressure variation at regenerator inlet, pulse tube and reservoir has been made with the experimental result. A fairly good qualitative agreement has been observed.

References

1. Gifford WE, Longsworth R (1964) Pulse-tube refrigeration. J Eng Industry 86(3):264–268
2. Yuan J, Maguire J (2005) Development of a single stage pulse tube refrigerator with linear compressor. In: Cryocoolers 13. Springer, pp 157–163
3. Fujimoto S, Kang Y, Matsubara Y (2002) Development of a 5 to 20 w at 80 k gm pulse tube cryocooler. In: Cryocoolers 10. Springer, pp 213–220
4. Ravex A, Poncet J, Charles I, Bleuze P (1998) Development of low frequency pulse tube refrigerators. In: Advances in cryogenic engineering. Springer, pp 1957–1964
5. Kumar P, Gupta AK, Sahoo RK, Jena DP Numerical investigation of a 3D inertance pulse tube refrigerator from design prospective. Cryogenics 98:125–138 (2019)
6. Flake B, Razani A (2004) Modeling pulse tube cryocooler with CFD. Adv Cry Eng AIP Proc 49:1493–1499
7. Yarbrough A, Flake BA, Razani A (2004) Computational fluid dynamic modeling of pressure drop through wire mesh. AIP Proc Adv Cry Eng 49:1338–1345
8. Jiao A, Jeong S, Ma HB (2004) Heat transfer characteristics of cryogenic helium gas through a miniature tube with a large temperature difference. Cryogenics 44:859–866
9. Cha JS, Ghiaasiaan SM, Desai PV, Harvey JP, Kirkconnell CS (2006) Multidimensional flow effects in pulse tube refrigerators. Cryogenics 46:658–665, 185
10. Zhang X-B, Gan Z-H, Qui L-M (2008) Computational fluid dynamics simulation of an interphasing pulse tube cooler. J Zhejiang Univ Sci A 9(1):93–38

11. Kumar P, Gupta AK, Sahoo RK Approximation techniques for solving cooling capacity of orifice pulse tube cryocooler. IOSR J Mech Civil Eng (IOSR-JMCE), pp 15–19, e-ISSN: 2278-1684, p-ISSN: 2320-334X
12. Banjare YP, Sahoo RK, Sarangi SK CFD Simulation of inertance tube pulse tube refrigerator. In: 19th national and 8th ISHMT-ASME heat and mass transfer conference JNTU College of Engineering Hyderabad, India January 3–5, 2008 Paper No. EXM-7, PP34
13. Banjare YP, Sahoo RK, Sarangi SK CFD simulation of orifice pulse tube refrigerators. In: International conference on "recent trends in mechanical engineering IRCTME 2007, Dept. of Mech. Engg. Ujjain Engg College Ujjain October 4–6, 2007, Paper no. HT1, pp 235–245
14. Roach PR, Kashani A (1995) Theoretical analysis of a pulse tube regenerator. Adv Cryogenic Eng 41:1357
15. de Waele ATAM, Hooijkass HWG, Steijaert PP, Benschop AAJ (1998) Regenerator dynamics in the harmonic approximation. Cryogenics 38:995–1006
16. Wheatly JC, Hofler T, Swift GW, Migliori A (1985) Understanding some simple phenomenon in thermo acoustics with application to acoustical heat engines. Am J Phy 53:147–162
17. Jin T, Chen GB, Shen Y (2001) A thermo acoustically driven pulse tube refrigerator capable of working below 120 K. Cryogenics 41:595–601
18. Tua Q, Li Q, Guo F, Wu J, Liu J (2003) Temperature difference generated in thermo-driven thermo acoustic refrigerator. Cryogenics, 43:515–522
19. 2-D line plot—MATLAB plot—MathWorks India
20. Yusoff N, Ramasamy M, Yusup S (2011) Taguchi's parametric design approach for the selection of optimization variables in a refrigerated gas plant. Chem Eng Res Des 89:665–675

Entropy Generation Minimization of Vapour Absorption Heat Transformer

S. Sekar, P. Chandrasekar, S. Kumar, Abraham J. S. Jospher, R. Sheeja, T. N. Valarmathi, and G. M. Lionus Leo

Nomenclature

VAHT	Vapour absorption heat transformer
COP	Co-efficient of performance
p	Pressure
t	Temperature
CR	Circulation ratio
$\dot{m}$	Mass flow rate
GTL	Gross Temperature Lift
$\dot{S}$	Entropy generation
X	Solution concentration
$\dot{Q}$	Heat transfer rate

S. Sekar (✉) · P. Chandrasekar · S. Kumar · A. J. S. Jospher
Department of Mechanical Engineering, Rajalakshmi Engineering College, Chennai 602105, India
e-mail: sekar.s@rajalakshmi.edu.in

R. Sheeja
Department of Civil Engineering, Easwari Engineering College, Chennai 600089, India

T. N. Valarmathi
School of Mechanical Engineering, Sathyabama Institute of Science and Technology, Chennai 600119, India

G. M. Lionus Leo
Department of Mechanical Engineering, St. Joseph's College of Engineering, Chennai 600119, India

M. Palanisamy et al. (eds.), *Theoretical, Computational, and Experimental Solutions to Thermo-Fluid Systems*, Lecture Notes in Mechanical Engineering,
https://doi.org/10.1007/978-981-33-4165-4_25

Suffix

A	Absorber
E	Evaporator
G	Generator
C	Condenser
ws	Weak solution
ss	Strong solution
r	Refrigerant
gen	Generation
SHX	Solution heat exchanger

1 Introduction

Several systems use heat energy for converting into power, processes, etc. Most of these systems reject waste heat into the atmosphere. 72% of the global primary energy is lost after conversion [1]. This not only wastes energy but also pollutes the environment. If some of the heat energy is upgraded and used for useful purposes efficient use of resources will improve. It is not only important to save energy for economical, also to reduce CO_2 for environment protection [2]. Recovering such waste heat can provide power, heat or cooling output without extra energy input. This increases the energy utilizing efficiency and is considered to be a significant "technology wedge" with the potential to contribute a particular figure for emission reduction [3]. Up to 50% of the waste heat may typically be recovered [4].

The Intergovernmental Panel on Climate Change (IPCC) finds industrial waste heat recovery as one of the tools for CO_2 mitigation [5]. By increasing the temperature of these sources economically, i.e., upgrading the waste heat from any industry and energy sources like solar and geothermal energy, they can be utilized for the processes which need relatively higher temperature heat energy for their operations. This option is an attractive one, as the use of fossil fuels is reduced, and global warming is decreased. Heat transformers are thermally operated systems used for upgrading any low-temperature waste heat or solar or geothermal energy. With growing interest in waste heat recovery, absorption technology is experiencing resurgence [6]. Growing energy demand can be tackled by the exploitation of low-temperature waste heat with improvement in efficiency and reduction in cost and emissions. Waste heat utilization is becoming important and use of heat operated absorption heat transformers is experiencing a resurgence. The absorption heat transformer is a type II heat pump. In type I heat pump, low temperature heat is upgraded to higher temperature heat by the application of a heat input at a temperature higher than the output from the heat pump, whereas in type II heat pump, a medium temperature heat source is upgraded to higher temperature as shown in Fig. 1.

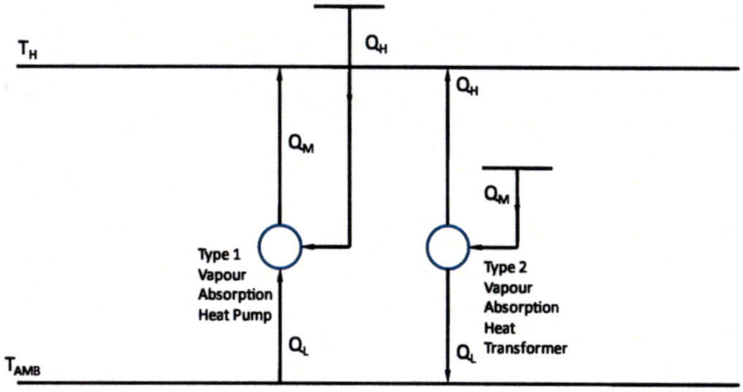

Fig. 1 Type I and type II absorption heat pumps

Thermodynamic analysis and improvement of power systems have two options, energy or first law based analysis and second law or exergy based analysis. First law analysis is based on quantitative analysis, where energy efficiency is calculated whereas in second law analysis qualitative analysis is carried out [7]. Recently, second law based methods have become very attractive [8]. Several models such as exergy analysis [9], irreversibility analysis [10], and entropy balance method [11] are available and some of these have been applied for absorption systems. Entropy generation is another method of second law analysis, which will indicate where and how, there are losses due to irreversibilities. Entropy generation analysis has been carried out for several thermal applications, namely thermoelectric systems for electronic cooling [12], evaporator in microscale refrigeration cycle [13], desalination [14], supercritical water reactor [15], falling film type plate-fin condenser/re-boilers [16], etc. Though entropy generation principle is applied for vapour absorption systems such as energy and entropy-based optimization for a single-stage refrigeration [17], entropy generation based thermodynamic analysis of absorption refrigeration system [18], second law analysis and entropy generation minimization of an absorption chiller [19], etc., entropy generation and minimization studies for vapour absorption heat transformer have not been found in the literature. So, in this work, an attempt has been made to analyze the vapour absorption heat transformer for entropy generation and minimization. A mathematical model has been developed and the results are presented in graphical form.

2 Working Principle

A VAHT works with two pressure and three temperature levels has a generator, a condenser, an evaporator, an absorber and a solution heat exchanger. Figure 2 shows a simple sketch of the VAHT. Waste heat at an intermediate temperature is added to the generator which vaporizes the working fluid (refrigerant), from the weak salt

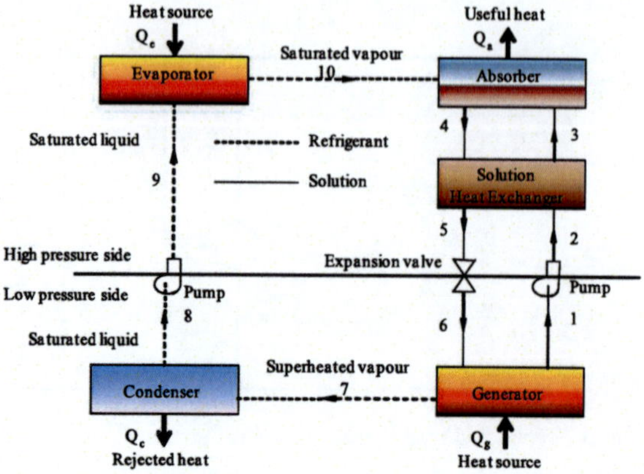

Fig. 2 Vapour absorption heat transformer

solution containing a low concentration of absorbent. The vaporized refrigerant flows to the condenser where it is condensed by cooling water and delivers some amount of heat at low-temperature level. The liquid refrigerant from the condenser is pumped to the evaporator. At the evaporator, the refrigerant is evaporated by the waste heat. The vaporized refrigerant then flows to the absorber, where the strong salt solution coming from the generator absorbs the vapour and heat is delivered at a usable temperature level.

Absorption of refrigerant vapour into solution is an exothermic reaction and generates useful heat at the highest temperature in the absorber. The absorption of refrigerant vapour into a strong solution makes it weak in absorbent. This weak solution from the absorber is fed to the generator. To increase the performance of the VAHT, a solution heat exchanger is provided between the generator and absorber.

3 Mathematical Modeling

3.1 Assumptions Used in the Analysis

The analysis has been carried out with the following assumptions:

1. The flow through the components is under steady state.
2. The working fluid at the exit of the generator and the absorber, the condenser and the evaporator are all assumed to be saturated.
3. The heat losses and gains through the various components and piping are neglected.
4. The pressure drops through the components are neglected.
5. Pure refrigerant vapour is liberated at the generator as the difference between the boiling point of lithium bromide (1265 °C) and water is very high.
6. The effectiveness of solution heat exchanger is 0.7.

7. Gross temperature lift is 20 °C.
8. The analysis is carried out for a generator heat input of 5 kW.
9. Refrigerant expansion is isenthalpic.
10. Pumping work is neglected.

Pressures in the main components,

$$pc = pg < pe = pa \tag{1}$$

Temperatures of the main components,

$$tc < tg = tc < ta \tag{2}$$

Circulation ratio (CR) is an important parameter used to design and optimize the vapour absorption transformer as is directly related to the size and cost of the components of the system. It is defined as the ratio of the mass flow rate of the solution leaving the absorber to the generator ($\dot{m}_4$) to the mass flow rate of the working fluid ($\dot{m}_7$).

$$CR, f = \frac{\dot{m}_{ws}}{\dot{m}_r} = \frac{\dot{m}_4}{\dot{m}_7} \tag{3}$$

The difference between the solution concentrations at the generator outlet and the absorber outlet is concentration differential,

$$\mathrm{d}X = \frac{X_{ss} - X_{ws}}{X_{ss}}(100) \tag{4}$$

Mass balance,

$$\dot{m}_{ws} = \dot{m}_{ss} + \dot{m}_r \tag{5}$$

Also, Mass balance is

$$\dot{m}_6 = \dot{m}_1 + \dot{m}_7 \tag{6}$$

Coefficient of Performance,

$$\mathrm{COP} = \frac{\dot{Q}_A}{\dot{Q}_G + \dot{Q}_E} \tag{7}$$

Gross Temperature Lift,

$$\mathrm{GTL} = T_A - T_G \tag{8}$$

Entropy generation in the absorber,

$$\dot{S}_{genA} = \left(\sum \dot{S}_{out} - \sum \dot{S}_{in}\right)_A \tag{9}$$

$$= \left(\frac{\dot{Q}_A}{T_A} + \dot{S}_4\right) - (\dot{S}_{10} + \dot{S}_3) \tag{10}$$

Entropy generation in the generator,

$$\dot{S}_{genG} = \left(\sum \dot{S}_{out} - \sum \dot{S}_{in}\right)_G \tag{11}$$

$$= (\dot{S}_1 + \dot{S}_7) - \left(\dot{S}_6 + \frac{\dot{Q}_G}{T_G}\right) \tag{12}$$

Entropy generation in the condenser,

$$\dot{S}_{genC} = \left(\sum \dot{S}_{out} - \sum \dot{S}_{in}\right)_C \tag{13}$$

$$= \left(\frac{\dot{Q}_C}{T_C} + \dot{S}_8\right) - \dot{S}_7 \tag{14}$$

Entropy generation in the evaporator,

$$\dot{S}_{genE} = \left(\sum \dot{S}_{out} - \sum \dot{S}_{in}\right)_E \tag{15}$$

$$= (\dot{S}_{10}) - \left(\dot{S}_9 + \frac{\dot{Q}_E}{T_E}\right) \tag{16}$$

Entropy generation in the solution heat exchanger,

$$\dot{S}_{genSHX} = \left(\sum \dot{S}_{out} - \sum \dot{S}_{in}\right)_{SHX} \tag{17}$$

$$= (\dot{S}_3 + \dot{S}_5) - (\dot{S}_2 + \dot{S}_4) \tag{18}$$

Total entropy generation of all the components,

$$\dot{S}_{gen} = \dot{S}_{genA} + \dot{S}_{genG} + \dot{S}_{genC} + \dot{S}_{genE} + \dot{S}_{genSHX} \tag{19}$$

In the thermodynamic mathematical modelling, the properties of water and steam are taken from standard steam tables and properties of lithium bromide solutions are calculated from the equations provided by Kaita [20]. Typical property values at various points of the cycle at generator and evaporator temperature of 60 °C, condenser temperature of 30 °C, GTL and solution heat exchanger effectiveness of 0.7 are presented in Table 1.

Table 1 Typical thermodynamic properties of each state point

State point	Chemical composition	Temperature (°C)	*X* (%)	$\dot{m}$ (kg/s)	*h* (kJ/kg)	*s* (kJ/kg K)
1	Strong solution	60	54.34	0.011	134	0.3488
2	Strong solution	60	54.34	0.011	134	0.3488
3	Strong solution	74	54.34	0.011	165.67	0.4276
4	Weak solution	80	45.67	0.0131	172	0.607
5	Weak solution	66	45.67	0.0131	145.4	0.513
6	Weak solution	66	45.67	0.0131	145.4	0.513
7	Water vapour	60	–	0.002	2607.11	7.908
8	Liquid water	30	–	0.002	125.7	0.297
9	Liquid water	30	–	0.002	125.7	0.297
10	Water vapour	60	–	0.002	2610.3	7.92

4 Results and Discussion

Simulations were carried by varying operating temperatures such as condenser temperature from 20 to 40 °C, heat source temperature or generator temperature from 60 to 80 °C with gross temperature lift of 20 °C and solution heat exchanger effectiveness of 0.7.

4.1 Effect of Heat Source Temperature

The effect of the heat source temperature on the COP at a condenser temperature of 20 °C, GTL of 20 °C and a solution heat exchanger effectiveness of 0.7, is presented in Fig. 3. The CR decreases with an increase in the heat source temperature. The decrease in the CR causes the increase in concentration ratio, which in turn leads COP to increase. Figure 4 depicts the variation of entropy generation with a heat source or generator temperature. When the generator temperature increases, the entropy generation rates at absorber and evaporator almost remain the same, whereas the entropy generation and total entropy generation values increase. This is due to the fact that the condenser is most affected by the generator temperature and corresponding pressure. As the entropy generation rates of evaporator and absorber are almost constant, the entropy generation of the condenser and total entropy generation show the same trend of variation.

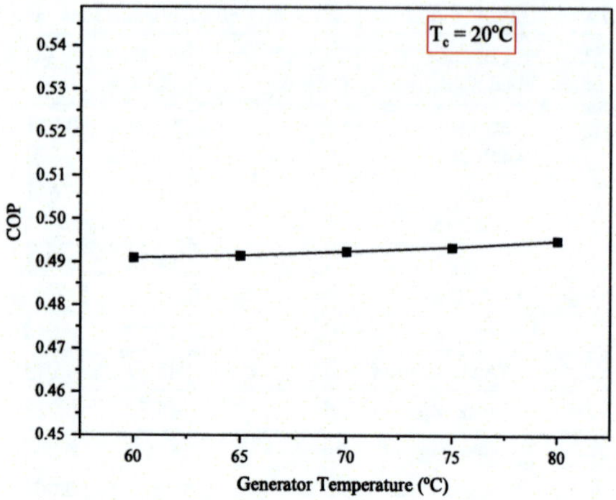

Fig. 3 Variation of COP with generator temperature

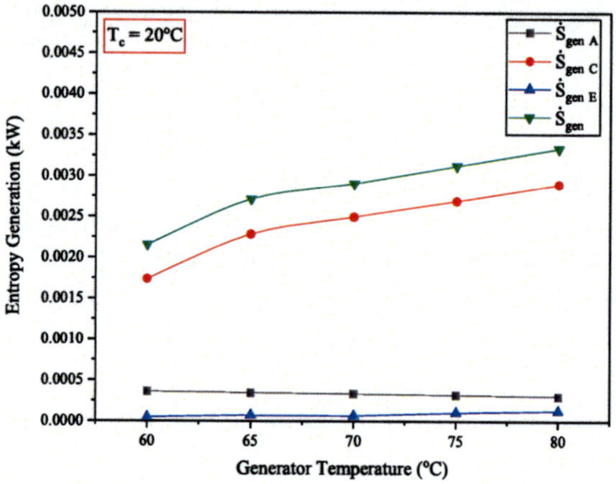

Fig. 4 Variation of entropy generation with generator temperature

4.2 *Effect of Condenser Temperature*

The influence of the condenser temperature on the COP is presented in Fig. 5, at a fixed generator and evaporator temperatures of 60 °C, the GTL of 20 °C and a solution heat exchanger effectiveness of 0.7. The condenser temperature was varied from 20 to 40 °C. With increasing Tc, the circulation ratio increases and the concentration differential decreases. This decrease in concentration differential lowers the COP. The decrease in COP is at a high rate after condenser temperature of 35 °C. The operation of vapour absorption heat transformer above this temperature will obviously lower the COP. Figure 6 shows the variation of entropy generation by varying the condenser temperature at a constant generator temperature of 60 °C, GTL of 20 °C and solution exchanger effectiveness of 0.7. Similar to the sudden change in the value of COP, after the condenser temperature of 35 °C, there is rapid change in the entropy generation

Fig. 5 Variation of COP with condenser temperature

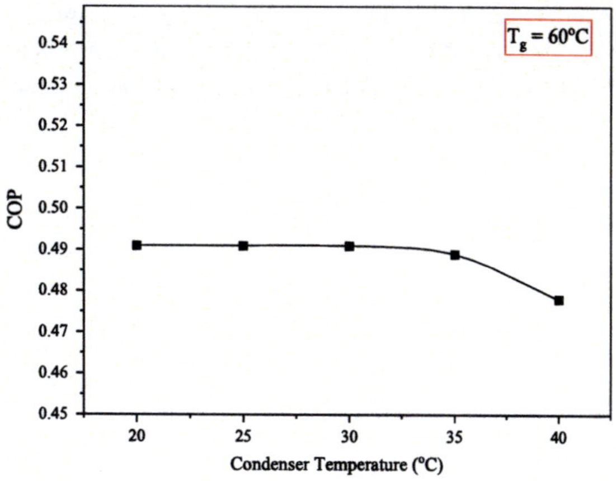

Fig. 6 Variation of entropy generation with condenser temperature

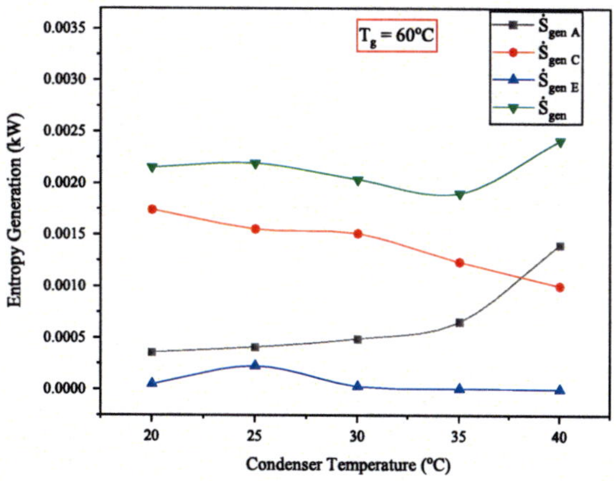

values. Entropy generation at the absorber increases to a large value causing an increase in the total entropy generation. This is again caused by the increase in circulation ratio at higher condenser temperatures.

5 Conclusion

Second law analysis with entropy generation at the components of vapour absorption heat transformer operating with lithium bromide working fluid has been performed. A mathematical was developed and the results have been presented. Among the working conditions considered, minimum entropy generation has arrived at generator temperature of 65 °C and condenser temperature of 40 °C, maximum COP has been obtained at generator temperature of 70 °C and condenser temperature of 35 °C and that is also best combination with lower entropy generation and higher COP. From the analysis, it is concluded that entropy generation analysis is a useful tool to find the qualitative performance of thermal systems and is suitable for optimizing the working condition for the best performance of the system.

References

1. Forman C, Muritala IK, Pardemann R, Meyer B (2016) Estimating the global waste heat potential. Renew Sustain Energy Rev 57:1568–1579
2. Rivera W, Best R, Cardoso MJ, Romero RJ (2015) A review of absorption heat transformers. Appl Therm Eng 91:654–670
3. Xu ZY, Wang RZ, Yang C (2019) Perspectives for low-temperature waste heat recovery. Energy 176:1037–1043
4. Donnellan P, Cronin K, Byrne E (2015) Recycling waste heat energy using vapour absorption heat transformers: a review. Renew Sustain Energy Rev 42:1290–1304
5. Miró L, Brückner S, Cabeza LF (2015) Mapping and discussing industrial waste heat (IWH) potentials for different countries. Renew Sustain Energy Rev 51:847–855
6. Roos: an overview of industrial waste heat recovery technologies for moderate temperatures less than 1000 °F. WSU Ext Energy Progr (2013)
7. Adrian B (2002) Fundamentals of exergy analysis, entropy generation minimization, and the generation of flow architecture. Int J Energy Res 26:545–565
8. Sciacovelli A, Verda V, Sciubba E (2015) Entropy generation analysis as a design tool—a review. Renew Sustain Energy Rev 43:1167–1181
9. Abdul K, Rajesh K (2008) Exergy analysis of double effect vapor absorption refrigeration system. Int J Energy Res 32:161–174
10. Emma BB, Ali F, Ammar BB, Fethi A, Michel F (2016) Thermodynamic analysis of the irreversibilities in solar absorption refrigerators. Entropy 18(4):107
11. Huang CLD, Fartaj SA, Fenton DL (1992) Analysis of thermal systems using the entropy balance method. Int J Energy Res 16:189–202
12. Yang C, Wei-Wei W, Wen-Tao D, Guo-Biao Y, Di L, Fu-Yun Z (2019) Entropy generation minimization of thermoelectric systems applied for electronic cooling: parametric investigations and operation optimization. Energy Convers Manage 186:401–414
13. Goker T, Tuba OO (2015) Entropy generation analysis and dimensional optimization of an evaporator for use in a micro scale refrigeration cycle. Int J Ref 56:140–153
14. Karan HM, Ronan K, McGovern Gregory PT, Edward KS, Syed MZ, John HLV (2011) Entropy generation analysis of desalination technologies. Entropy 13:1829–1864
15. Valujerdi MM, Talebi S (2020) Entropy generation study for a supercritical water reactor (SCWR). Prog Nucl Energy 118:103–129
16. Yuanyuan Z, Jianlin Y (2019) Optimization design of falling film type plate-fin condenser/reboilers by minimizing specific entropy generation rate. Cryogenics 99:25–31
17. Samiran S, Dipankar NB (2016) Energy and entropy-based optimization of a single-stage water-lithium bromide absorption refrigeration system. Heat Transf Eng 37(2):232–241
18. Omer K, Recep Y (2007) Thermodynamic analysis of absorption refrigeration system based on entropy generation. Curr Sci 92(4):472
19. Aung M, Kyaw T, Young-Deuk K, Chakraborty A, Won GC, Kim CN (2011) A second law analysis and entropy generation minimization of an absorption chiller. App Therm Eng 31:2405–2413
20. Kaita Y (2001) Thermodynamic properties of lithium bromide—water solutions at high temperatures. Int J Ref 24:374–390

Exergy Studies on a Hybrid Desalination and Cooling Plant

Chalasani Chiranjeevi, Yendaluru Raja Sekhar, Muthuswamy Natarajan, Tangellapalli Srinivas, Mehran Hashemian, and Vinod Aditya

Nomenclature

Symbols

LPH	Liters per hour
ν	Humidity ratio
H	Heat supplied/absorbed
χ	Exergy
Ω	Exergy destroyed
Π	Exergy efficiency

Subscripts

hw	Hot water
wv	Water vapour
cw	Circulating water
chw	Chilled water

C. Chiranjeevi (✉) · Y. Raja Sekhar · M. Natarajan · V. Aditya
School of Mechanical Engineering, Vellore Institute of Technology, Vellore, TN, India
e-mail: chiranjeevi.chalasani@gmail.com

T. Srinivas
Department of Mechanical Engineering, Dr. B R Ambedkar NIT, Jalandhar, Punjab, India

M. Hashemian
Department of Mechanical Engineering, Urmia University, Urmia, Iran

M. Palanisamy et al. (eds.), *Theoretical, Computational, and Experimental Solutions to Thermo-Fluid Systems*, Lecture Notes in Mechanical Engineering,
https://doi.org/10.1007/978-981-33-4165-4_26

sp1, sp2 Hot water absorbed in humidifier1, humidifier2
ds1, ds2 Desalinated water from dehumidifier1, dehumidifier2

1 Introduction

The scarcity of fresh water due to natural and unnatural actions is imminent. Water is a primary resource which is used in occupations of huge and varying spectrum. Water is a necessary resource for humans and is taken for granted since Earth seems to have it in excess. Desalination is a process of removing the salts and certain minerals from the sea water making the product of the process usable. There are different methods of performing desalination. Here, the principle of humidification and dehumidification is used to obtain desalinated water. This project aims on conducting an exergy analysis on two-stage HDH desalination system with an integrated cooling system.

The irreversibility analysis is applied to HDH desalination system to identify the irreversibility prone areas. This observation was further used to improve cycles and components individually. The system can lead to peak performance by diminishing the specific irreversibility [1]. Thermodynamic studies by applying second law to evaluate minimum work requirement for dehumidification process in HDH desalination cycle assuming the dead state as dry ambient and final state as to be saturated air. At constant air temperature, rise in relative humidity by 10% can lead to increment in work by 2.4–32% based on the air temperatures. Similarly, at constant relative humidity, as the air temperature rises, minimum work required to condense the humid air increases [2]. The effect of different operating conditions and design parameters on HDH desalination system operating with solar water heater was studied and identified that the productivity rises with mass flow rate of air up to an optimal value then drops. The study depicts that double pass heater is better than single pass solar air heater and effect the performance of system as it increased the productivity of system by 10%. Rate of airflow also affects the productivity of system, with the increase in flow rate by 200% productivity rises to 400% increment [3]. The characteristics of HDH desalination system as a function of flow rate of water and air, temperature of hot water and temperature of cooling water were studied [4]. The study reveals water temperature plays a major role in output water production and high water yield can be achieved by increase of air flow rate and reduced cooling fluid temperature. A source and sink HDH desalination model to identify the exergy losses was developed by [5]. Study reveals that the highest irreversibility was determined in the heater, further the investigation tells that for a HDH process, exergy destruction of system is not affected by phenomena of mass transfer. Higher temperature part of system plays an immense role in exergy destruction.

A two-stage HDH desalination system with cooling system integration was designed, developed and carried out experiments by [6–11]. They carried out thermodynamic, simulation and experimental studies for the maximization of energy utilization. Studies reveal that the degree of humidification depends on the effectiveness humidifier, first law efficiency and hot water inlet temperature. Desalinated water yield depends on the temperature of cooling water circulated in dehumidifier. Approximately 1.5 LPH of fresh water and 150–200 W of cooling effect with an EUF of 0.33 and 0.58 was observed for plant and cycle respectively. Exergy analysis on solar multi-effect HDH desalination process was carried out by [12] and studies reveal that the solar collector has minimum exergy efficiency. To enhance fresh water output, the energy and exergy efficiency should be improved and rejected water should be reused to obtain fresh water. Thermodynamic analysis of various psychrometric processes was carried out and was concluded that second law efficiency is a function of mass flow rate, temperature and relative humidity of air [13]. Experimental studies were carried out to enhance the productivity of desalinated water by using various type of inserts in the air pre heater and different packing materials in the humidifier. Energy and exergy analysis was conducted to determine the EUF and compare the various changes been made in components. The improved system was able to develop 45% increase in water yield when compared to conventional system. It was observed that exergy efficiency increases with higher turbulence. Using twisted tape in air heater and gunny bags in the humidifier resulted in maximum obtained efficiency. In the modified system energy efficiency jumped from 20 to 44% and exergy efficiency from 15 to 38% [14].

The effect of various operating conditions on the performance of solar HDH desalination system was investigated experimentally. The results that by increasing the feed water mass flow rate, productivity of the system can be increased. However, air mass flow rate does not have any major effect on the productivity of the system and double pass solar air heater plays a key role and productivity was decreased by 15% without solar air heater [15]. Energy and exergy analysis was carried out on a HDH solar desalination system and found that maximum energy and exergy efficiency was obtained as 31.54% and 1.87% respectively with fresh water production of 1.11 kg/h [16]. Various methods of exploiting exergy as an influential tool for the generation of better energy strategy for various applications were described [17]. Different thermodynamic studies on HDH cycle was studied to improve the performance. These include vapour compression cycles, multi-extraction cycles and multi-pressure cycles. On the basis of study, the use of novel air heater to increase the efficiency and usage of vacuum operations was recommended [18].

Exergy analysis is generally carried out to make the thermal system more economic in terms of available energy. The losses in exergy flow also known as exergy destruction is responsible for irreversibility occurrence associated with heat addition processes and finite temperature variations. Thus, this analysis of available energy or exergy plays an inevitable role in determination of losses in energy system and later in refining the outputs in components.

2 Methodology

The line diagram of the proposed integrated plant with different components is shown in Fig. 1. A rotameter is provided at the inlets of the components to verify the flowrate is near the supposed value or not. Rotameter is used to measure the flowrates of hot water to air preheater (APH)s, circulating water and chilled water to dehumidifiers and constant flow rates of 1150, 300 and 125 L/h is maintained to the respective devices. Similarly, air flow rate was kept constant at 15 m^3/h. As this a study for determining the variation in the exergetic and performance parameters with respect to varying hot water supply rate as well as inlet water temperature to the humidifiers in the systems, rotameters were utilized to maintain the flow rates at 100, 125, 150 and 200 L/h at different instances of time to humidifiers. The inlet temperatures considered are at 46, 48 and 51 °C.

Figure 1 emphasizes on the experimental setup. The water is heated using solar flat plate collector. The heated water is supplied to APHs and humidifier for obvious purposes. The heated water flow rate for both the preheaters are kept constant and is kept variable for both the humidifiers. The air from the atmosphere is passed to APH with the help of a blower in order to maintain a constant air flow rate. The air flowing through preheater undergoes sensible heating i.e. rise in temperature of air without any changes in specific humidity. This decreases the heating load for humidifier. The hot water is sprayed from the top side of the humidifier while preheated air enters from below. This increases the humidity of air. The humidified air is then passed through water cooled dehumidifier. This leads to the end of first stage of the HDH system. Similar procedure is followed as above in second stage except it uses chilled water from a water cooler. After going through the second dehumidifier, distilled water is obtained at the bottom of dehumidifier and the chilled air can be used for centralized air conditioning in hotter environment.

The following formulae were used to evaluate different parameters. Mass of dry air is to be considered for exergy balancing of components. The estimation of mass of dry air can be done by the following equation:

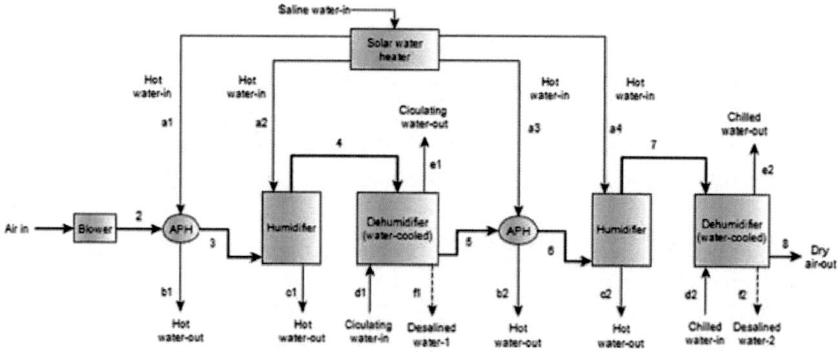

Fig. 1 Two Stage HDH system with cooling effect

$$m_{\text{dry\_air}} = \frac{m_{\text{atm}}}{1+\nu} \tag{1}$$

The air flows through the annular tube heat exchanger APH and gains heat via conduction and forced convection from the hot water coming heated up by solar flat plat collector but there is mass exchange involved in this process as it is indirect heat exchanging process. So, inlet and outlet mass flow rate of dry air in the APH is similar.

The mass of water vapour absorbed in the humidifiers by the air stream is as follows:

$$m_{\text{sp1}} = m_{\text{dry\_air}}(\nu_4 - \nu_3) \tag{2}$$

$$m_{\text{sp2}} = m_{\text{dry\_air}}(\nu_7 - \nu_6) \tag{3}$$

The addition of water vapour leads to change in the humidity ratio of the air stream passing out of the humidifier as well as the mass of the stream.

$$m_{\text{ds1}} = m_{\text{dry\_air}}(\nu_6 - \nu_7) \tag{4}$$

$$m_{ds2} = m_{dry\_air}(\nu_8 - \nu_7) \tag{5}$$

The above equations are used to obtain the mass of condensate in each stage. The motive of preheater is to increase the absorption capacity of air before undergoing the humidification. The air stream undergoes only sensible heating, hence there is no change in the humidity ratio of the air leaving the preheater.

$$\text{H}_{\text{aph1}} = m_{\text{hw1}} C_{\text{p\_hw}} \Delta T_{a1-b1} \tag{6}$$

$$\text{H}_{\text{aph2}} = m_{\text{hw1}} C_{\text{p\_hw}} \Delta T_{a3-b2} \tag{7}$$

The above equations signify the heat transferred by the hot stream from the collector to the atmospheric air passing through the APHs.

The heating of air in humidifiers can be negligible. The cooling offered by the cooling water in respective dehumidifiers can be given below.

$$H_{\text{dehum1}} = m_{\text{cw}} C_{\text{p\_cw}} \Delta T_{e1-d1} \tag{8}$$

$$H_{\text{dehum2}} = m_{\text{chw}} C_{\text{p\_cw}} \Delta T_{e2-d2} \tag{9}$$

Exergy is the maximum amount of usable energy in a stream with reference to the environment as the dead state, that is, available energy. The specific exergy of a stream at a given point, neglecting the mechanical and chemical oriented terms can be expressed as below:

$$\chi = \hbar - \hbar_0 - T_0(\kappa - \kappa_0) \quad (10)$$

The following equations are used in determining the irreversibility of each component from each stage of the desalination system:

$$\Omega_{\text{aph1}} = m_{\text{dry\_air}}(\Delta\chi_{\text{dry\_air}}^{3-2}) + m_{\text{hw1}}(\Delta\chi_{\text{hw1}}^{a1-b1}) \quad (11)$$

$$\Omega_{aph2} = m_{dry\_air}(\Delta\chi_{dry\_air}^{5-6}) + m_{hw1}(\Delta\chi_{hw1}^{a3-b2}) \quad (12)$$

$$\Omega_{hum1} = m_{dry\_air}(\Delta\chi_{dry\_air}^{3-4}) + m_{hw2}(\Delta\chi_{hw1}^{a2-c1}) \quad (13)$$

$$\Omega_{\text{hum2}} = m_{\text{dry\_air}}(\Delta\chi_{\text{dry\_air}}^{6-7}) + m_{\text{hw2}}(\Delta\chi_{\text{hw1}}^{a4-c2}) \quad (14)$$

$$\Omega_{\text{dehum1}} = m_{\text{dry\_air}}(\Delta\chi_{\text{dry\_air}}^{4-5}) + m_{\text{cw}}(\Delta\chi_{\text{cw}}^{d1-e1}) \quad (15)$$

$$\Omega_{\text{dehum2}} = m_{\text{dry\_air}}(\Delta\chi_{\text{dry\_air}}^{7-8}) + m_{\text{cw}}(\Delta\chi_{\text{cw}}^{d2-e2}) \quad (16)$$

Exergy Efficiency is an entity that can be used to demonstrate the degree of efficiency of a transformation process of a type of energy to another. In order to determine such efficiencies, the following equations were used [17]:

$$\prod_{\text{aph}} = 1 - \frac{\Omega_{\text{aph}}}{m_{\text{dry\_air}}\chi_{\text{dry\_air}}^{\text{in}} + m_{\text{hw1}}\chi_{\text{hw1}}^{\text{in}}} \quad (17)$$

$$\prod_{\text{hum}} = 1 - \frac{\Omega_{\text{hum}}}{m_{\text{dry\_air}}\chi_{\text{dry\_air}}^{\text{in}} + m_{\text{hw2}}\chi_{\text{hw2}}^{\text{in}}} \quad (18)$$

$$\prod_{\text{dehum1}} = \frac{m_{\text{cw}}(\Delta\chi_{\text{cw}})}{m_{\text{dry\_air}}(\Delta\chi_{\text{dry\_air}})} \quad (19)$$

$$\prod_{\text{dehum2}} = \frac{m_{\text{chw}}(\Delta\chi_{\text{chw}})}{m_{\text{dry\_air}}(\Delta\chi_{\text{dry\_air}})} \quad (20)$$

Gained Output Ratio (GOR) a commonly used performance parameter for evaluating a desalination system is expressed by;

$$\text{GOR} = \frac{m_{\text{ds}}\omega_{\text{ds}}}{Q^{\text{in}}} \quad (21)$$

The GOR of each stage of the HDH system can be estimated from the above equation is used to compare the performance of each dehumidifier. Effect of cooling is seen via this performance parameter.

3 Results and Discussion

Exergy destruction, exergy efficiency, stage-wise GOR comparison and total yield comparison among the stages of the system are estimated in this study. Highest irreversibility was observed during sensible heating process, while humidification and dehumidification process had lowest, and humidification being the least. Hence exergy destruction for APHs at various configurations was studied.

Figure 2 show the exergy destruction in APHs for two stages has similar trend except for 100 LPH and 48 °C. The exergy destruction increases linearly with raise in inlet water temperature for 100 and 150 LPH flowrate, while for 200 LPH it decreases. This would be due to increased heat loss to ambient.

The variation of exergy efficiency of 1st and 2nd stage humidifiers is shown in Fig. 3. It is observed that efficiency of humidifier 1st stage remains constant but tends to slightly increase at higher inlet temperature of hot water except for 150 LPH configuration. In humidifier 2nd stage exergy efficiency increases with inlet water temperature having lowest for 100 LPH. However, decreasing trend was observed for 125 LPH. It can be concluded that exergy efficiency of humidifier increases with increase in temperature and flow rate of water with 2nd stage acquiring higher exergetic efficiency than 1st stage humidifier. It is observed that highest efficiency of 80 and 90% was observed at 200 LPH flow rate and 51 °C inlet temperature of water.

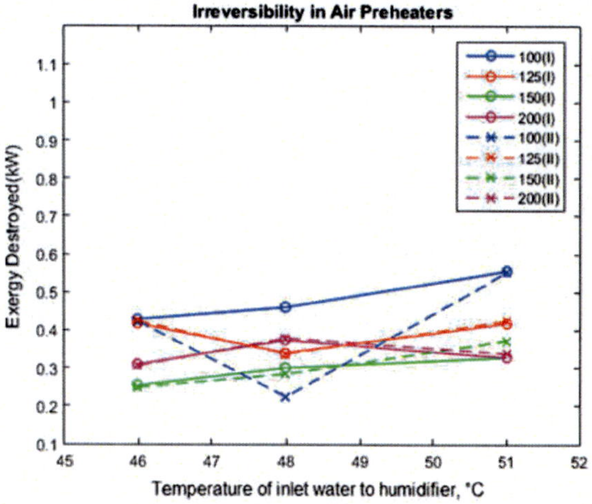

Fig. 2 Variation of exergy destroyed in 1st and 2nd stage air preheaters

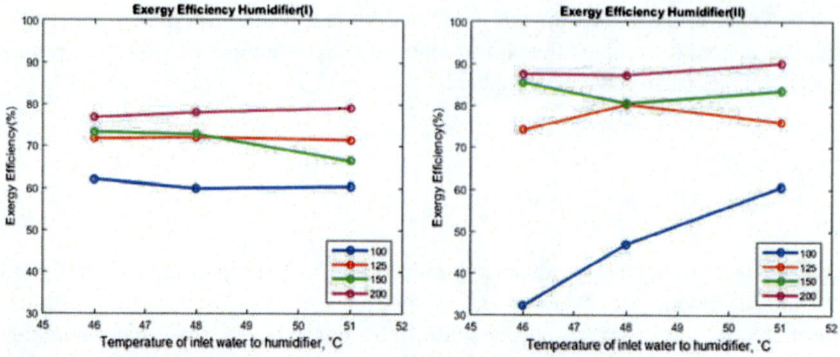

Fig. 3 Exergy efficiency variation in 1st and 2nd stage humidifier

Variation of exergy efficiency of 1st and 2nd stage dehumidifiers at different inlet hot water temperature is shown in Fig. 4. Exergetic efficiency decrease with increase in inlet hot water temperature and decrease as hot water flowrate increase. However, at dehumidifier second stage, efficiency increases with hot water inlet temperature and flowrate. Best configuration for first stage dehumidifier is observed for 100 LPH and 46 °C with exergy efficiency of 47% while for second stage, best condition is noted for 200 LPH and 51 °C with an efficiency of 55%.

Figure 5a shows the fresh water yield obtained for the two stages of dehumidifier, it can be concluded that in second stage of the HDH system is always higher than the first stage of the system. Hence, cooling effect does increase with the yield of the respective stages. At 48 °C, the total yield decreases irrespective of inlet hot water flow rate and an increase with increase in temperature. Highest yield is obtained for 125 LPH at 51 °C of 1.9 kg/h at 200 LPH and 51 °C with second stage being the better. Figure 5b shows the GOR of the two-stage HDH desalination system. Considering the system GOR for the two stages, second stage has lower value than first stage in all

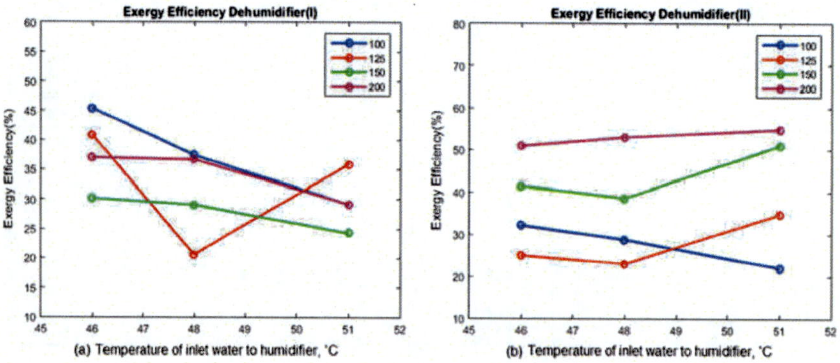

Fig. 4 Exergy efficiency variation in 1st and 2nd stage dehumidifiers

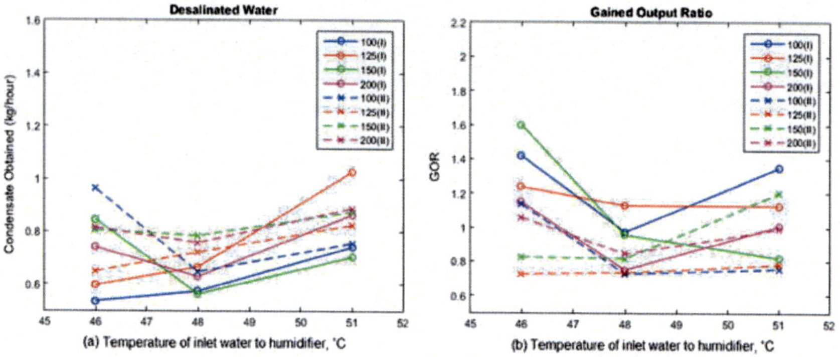

Fig. 5 Condensate obtained and GOR in 1st and 2nd stage dehumidifiers

the configurations considered in this study. For achieving higher performance, low flowrate and higher inlet water temperature to humidifier are suggested for first and second stage, respectively, with two-stage humidifier yielding better performance.

4 Conclusions

Exergy studies were carried out on an integrated two-stage HDH desalination with the cooling system is analyzed in this study. It is observed that Exergy destruction during sensible heating reduces with decreasing inlet hot water temperature. By reducing hot water inlet temperature, exergetic efficiency of dehumidifier as well as GOR of the system can increase but humidifier's exergetic efficiency will decrease. Therefore, increasing dehumidifier efficiency will increase GOR of the system. The configuration of 200 LPH and 51 °C can be said as the best configuration in terms of exergetic efficiency, 125 LPH and 51 °C for higher yield of condensate and 150 LPH and 46 °C configuration for lesser irreversibility during sensible heating.

References

1. Mistry KH, Zubair SM (2010) Effect of entropy generation on the performance of humidification-dehumidification desalination cycles. Int J Therm Sci 49(9):1837–1847
2. Alhazmy MM (2007) Minimum work requirement for water production in humidification—dehumidification desalination cycle. Desalination 214(1–3):102–111
3. Tiwari A, Sachdev T (2012) Conceptual analysis of desalination system working on humidify and dehumidify technique using solar air heater. In: International conference on mechanical and robotics engineering (ICMRE' 2012) May 26–27, 2012 Phuket
4. Al-Enezi G, Ettouney H, Fawzy N (2006) Low temperature humidification dehumidification desalination process. Energy Convers Manage 47(4):470–484

5. Ashrafizadeh SA, Amidpour M (2012) Exergy analysis of humidification–dehumidification desalination systems using driving forces concept. Desalination 285:108–116
6. Chiranjeevi C, Srinivas T (2014) Combined two stage desalination and cooling plant. Desalination 345:56–63
7. Chiranjeevi C, Srinivas T (2015) Experimental and simulation studies on two stage humidification–dehumidification desalination and cooling plant. Desalination 376:9–16
8. Chiranjeevi C, Srinivas T (2016) Influence of vapor absorption cooling on humidification-dehumidification (HDH) desalination. Alexandria Eng J 55(3):1961–1967
9. Marale SC, Chiranjeevi TS, ThundilKaruppa Raj R (2017) Experimental and computational fluid dynamics studies on dehumidifier in a combined cooling and desalination plant. J Thermal Sci Eng Appl 9:1–10
10. Chiranjeevi C, Srinivas T (2017) Augmented desalination with cooling integration. Int J Refrig 80:106–119
11. Chiranjeevi C, Srinivas T, Shankar R (2019) Experimental investigation on a hybrid desalination and cooling unit using humidification-dehumidification technique. Desalin Water Treatment 156:148–160
12. Hou S, Zeng D, Ye S, Zhang H (2007) Exergy analysis of the solar multi-effect humidification–dehumidification desalination process. Desalination 203(1–3):403–409
13. Qureshi BA, Zubair SM (2003) Application of exergy analysis to various psychrometric processes. Int J Energy Res 27(12):1079–1094
14. Muthusamy C, Srithar K (2015) Energy and exergy analysis for a humidification–dehumidification desalination system integrated with multiple inserts. Desalination 367:49–59
15. Yamalı C, Solmus İ (2008) A solar desalination system using humidification–dehumidification process: experimental study and comparison with the theoretical results. Desalination 220(1–3):538–551
16. Deniz E, Çınar S (2016) Energy, exergy, economic and environmental (4E) analysis of a solar desalination system with humidification-dehumidification. Energy Convers Manage 126:12–19
17. Dincer I, Rosen MA (2012) Exergy: energy, environment and sustainable development. Newnes
18. Narayan GP, Sharqawy MH, John H, Lienhard V, Zubair SM (2010) Thermodynamic analysis of humidification dehumidification desalination cycles. Desalin Water Treatment 16(1–3):339–353

Performance Analysis of a Desert Cooler for Different Climatic Conditions in India

Prabhat Ranjan Mishra, Aneesh Somwanshi, and Vivek Kumar Gaba

Nomenclature

C_{pa}	Specific heat of the air, (J/Kg °C)
C_w	Specific heat of water, (J/Kg °C)
h_c	Convective heat transfer coefficient between water and air, $(\mathrm{W/m^2\,°C})$
h_e	Coefficient of mass transfer between air and water, $(\mathrm{Kg/m^2s})$
M_t	Mass of water at the cooler tank, (Kg)
$\dot{m}_w$	Water mass flow rate, $(\mathrm{Kg/s})$
P_a	Vapour pressure of saturated water in the air, $(\mathrm{N/m^2})$
P_w	Vapour pressure of saturated water in the air at the temperature of water, $(\mathrm{N/m^2})$
t	Time, (s)
$\overline{T}_a$	Mean temperature of air, (°C)
T_{ae}	Temperature of exit air, (°C)
T_{ai}	Temperature of inlet air, (°C)
$\langle T_a(\xi)\rangle$	Mean exit air temperature, (°C)
$\langle T_w(\xi)\rangle$	Mean temperature of water flowing through the cooler pad, (°C)
y_0	Pad breadth, (m)
x_0	Pad thickness, (m)
z_0	Pad height, (m)
ρ_a $(\mathrm{Kg/m^3})$	Air density
γ	Relative humidity of the air

P. R. Mishra (✉) · V. K. Gaba
National Institute of Technology, Raipur, Chhattisgarh, India
e-mail: prabhat.rjnmishra@gmail.com

A. Somwanshi
School of Engg. & I.T, MATS University, Raipur, Chhattisgarh, India

M. Palanisamy et al. (eds.), *Theoretical, Computational, and Experimental Solutions to Thermo-Fluid Systems*, Lecture Notes in Mechanical Engineering,
https://doi.org/10.1007/978-981-33-4165-4_27

1 Introduction

Desert cooler commonly known as air cooler works on the principle of evaporation is an economical and eco-friendly alternative to a conventional air conditioner. Desert cooler uses natural energy source (water) and hence it is more energy efficient than vapour compression-based air conditioners. The currently available evaporative cooling system includes both direct and indirect types. Direct evaporative cooling system in which the dry air directly comes in contact with water can lower the temperature of air by using latent heat of water evaporation. As a result, warm dry air is changed into cool and moist air. Indirect evaporative cooling systems are able to lower the air temperature by avoiding addition of moisture to the air. Desert coolers utilizes the direct evaporative cooling it consists of a cubical box made up of mild steel or PVC enclosed by evaporative pad from three sides in addition to which an exhaust fan is fitted in front side to induce dry outside air into the cooling pad. The induced air comes in contact with water flowing from the top of the cooler pad. Water stored in the box (cooler tank) is pumped into the top of the cooler pad by a pump, water comes in contact with dry air evaporates and as a result heat transfer air gets cooled. The cooled air commonly known as washed air thus can be utilized for space conditioning.

Unfortunately, evaporative air cooling requires an abundant water source and is effective only in hot and dry climate, this is one of the reasons for the decline in the popularity of evaporative air conditioning. However, the rising cost of electricity and detrimental effect on the environment due to air conditioners has created a need to develop a clean environmental friendly alternative to conventional air conditioners.

Various research work has been carried over in the past to design and develop the evaporative air cooler with enhanced effectiveness pad materials and thermal comfort sizing, etc. Sawhney et al. [1] introduced two new parameters to define the cooling capacity of the cooler, they obtained optimum values of the defined variables in different environmental conditions. Singh et al. [2] determined optimum values of operating parameters viz. pad area, packing factor and airflow rate for minimum TSI (Tropical Summer Index) considering climate of Delhi, India. Sodha et al. [3] computed the performance of a desert cooler and they have given the 'Discomfort index' (DI) for three climatic zones found in India; moreover, they determined the effect of different parameters on discomfort index. Sodha et al. [4] introduced a thumb rule for the sizing of desert cooler in terms of the size of the floor region to be cooled for two atmospheres to be specific hot and dry as well as composite. To determine the performance of desert cooler Camargo et al. [5] developed a mathematical model, the model has been experimentally validated. Wu et al. [6] developed a simple mathematical model to predict the thermal performance of desert cooler and the effect of operating parameters on cooling efficiency has been investigated. Fouda and Melikyan [7] developed and validated a simplified mathematical model to describe the heat and mass transfer between air and water in a direct evaporative cooler. Sachdeva et al. [8] analyzed the performance of direct evaporative cooler by the heat and mass transfer. They considered cooling pads of various thicknesses and decided the impacts of varieties in some significant parameters like wet bulb saturation effectiveness, COP, Nu, NTU, HTC, CC and consumption of water with different face velocities. Kabeel and Bassuoni [9] presented a theoretical model to lessen the water utilization rate. The saline water at various concentrations is utilized as feeding water to the introduced model. They probed a direct evaporative cooler and found that at 200,000 PPM saltiness proportion, a decrease in water utilization by about 1.5 L/h. Dhamneya et al. [10] presented the thermodynamic performance analysis of a direct evaporative cooling system by increasing the sides of the cooling pads. They analyzed the performance of cooling by considering the different configurations of cooling pads and found that the maximum saturation efficiency of the triangular configuration was about 97%. Sodha et al. [11] built up a mathematical model considering about the variable water temperature to investigate the desert cooler, the proposed model is valuable to anticipate the transient temperature of tank water, they proposed to use the cooled tank water to cool outside objects. Bishoyi et al. [12] analyzed the thermal performance of direct evaporative cooler with respect to cooling capacity, power consumption and energy efficiency ratio considering different cooling pad materials for the composite climate. With reference to the above study it is particularly evident that the exhibition of direct evaporative cooling for various climatic conditions is yet required.

In the present paper, authors investigated the performance of direct evaporative cooler in three different climates viz. hot and dry, composite and warm and humid climate in India considering honeycomb cooling pads. The theoretical temperature of washed air (exit air) is computed for each hour in a day for different climates. Hourly effectiveness of cooler has been computed. Hourly average of ambient temperature and relative humidity is considered for numerical computations. Exit air temperature is computed considering the mathematical model presented by Sodha et al. [13]. The model presented by Sodha and Somwanshi is experimentally validated for the climate of Raipur (Chhattisgarh) by conducting laboratory experiments in stable conditions.

2 Mathematical Analysis

Following an earlier mathematical model proposed by Sodha et al. [11], considering a cooler pad (Fig. 1), the water streams from top to bottom in z-direction and flow of air is assumed to be in the x-direction.

Following earlier work, the temperature of the air in the cooling pad is given by

$$\frac{T_a - T_w}{T_{ai} - T_w} = \exp(-\alpha x), \tag{1}$$

where $\alpha = h_c F_p / \rho_a v_a$.

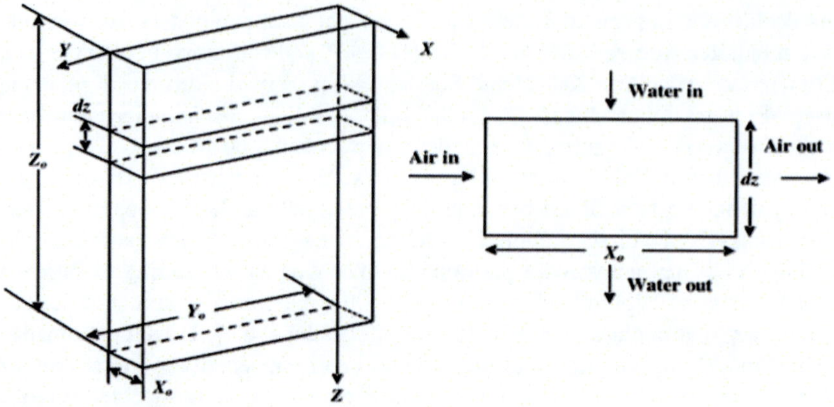

Fig. 1 Pad profile and elemental volume [11]

The term T_{ai} is the temperature of the inlet air,

From Eq. (1), the temperature of air coming out of cooler and average temp. of air (x-direction) given by,

$$T_{ae} = T_w\{1 - \exp(-\alpha x_o)\} + T_{ai}\exp(-\alpha x_o) \tag{2}$$

$$\overline{T}_a = T_w + \{T_{ai} - T_w\}\{1 - \exp(-\alpha x_o)\}/\alpha x_o \tag{3}$$

Taking a small element of dz thickness (Fig. 1), the equation of energy balance for water may be expressed as,

$$\dot{m}_w c_w \frac{dT_w}{dz}dz + \dot{Q}_L dz + \dot{Q}_S dz = 0 \tag{4}$$

The latent heat transfers per unit area are given by Tiwari [13]

$$\dot{q}_L = C_n h_c (P_w - \gamma P_a) \tag{5}$$

C_n is a constant and it is given by,

$C_n = \frac{M_w L}{RT\rho_a c_{pa} Le^{2/3}}$, $\rho_a = \frac{P_a M_a}{RT}$, $C_n = \frac{L}{c_{pa}}\frac{M_w}{M_a}\frac{1}{P_T Le^{2/3}}$. Here $P_a = P_T$, $Le = \frac{\alpha'}{D_{ab}}$.

Remembering that the evaporating surface in the element $x_o y_o$dz is $F_p x_o y_o$dz.

$$\dot{Q}_L = \dot{q}_L F_p x_o y_o = C_n h_c (P_w - \gamma P_a) F_p x_o y_o \tag{6}$$

Further,

$$\dot{Q}_S = h_c (T_w - T_a) F_p x_o y_o \tag{7}$$

The saturation vapor pressure of water can be given by,

$$P = R_1T^2 + R_2T + R_3 \tag{8}$$

where $R_1 = 6.36\,\mathrm{N\,m^{2\circ}C^{-2}}$, $R_2 = -112.8\,\mathrm{N\,m^{2\circ}C^{-1}}$ and $R_3 = 1890\,\mathrm{Nm^{-2}}$.

From Eqs. (3), (4), (5), (6), (7) and (8) one obtains,

$$\frac{\mathrm{d}T_w}{\mathrm{d}z} = -AT_w^2 + BT_w + C \tag{9}$$

where,

$N_1 = \frac{\dot{m}_w c_w}{h_c F_p x_o y_o}$, $K_1 = \{1 - \exp(-\alpha x_o)\}/\alpha x_o$, $A = R_1C_n/N_1$, $B = (R_2C_n - K_1)/N_1$ and $C = [C_nR_1\gamma T_{ai}^2 + T_{ai}(K_1 - \gamma C_nR_2) + C_nR_3(\gamma - 1)]/N_1$.

Integrating Eq. (9) one obtains,

$$\frac{T_w - (B/2A) - C_1}{T_w - (B/2A) + C_1} = \beta\exp(-2AC_1z) \tag{10}$$

where,

$\beta = \frac{[T_{wi} - (B/2A) - C_1]}{[T_{wi} - (B/2A) + C_1]}$ and $C_1 = \sqrt{(B/2A)^2 + C/A}$.

where T_{wi} is the temperature of water at the top of the cooling pad ($z = o$).

From Eq. (9) the exit and the mean (over z) temp. of water in the cooler pad and mean exit air temperature is given by,

$$T_{we} = (B/2A) + C_1\left\{\frac{(1 + \beta\exp(-2AC_1z_o)}{(1 - \beta\exp(-2AC_1z_o)}\right\} \tag{11}$$

$$\langle T_w\rangle = \frac{B}{2A} + \frac{1}{Az_o}\ln\left\{\frac{\exp(2AC_1z_o) - \beta}{1 - \beta}\right\} - C_1 \tag{12}$$

$$\langle T_a\rangle = \exp(-\alpha x_o)\{T_{ao} - \langle T_w\rangle\} + \langle T_w\rangle \tag{13}$$

3 Experimental Validation

To validate the proposed model experiment was performed in the climate of Raipur, Chhattisgarh, India (21.25°N, 81.62°E), for a typical day in the month of April (27/04/2018). The experimental cooler consists of a cooler tank (90 L) provided with cellulose evaporative pads (0.76 × 0.60 × 0.10) m^3 dimensions in three different faces. The front face consists of an exhaust fan (400 W) to induce the outside dry air into the cooler pad. A pump (18 W) is placed into the cooler tank to pump water

Fig. 2 Experimental setup

into the cooler pads. Pictorial view of the experimental setup is shown in Fig. 2. The temperature of an air inlet and exit is measured by the help of a calibrated K-type temperature of accuracy (±0.1 °C), humidity of the exit air and inlet air is measured by the help Maxtech humidity meter of accuracy (±5%).

To have controlled experiments cooler is fitted in a small room with exit air going out of the window. Before starting the experiment, experimental cooler is allowed to run for 15 min to make pads uniformly wet. Average exit air temperature, inlet air temperature, inlet water temperature, and exit water temperature are recorded at a time interval of 15 min. The mass flow rate of water flowing through the cooling pad is measured by measuring the time taken to fill the tank of the cooler of known volume. The theoretical temperature of exit air and water is computed by using Eqs. 10 and 12. Computed values are compared with recorded values.

The theoretical and experimental temperature of exit air is computed and shown in Fig. 3, Inlet parameters are as follows: ($T_{ai} = 35.5$ °C, Relative humidity = 0.30, $T_{w0} = 32.5$ °C, $m_w = 0.15$ kg/s, $V_a = 1.5$ m/s, $C_w = 4200$ J/kg k, $F_p = 370$ m^2/m^3, $X_0 = 0.10$ m, $y_0 = 0.45$ m, $Z_0 = 0.5$ m, $M_t = 97$ kg).

4 Determination of Closeness of Experimental and Theoretical Readings

4.1 Root Mean Square of Percentage Deviation (e)

Closeness of theoretical to experimental values can be presented in terms of root mean square of percent deviation (e) and it is given by,

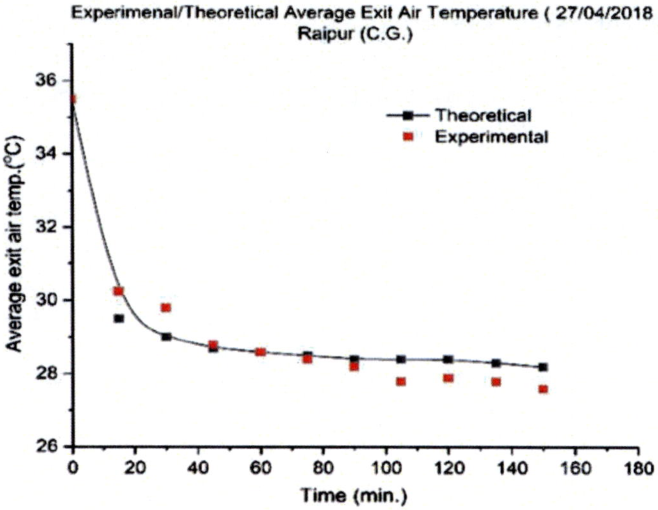

Fig. 3 Theoretical and experimental exit air temperature

$$e = \sqrt{\frac{\sum (e_i)^2}{n}} \tag{14}$$

$$e_i = \left[\frac{X_{\text{pre}(i)} - X_{\exp t(i)}}{X_{\text{pre}(i)}}\right] \times 100 \tag{15}$$

4.2 *Coefficient of Correlation* (r)

The relation between the theoretical and experimental values is exhibited by a coefficient called coefficient of correlation (r). The coefficient of correlation can be assessed with the assistance of the following equation:

$$r = \frac{N\sum X_{\text{pre}}X_{\exp} - (\sum X_{\text{pre}})(\sum X_{\exp})}{\sqrt{N\sum X_{\exp}^2 - (\sum X_{\exp})^2}\sqrt{N\sum X_{\text{pre}}^2 - (\sum X_{\text{pre}})^2}} \tag{16}$$

Here N is the number of observations. The experimental and theoretical values are said to be in a strong correlation, if estimation of r is near 1.

The values of the root mean square of percentage deviation (e) and the coefficient of correlation (r) is computed by the relations discussed above. It is seen that the value of e is 1.15% and the value of r is 0.97 for the degree of freedom 10. The values seem to be reasonably good, the small discrepancy may be due to small variations in room conditions.

5 Performance of Cooler

Daily performance of cooler is computed for summers (average data of April, May and June) for the three different climatic zones in India viz. hot and dry climate, composite climate and warm humid climate. The different cities corresponding to the three different climatic zones are selected for numerical computations. Hourly average of ambient temperature and humidity is considered for numerical computations. Hourly degree of cooling ($T_{ai} - T_{ae}$) is computed for the different climates. The velocity of inlet air is considered as 1.5 m/s, mid-size cooler (90 L) capacity with cellulose pad of area (400 m^2/m^3) and dimensions (0.1, 0.5, 0.7 m) is considered for numerical computations.

The degree of cooling will be the temperature difference between inlet air temperature and the temperature of air coming out of the cooler (exit air). The degree of cooling is plotted with time for the different climatic conditions and shown in Figs. 4, 5 and 6. It is seen that the degree of cooling for April month is maximum for the climate of Jodhpur and minimum for the climate of Kolkata. For the month of May, it is seen that the degree of cooling is more for Jodhpur climate up to time 5.00 pm in evening after that the degree of cooling is more for the composite climate of Delhi the reason could be due to increase in relative humidity in evening hours for the

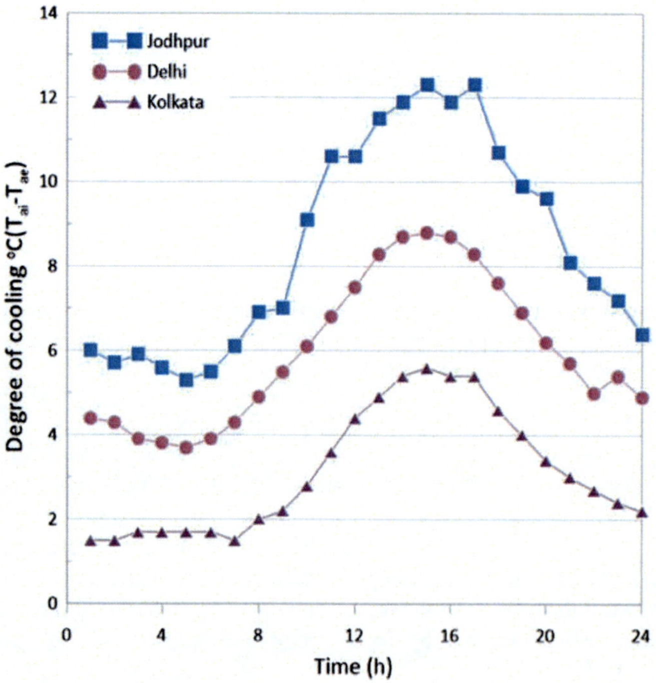

Fig. 4 Degree of cooling for April month

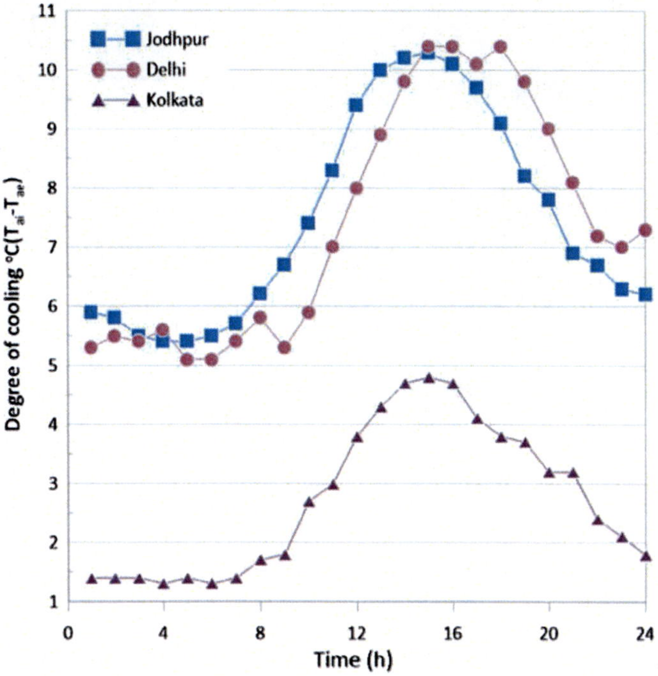

Fig. 5 Degree of cooling for May month

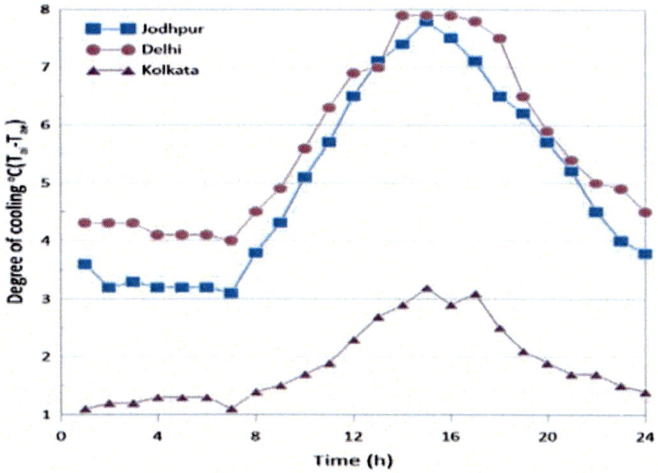

Fig. 6 Degree of cooling for June month

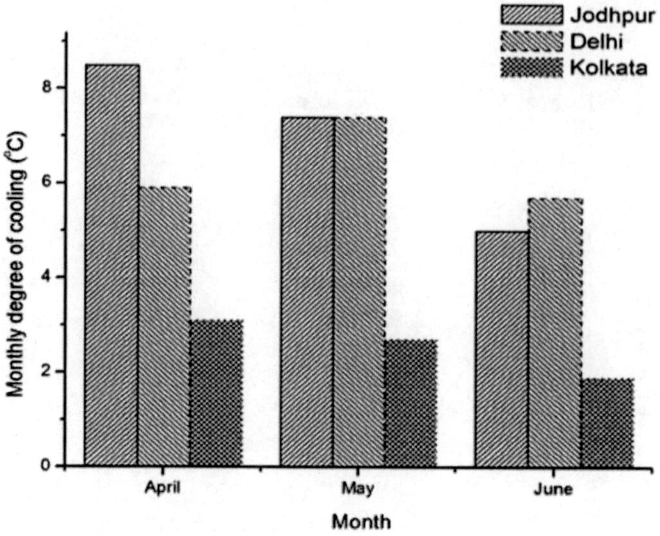

Fig. 7 Monthly average degree of cooling

climate of Jodhpur. For June, the degree of cooling is almost the same for the climate of Jodhpur and Delhi whereas for humid climate it is very less.

The average degree of cooling for the month of April, May, and June for all three climatic zones are shown in Fig. 7. It is seen that the average degree of cooling for the month of April is 8.5 °C, 5.9 °C and 3.1 °C for climates of Jodhpur, Delhi, and Kolkata respectively. For the month of May, the average degree of cooling for Jodhpur, Delhi and Kolkata are 7.4, 7.3, and 2.7 °C. It is seen that the average degree of cooling is almost the same for Jodhpur and Delhi climate. The degree of cooling for day time is more for Jodhpur and for Delhi, degree of cooling is more for evening and night. The degree of cooling for June month is very less for all the climates.

6 Result and Conclusion

The proposed mathematical model can be implemented to analyze the performance of desert coolers for different climatic conditions. Various designs and dimensions can be considered to get the cooling performance. The model will be helpful to design a cooler to give maximum efficiency. In the present work daily performance of an evaporative cooler has been computed for summer months in the three different climatic conditions found in India. It is clearly seen that like other devices based on the evaporative cooling, the desert cooler performs badly for the places with higher humidity. The cooler works reasonably well for climates with low humidity and high ambient temperatures. The average degree of cooling for summers (April, May, and

June) is a maximum of about 6.9 °C for the climate of Jodhpur. For the climate of Kolkata, its value is a minimum of about 2.6 °C. For the climate of Delhi, the average degree of cooling for summers is about 6.3 °C.

References

1. Sawhney RL, Singh SP, Bansal NK, Sodha MS (1987) Optimization of an evaporative cooler for space cooling. Housing Sci 11(3):225–231
2. Singh SP, Sawhney RL, Bansal NK, Sodha MS (1987) Sizing of an evaporative cooler for thermal comfort inside a room. Housing Sci 11(2):141–148
3. Sodha MS, Singh SP, Sawhney RL (1990) Evaluation of discomfort in a room with desert cooler. Int J Energy Res 14:745–756
4. Sodha MS, Singh SP, Sawhney RL (1995) Evolution of design patterns for direct evaporative coolers. Build Environ 30(2):287–291
5. Camargo JR, Ebinuma CD, Silveria JL (2005) Experimental performance of a direct evaporative cooler operating during summer in a Brazilian city. Int J Refrig 28:124–1132
6. Wu JM, Huang X, Zhang H (2009) Numerical investigation on the heat and mass transfer in a direct evaporative cooler. Appl Therm Eng 29:195–201
7. Fouda A, Melikyan Z (2011) A simplified model for analysis of heat and mass transfer in direct evaporative cooler. Appl Therm Eng 31:932–936
8. Sachdeva A, Rajput SPS, Kothari A (2015) Performance analysis of direct evaporative cooling in Indian Summer. Int J Res Aeronaut Mech Eng 3(12):131–141
9. Kabeel AE, Bassuoni MM (2017) A simplified experimentally tested theoretical model to reduce water consumption of a direct evaporative cooler for dry climates. Int J Refrig. https://doi.org/10.1016/j.ijrefrig.2017.06.010
10. Dhamneya AK, Rajput SPS, Singh A (2018) Thermodynamic performance analysis of direct evaporative cooling system for increased heat and mass transfer area. Ain Shams Eng J 9:2951–2960
11. Sodha MS, Somwanshi A (2012) Variation of water temperature along the direction of flow: effect on performance of an evaporative cooler. J Fundamentals Renew Energy Appl 2. https://doi.org/10.4303/jfrea/R110301
12. Bishoyi D, Sudhakar K Experimental performance of a direct evaporative cooler in composite climate of India. Energy Build https://doi.org/10.1016/j.enbuild.2017.08.014
13. Tiwari GN (2002) Solar Energy fundamentals. Modelling and applications, Narosa Publication, Design

Energy and Environmental Analyses of Active Solar Dryer for Medicinal Herbs Drying

D. V. N. Lakshmi and Muthukumar Palanisamy

1 Introduction

A large variety of spices and herbs such as turmeric, cardamom, black pepper, black ginger, sugandamantri, stevia, basil, giloy, etc., are cultivated in India, and being exported to various parts of the world. The diverse climatic condition of India is the prime reason for the cultivation of good quality and large variety of spices and herbs. Drying is the most common method to store the cultivated products for longer time. The quality of the product depends on the type of drying conditions (normal, blanched and chemically pre-treated) and mode of drying (open sun drying, artificial drying, and solar drying). The artificial dryers consume fossil fuels and emit harmful greenhouse gases to the environment. The cost of these dryers is very high and unaffordable by the small scale farmers. In tropical places, sun drying is a normal practice, which impacts the medicinal qualities of the herbs and spices. As a result, the product's price decreases in the international market. Solar dryers are of low cost as compared to conventional dryers and they can be fabricated with locally available materials. In India, the most common practice for preserving stevia leaves is by drying them under the open sun. The product quality of open sun-dried (OSD) products is not up to the mark because of the uneven drying, dust and sudden rains. Further, long drying hours in open sun spoils the nutritional and medicinal value of the product [1]. For drying of medicinal herbs, the temperature between 40 and 60 °C is recommended [2]. Artificial dryers (Fluidized bed dryers, infrared dryers, etc.)

D. V. N. Lakshmi
Faculty of Engineering, Usha Martin University, Ranchi, Jharkhand 834003, India

M. Palanisamy (✉)
Department of Mechanical Engineering, Indian Institute of Technology Guwahati, Guwahati, Assam 781039, India
e-mail: pmkumar@iitg.ac.in

M. Palanisamy et al. (eds.), *Theoretical, Computational, and Experimental Solutions to Thermo-Fluid Systems*, Lecture Notes in Mechanical Engineering,
https://doi.org/10.1007/978-981-33-4165-4_28

are the other available options that are used to fulfil the purpose. The high initial investment and operational costs, requirement of skilled operators and emission of harmful gases are the drawbacks of the artificial dryers. In contrast to this, solar-based dryers are better alternates to address the above-cited problems. The fresh products are loaded in an insulated closed cabinet known as dryer while the hot air from the solar collector is allowed to pass around the product and allowed to dry till the desired moisture content is reached. This method preserves the medicinal and nutritional values of the herbs as compared to conventional sun drying [3]. Because of progress and changes in life style, majority of individuals are with threats like hypertension, obesity, cancer, diabetes, etc. Roughly 463 million grown-ups (20–79 years) are living with diabetes,by 2045 this will soar to 700 million. Usage of zero-calorie foods is getting popular to maintain a healthy lifestyle. The most common use of artificial sweeteners is Acesulfame potassium, sucralose, etc. These sweetners create severe health issues like abdominal and cardiac issues. Stevia is a natural sweetner and its sweetness is 300 times more than sugar [4]. Stevia has a low glycaemic index and non-cariogenic. Since, it does not influence blood sugar level, it is more suitable for diabetic patients. Fresh stevia leaves are not available throughout the year and they are generally consumed in dried form. In India, the most common practice is drying under the open sun. This method degrades the quality in terms of colour and medicinal values. Solar drying is a renewable energy-based technique to dry the stevia products in rural places. This technique preserves the nutritional and medicinal values of the stevia compared to conventional drying. The solar air heater with multi-pass was developed for screw pine leaves drying [5]. Sensible heat storage materials like granite was placed below the absorber plate to reduce the drying time by 50%. Similarly, the drying behaviour of the fresh rosemary leaves was investigated in an indirect solar dryer with an auxiliary heating system [6]. It was observed that the drying time was less at higher operating temperature. A mixed mode solar dryer with paraffin-based thermal energy storage was developed for studying the drying behaviour of medicinal products like Curcuma caesia and Curcuma zedoaria [7]. The drying kinetics of the dried samples were analyzed under two different conditions. Lemus-Mondaca et al. [8] investigated the effect of drying on stevia leaves by using freeze, infra-red, vacuum, shade and convective drying methods, and found that the phenolic and flavonoid contents were more in the freeze-drying, as compared to the other drying techniques. Direct and indirect solar dryers, with and without mesh were developed for stevia leaves drying [9]. Experiments were performed by placing 20 g of samples in the drier. The drying kinetic analysis was performed and it was observed that the Weibull model was the suitable kinetic model. Recently, a mixed mode active solar dryer was developed for stevia leaves drying [10]. Exergy analysis of the dryer and quality analysis of the samples were performed and observed that DPPH, total phenolic and total flavonoid content of the samples dried in the solar dryer were better compared to the OSD. The discolorations of the solar-dried samples were high due to direct exposure of sunlight in the drying chamber.

Limited works have been reported on drying of medicinal herbs like stevia, neem and giloi in solar dryer till date [9, 10]. It is also seen that there is a lack of profound experimental studies on drying of stevia leaves and performance investigations of solar dryer for drying stevia leaves. The present research work focuses on (i) development of an indirect active solar dryer for the climatic conditions of North East India,and (ii) to perform experiments on drying of stevia for evaluating the performance of the solar dryer in terms of dryer efficiency and pick up efficiency. Further, quality analysis of the samples and carbon mitigation potential of the dryer are presented.

2 Experimental Setup and Experimental Procedure

The experimental setup is fabricated as per the following design procedure. The setup is located on the top of Thermal Science Lab, Department of Mechanical Engineering, IIT Guwahati, India. The system consists of a solar dryer (insulated and parallel flow type); a double pass solar air heater (counter flow) and a blower. The schematic representation of the setup is shown in Fig. 1. The solar air heater is 1.01 m in width and 2.01 m in length. The absorber plate (0.85 mm thick GI plate) is coated with black colour to absorb the maximum solar heat. To reduce heat losses, the absorber plate is covered with plexiglass (4 mm thickness). A base plate of 1 mm thick aluminum sheet is used to reduce the re-radiation losses from the absorber plate. The dryer consists of a diverging section at the entry for uniform air flow inside the chamber. The designed solar dryer can accommodate six trays and one sample tray for holding the products. The blower is placed between the solar air heater and the chamber. The fresh air enters the solar air heater, gets heated up and then enters into the dryer. To vary the air flow rate, a control valve is placed before the blower and the hot air is pumped into the drying chamber through the blower. The air heater was placed 25° south to receive the maximum solar radiation.

During the experiments, the flow rate was maintained constant. To achieve a steady-state condition, the blower was operated 1 h prior to the loading of products. Experiments were conducted in the month of October 2017, from 8 AM to 4 PM.

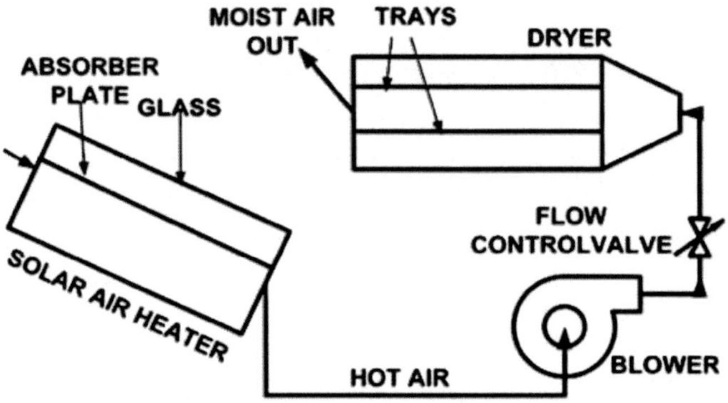

Fig. 1 Schematics of forced convection solar dryer

Table 1 Details of the instruments and their accuracy

S.No.	Parameter	Instrument	Accuracy
1	Solar radiation (I)	Pyranometer	±1 W/m^2
2	Temperature (T)	T type thermocouples	±0.2 °C
3	Wet-bulb temperature	Hygrometer	±0.5 °C
4	Velocity (V)	Hotwire anemometer	±0.1 m/s
5	Weight of samples (W)	Weighing balance	±0.01 g

The fresh samples (stevia) were collected and cleaned properly with running water before drying. The dryer was loaded with full capacity (10 kg) to study the dryer performance. Temperature, humidity, velocity of air and solar radiation intensity at various positions were measured. The details of instrumentation are presented in Table 1. Thermocouples and pyranometer were connected to the data acquisition system (Agilent—34972A).

3 Performance Analysis

The amount of water dehydrated from the product (y_w) was calculated by applying Eq. 1,

$$y_w = \frac{m_{pr}(X_{in} - X_{fn})}{100 - X_{fn}} \tag{1}$$

The dryer efficiency is defined as the ratio of energy required to remove moisture from the product to energy input to the dryer. Input energy to the dryer is given by using Eq. 3, where h_{fg} is latent heat of vaporization (kJ/Kg), I is the intensity of solar radiation (W/m^2), BP is energy consumed by the blower (Kwh), A_{SAH} is area of solar air heater (m^2), and α and τ are the absorptivity and transmissivity of the SAH.

$$\eta_{ISD} = \frac{y_w * h_{fg}}{E_{ISD}} \tag{2}$$

$$E_{ISD} = [A_{SAH} * I * \alpha * \tau + BP] * t_{ISD} \tag{3}$$

Pick up efficiency is another important parameter used for evaluating the dryer performance. The moisture removal capacity from the product inside the solar dryer can be estimated by using Eq. (4) [11, 12],

$$\eta_{\text{pickup}} = \frac{m_{\text{wr}}}{v\rho t_{\text{ISD}}(h_{\text{asdi}} - h_{\text{di}})} \quad (4)$$

where, v is volumetric flow rate of drying air entering into the chamber (m^3/s) and h_{asdi} is absolute humidity of the air entering the dryer at the point of adiabatic saturation (%) and h_{di} is absolute humidity of drying air at inlet (%). The effective moisture diffusivity ($D_{\text{effective}}$) is another parameter considered in the design of a solar dryer, and this value is estimated by using Fick's diffusion equation,

$$\text{MRR} = \frac{8}{\pi^2}\sum_{n=0}^{\infty}\frac{1}{(2n+1)^2} \times \exp\left[-\frac{(2n+1)^2\pi^2 D_{\text{eff}} t_{\text{drying}}}{4(h^*)^2}\right] \quad (5)$$

Considering the first term in the above series, Eq. 5 can be written as:

$$\text{MRR} = \frac{8}{\pi^2}\exp\left[-\frac{\pi^2 D_{\text{eff}} t_{\text{drying}}}{4(h^*)^2}\right] \quad (6)$$

From the experimental data, effective moisture diffusivity is obtained. By plotting ln(MR) versus time and from the slope of this line, $D_{\text{effective}}$ is estimated as:

$$D_{\text{eff}} = \frac{\text{Slope} \times 4(h^*)^2}{\pi^2} \quad (7)$$

The economic study and environmental analysis of the dryers are very important in terms of acceptability. The first one includes cost and feasibility of the system and the second one explains about the CO_2 emissions, carbon credit and energy payback time (*EPBT*). The amount of energy required to produce the final end user components is called as embodied energy (E_{m}). The consumption of energy produces carbon emissions to the environment. From embodied energy analysis, one can evaluate how much carbon will be emitted by the dryer. This value can be minimized by selecting materials which have less E_{m} value for reducing the CO_2 emissions.

Energy payback time (*EPBT*) is defined as the ratio of total embodied energy of the system (E_{m}) to the annual thermal energy output (E_{annual}).

$$EPBT = \frac{E_{\text{m}}}{E_{\text{annual}}} \quad (8)$$

The thermal output of the dryer annually (E_{annual}) is evaluated from the product of thermal output daily (E_{daily}) and number of dryer operated days (T_{b}). The daily thermal output evaluated as follows:

The thermal output daily (kWh) of the solar dryer is given by,

$$E_{\text{daily}} = \frac{y_{\text{w}} \times h_{\text{fg}}}{3.6 \times 10^6} \quad (9)$$

Thermal output of the dryer annually is given by,

$$E_{\text{annual}} = E_{\text{daily}} \times T_{\text{b}} \tag{10}$$

The country's socio-economic development depends on energy consumption per capita. Industrialization, lifestyle and employment are the other factors that influence the economic development of any country. The transmission losses are taken as 45% and loss of energy by consumers as 25% in India.

$$\begin{aligned} CO_2 \text{ emission per year} &= \frac{E_{\text{in}} \times 2.042}{n_{\text{lifespan}}} \\ \frac{1}{1-l_a} * \frac{1}{1-l_T} * 0.98 &= \frac{1}{1-0.2} * \frac{1}{1-0.4} * 0.98 = 2.042 \end{aligned} \tag{11}$$

where, E_{in} is embodied energy, l_t is transmission losses and l_a is domestic appliances losses. The net CO_2 mitigation over the lifetime of a system is evaluated as follows:

The net CO_2 mitigation over the lifetime of a system = Total CO_2 mitigation − Total CO_2 emission.

$$\text{Net } CO_2\text{mitigation (in tons)} = (E_{\text{annual}} \times n - E_{\text{m}}) \times 2.042 \times 10^{-3} \tag{12}$$

Carbon credits can be used to finance carbon reduction schemes. The carbon credit is estimated by using Eq. 13,

$$\text{Carbon credit} = \text{Net } CO_2\text{mitigation} * \text{ cost of carbon credit per ton (in USD)} \tag{13}$$

4 Results and Discussion

The dryer was loaded fully (10 kg) to study the performance. Using AOAC method, the moisture content of the stevia samples was estimated in the hot air oven. Experiments were performed in October 2017. A sample of 100 g was placed inside the active solar dryer to study the drying behaviour of stevia leaves and the same amount of sample was placed under open sun. The variation of air temperature inside the dryer and the ambient temperature is shown in Fig. 2. Temperature inside the IASD was varied from 27.2 to 55.4 °C, whereas the average ambient temperature of air was 25.9 °C. Temperature of air plays a significant role on the drying time, wherein higher the temperature, lesser is the drying time. The moisture ratio (MR) variation with time is shown in Fig. 3. The stevia leaves reached the final moisture content moisture content of 0.053 (d.b) from initial moisture content 4.36 (d.b) within 7 h in IASD condition. Under OSD, the final moisture content of leaves was achieved

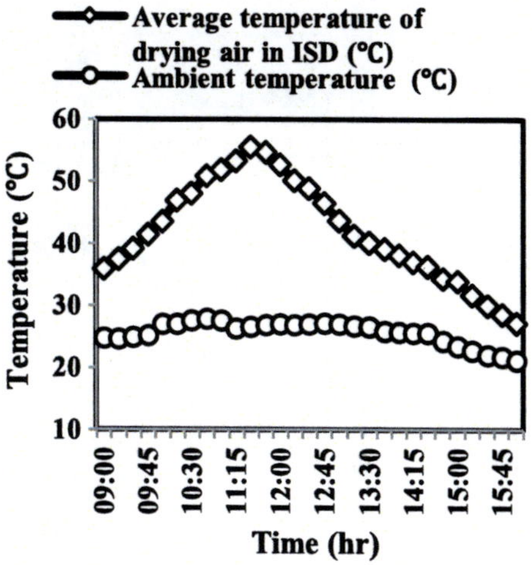

Fig. 2 Temperature of air inside the dryer and ambient conditions

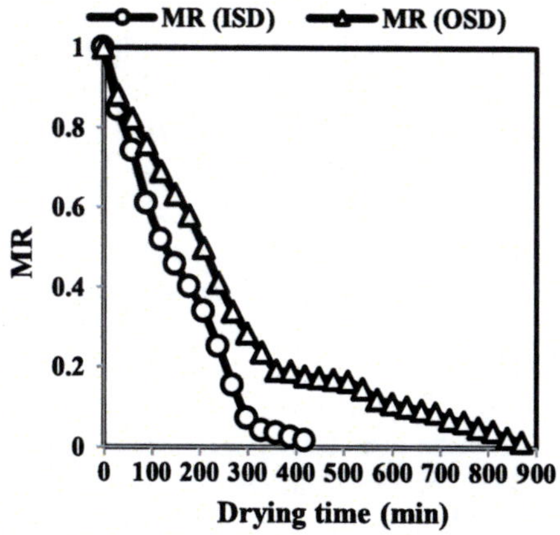

Fig. 3 Variation of moisture ratio of stevia leaves in IASD and OSD

in about 14.5 h. From Fig. 3, it was observed that the drying was taking place in falling rate drying period. In OSD, the moisture content reduced slowly due to low temperature and high humidity. As a result, the medicinal values (total phenolic, total flavonoid) of dried leaves under sun-drying condition was low compared to the leaves dried in the IASD.

Effective moisture diffusivity is vital parameter to evaluate the dryer performance. This value is determined by using the method of slopes. The relationship of ln(MR) versus time is shown in Fig. 4. Using Eq. 8, the diffusivity value was estimated for samples dried in the solar dryer and OSD which were 2.774×10^{-9} m^2/s and 1.254×10^{-10} m^2/s, respectively. The dryer performance depends on the amount of water evaporated from the product and the time of drying. Efficiency of the dryer is defined as the ratio between the energy required to remove the water from the product

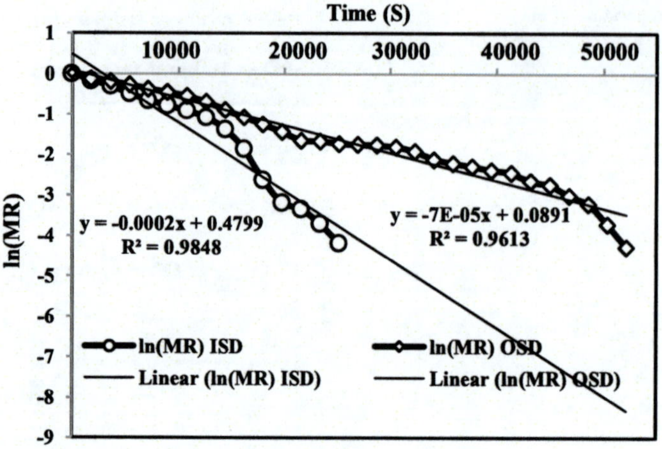

Fig. 4 Effective moisture diffusivity of stevia leaves in IASD and OSD

to the energy input to the dryer. The input energy to the dryer is mainly from the solar collector. The energy needed to dehydrate the product depends on the shape, thickness and initial moisture content of the product. The efficiency of the dryer is estimated by using Eq. 3 ($I_{avg} = 535.4$ W/m^2, $A_{SAH} = 2.1$ m^2, $y_w = 7.9$ kg) and found that the overall thermal efficiency of the dryer was 31.5%. Pick-up efficiency is the ability of drying air to remove the moisture from product. Temperature of the drying air plays a significant role in removing the moisture. This value is high during the initial drying hours and decreases when the drying advances. This is mainly due to higher moisture evaporation in initial drying period, which decreases as drying progress. The dryer average pick up efficiency was estimated as 73.5%.

Environmental analysis of the active solar dryer was estimated, as the carbon credit and the net carbon dioxide mitigation potential were found by using Eqs. 9–14. The materials used for manufacturing the components have some energy densities [1, 13]. This energy density was multiplied with the amount of material that gave the embodied energy value. Table 2 shows the energy densities, weight of the materials used for manufacturing the solar dryer and the embodied energy for each component.

The yearly thermal output of the dryer was found to be 1685.7 kWh/year. The EPBT of the active solar dryer was estimated using Eq. 16. The total embodied energy of the dryer was 876.6 kWh, while the annual energy output from the developed dryer was 2767.5 kWh/year and the EPBT value was found as 0.5 year. The value is very low as compared to the other developed solar dryers [14]. The developed solar dryer is environmentally energy efficient and environment friendly. The net CO_2 mitigation potential of the system is evaluated by using Eq. 13, and it was found as 55.06 tonne for the 10-year life span of the dryer. If CO_2 emission is being traded at 13 US $ per tonne of CO_2 mitigation, then the carbon credit earned by the system can be obtained. The earned carbon credit varies from ₹12,561.00 to ₹50,245.49, which is quite high.

Table 2 Energy densities and quantities of solar dryer components for evaluating the embodied energy

S. No.	Material	Quantity (kg)	E_c (kWh/kg)	E_{total} (kWh)
Solar air heater				
1	Glass cover	5.16	7.28	37.56
2	Absorber plate (GI sheets)	2.08	13.88	28.87
3	Aluminum base plate 1 mm thickness	0.8	55.28	44.22
4	Plywood bottom cover and base	16.8	2.88	48.38
5	Glass wool for insulation	2.5	4.044	10.11
6	Al. Angles	2	55.28	110.56
7	Black paint	1	25.11	25.11
(a) Embodied energy for solar air heater				*304.8 kWh*
Indirect solar drying chamber				
8	Drying chamber (M.S. Sheet thickness 1 mm	45	8.8	396
9	Wooden stand	10	2.89	28.9
10	Wire mesh	4.2	8.89	37.33
11	PVC pipe	2	19.4	38.8
(b) Embodied energy for construction of solar dryer				*501 kWh*
Blower				
12	Motor body casing	10	6.1	61
13	Copper wire	0.05	19.62	0.981
14	Fittings	1	8.89	8.89
(c) Embodied energy for blower				*70.8 kWh*
Total embodied energy of solar dryer (a) + (b) + (c)				876.6 kWh

The total flavonoid (TFC) and total phenolic content (TPC) of the raw samples and dried samples (solar and open sun) were evaluated by standard testing methods and are reported in Table 3 [7]. The phenolic and flavonoid content values of solar-dried samples were high due to less drying time.

Table 3 Total flavonoid and total phenolic content of raw and dried stevia samples

	TPC (mg GAE per gm of sample)	TFC (mg quercetin per gm of sample)	Antioxidant activity (μ mol of TE per gm of sample)
Raw	48.40 ± 1.27	1.98 ± 0.03	49.85 ± 2.78
IASD	62.85 ± 5.606	28.25 ± 2.19	92 ± 7.29
OSD	33.25 ± 1.01	20.69 ± 1.14	3.01

5 Conclusions

In the present experimental study, active solar dryer indirect type was designed and experiments have been carried out at full load condition to study the performance of the dryer using stevia leaves as drying sample. Major conclusions drawn are as follows:

- The time taken the fresh stevia leaves to reach the desired moisture content of 6.5% w.b was 420 min and 870 min, respectively, in indirect solar dryer and OSD to dry till the desired moisture content was achieved.
- The overall dryer efficiency and pickup efficiency of the active solar dryer were respectively 31.5% and 73.5%.
- The developed dryer is economically viable and can compete with commercial dryers.
- The estimated effective moisture diffusivities of stevia leaves dried in IASD and OSD were 2.774×10^{-9} m^2/s and 1.254×10^{-10} m^2/s, respectively.
- The TFC and TPC of the open sun-dried samples were low compared to the samples dried in IASD.
- The earned carbon credit varies from ₹12,561 to ₹50,245, which is quite high.
- The energy payback time of the dryer was estimated as 0.52 year.

References

1. Tomar V, Tiwari GN, Norton B (2017) Solar dryers for tropical food preservation: thermo physics of crops, systems and components. Sol Energy 154:2–13
2. Cipliene A, Novosinskas H, Raila A, Zvicevicius E (2015) Usage of hybrid solar collector system in drying technologies of medical plants. Energy Convers Manage 93:399–405
3. Murthy MVR (2008) Review of new technologies, models and experimental investigations of solar driers. Renew Sustain Energy Rev 13:835–844
4. Periche A, Castelló ML, Heredia A, Escriche I (2015) Influence of drying method on steviol glycosides and antioxidants in Stevia rebaudiana leaves. Food Chem 172:1–6
5. Kareem MW, Khairul H, Sopian K, Ruslan MH (2017) Multi-pass solar air heating collector system for drying of screw-pine leaf (Pandanus tectorius). Renew Energy 112:413–424
6. Mghazli S, Ouhammou M, Hidar N, Lahnine L, Idlimam A, Mahrouz M (2017) Drying characteristics and kinetics solar drying of Moroccan rosemary leaves. Renewable Energy 108:303–310
7. Lakshmi DVN, Muthukumar P, Apurba L, Nayak PK (2018) Drying kinetics and quality analysis of black turmeric (Curcuma caesia) drying in a mixed mode forced convection solar dryer integrated with thermal energy storage. Renew Energy 120:23–34
8. Lemus-Mondaca R, Vega-Gálvez A, Rojas P, Stucken K, Delporte C, Valenzuela-Barra G, Jagus RJ, Agüero MV, Pasten A (2018) Anti-oxidant, antimicrobial and anti-inflammatory potential of Stevia rebaudiana leaves: effects of different drying methods. J Appl Res Med Aromatic Plants 11:37–46
9. Téllez MC, Figuero IP, Téllez BC, Vidaña ECL, Ortiz AL (2019) Solar drying of Stevia (Rebaudiana Bertoni) leaves using direct and indirect technologies. Sol Energy 159:898–907
10. Lakshmi DVN, Muthukumar P, Apurba L, Nayak PK (2019) Performance analyses of mixed mode forced convection solar dryer for drying of stevia leaves. Sol Energy 188:507–518

11. Fudholi A, Sopian K, Alghoul MA, Ruslan MH, Othman OY (2015) Performances and improvement potential of solar drying system for palm oil fronds. Renew Energy 78:561–565
12. Akpinar EK (2010) Drying of mint leaves in a solar dryer and under open sun: modelling, performance analyses. Energy Convers Manage 51:2407–2418
13. Shrivastava V, Kumar A (2017) Embodied energy analysis of the indirect solar drying unit. Int J Ambient Energy 38:280–285
14. Prakash O, Kumar A, Sharaf-Eldeen YI (2016) Review on Indian solar drying status. Current Sustain Renew Energy Reports 3:113–120

Modelling and Trade-Off Analysis of Performance Parameters for Counter Flow Packed Bed Liquid Desiccant air Dehumidifier

Mrinal Bhowmik, Muthukumar Palanisamy, and R. Anandalakshmi

Nomenclature

$CaCl_2$	Calcium chloride
CR	Condensation rate (g/s)
ε_h	Enthalpy effectiveness
ε_m	Moisture effectiveness
HFCs	Hydrofluorocarbons
LDAC	Liquid desiccant dehumidification system
LiBr	Lithium bromide
LiCl	Lithium chloride
$\dot{m}$	Mass flow rate (kg/s)
NOAA	National Oceanic and Atmospheric Administration
P_{atm}	Atmospheric pressure (kPa)
R^2	Correlation coefficient
RMSE	Root mean square error
S/A	Solution to air flow mass rate
SWR	Stepwise regression
T	Temperature (°C)
VCS	Vapor compression system

M. Bhowmik · M. Palanisamy (✉)
Centre for Energy, Indian Institute of Technology Guwahati, Guwahati, India
e-mail: pmkumar@iitg.ac.in

M. Palanisamy
Department of Mechanical Engineering, Indian Institute of Technology Guwahati, Guwahati, India

R. Anandalakshmi
Department of Chemical Engineering, Indian Institute of Technology Guwahati, Guwahati, India

M. Palanisamy et al. (eds.), *Theoretical, Computational, and Experimental Solutions to Thermo-Fluid Systems*, Lecture Notes in Mechanical Engineering,
https://doi.org/10.1007/978-981-33-4165-4_29

ω	Specific humidity (kg of water vapor by kg of dry air (kg_{wv}/kg_{da}))
h	Enthalpy of fluid (kJ/kg)
X	Solution concentration (wt.%)
z	Number of experimental runs
λ	Data point

Subscript

exp	experiment
e	equilibrium
i	inlet
a	air
s	solution
o	outlet

1 Introduction

National Oceanic and Atmospheric Administration (NOAA) had prescribed the year 2015 as the warmest year from the time when records began in 1880 [1]. The increased number of scorching hot summer days are record-breaking in various zones of the globe. Therefore, the demand for heating, ventilation and air conditioning systems are annually expected to hike by 6.2% [2]. The requirements of human comfort indoor conditions are elucidated in terms of regulated temperature (sensible load) and also by control of humidity (latent load), especially for hot and humid geographic regions. In order to control the required humidity level, the conventional refrigerant aided vapour compression systems (VCS) are condensed out the water vapour present in air by lowering the air dry-bulb temperature below its dew point temperature. Subsequently, the dehumidified air is subjected to reheating to reach the desired indoor temperature range that eventually consumes more power in terms of electricity. For example, in order to extract the air specific humidity of ~0.07 kg_{wv}/kg_{da}, the air needs to be cooled to nearly 9 °C [3]. Most often, the electricity generation involves the fossil-fuelled driven power plant that is having a resultant emission of CO_2 into the lower atmosphere.

Emerging nations like China and India are still relying on some strong based refrigerants for air conditioning. However, European Commission Regulation 2037/2000, articulates the usage of all HFCs would be phasing out from the year of 2015 [4]. Even so, environmental concern seems to revoke when the level of comfort, luxury and standard of living associated with humans. Many convenient protocols, such as Montreal Protocol are directed to some significant changes in the production of refrigerants and their control consumption. Since then, different initiatives have been considered to minimise the use of HFCs. In this perspective, a holistic approach needs to be adopted to maintain the energy-intensive humidity level. Recently, sorption technology has taken significant attention as an alternative process to standardise mechanical air conditioning systems for dehumidification of air without cooling below its dew point temperature. Liquid desiccant dehumidification systems (LDAC) are one of the sorption technologies, where a synthetic or natural desiccant is used as a hygroscopic material to absorb the moisture of the surrounding air.

Several experimental investigations have been conducted on dehumidifiers by different salt solutions as desiccants. Many researchers used aqueous lithium chloride (LiCl) as a desiccant [5–7]. Other investigations using calcium chloride ($CaCl_2$) and lithium bromide (LiBr) as desiccants in a dehumidifier were also carried out [3, 8, 9]. Moreover, it is noticed that LiCl becomes more feasible desiccant compared to other pure desiccants because of its inherent characteristics as well as its stability feature (it does generally not vaporise into the air) at ambient conditions [10]. Chung and Gosh [11] experimentally studied dehumidifier with LiCl aqueous desiccant and concluded that the structured packed bed liquid desiccant dehumidifier yielded better performance compared to the random packed based dehumidifier. Fumo and Goswami [10] studied LiCl potentiality as an aqueous desiccant in liquid desiccant dehumidification systems and evaluated the performance of dehumidifier by estimating the condensation /evaporation rate as performance characteristics. Naik and Muthukumar [7] experimentally studied the desiccant dehumidifier with the help of LiCl desiccant and patterned the dehumidifier performance characteristics at different operating conditions. In the recent past, reasonable amount of effort has been oriented towards estimation of performance characteristics of the desiccant dehumidifier. However, no attention was devoted toward the investigation of trade-off analysis, which is essential for decision-making exercises, among various dehumidifier performance characteristics to get optimum input parameters.

On the other hand, it is essential to investigate the underlying physicochemical properties of desiccants, which are most prominent in the system. Mainly, in desiccant dehumidification system, the equilibrium air humidity, which is a function of vapour pressure difference between the liquid desiccant and process air, at the desiccant surface that is the most direct impelling properties on the system's effectiveness [12]. The approach for estimation of equilibrium air humidity at the desiccant surface was a subject of earlier research for specific spectra of interest in different technological and scientific applications, like metrology as well as for water distillation. In these mentioned fields, the use of vapour pressure and their corresponding study mostly fall either to intermediate temperatures like 28–35 °C or to relatively higher temperatures 35–45 °C. However, the vapour pressure (driving force of the water vapour) at low temperature (15 °C) levels up to high temperature (50 °C) is crucial for certain other industrial domains, like drying and air conditioning, to estimate a precise prediction of heat and mass transfer activity. In the present study, the concentration range and temperature range has chosen between 20 and 40 wt.% and 20°–50 °C, respectively. The concentration and temperature are maintained below 40% and less than 20 °C, respectively, to avoid crystallisation of desiccant, which might block the packing material pores. Besides, in order to achieve adequate moisture absorption rate from the air, the temperature and concentration are maintained below 50 °C and above 20 wt.%, respectively. Even though a significant number of experimental studies were conducted to establish some empirical correlations for vapour pressure estimation, they are represented in a complicated form and sporadically represented in the literature.

In this perspective, the objectives of the present study are to develop a simple correlation to predict the vapour pressure as a function of desiccant solution concentration and temperature for LiCl desiccant solution at temperature range between 20° and 50 °C. In order to ensure that, the data of the corresponding properties are compiled from available literature. With the help of vapour pressure and desiccant temperature, the equilibrium specific humidity is also demonstrated in the form of Othmer Chart (as a function of concentration and temperature). An investigation is made to analyse the trade-off among performance parameters based on different liquid to air ratios. Subsequently, empirical regression models are also developed for prediction of performance parameters and compared with previous published correlations for modelling of dehumidifier testing paradigms. Experimental data from Chung and Ghosh [11] were used to analyse this present theoretical analysis for LiCl solution.

2 Evaluation of Equilibrium Air Humidity

2.1 Vapour Pressure of LiCl

In order to calculate the vapour pressure of LiCl, the vapour pressure data were collected from several literatures [12–14] at different ranges of temperatures and concentrations. Moreover, the existing procedure [13, 14] is used to calculate the vapour pressure of the LiCl, as follows:

$$p_{\mathrm{sol}} = p_{\mathrm{H_2O}}[\Psi f] \tag{1}$$

where $f = A + B\theta$; $A = 2 - \left[1 + \left(\frac{X}{\pi_0}\right)^{\pi_1}\right]^{\pi_2}$

$$B = \left[1 + \left(\frac{X}{\pi_3}\right)^{\pi_4}\right]^{\pi_5} - 1;\ \theta \equiv \frac{T_s}{T_{c,H_2O}}$$

and $\Psi = 1 - \left[1 + \left(\frac{X}{\pi_6}\right)^{\pi_7}\right]^{\pi_8} - \pi_9 e^{-\frac{(X-0.1)^2}{0.005}}$;

$$In\left(\frac{p_{\mathrm{H_2O}}}{p_{c,\mathrm{H_2O}}}\right) = \frac{A_0\tau + A_1\tau^{1.5} + A_2\tau^3 + A_3\tau^{3.5} + A_4\tau^4 + A_5\tau^5}{1-\tau};$$

$$\tau = 1 - \frac{T_s}{T_{c,\mathrm{H_2O}}};$$

The constants of Eq. 1 are represented in Table 1. A simple straight-forward polynomial equation is proposed form the extracted vapour pressure data. It can be used to represent the vapour pressure as a function of concentration and temperature as follows:

$$P_v\,(\mathrm{kPa}) = A_o + A_1\frac{X}{T_s} + A_2\left(\frac{X}{T_s}\right)^2 + A_3\left(\frac{X}{T_s}\right)^3 \tag{2}$$

where, A_0, A_1, A_2 and A_3 are constants, and their corresponding values are 16.08, –26.533, 15.039, and −2.801, respectively, for LiCl solution.

2.2 *Air Equilibrium Specific Humidity at Desiccant Surface (ω_e)*

Desiccants absorb moisture due to vapour pressure difference between the surface of the desiccant solution and surrounding air. The dehumidification phenomena are said to happen when vapour pressure value of air is more compared to that at surface of the desiccant solution and expected to be continued till the vapour pressure of desiccant is in equilibrium with air. The parameter which conveys the feasibility of dehumidification process is equilibrium air specific humidity, which is the air specific humidity at the desiccant solution surface at local solution concentration and temperature, calculated by Eq. (3)

$$\omega_e = 0.622\left(\frac{P_v(X, T_s)}{P_{\mathrm{atm}} - P_v(X, T_s)}\right); \tag{3}$$

where, $P_{\mathrm{atm}} = 101.325$ kPa.

In Fig. 1, the lines of constant aqueous salt concentration (isosteres) (40–20 wt.%, from top to bottom) of LiCl are depicted in an Othmer chart. This chart demonstrates the equilibrium humidity ratio of air, $\omega_e = f\,(X, T_s)$ at the desiccant solution surface as a function of solution temperature and concentration. It is observed that humidity ratio decreases with temperature for a fixed salt concentration ($X = m_{\mathrm{salt}}/m_{\mathrm{sol.}}$). However, the humidity ratio decreases with increasing concentration for a constant temperature. It could be derived from Othmer chart that desiccant absorbs more moisture at lower temperatures and higher concentration. A straightforward correlation is developed as function of solution temperature and concentration (Eq. 4).

$$\omega_\mathrm{e} = \mathrm{A_1} + \mathrm{A_2T} + \mathrm{A_3T^3} + \mathrm{A_4X} + \mathrm{A_5X\,T^3} \tag{4}$$

Table 1 Parameters for the vapour pressure equations of LiCl–H_2O [13, 8]

Constants	π_0	π_1	π_2	π_3	π_4	π_5	π_6	π_7	π_8	π_9
Value	0.28	4.3	0.6	0.21	5.10	0.49	0.362	−4.75	−0.4	0.03
Constants	A_0	A_1	A_2		A_3	A_4	A_5		$p_{c,H2O}$	$T_{c,H2O}$
Value	−7.85823	1.839910	−11.781100		22.67050	−15.9393	1.775160		22.064×10^3	647.3 K

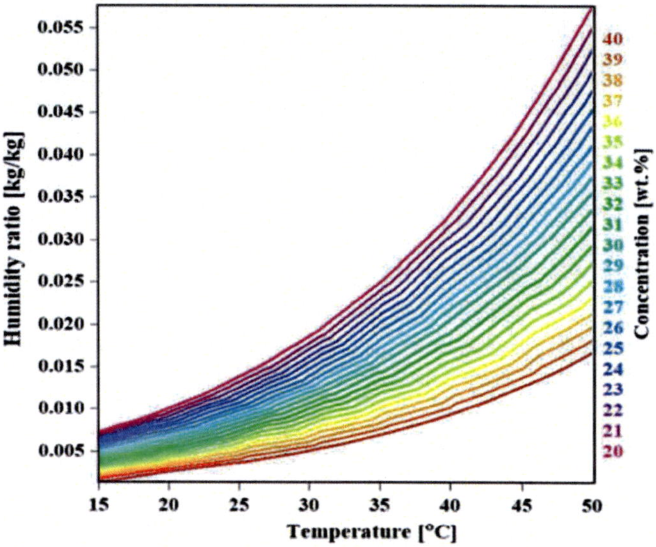

Fig. 1 Othmer chart for LiCl aqueous solution

where, A_1, A_2, A_3, A_4 and A_5 are constants and their corresponding values are 9.69E−03, 1.08E−04, 6.63E−07, −2.57E−04 and −1.43E−08, respectively, for LiCl solution.

3 Performance Indices

3.1 Condensation Rate

It expresses the rate of change of air specific humidity at a particular mass flow rate. It is calculated as

$$CR\,(\text{kg/s}) = \dot{m}_a\left(\omega_a - \omega_{a,o}\right) \tag{5}$$

3.2 Moisture Effectiveness

This expresses the air specific humidity variation across the dehumidifier with respect to maximum possible humidity variation.

$$\varepsilon_m = \frac{\omega_a - \omega_{a,o}}{\omega_a - \omega_e} \tag{6a}$$

$$\omega_e = 0.622\left(\frac{P_V(X, T_s)}{P_{atm} - P_V(X, T_s)}\right) \tag{6b}$$

3.3 Enthalpy Effectiveness

This performance parameter expresses the actual enthalpy variation of air during dehumidification operation compared to the maximum possible enthalpy variation.

$$\varepsilon_h = \frac{h_a - h_{a,o}}{h_a - h_e} \tag{7}$$

4 Results and Discussion

4.1 Experimental Data Analysis

Chung and Ghosh [11] experimental data were collected from Koronaki et al. [15] (adapted with permission from Elsevier, copyright 2019, Order Number: 4701141019796) for the present study. The experiment was conducted using LiCl liquid desiccant in an adiabatic counter flow dehumidifier with packing density of 223 m^2/m^3. T_a, T_s, X, ω_a, $\dot{m}_a$ and $\dot{m}_s$ are inlet parameters and T_{ao}, T_{so} and ω_{ao} represent the outlet conditions of the dehumidifier, respectively [15].

Table 2 provides the experimental data for a packed column dehumidifier with PVC structure packing. In the case of air dry bulb temperature, the output results showed that the process air is being cooled during dehumidification process. It is due to the fact that the solution inlet temperature is much below the air inlet temperature.

Though mass transfer occurs between the air and desiccant, the outlet air temperature becomes lower compared to the inlet air temperature. It is due to sensible cooling domination over sorption heating. The maximum air specific humidity difference between the inlet and outlet was 0.0074 kg_{wv}/kg_{da}. It can also be observed that the outlet solution temperature is higher compared to the inlet solution temperature and this is due to sensible and sorption heat transfer during dehumidification. The performance parameters are calculated using Eqs. (5–7) and are represented in Table 3.

Table 2 Experimental values of Chung and Ghosh [11] obtained from Koronaki et al. [15] (adapted with permission from Elsevier, copyright 2019, Order Number: 4701141019796)

S. No.	Inlet condition						Outlet condition		
	Air			Solution			Air/Solution		
	$\dot{m}_a$ (kg/s)	T_a (°C)	ω_a (kg_{wv}/kg_{da})	$\dot{m}_s$ (kg/s)	T_s (°C)	X (wt.%)	T_{ao} (°C)	ω_{ao} (kg_{wv}/kg_{da})	T_{so} (°C)
1	0.0231	21.8	0.0112	0.3071	16.8	31	17.8	0.0064	16.9
2	0.0260	21.6	0.0111	0.3071	16.5	31	17.6	0.0063	16.6
3	0.0289	20.4	0.0115	0.3071	16.4	31	17.3	0.0066	16.6
4	0.0318	21.0	0.0117	0.3071	17.0	31	17.8	0.0067	17.2
5	0.0318	24.1	0.0149	0.1919	16.9	31	19.1	0.0080	18.3
6	0.0318	23.4	0.0147	0.2303	17.5	31	19.1	0.0077	18.6
7	0.0318	23.3	0.0147	0.2687	17.7	31	19.1	0.0075	18.5
8	0.0318	24.1	0.0147	0.3071	17.1	31	18.9	0.0073	18.0
9	0.0231	23.3	0.0127	0.3071	19.5	37	20.6	0.0058	19.8
10	0.0260	22.9	0.0126	0.3071	18.9	37	19.9	0.0058	19.0
11	0.0289	20.9	0.0101	0.3071	19.3	37	19.8	0.0051	19.4
12	0.0318	20.8	0.0105	0.3071	19.1	37	19.5	0.0052	19.2
13	0.0318	21.1	0.0103	0.1919	17.6	38	18.9	0.0049	18.4
14	0.0318	20.9	0.0103	0.2303	17.5	38	18.6	0.0047	18.2
15	0.0318	20.8	0.0103	0.2687	17.5	38	18.5	0.0046	18.1
16	0.0318	20.7	0.0103	0.3071	17.7	38	18.4	0.0045	18.0

Table 3 Calculated performance parameters based on Table 2

S. No.	CR (g/s)	ε_h	ε_m
1	0.111	0.724	0.692
2	0.125	0.724	0.697
3	0.142	0.694	0.667
4	0.159	0.707	0.678
5	0.219	0.667	0.651
6	0.223	0.701	0.684
7	0.229	0.724	0.708
8	0.235	0.729	0.715
9	0.159	0.728	0.720
10	0.177	0.722	0.707
11	0.145	0.720	0.713
12	0.169	0.723	0.709
13	0.172	0.690	0.692
14	0.178	0.718	0.716
15	0.181	0.733	0.729
16	0.184	0.758	0.747

4.2 Trade-Off Analysis of Performance Parameters

Trade-off analysis among the performance parameters (ε_h-ε_m-CR) is investigated to rationalise the effectiveness of the liquid desiccant dehumidifier. The entire ε_h-ε_m-CR footprint of the experimental endeavour is encapsulated in Figs. 2 and 3. Figure 2 corresponds to S/A of 6.7, 8.5 and 9.5 operating operations. The loci 6–7–8.5–9.5 in Fig. 2 represents the S/A and the corresponding ε_m and ε_h are represented in X-axis, and Y-axis, respectively. CR are represented in colour code. Dehumidifier operation yielded a persistent increase in ε_{h}, ε_{m}, and CR for higher rates of solution enrichment.

However, beyond S/A of 8.5, the air-desiccant operation registered a significant decrease of ε_h, ε_m, and CR. It is noted that ε_{h}, ε_{m} and CR trade-off footprint of dehumidifier operation is higher for lower S/A value (8.5) when compared to higher S/A values (10.5–13.5), as shown in Figs. 2 and 3. However, ε_h, ε_m and CR value increase with S/A from 10.5 to 13.5 (Fig. 3). S/A of 8.5 registered 77.1% increase in CR, 29.2% and 26.1% lower in ε_m and ε_h, respectively, as compared to S/A of 13.5. It is due to more desiccant participation at higher S/A, which blocks packing material pores and reduces the contact area in the dehumidifier chamber paradigm.

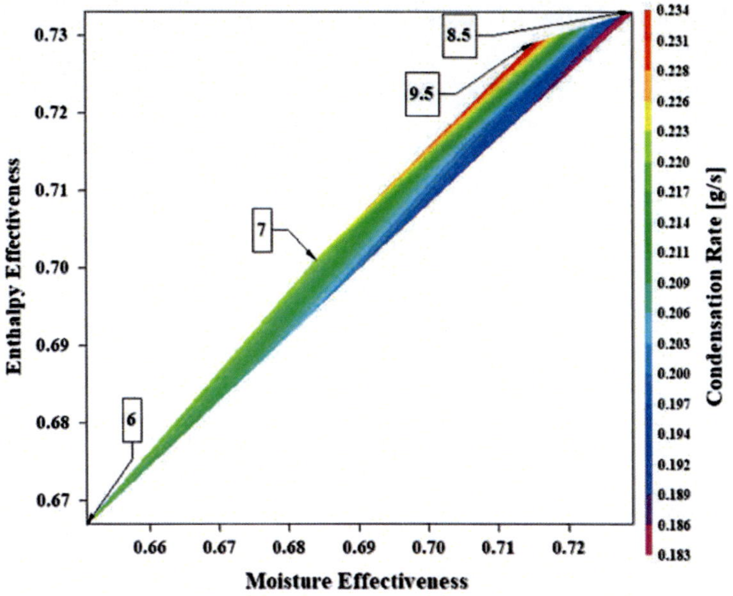

Fig. 2 Trade-off map for dehumidifier performance parameters at S/A: 6–8.5

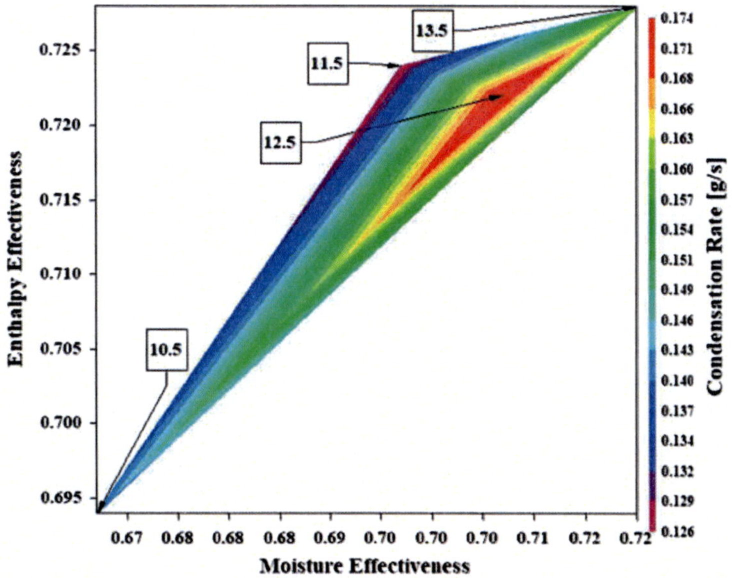

Fig. 3 Trade-off map for dehumidifier performance parameters at S/A: 10.5–13.5

5 Modelling Using Stepwise Regression (SWR) Approach

Regression analysis can be used to identify a trend in the data, whether it is linear or non-linear form. The heat and mass transfer process is complicated in nature, so, all the input parameters are considered for modelling. T_a, ω_a, X, T_s, $\dot{m}_a$ and $\dot{m}_s$ are selected as inlet parameters, as mentioned in Eq. (8).

$$y_i = f(T_a, \omega_a, T_s, X, \dot{m}_a, \dot{m}_s) \tag{8}$$

where y_i is the performance parameters of the model, i.e., CR and ε_m. In the dehumidification process, the T_{ao} is varying for different composition/operating conditions and this is due to different rate of sorption and sensible heat release. In some operation/composition, the sorption heat is dominated over sensible heat and in some cases, sensible heat between the air and desiccant is dominated. Hence, CR and ε_m model shows a non-linear trend. Therefore, SWR model is chosen to build the empirical models. The developed models for the performance parameters are shown in Eqs. (9–10).

$$\begin{aligned} CR = {} & 5.893\,\omega_a T_s + 0.2897\,\dot{m}_a T_a + 0.006869\,T_a \dot{m}_s + 0.001685\,T_a X - 0.1621 \\ & - 0.003356\,T_a T_s - 2.703\,\omega_a X \end{aligned} \tag{9}$$

$$\begin{aligned} \varepsilon_m = {} & 0.4378 + 0.09562X + 0.0162\,T_a T_s + 0.663\,\sin(\sin(\dot{m}_a)) + 0.04326\,T_a \dot{m}_s^2 \\ & + 0.07573\,\omega_a / \sin(\sin(2.737/\omega_a)) - 0.3526\,T_s - 0.0001918\,X T_a^2 \end{aligned} \tag{10}$$

The adequacy of fitted regression models, R-square value is 99.93% for CR, and 97.7% for ε_m, which indicates that the regression models are highly significant (Fig. 4). Details of correlation coefficient (R^2) can be found from literature [16]. This

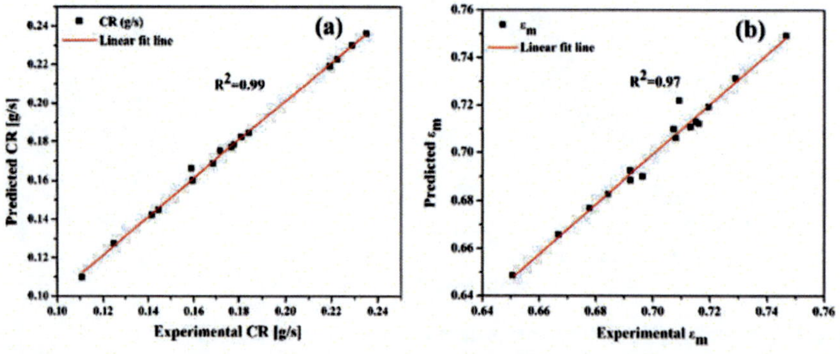

Fig. 4 Comparison of experimental and SWR predicted results **a** condensation rate (g/s), **b** moisture effectiveness

Table 4 Comparison of R-square and RMSE for present and Koronaki et al. [15] models

	Statistical parameters	(ε_m)	(CR)
Koronaki et al. [15]	R-square	0.779	0.8663
	RMSE	0.246	0.0880
Present study	R-square	0.977	0.9971
	RMSE	0.006	0.0020

also expresses the performance prediction capability of present model, which seems to be in line with the trends of the experimental data, as seen from Fig. 4. Further, present models corroborated that the effects of all independent parameters can also be well predicted.

6 Model Comparison with Literature

The models developed in the current study are compared with Koronaki et al. [15] models for the same experimental datasets. The performance of developed models are compared based on two statistical criteria, RMSE (Root Mean Square Error) and R^2 (correlation coefficient) for the same datasets (Eqs. 11, 12). Preliminary studies suggest that the SWR model could provide excellent predictions throughout the region of interest. Accordingly, the developed models are found to be reasonably consistent and produces stable responses at all points of interest. It is observed from Table 4 that the present models exhibit better prediction compared to Koronaki et al. [15] models. The adequacy of fitted regression models, R-square and RMSE values are 99.93% and 0.0020 for CR and 97.7% and 0.006 for ε_m, respectively (Table 4).

$$R^2 = 1 - \left(\frac{\sum_{i=1}^{z} \left(\lambda_{ex} - \lambda_p\right)^2}{\sum_{i=1}^{z} \lambda_{ex}^2} \right) \tag{11}$$

$$\text{RMSE} = \sqrt{\frac{1}{z}\left(\sum_{i=1}^{z} \left(\lambda_{ex} - \lambda_p\right)^2\right)} \tag{12}$$

7 Parametric Study

In this section, the effects of independent parameters on the dehumidifier performance parameter are analysed using the developed empirical correlations. It is well known that dehumidifier operation is performed to remove moisture from the air. So, the condensation rate becomes a crucial factor. Therefore, in the present study, the effects of independent parameters on the condensation rate (CR) is analysed in detail. The influences of the four different inlet parameters on the condensation rate are shown

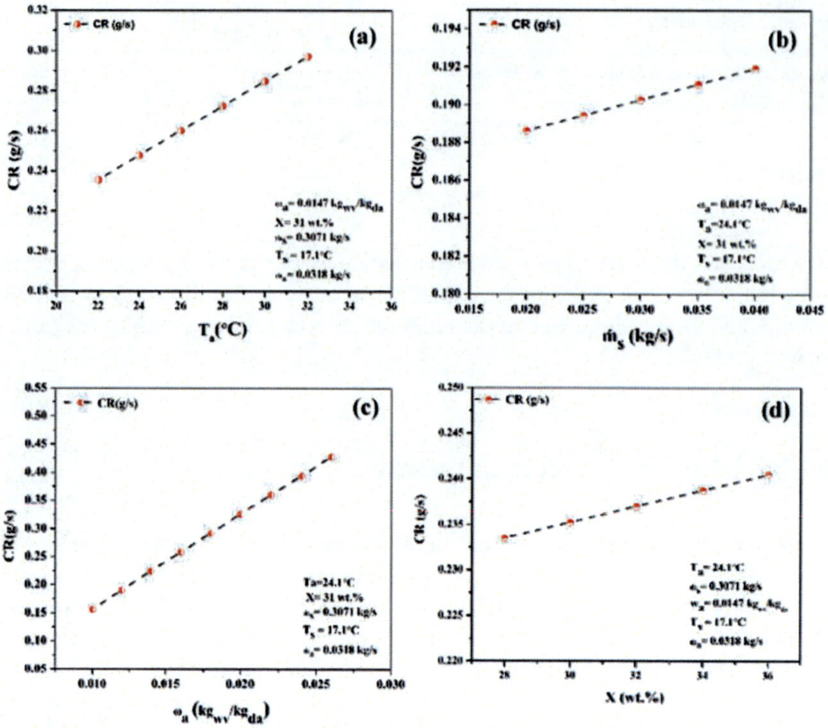

Fig. 5 Influences of inlet parameters on the condensation rate: **a** air temperature, **b** desiccant mass flow rate, **c** air specific humidity, **d** desiccant concentration

in Fig. 5. All of the four parameters have a noticeable influence on the condensation rate, as shown in Fig. 5. It can be seen from Fig. 5 that the inlet desiccant flow rate, air humidity ratio, air temperature and desiccant concentration have a positive influence on the condensation rate. Increasing the desiccant mass flow rate leads to an increasing mass transfer coefficient between the air and the desiccant, as mentioned by Zhang et al. [17]. As a result, the condensation rate increases with a rising flow rate of desiccant. Further, increasing the air temperature can result in a higher surface vapour pressure difference between the air and desiccant, which increases the mass transfer potential between the air and the desiccant and then increase the condensation rate. The reason for the increase of the condensation rate with the desiccant inlet concentration is due to the decrease in the surface vapour pressure of the desiccant. However, a higher surface tension caused by a higher concentration would reduce the wettability of the packing [18]. This effect counteracts the increase of the mass transfer potential; positive exponent represents the increasing trend while the negative exponent represents the decreasing trend. Finally, it is found from the prediction curve (Fig. 5) that the positive exponent has a more significant influence than the negative exponent.

8 Conclusions

In the present study, LiCl solution is used as a desiccant material to analyse the dehumidifier performance. Based on the above study, the following major conclusions can be drawn:

- A polynomial correlation is developed to calculate the vapour pressure of LiCl solution for the temperature range between 20 and 50 °C within the desiccant concentration between 20 and 40 wt.%.
- Othmer chart (ω_e-T_s-X diagram) is constructed in a concentration range between 20 and 40% within the temperature range between 15 and 50 °C.
- The trade-off analysis showed that the solution to airflow rate (S/A) ratio of 8.5 is best compared to the other S/A ratios and the condensation rate, moisture effectiveness, and enthalpy effectiveness is 0.234 g/s, 71.5% and 72.5%, respectively at S/A of 8.5.
- Empirical regression models of CR and ε_m are developed for prediction of performance parameters and compared with existing literature correlations. It reveals that the developed models are robust for surrogate modelling in dehumidifier testing paradigms.
- From the parametric study, it is noticed that the inlet desiccant flow rate, air humidity ratio, air temperature and desiccant concentration have a positive influence on the condensation rate.

References

1. HVAC&R Industry. A Weekly ENewsletter from ASHRAE, 21 Jan 2016, 15, 3
2. Rafique MM, Gandhidasan P, Rehman S, Al-Hadhrami LM (2015) A review on desiccant based evaporative cooling systems. Renew Sustain Energy Rev 45:145–159
3. Ahmed CSK, Gandhidasan P, Al-Farayedhi AA (1997) Simulation of a hybrid liquid desiccant based air-conditioning system. Appl Therm Eng 17(2):125–134
4. Sarbu I, Sebarchievici C (2013) Review of solar refrigeration and cooling systems. Energy Build 67:286–297
5. Bhowmik M, Muthukumar P, Anandalakshmi R (2018) Numerical study on dehumidification performance of a cross-flow liquid desiccant adiabatic dehumidifier with various halide salt desiccants. International Conference on Computational Methods for Thermal Problems, THERMACOMP 2018, issue 223309, pp: 95-98
6. Chen Y, Zhang X, Yin Y (2016) Experimental and theoretical analysis of liquid desiccant dehumidification process based on an advanced hybrid air-conditioning system. Appl Therm Eng 98:387–399
7. Naik BK, Muthukumar P (2019) Experimental investigation and parametric studies on structured packing chamber based liquid desiccant dehumidification and regeneration systems. Build Environ 149:330–348
8. Lazzarin RM, Gasparella A, Longo GA (1999) Chemical dehumidification by liquid desiccants: theory and experiment. Int J Refrig 22(4):334–347
9. Bhowmik M, Haldar S, Dharmalingam K, Muthukumar P, Anandalakshmi R (2020) Evaluation of thermo-kinetic and absorption characteristics of pure desiccants and desiccant mixtures. Mater Today Proc, 26 (2):1967-1971

10. Fumo N, Goswami DY (2002) Study of an aqueous lithium chloride desiccant system: air dehumidification and desiccant regeneration. Sol Energy 72(4):351–361
11. Chung TW, Ghosh TK, Hines AL (1996) Comparison between random and structured packings for dehumidification of air by lithium chloride solutions in a packed column and their heat and mass transfer correlations. Ind Eng Chem Res 35(1):192–198
12. Sahlot M, Riffat SB (2016) Desiccant cooling systems: a review. Int J Low-Carbon Technol 11(4):489–505
13. Conde MR (2004) Properties of aqueous solutions of lithium and calcium chlorides: formulations for use in air conditioning equipment design. Int J Therm Sci 43(4):367–382
14. Gurubalan A, Maiya MP, Geoghegan PJ (2019) A comprehensive review of liquid desiccant air conditioning system. Appl Energy 254(July):113673
15. Koronaki IP, Christodoulaki RI, Papaefthimiou VD, Rogdakis ED (2013) Thermodynamic analysis of a counter flow adiabatic dehumidifier with different liquid desiccant materials. Appl Therm Eng 50(1):361–373
16. Bhowmik M, Muthukumar P, Anandalakshmi R (2019) Experimental based multilayer perceptron approach for prediction of evacuated solar collector performance in humid subtropical regions. Renew Energy 143:1566–1580
17. Zhang L, Hihara E, Matsuoka F, Dang C (2010) Experimental analysis of mass transfer in adiabatic structured packing dehumidifier/regenerator with liquid desiccant. Int J Heat Mass Transf 53(13–14):2856–2863
18. Moon CG, Bansal PK, Jain S (2009) New mass transfer performance data of a cross-flow liquid desiccant dehumidification system. Int J Refrig 32(3):524–533

Design of Solar Photovoltaic/Thermal System (PVT) with Thermal Energy Storage for Air-Conditioning Applications

A. Sai Kaushik, Pulla Rao Muvvala, and Satya Sekhar Bhogilla

List of symbols

A	Area of the collector (m^2)
$\dot{Q}_l$	Heat lost by the panel (W)
G	Incident solar radiation (W/m^2)
η_{th}	Thermal efficiency (%)
$\dot{m}$	Mass flow rate (kg/s)
$c_{cp(cf)}$	Specific heat capacity of the coolant fluid (kJ/kg.K)
T_o	Outlet temperature of the coolant fluid (K)
T_i	Inlet temperature of the coolant fluid (K)
η_e	Electrical efficiency (%)
P_g	Power generated by the solar cell (W)
$\dot{Q}_e$	Heat absorbed by the evaporator (W)
$\dot{Q}_g$	Heat supplied to the generator (W)
m_{PCM}	Mass of PCM material (kg)
L_{PCM}	Latent heat of fusion of lauric acid (kJ/kg)
$\dot{Q}_{PCM}$	Heat transferred to the PCM (W)
γ	Circulation ratio
$\in$	Concentration ratio
E_{in}	Amount of exergy that is received by the system (W)
T_a	Ambient temperature (K)

A. Sai Kaushik · P. R. Muvvala
IIITDM Kurnool, Kurnool, India

S. S. Bhogilla (✉)
Indian Institute of Technology Jammu, Jammu, India
e-mail: satya.bhogilla@iitjammu.ac.in

M. Palanisamy et al. (eds.), *Theoretical, Computational, and Experimental Solutions to Thermo-Fluid Systems*, Lecture Notes in Mechanical Engineering,
https://doi.org/10.1007/978-981-33-4165-4_30

T_s Sun temperature (K)
E_{out} Amount of exergy obtained as output (W)

1 Introduction

Solar energy is one of the most abundant and promising renewable sources of energy with a lot of potential and scope for development in generation of power for useful work. Solar energy systems are becoming increasingly popular due to the current global trend shifting towards adopting cleaner and eco-friendly sources of energy for power generation in an ode to counteract the harmful effects that are being caused due to excessive use of fossil fuels and other non-renewable sources of energy for the same. As a result of this, a lot of research and studies are being performed to increase the efficiency or the performance of these systems to make them more practically feasible for a wider scale of implementation.

A large fraction of the solar heat that is incident on to the solar panel is dissipated in the form of thermal energy or waste heat and only a mere fraction of which is actually being utilized by the solar panel. Since the electrical conversion efficiency of the solar panel is very low with an average range of efficiency ranging from 15% to 18%, a hybrid system involving an additional thermal extraction cycle to capture or absorb this dissipated heat is coupled with the solar cell panel to increase their overall efficiency. Co-generation of both electricity and thermal energy can be obtained from the same system. Such hybrid systems are becoming more common and are playing a pivotal role in extracting the maximum output from the given system under consideration and making them more diverse and flexible in terms of the different applications they can be used for. Optical devices, such as plane reflectors or parabolic dish collectors, are used as solar concentrators for capturing a large area of incident sunlight and focusing it onto the solar panel for operating the panel at its peak performance throughout its working cycle. As the efficiency of the solar cell decreases with increase in its temperature, the implementation of the thermal energy extraction cycle to the panel absorbs the excess heat and maintains the solar panel at its optimum temperature for best performance and to avoid damage to any of the components. This leads to an increased lifespan of the solar panel in addition to the extra thermal energy that is being harnessed [1, 2].

A pump is used to circulate the coolant fluid (primarily water) across the solar panel to absorb the heat and the thermal energy system employing phase-change materials to ensure that the coolant fluid passing through it receives enough thermal energy for producing useful work. The extent of using this system for various applications is determined by the amount of heat that is being absorbed by the thermal extraction cycle or the maximum outlet temperature that is obtained from the circulating coolant fluid. For low temperature ranges, this system can be employed for water heating or space heating and so on, and for medium temperature ranges, the scope of application of this system can be extended further to air-conditioning or cooling applications such as the vapor-absorption cycle, desalination systems, and water pumping. The applications for which this system can be used are diverse and many are based on the amount of heat that is being generated by the solar panel system [3]. Various studies have been performed in studying the feasibility of the waste heat obtained from the PVT module for different applications such as for heating and cooling applications in buildings [4] to improve the collector efficiency by about 20–30%, for water heating using PCM materials [5] for domestic applications, and for using the waste heat obtained from the system for solar cooling through the vapor-absorption cycle [6] to determine its coefficient of performance and cooling capacity. The applications for which this system can be used are diverse and many are based on the amount of heat that is being generated by the solar panel system. The primary advantage of using concentrators if installed is to focus the solar energy onto the system so that it reduces the number of photovoltaic cells required for the panel and makes it more viable and compact. This can solve the problem of deploying energy harvesting systems where the available space is the main constraint. This system is also more economical and practical to operate in the longer run with a faster pay-back period and also with lesser maintenance and service that is required.

2 System Modeling

In this model analysis, a coolant circuit made up of copper tubes passing underneath the solar panel so as to absorb the excessive heat that is being produced by the panel is considered. Water is pumped through these copper tubes as the coolant fluid to transfer the thermal energy from the PV module to the source of application of the waste heat. The concept of total system is shown in Fig. 1. The heat that is absorbed by the water passing through the solar module is then split into two channels, with one of them exchanging the absorbed heat with the phase-change material system (thermal energy storage) and the other exchanging the absorbed heat to the vapor-absorption cycle. A 50:50 split of the absorbed heat is assumed for both the channels

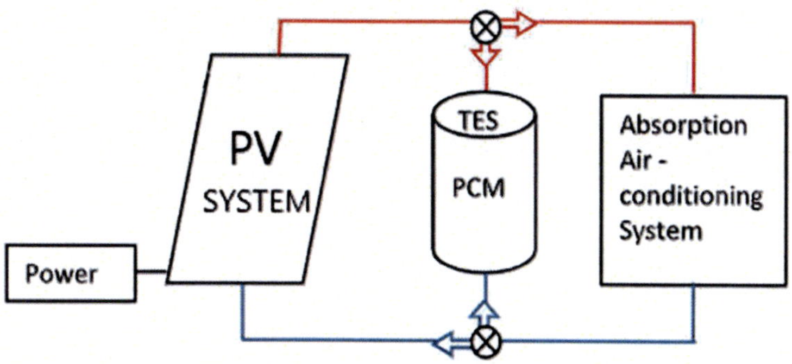

Fig. 1 Concept of total system

of coolant circuit with both the channels receiving half the amount of the total waste heat that is absorbed.

Stearic acid is selected as the primary phase-change material for the thermal energy storage system as it exhibits the most suitable properties with a high latent heat of fusion capacity for optimum energy storage at the required temperatures.

2.1 *Photovoltaic Module with Coolant Circuit*

The PV module under consideration for our analysis is a 250 W panel with a short circuit current (I_s) of 8.92 A and an open-circuit voltage (V_0) of 38.19 V. The effective collector area of the solar panel is 1.623 m^2 (length = 166.32 cm, width = 99 cm). The thermal efficiency of the flat-plate solar panel/collector is given by the following equation:

$$\eta_{\text{th}} = \frac{\dot{Q}_{\text{l}}}{A \times G} \tag{1}$$

where $\dot{Q}_{\text{l}}$ is the heat lost by the panel to the surroundings, A is the area of the collector, and G is the incident solar radiation. This amount of heat that is lost by the solar cell can be transferred to the coolant fluid and to the phase-change material in the thermal energy storage system by the following equation:

$$\dot{Q}_{\text{l}} = \dot{m}_{\text{cf}} c_{\text{cp(cf)}} \Delta T \tag{2}$$

where $\dot{m}_{\text{cf}}$ is the mass flow rate of the coolant fluid, $c_{\text{cp(cf)}}$ is the specific heat capacity of the coolant fluid at constant pressure which is taken as the constant value of 4.187 kJ/kg.K and ΔT is the temperature difference between the outlet temperature (T_{o}) and inlet temperature of the coolant fluid (T_{i}) passing through the solar panel. The electrical efficiency of the solar cell is given by the equation:

$$\eta_{\text{e}} = \frac{P_{\text{g}}}{A \times G} \tag{3}$$

where P_{g} is the amount of power generated by the solar cell, A is the area of the collector, and G is the incident solar radiation. The electrical power generated by the solar cell is given by the equation:

$$P_{\text{g}} = V.I \tag{4}$$

where V is the voltage output obtained from the solar panel, and I is the current output obtained from the solar cell. The photovoltaic panel efficiency is lesser than the efficiency of the individual cell due to current mismatch loss and coupling losses associated with the series of the PV panels that have been joined. The auxiliary work that is supplied to the pump for circulating the coolant fluid is neglected in our analysis as the working fluid considered in our analysis is water which would require minimum pump work for effective circulation. The coefficient of performance (COP) of the vapor-absorption cycle is given by:

$$\mathrm{COP} = \frac{\dot{Q}_{\mathrm{e}}}{\dot{Q}_{\mathrm{g}}} \quad (5)$$

where $\dot{Q}_{\mathrm{e}}$ is the heat absorbed by the evaporator to cool the working space to the desired temperature values and $\dot{Q}_{\mathrm{g}}$ is the heat supplied to the generator that is obtained through one of the channels of the absorbed heat from the solar panel.

3 Model Analysis and Observations

The irradiation data and temperature data are taken from the reference [7]. Tables 1 and 2 show the various calculated parameters for different mass flow rates. The same values of incident radiation have been considered for the cases of different mass flow rates to study the trend in the coolant output temperature values obtained.

Average solar radiation intensity over a time period of 6 h (21,600 s): 730.84 W/m^2

Average thermal efficiency of the system: 66.7%

Average waste heat absorbed by the coolant fluid = (730.84 × 1.623) × 0.73 = 791.2 W.

Assuming 50% distribution to the PCM material: $\dot{Q}_{\mathrm{PCM}}$ (avg) = 395.6 W

Density of stearic acid = 840 kg/m^3

Amount of heat transferred to the PCM = $\dot{Q}_{\mathrm{PCM}}$ (avg) × 21,600 s = 8.545 MJ

$$Q_{\mathrm{PCM}}(\mathrm{avg}) = m_{\mathrm{PCM}} x L_{\mathrm{PCM}} \quad (6)$$

where m_{PCM} is the mass of PCM material required for the analysis and L_{PCM} is the latent heat of fusion of stearic acid = 203 kJ/kg

Hence, $m_{\mathrm{PCM}} = 42.1$ kg

Table 3 shows different PCM material for comparison. Stearic acid is selected as the primary PCM material in the analysis as compared to other materials because of its relatively higher melting point. As a result of this, it can store energy at higher temperatures and can supply good quality heat to the vapor-absorption cycle at the desired temperatures for obtaining the best performance from the system. Also, stearic acid can handle higher fluctuation of loads when acting as a standalone supplier of energy due to its better quality of energy stored that can supply the required amount of varied thermal heat requirements over a wider range of operation.

Table 1 PVT module analysis

$\dot{V}_{cf}$ (m^3/min)	$\dot{m}_{cf}$ (kg/s)	Time (h)	G (W/m^2)	T_o (K)	T_i (K)	$\dot{Q}_1$ (W)	η_{th} (%)
0.0005	0.00834	10	376	361.140	350.15	383.900	62.91
		11	737	370.120	350.15	697.350	58.30
		12	853	378.536	350.15	991.240	71.60
		13	937	382.900	350.15	1143.756	75.21
		14	948	380.670	350.15	1065.800	69.27
		15	699	371.710	350.15	752.720	66.35
		16	369	365.350	350.15	530.910	63.20
0.001	0.01667	10	376	355.650	350.15	383.900	62.91
		11	737	360.140	350.15	697.350	58.30
		12	853	364.351	350.15	991.240	71.60
		13	937	366.530	350.15	1143.760	75.21
		14	948	365.420	350.15	1065.800	69.27
		15	699	360.934	350.15	752.720	66.35
		16	369	357.756	350.15	530.910	63.20
0.0015	0.025	10	376	353.810	350.15	383.900	62.91
		11	737	356.812	350.15	697.350	58.30
		12	853	359.620	350.15	991.240	71.60
		13	937	361.076	350.15	1143.760	75.21
		14	948	360.332	350.15	1065.800	69.27
		15	699	357.341	350.15	752.720	66.35
		16	369	355.222	350.15	530.910	63.20
0.002	0.0334	10	376	352.895	350.15	383.900	62.91
		11	737	355.136	350.15	697.350	58.30
		12	853	357.238	350.15	991.240	71.60
		13	937	358.328	350.15	1143.750	75.21
		14	948	357.771	350.15	1065.800	69.27
		15	699	355.530	350.15	752.720	66.35
		16	369	353.946	350.15	530.910	63.20

The operating parameters and conditions of the vapor-absorption cycle along with the thermodynamic properties at different states of the cycle considered in the below tables are taken from the Kaushik et al. [8].

Assuming 50 percentage distribution of heat transferred to the vapor-absorption cycle: $\dot{Q}_{VAC}$ (avg) = 395.6 W.

This heat is provided as the input to the generator of the cycle:

$$\dot{Q}_{VAC}(\text{avg}) = \dot{Q}_g(\text{avg}) = \dot{Q}_3(\text{avg})$$

Table 2 Efficiency calculation

Time (h)	Input Power (P_{in}) (W)	P_{el} (W)	$\dot{Q}_1$ (W)	η_{el} (%)	$\eta_{overall}$ (%)
10	610.248	83.287	383.900	13.64	76.55
11	1196.200	126.78	697.356	10.60	68.89
12	1384.420	140.765	991.244	10.16	81.70
13	1520.760	150.887	1143.756	09.90	85.13
14	1538.600	152.213	1065.800	09.89	79.16
15	1134.477	122.2	752.725	10.77	77.12
16	598.887	82.44	378.500	13.70	76.90

Table 3 PCM material comparison

PCM material	Heat supplied to the PCM (MJ)	Density (kg/m^3)	Melting point (°C)	Latent heat of fusion (kJ/kg)	Mass of PCM required (kg)	Volume occupied by PCM (m^3)
Lauric acid	9.441	880	43.20	228.29	41.245	0.047
Decanoid acid		893	31.60	179.13	52.700	0.060
1-Tetradecanol		824	38.00	259.44	36.390	0.044
Stearic acid		840	70.00	203.00	46.500	0.055

4 Exergy Analysis

The total energy that has been extracted by the heat recovery system is equal to the net exergy loss in the complete process.

The amount of exergy that is received by the system as input is given by:

$$E_{in} = A\ x\ G\ x \left[1 - \frac{4}{3}\left(\frac{T_a}{T_s}\right) + \frac{1}{3}\left(\frac{T_a}{T_s}\right)^4\right] \tag{7}$$

where A is the area of the collector plate of the solar panel, G is the amount of incident radiation onto the collector plate, T_a is the ambient temperature, and T_s is the sun temperature.

The effective collector area of the solar panel is 1.623 m^2 (length = 166.32 cm, width = 99 cm).

Incident radiation values of different days of experimental observations for various mass flow rates have been considered for the exergy analysis as the temperature of the solar cell is dependent on the incident radiation conditions. The electrical efficiency of the PV module has been considered as a function of incident radiation.

The amount of exergy obtained as output is given by:

$$E_{out} = E_{th} + E_{pv} \tag{8}$$

where E_{th} is the amount of thermal exergy of the thermal extraction cycle and E_{pv} is the photovoltaic thermal exergy. The thermal exergy is given by:

$$E_{th} = \dot{Q}_l\left[1 - \frac{T_a}{T_c}\right] \tag{9}$$

where $\dot{Q}_l$ is the amount of heat extracted, T_a is the ambient temperature, and T_c is the temperature of the photovoltaic cell. The photovoltaic exergy is given by:

$$E_{pv} = \eta_{el} x\ A\ x\ G\ x\left[1 - \frac{4}{3}\left(\frac{T_a}{T_s}\right) + \frac{1}{3}\left(\frac{T_a}{T_s}\right)^4\right] \tag{10}$$

where η_{el} is the electrical efficiency of the cell, A is the area of the collector plate of the solar panel, G is the amount of incident radiation onto the collector plate, T_a is the ambient temperature, and T_s is the sun temperature. The exergic efficiency of the system is given by:

$$\eta_{exergic} = \frac{E_{out}}{E_{in}} \tag{11}$$

As seen in Figs. 2 and 3, the exergic efficiency and the exergic work output depend upon the levels of incident radiation and the amount of flow rate of the coolant fluid.

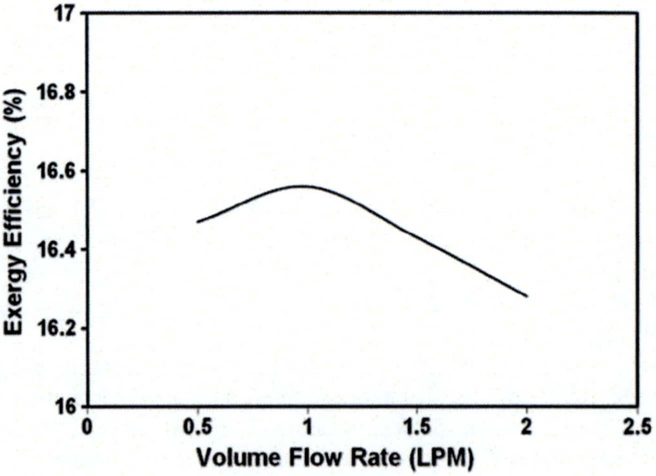

Fig. 2 Effect of volume flow rate on exergic efficiency

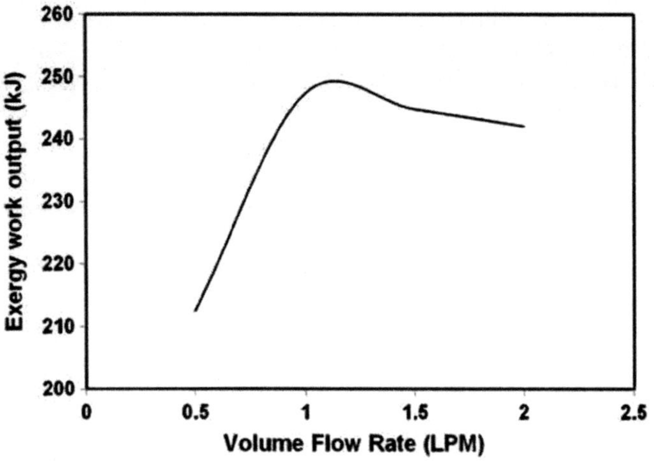

Fig. 3 Effect of volume flow rate on exergic work output

Exergic work output increases with higher levels of incident radiation on the panels due to increased heat dissipation and electric work produced by the system. On the contrary, the exergic work output decreases as the flow rate of the coolant fluid increases due to higher irreversibility. As the entropy of the system increases due to increased flow rate, there are higher losses or disturbances in the energy flow of the system that disrupts and reduces the amount of exergy available from the system to do the useful work. The system becomes relatively unstable with increasing flow rate of the system that reduces its exergic capabilities. Also, as the flow rate of the working fluid increases, the thermal capacity of the fluid to do the useful work decreases due to a higher amount of energy required to raise the energy of the coolant fluid. A maximum exergic output of 247.35 W and exergic efficiency of 16.56% at a flow rate of 1 LPM is obtained from the system at a time of 12 h.

5 Conclusions

In this analysis, the application of waste heat dissipated by the solar panel to be used in the thermal energy system with stearic acid as the PCM material and the vapor-absorption cycle has been carried out.

- It has been found that higher levels of radiation increases the electrical output produced by the solar panel with a maximum output of 152 W at 14 h (incident radiation = 948 W/m^2) but the electrical efficiencies of panel show a decline due to the increased operating temperature of the system with a maximum electric efficiency obtained being 13.7% at 16 h (incident radiation = 369 W/m^2).

- It is observed that with increasing mass flow rate of the liquid water circulating fluid, there is a decrease in the temperature at the outlet of the coolant flow, and the maximum change in temperature of 32.75 °C is observed when the flow rate of the water is set at 0.5 LPM and at a time of 13 h.
- The thermal output of the system increases as time progresses, reaching a maximum value of 1143 W at 13 h due to higher levels of incident solar radiation with a maximum thermal efficiency obtained from the heat extraction cycle of 75.21%.
- The overall efficiency of the system varied from values ranging from 68.89 to 85.13%.
- The exergic efficiency and the exergic work output are dependent on the flow rate of the coolant fluid and the incident radiation on the solar panels. A maximum exergic output of 247.35 W at a maximum exergic efficiency of 16.56% is obtained at a flow rate of 1 LPM at a time of 12 h.

References

1. Riffat SB, Cuce E (2011) A review on hybrid photovoltaic/thermal collectors and systems. Int J Low-Carbon Technol 6:212–241
2. Javed M, Leila L (2016) Optimal management of a solar power plant equipped with a thermal energy storage system by using dynamic programming method". Proc Mech Eng Power Energy 230:219–233
3. Nick J, Thomas S (2015) Concentrated solar power: Recent developments and future challenges. Proc Mech Eng Power Energy 229:693–713
4. Arkin H, Navon R, Burg I (1997) HVAC with thermal energy storage: optimal design and optimal scheduling". Proc Mech Eng Power Energy 18:31–38
5. Cocco D, Cau G (2015) Energy and economic analysis of concentrating solar power plants based on parabolic trough and linear Fresnel collectors. Proc Mech Eng Power Energy 229:677–688
6. Shouquat HM, Abd RN, SelvarajJeyraj AL, Kumar PA (2019) Experimental investigation on energy performance of hybrid PV/T-PCM system. In: ICEES 2019 5th international conference on electrical energy systems
7. Hossain MS, Pandey AK, Selvaraj J, Rahim NA, Islam MM, Tyagi VV (2019) Two side serpentine flow based photovoltaic-thermal-phase change materials (PVT-PCM System): energy, exergy and economic analysis. Renew Energy 136:1320–1336
8. Kaushik S, Singh S (2014) Thermodynamic analysis of vapor absorption refrigeration system and calculation of COP. Int J Res Appl Sci Eng Technol 2:1–8

Measurement of Temperature Distribution Using Liquid Crystal Thermography Technique Over the Absorber Plate of Solar Air Heater

Dheeraj Kumar, Amit Kumar, and Apurba Layek

Nomenclature

A	Thermal diffusivity, (m^2/s)
D	Diameter of the jet, (m)
HSI	Hue, saturation, intensity
$h\,(X, Z)$	Local heat transfer coefficient, (convective) $W/(m^2K)$
K_a	Thermal conductivity of air, (W/m K)
LCT	Liquid crystal thermography
Nu	Nusselt number, dimensionless
Nu (X, Z)	Local Nusselt number, $h\,(X, Z)\,D/k_a$
$R'G'B'$	Red, green, blue primaries
T	Thickness of the plate, (m)
THPP	Thermohydraulic performance
TLCs	Thermochromic liquid crystals
T_i	Air inlet temperature, (K)
T_j	Exit temperature, (K)
Z	Distance along streamwise direction, (m)
$T\,(X, Z, t)$	Wall temperature on the surface at any given time, (K)

D. Kumar (✉) · A. Kumar · A. Layek
Department of Mechanical Engineering, National Institute of Technology Durgapur, Durgapur, West Bengal 713209, India
e-mail: dknitdgp1@gmail.com

M. Palanisamy et al. (eds.), *Theoretical, Computational, and Experimental Solutions to Thermo-Fluid Systems*, Lecture Notes in Mechanical Engineering,
https://doi.org/10.1007/978-981-33-4165-4_31

1 Introduction

In the past two to three decades, application of liquid crystals has become quite popular in the field of heat transfer engineering. LCT uses organic cholesteric liquid crystals as a sensing agent to visualize the surface temperature [1] in the form of color pattern. LCT sheets reflect various color patterns depending upon their surface temperature. This advanced technique of LCT is usually being applied over the surface of the absorber plate to temperature measurement. The operating temperature range of the TLC lies between −30 and 150 °C. It is desired to have a range of bandwidth; sometimes, it is also known as range of an active temperature which lies in between 1 and 2 °C [2]. In the entire temperature range of operations, this method can be applied for measuring the temperature of the absorber plate by visualizing the color pattern of the TLC. Smoothly the color pattern varies from red to blue color, and the reply timing of thermochromic liquid crystal is about 3 ms. It has been noticed that when pure organic materials are being directly exposed to an environment of chemical contamination and UV lights, degradation starts quickly of TLC sheets. However, currently, this degradation process can be minimized with the application of microencapsulation process. This process enables the TLC to be more stable and capable for the ease of handling.

2 Prior Work for the Calibration of TLC and Uncertainty Measurement

The TLC thermography technique is capable of giving the feedback with the color response for the temperature measurement having intricate structured heat transfer absorber surface. However, before the application of TLC sheets, it becomes mandatory to perform calibration test of hue–temperature. In the last two eras, many research works have been performed for the calibration of TLC material, and the experimental work has also been performed for the calibration of TLCs. Effect of parameters, like angle of lighting and view perspective, on the calibration curve of hue–temperature of TLC has been studied. Abdullah et al. [3] observed the effects of coating thickness on the calibration curve through an experimental approach. It is inferred from the above research work that the calibration curve of TLC hue–temperature can be significantly affected by the angle of view, lighting angle, and thickness of coating material.

3 Parametric Analysis for the Uncertainty Measurement of TLCs

In order to demonstrate the uncertainty of TLC thermography technique, many factors are essential that govern its uncertainty measurement. Those parameters can be enlisted as coating thickness, the bandwidth of TLC, lighting angle, and quality of TLC coating. An experimental setup for calibration has been shown in Fig. 1.

3.1 *Effect of Bandwidth on TLC*

TLC can be commonly divided into two categories as narrow-band TLC and wide-band TLC. It is categorized as per the active temperature range, which is of 1 °C for narrow-band and 5, 10, 20 °C for wide-band TLC, respectively. An increment in the sensitivity of hue–temperature with a decrement in the bandwidth of TLC occurs. Uncertainty in the TLC is observed up to ±0.1 °C [3], and ±0.1– 0.3 °C for the bandwidth of 5 and 10 °C, respectively. An uncertainty in the measurement of TLC of ±0.4–0.5 °C has been observed for a bandwidth of 20 °C.

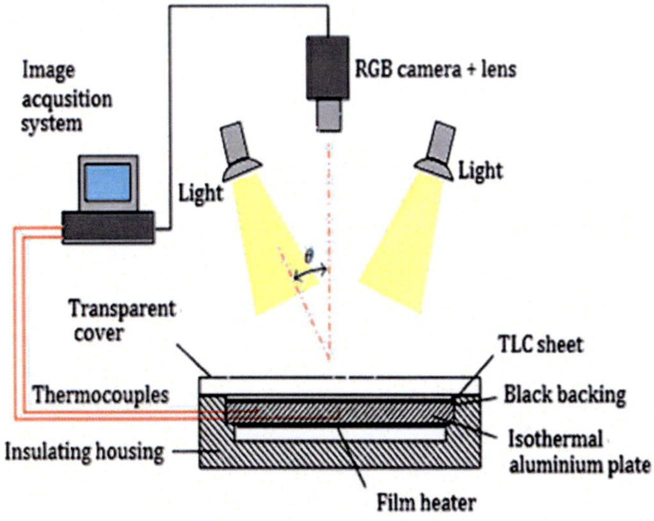

Fig. 1 Representation of the TLC calibration experimental setup

3.2 Effect of Lighting Angle

Calibration process has been conducted at different angles with the TLC sheet. It has been inferred from the experimentation that lighting angle directly affects the hue–temperature calibration curve. It shifts in the upward direction, as there is an increment in lighting angle. The experimental result shows that when the lighting angle shifts from 20 to 34° the average value of ambiguity in the temperature increases by 25% [4]. The reason behind is the concept of reflection of color signals, which leads to a higher level of uncertainty.

3.3 Effect of Coating Thickness

This parameter has a significant influence on the level of uncertainty measurement. Figure 2 shows the hue curve for the different thickness of TLC coatings 10, 20, 25, 30, and 40 μm. It can be observed from Fig. 2 that the hue curve moves upward as the thickness of the coating material increases. In the case of coating thickness having 10 μm levels of uncertainty in the measurement is of higher degree. It is only

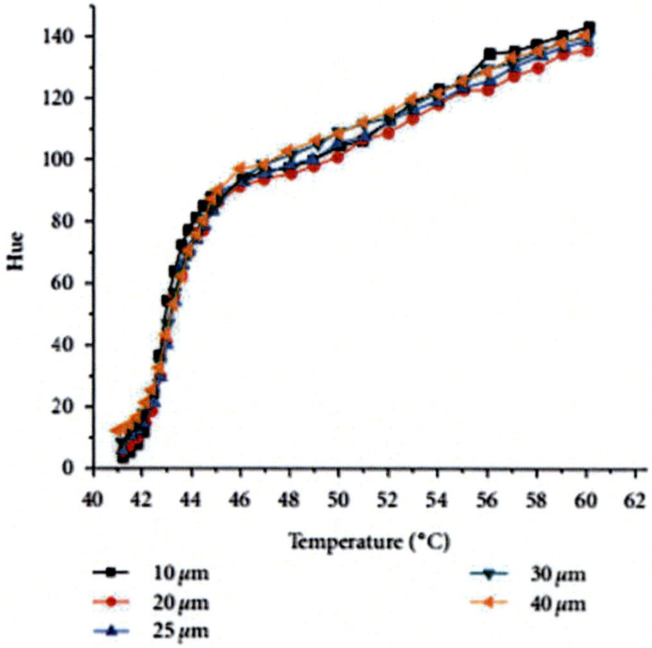

Fig. 2 Effect of coating thickness of TLC on hue curve

due to the very thin coating layer. It results in a frailer reflection of color response, and an advanced level of noise in hue value.

3.4 Effect of Coating Material Excellence

Experiments were conducted to scrutinize the consequence of quality of the coating material on the hue curve calibration and uncertainty in the measurement. Two coating samples have been prepared of the same thickness, but one has a rough surface and another of the smooth surface. The fine coating surface of TLC shows its behavior of broader range of hue value. It means fine coating surface is more sensitive to hue–temperature. It also has higher level of resolutions toward the measurement [5]. However, for the application of temperature, measurement for fine coating surface should be used.

4 Prior Work in LCT for the Temperature and Nu Number Distribution

Numerous works have been completed to identify the Nusselt number and pattern of the temperature distribution in solar air heater. The technique of liquid crystal thermographic has been evolved as an effective technique to execute the operation. It has been applied as the technique of liquid crystal thermography over the plate surface. The additional effect of artificial rib roughness on the heat transfer in a rectangular duct arrangement has been studied. Wang et al. [6] performed experimentation with the LCT on the absorber plate having truncated and continuous type rib roughness. The entire distribution pattern of Nusselt number for rib surface has been obtained. The result confirms that the index value of thermohydraulic performance (THPP) is supreme for the absorber plate having continuous ribs as compared to discrete ribs. Tariq et al. [7] performed experiments with the roughness type of chamfered ribs over the absorber plate. The pattern and behavior of fluid flow and Nusselt number distribution have been studied. The author viewed the variation in different parameters influencing the fluid flow with the assistance of liquid crystal thermographic technique. The assumed parameters were thickness of the thermal boundary layer, skin friction coefficient, distribution of temperature profile, and velocity profile. Reddy et al. [8] used the LCT technique to observe the augmentation of heat transfer in the environment of vertical rectangular fin. Steady-state condition has been assumed for the investigation of heat transfer. Natural convection phenomena have been retrieved with the developed algorithm of golden section search method. Kumar et al. [9] did experimental work to find out the Nusselt number distribution with the heated absorber plate having twisted ribs over the absorber surface. Maurer et al. [10] approved out a study to find out the characteristics behavior of the locally heat transfer coefficient over the roughened surface having V-shaped ribs. TLC thermography technique has been implemented in steady-state mode of application along with the stainless-steel foil-type heater.

5 Evaluation of Heat Transfer Coefficients

In Fig. 3, the flow diagram has been shown, which helps to predict the process paths to find out the coefficient of heat transfer by means of the liquid crystal thermography technique. One-dimensional equation of transient heat transfer is being resolved to get the value of the coefficient of heat transfer. The captured images are processed using MATLAB software. The profile of the temperature contour obtained is used to discover the value of heat transfer coefficient. Evaluations are performed with the application of normalized temperature distribution, which is based on semi-infinite approximate solutions. Equation (1) executes the operation with a known value of temperature, and Eq. (2) stands for getting the local value of Nusselt number [11].

$$\frac{T(X,Z,t)-T_i}{T_j-T_i}=1-\left[\exp\left(\frac{h(X,Z)^2\alpha t}{k^2{}_s}\right)\right]\left[\mathrm{erfc}\left(\frac{h(X,Z)\sqrt{\alpha t}}{k_s}\right)\right] \quad (1)$$

$$\mathrm{Nu}(X,Z)=\frac{h(X,Z)\times D}{k_a} \quad (2)$$

Calibration of Thermochromic Liquid Crystals (TLCs)
The thermochromic liquid crystal (TLC) sheet (Hallcrest™ R40C5W) has the characteristics of activation temperature of 40 °C and bandwidth of 5 °C. This sheet

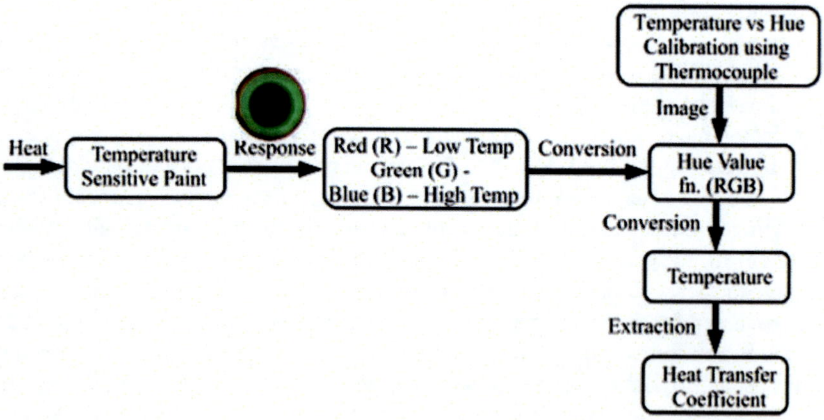

Fig. 3 Flowchart for the evaluation of the coefficient of heat transfer

has been placed over the surface of the roughened plate having twisted rib geometry. There are 24 thermocouples of T-type mounted on the absorber plate from the underneath of the rough surface. These thermocouples are also being connected to the data acquisition system. The isothermal condition has been maintained inside the rectangular duct by blocking both sides of the duct. An electric heater is being placed just above the back surface of the plate to heat the surface plate. Now the electric heater has been turned on to heat the plate surface to a temperature range just above the aforementioned active range of boundary. The heater is now switched off, and the surface is left for cooling by natural convection mode. The TLC sheet shows different color distribution pattern on its surface, which has been recorded with the help of a CCD camera. The image taken is transformed to a file of 24-bit JPG format. MATLAB programming converts the RGB values to its corresponding HSI values. Some relevant equations [3–7] used in MATLAB programming to execute the conversion process are listed as follows [12]:

$$\cos\theta = \left\{ \frac{0.5[(R' - G') + (R' - B')]}{(R' - G') + [(R' - B')(G' - B')^{0.5}]} \right\} \tag{3}$$

$$H = \theta \text{ if } B' \leqslant G \tag{4}$$

$$H = 360 - \theta \text{ if } B' \geqslant G \tag{5}$$

$$S = 1 - \left[\frac{\text{Min}\{\, R', G', B' \}}{I} \right] \tag{6}$$

$$I = \frac{R' + G' + B'}{23} \tag{7}$$

In Fig. 4 the variation in the hue value and its corresponding temperature changes of the TLC sheet have been shown. Here, the hue value shows the best analogy with the temperature range of TLC sheet as compared to other parameters such as saturation and intensity. A correlation has been obtained for the hue value and temperature of TLC sheet.

$$T_{\text{TLC}} = 35.788H^3 - 36.866H^2 + 36.467H + 310.85 \tag{8}$$

It has been noticed that the temperature value corresponding to the color response shown in TLC sheet is quite similar to the mean temperature of the absorber plate having artificial roughness on its surface.

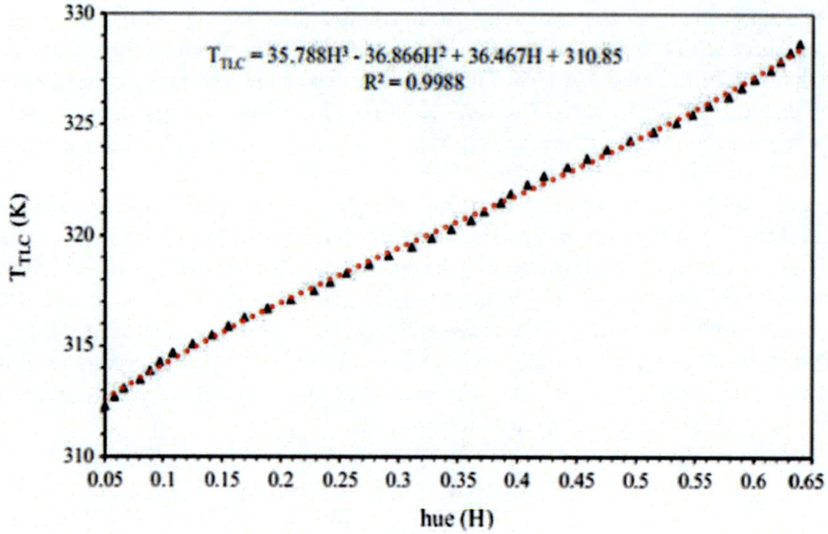

Fig. 4 Calibration curve of TLC sheet [9]

5.1 Precautions During Image Capture

i. The glowing intensity of the white lamp above the test section needs to be maintained constant for the entire time period of experimentation.
ii. There is a possibility of infrared heating of the TLC sheets. So, to minimize this effect, lights are swapped on only at the stint of image catching.
iii. The pixel resolution of the camera must be as high as (1368 × 768) so that the image formation will be clear as much as possible.
iv. The setup of the camera for the image acquisition should be in the normal direction to the TLC sheet.
v. A detachment between the TLC sheet and the CCD camera nearly 230 mm should be maintained.

6 Conclusions

A methodical approach for the introduction of liquid crystal thermography (LCT) and thermochromic liquid crystals has been presented in this paper. Application of TLC for the measurement of heat transfer and procedures for the calibration of TLC have been appropriately discussed. The different parameters influencing the uncertainty analysis for the measurement of TLC temperature have also been investigated. This study also deals with the detailed discussion of the effect of thickness of the coating material, lighting angle, quality of coating, and bandwidth of the TLC on the hue–temperature calibration curvature. The technique of LCT provides the information regarding the flow behavior over the roughened surface of an absorber plate. This technique not only gives the characteristics of fluid flow but also shares the dispersal of Nusselt number over the absorber plate.

Acknowledgements The first author thankfully acknowledges the guidance and consistent support by Dr. Apurba Layek (Professor, National Institute of Technology, Mechanical engineering). The first author further records his thankfulness for financial support by the Ministry of Human Resource and Development (MHRD).

References

1. Hallcrest (1991) Handbook of thermochromic liquid crystal technology
2. Rao Y, Zang S (2009) Calibrations and the measurement uncertainty of wide-band liquid crystal thermography. Meas Sci Technol 21(1):015105
3. Abdullah N, Saiah HR, Jaafar AA, Salleh MA (2009) Film thickness effects on calibrations of a narrowband thermochromic liquid crystal. Exp Therm Fluid Sci 33(4):561–578
4. Rao Y, Xu Y (2012) Liquid crystal thermography measurement uncertainty analysis and its application to turbulent heat transfer measurements. Adv Condens Matter Phys
5. Cooper TE, Field RJ, Meyer JF (1975) Liquid crystal thermography and its application to the study of convective heat transfer. J Heat Trans 97(3):442–450
6. Wang L, Sunden B (2005) Experimental investigation of local heat transfer in a square duct with continuous and truncated ribs. Exp Heat Transf 18:179–197
7. Tariq A, Swain SK, Panigrahi PK (2002) An experimental study of convective heat transfer from flat and ribbed surface. Indian J Eng Mater Sci 9:464–471
8. Reddy BK, Balaji C (2012) Estimation of temperature dependent heat transfer coefficient in a vertical rectangular fin using liquid crystal thermography. Int J Heat Mass Transf 55:3686–3693
9. Kumar A, Layek A (2018) Thermo-hydraulic performance of solar air heater having twisted rib over the absorber plate. Int J Therm Sci 133:181–195
10. Maurer M, von Wolfersdorf J, Gritsch M (2007) An experimental and numerical study of heat transfer and pressure loss in a rectangular channel with V-shaped ribs. J Turbomach 129(4):800–808
11. Ansu U, Godi SC, Pattamatta A, Balaji C (2017) Experimental investigation of the inlet condition on jet impingement heat transfer using liquid crystal thermography. Exp Therm Fluid Sci 1(80):363–375
12. Stasiek J, Stasiek A, Jewartowski M, Collins MW (2006) Liquid crystal thermography and true-colour digital image processing. Optics Laser Technol 38(4–6):243–256

Compressible Flow Through Convergent–Divergent Nozzle

Buddha Dev Das, Rajdeep Sardar, Sandip Sarkar, and Nirmal Kumar Manna

1 Introduction

A nozzle is a device designed especially in the form of a tube of varying cross-sectional area through which hot gas or liquid flows to generate thrust following Newton's third law of motion. The nozzle is often used to control the characteristics of fluid flow (specifically the rate of flow that emerges from the outlet), pressure and the direction of flow, and to enhance the velocity of a gaseous substance. In the area of compressible flow, the nozzles are typically categorized as a convergent nozzle and a convergent–divergent (CD) nozzle. Both types of nozzles have serious applications in industry and technology. In particular, the CD nozzle plays a vital role in the case of a supersonic version of the missile, jet engines, wind tunnel, ramjets, scramjets and rocket science as well [1]. A CD nozzle is used frequently to proselytize chemical energy into kinetic energy in a thermal chamber and vice versa [2].

In the present work, a typical geometry of converging–diverging (CD) nozzle (based on the standard literature [3]) is considered. For CD nozzle, the Mach number is less than 1.0 in the converging part (subsonic section) and Mach number is 1.0 at the throat section. In the diverging part (supersonic section), the Mach number

B. D. Das · R. Sardar · S. Sarkar · N. K. Manna (✉)
Department of Mechanical Engineering, Jadavpur University, Kolkata 700032, India
e-mail: nirmalkmannaju@gmail.com

B. D. Das
e-mail: dbuddhadev0@gmail.com

R. Sardar
e-mail: rajdeepmsclaret2014@gmail.com

S. Sarkar
e-mail: sandipsarkar.mech@jadavpuruniversity.in

M. Palanisamy et al. (eds.), *Theoretical, Computational, and Experimental Solutions to Thermo-Fluid Systems*, Lecture Notes in Mechanical Engineering,
https://doi.org/10.1007/978-981-33-4165-4_32

increases over unity; in other words, the supersonic flow is achieved particularly for which CD nozzle is designed. However, depending upon the outlet condition, Mach number can have different values at the exit, and a phenomenon of normal shock can develop the diverging part before the outlet. The main objective of the present work is to capture normal shocks under different exit conditions numerically using computational fluid dynamics (CFD). On the CD nozzle, the scarcity of works [4–6] is observed in the literature. There are a few works on nozzle flow [7–12] using CFD simulation; however, almost no work has reported earlier on the analysis of normal shock using the standard geometry [3]. To understand the shock phenomenon properly and to enrich the knowledge base in the area motivates us to this study. The dynamics of expansive flow in CD nozzle are analysed in terms of different involved parameters mainly pressure, temperature, velocity and Mach number. The abrupt changes of pressure and other parameters are captured and presented in terms of different contour plots and line curves. The present CFD simulation is performed using ANSYS FLUENT software.

2 Problem and Numerical Procedure

2.1 Theory

A CD nozzle is first invented by a Swedish scientist named De Laval, and now it is known as de Laval nozzle or converging–diverging nozzle. This type of nozzle is required to gain supersonic speed. Due to the certain values of backpressure [7] at the exit of the nozzle (or the overall pressure differential across the nozzle) under a supersonic flow condition, the supersonic flow suddenly transforms into a subsonic flow resulting in normal shock only in the divergent part of the nozzle. The expansion in the converging section continues as a subsonic flow and becomes sonic flow at the throat. Afterward, the flow becomes supersonic before the shock occurs. After the sudden shock, the flow becomes subsonic which continues to the exit of the nozzle. Consequently, a small continuous decrease in Mach number takes place in the diverging area of CD nozzle, but the density, temperature and pressure also increase after the shock at the exhaust section. According to the principle of energy conservation, the stagnation pressure and enthalpy does not vary across the normal shock.

2.2 Geometry and Mesh

This paper aims to simulate Anderson's geometry model [3] to capture the normal shocks at different exit conditions. The shape of the converging–diverging nozzle is specified by $A = A(x)$, considering a parabolic profile as given by

$$A(x) = 1.0 + 2.2(x - 1.5)^2 \quad (1)$$

where $0 \leq x \leq 3$ and $x \leq 1.5$ indicates the convergent section, $x = 1.5$ represents the throat of the nozzle, and the divergent section implies for $x \geq 1.5$ [1]. For the purpose of the simulation, the axisymmetric geometry is considered. The corresponding geometric model is shown in Fig. 1a in the platform of ANSYS FLUENT. The generation of computation mesh is depicted in Fig. 2, showing 2D structured grids. The structured mesh is employed for more accuracy of this simulation. The mesh consists of 14,271 nodes and 14,000 elements with refine mesh of 0.0001 sizing for all cases.

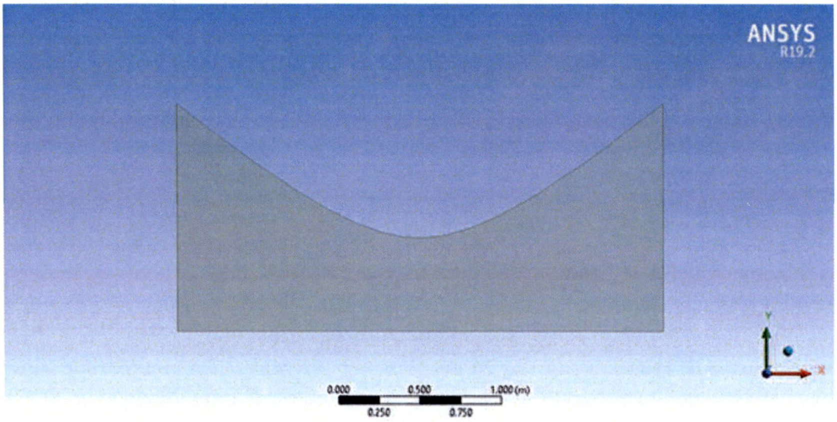

Fig. 1 Two-dimensional geometry of the converging–diverging nozzle

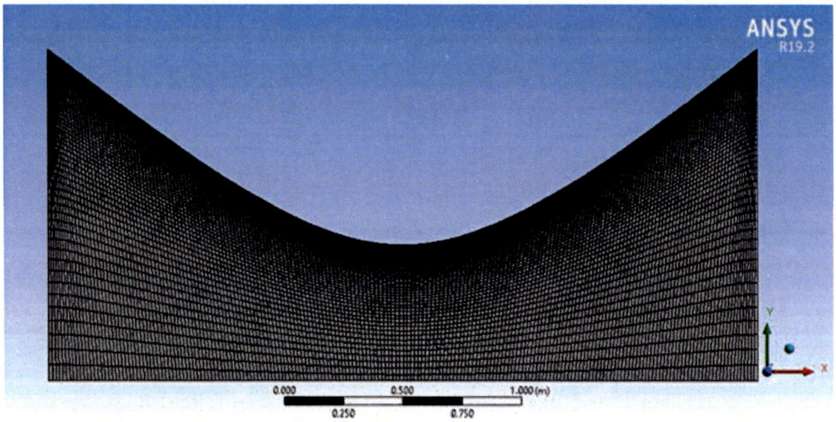

Fig. 2 Axisymmetric grid generation

2.3 Mathematical Formulation

As discussed earlier, in a CD nozzle, the normal shock occurs in the diverging portion; however, the location of the shock and the minimum pressure prevailed during the shock are the most significant questions regarding the same. To address the same, the present work is undertaken using the facility of ANSYS FLUENT. In the software, there are multifarious governing equations amalgamated with convergent–divergent nozzle, which should be taken into account along with the elementary theory of CFD. The set of equations is as follows:

The equation for an ideal gas is denoted by

$$\rho = \frac{P}{RT} \tag{2}$$

where P = pressure (Pa), T = temperature (K), R = universal gas constant (J/mol K).

As well as, the sound speed is given by

$$c = \sqrt{\gamma RT} \tag{3}$$

Mach number can be obtained by the following equation: $\frac{V}{C}$, where v is the flow velocity and c is the sonic velocity.

The conservation of mass and momentum equations for a Newtonian fluid is known as Navier–Stokes equations. For the present 2D compressible flow, considering steady-state condition and neglecting temporal term yields:

The expression of continuity equation is

$$\frac{\partial \rho u_j}{\partial x_j} = 0 \tag{4}$$

and the momentum equation is

$$\frac{\partial \rho u_i u_j}{\partial x_j} = -\frac{\partial p}{\partial x_j} + \frac{\partial \sigma_{ij}}{\partial x_j} + \Pi \partial_{i1,} \tag{5}$$

The steady flow energy equation becomes

$$\frac{\partial_{\rho u_j} T}{\partial x_j} = \frac{\partial \left(\rho \alpha \frac{\partial T}{\partial x_j} \right)}{\partial x_j} \tag{6}$$

where u_i is the velocity component in the ith direction, p is the thermodynamic pressure in Pa and ρ is the fluid density in kg/m^3, and σ_{ij} is the viscous stress.

3 Results and Discussion

In our work, the standard air is the working fluid. It flows at 1.0 atmospheric pressure (100 kPa) from the left-hand side of the CD nozzle and via throat section, finally exits from the right end. The simulations have been conducted under the pressure variation in the range of 10–68.740 kPa. It has been observed that from 10 to 30 kPa, the normal shock is not visible. However, when the pressure difference between the entry and exit becomes around 50–68.740 kPa, the shocks occur in the divergent region due to the over expansion. This may be attributed to the fact that the pressure is one of the significant parameters on which the formation of shockwave depends. All these matters are observed in simulation with the help of different contour and plot of various parameters, including density, velocity, pressure and Mach number, as illustrated in Figs. 2, 3, 4, 5, 6, 7, 8 and 9 for different situations.

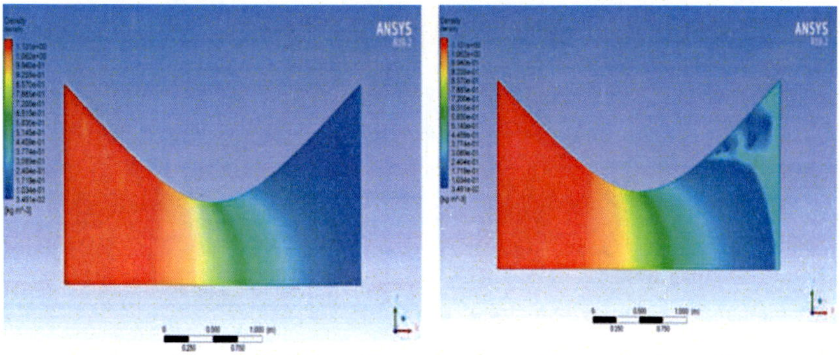

(a) (b)

Fig. 3 Density contours at exit pressures 10 kPa (**a**) and 30 kPa (**b**)

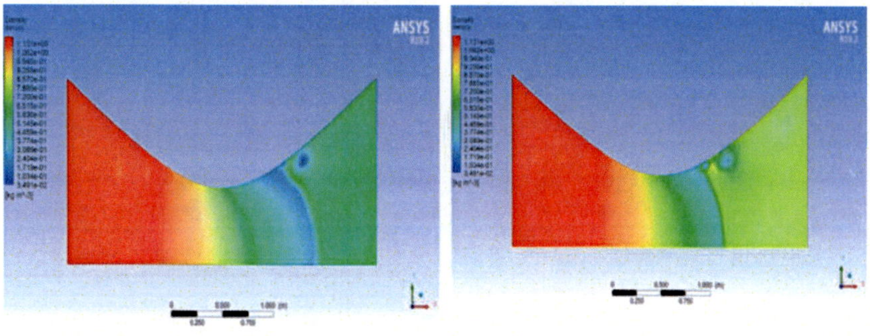

(a) (b)

Fig. 4 Density contours at exit pressures 50,000 Pa (**a**) and 68,740 Pa (**b**)

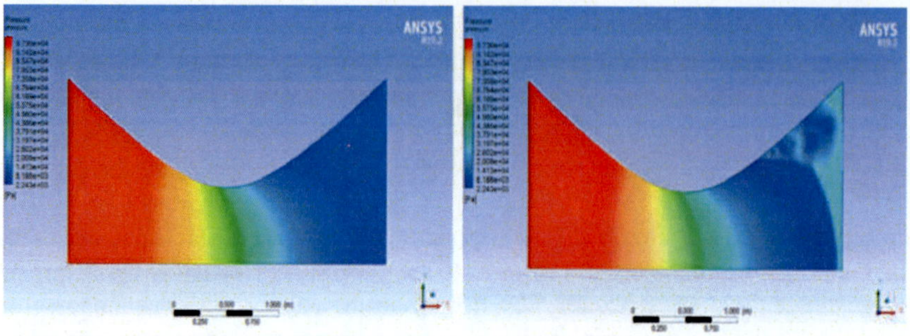

(a) (b)

Fig. 5 Pressure contours at exit pressures 10 kPa (**a**) and 30 kPa (**b**)

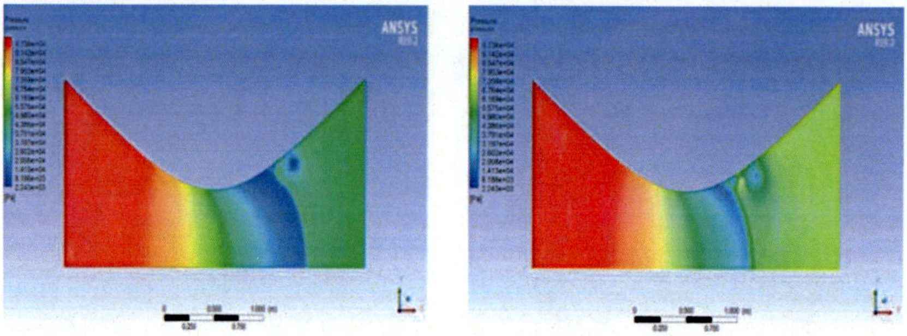

(a) (b)

Fig. 6 Pressure contours at exit pressures 50,000 Pa (**a**) and 68,740 Pa (**b**)

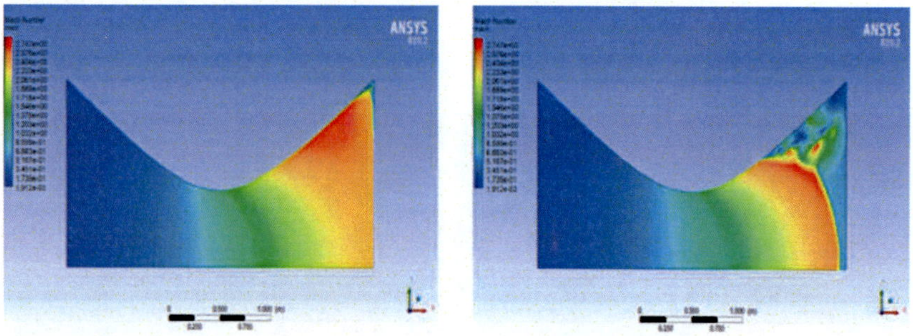

(a) (b)

Fig. 7 Mach numbers at exit pressures 10 kPa (**a**) and 30 kPa (**b**)

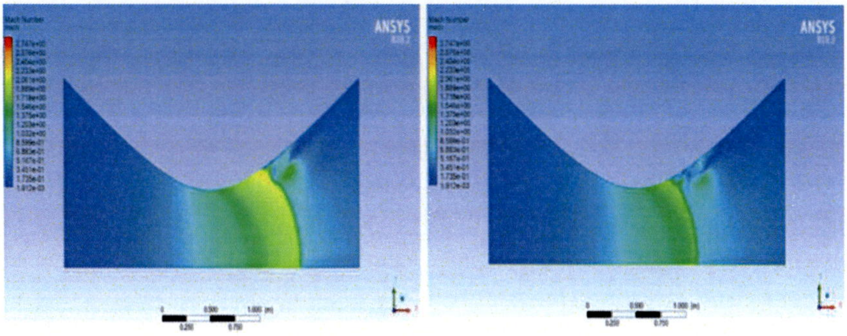

(a) (b)

Fig. 8 Mach numbers at exit pressures 50,000 Pa (**a**) and 68,740 Pa (**b**)

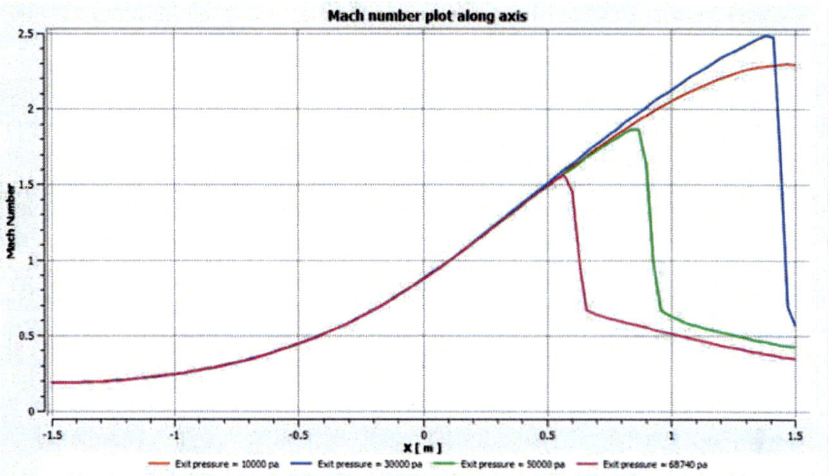

Fig. 9 Mach number variation for different outlet pressure along axis

The abrupt change in colour in the plot of Mach number (Figs. 7 and 8) indicates the region of shock, which starts to fall with a definite mutation. It is observed that Mach number increases up to throat section and afterward at the shock zone it starts to decrease, where the temperature is minimum. A schematic view of contours of Mach numbers at different pressure conditions is shown in Figs. 7 and 8.

The Mach number variation in the case of above CD nozzle, at different pressures at the nozzle exit, is depicted in Fig. 9, and the overall pressure variation along the axis, viz., the present of normal shock is illustrated in Fig. 10. As the exit pressure changes, the location of the normal shock changes. The normal shock shifts towards right when exit pressure decreases.

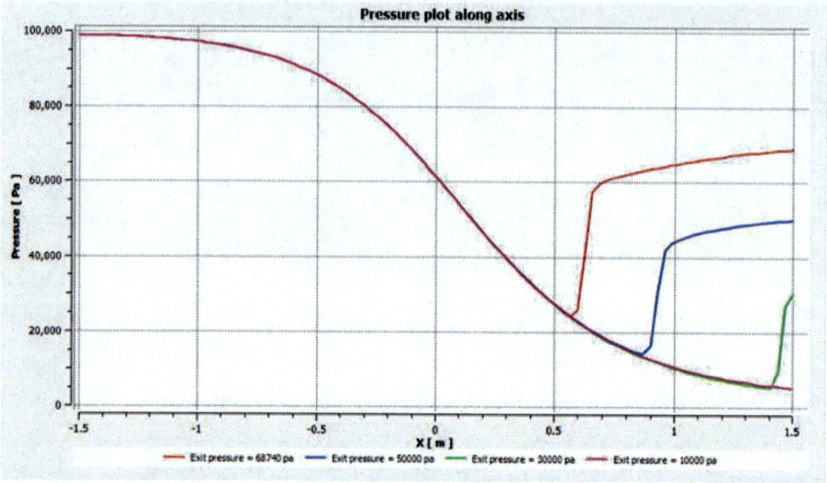

Fig. 10 Pressure variation along axis

4 Conclusions

In the aforementioned geometry, the maximum value of Mach number is in a place where pressure is minimum and at throat region pressure value should be near about backpressure. Up to 30,000 Pa Mach number curves are smooth just before the exit but increasing exit pressure, we have noticed shockwave occurs after throat, but before outlet boundary, as expected from theoretical understandings.

References

1. Md Iliyaas A, Karuppasamy K (2016) Computational analysis of CD nozzle for solid propellant rocket. Int J Innov Works Eng Technol (IJIWET) 2:314–324
2. Sudhakar BVVN, Sekhar BPC, Mohan PN, Md Ahmad T (2016) Modeling simulation of convergent-divergent nozzle using computational fluid dynamics. Int Res J Eng Technol (IRJET) 03:2395–0072
3. Anderson JR (1995) Computational fluid dynamics the basics with applications. McGraw Hill, Inc., United States of America
4. Malay SP, Mane SD, Raman M (2016) Concepts and CFD analysis of De-Laval nozzle. Int J Mech Eng Technol (IJMET) 7:221–240
5. Modesti D, Pirozzolic S, Grassoa F (2019) Direct numerical simulation of developed compressible flow in square ducts. Int J Heat Fluid Flow 76:130–140
6. Mohan KG, Dominic XF, Kumar RM (2013) Design and optimization of De Laval nozzle to prevent shock induced flow separation. Adv Aeros Sci Appl 3:119–124
7. Gopi G, Rao BR (2019) Effect of back pressure on nozzle flow. Int J Res Advent Technol. In: Conference proceedings of 3rd national conference on recent trends and innovations in mechanical engineering 15–16 Mar 2019

8. Kumar RR, Devarajan Y (2018) CFD simulation analysis of two-dimensional convergent-divergent nozzle. Int J Ambient Energy. doi: 10.1080/ 01430750. 2018.1517683
9. Yu Y, Shademan M, Barron RM, Balachandar R (2012) CFD study of effects of geometry variations on flow in a nozzle. Eng Appl Comput Fluid Mech 6:412–425
10. Madhu BP, Sameer S, Kalyana KM, Mahendra MG (2017) CFD analysis of convergent-divergent and contour nozzle. Int J Mech Eng Technol 8:670–677
11. Narayana KPSS, Reddy KS (2016) Simulation of convergent divergent rocket nozzle using CFD analysis. IOSR J Mech Civil Eng (IOSR-JMCE) 13:58–65
12. Aabid A, Chaudhary ZI, Khan SA (2019) Modelling and analysis of convergent divergent nozzle with sudden expansion duct using finite element method. J Adv Res Fluid Mech Therm Sci 63:34–51

A Comparative Study of Fluid Flow Characteristics of Dual Jet Using Different RANS-Based Turbulence Models

Sanjay Singh Rathore and Suresh Kant Verma

1 Introduction

The study of fluid flow characteristics of turbulent dual jet is an important research area because of its various engineering and industrial applications. Dual jets are often encountered in cooling, heating and mixing applications due to their superior heat and mass transfer capabilities. Some of the most general examples are cooling of gas turbine blades, cooling of electronic components and combustion chamber wall of boiler, air conditioning and heat exchanger, automobile exhaust stakes, fuel injection systems, waste water disposal and so on [1–7]. A dual jet is formed when a plane offset jet interacts with a parallel wall jet. The dual jet impingement over the wall can be characterized by three flow zones, namely the converging region, the merging region and the combined region as shown in Fig. 1. Initially, in converging region, the offset jet deflects toward the wall jet due to the presence of bottom impingement wall, which causes an asymmetric entrainment between jets and creates a sub-atmospheric pressure zone. Thus, the recirculation flow occurs in the converging region. The two jets then interact with each other at merge point (MP: $X_{\mathrm{mp}}, Y_{\mathrm{mp}}$) in merging region and subsequently behave like a single jet from combined point (CP: $X_{\mathrm{cp}}, Y_{\mathrm{cp}}$) in combined region. The offset ratio can be defined as the ratio of height of offset jet from impingement wall (H) to the nozzle width (w).

Wang and Tan [1] have experimentally studied the mean flow characteristics of a turbulent dual jet for jet exit Reynolds number 10,000 and offset ratio 2 using particle image velocimetry (PIV) measurement technique. Later on, Vishnuvardhanarao and Das [2] have investigated numerically the heat transfer phenomenon associated with a turbulent dual jet flow for different jet velocity ratios using standard k-ε turbulence model. A detailed numerical study for characterization of mean flow parameters of

S. S. Rathore (✉) · S. K. Verma
Department of Mechanical Engineering, NIT Patna, 800005 Patna, India
e-mail: sanjayrathoremech@gmail.com

M. Palanisamy et al. (eds.), *Theoretical, Computational, and Experimental Solutions to Thermo-Fluid Systems*, Lecture Notes in Mechanical Engineering,
https://doi.org/10.1007/978-981-33-4165-4_33

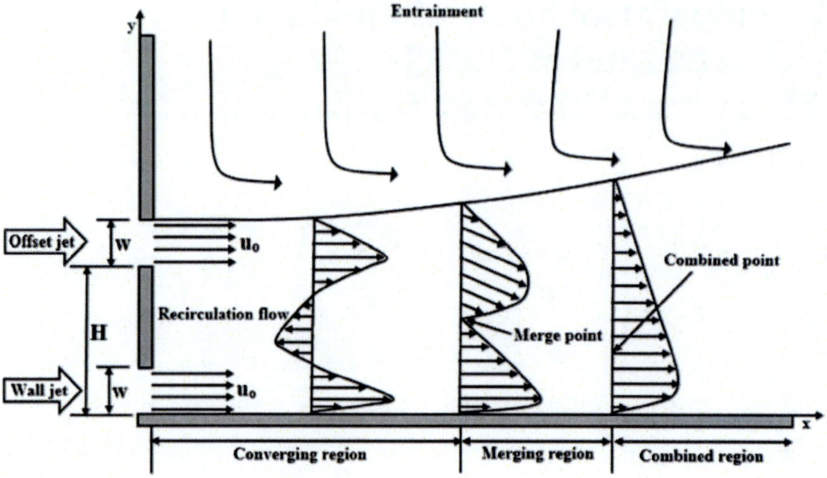

Fig. 1 Schematic diagram of a dual jet

turbulent dual jet has been carried out by Kumar and Das [3]. They used standard *k*-*ε* model as turbulence closer. The Reynolds number and offset ratio considered are 20,000 and 9, respectively. The effect of jet separation distance on steady and periodically unsteady flow behaviour of turbulent dual jet has been numerically investigated by Mondal et al. [4] with the help of standard *k*-*ε* turbulence model. An exhaustive numerical study of mean flow parameters of turbulent dual jet has been done by Kumar [5] for Reynolds number 15,000 and offset ratio between 3 and 15 at an interval of 2 using standard *k*-*ε* turbulence model. The effect of offset ratio on heat transfer characteristics of turbulent dual jet has been numerically studied by Hnaien et al. [6] using standard *k*-*ω* turbulence model. Recently, Assoudi et al. [7] carried out a numerical study to compare the fluid flow behaviour of a single turbulent offset jet and a dual jet.

On the basis of detailed literature review, the authors found that most of the numerical studies of dual jet flow have been carried out using high Reynolds number (HRN) standard *k*-*ε* turbulence model. The numerical study which explores the relative performance of different Reynolds averaged Navier–Stokes (RANS) based turbulence models for prediction of complex flow field of dual jet is very scarce in the literature. Thus, the objective of this work is to bridge this gap by carrying out a comparative study to judge the suitability of various RANS-based turbulence models for prediction of fluid flow characteristics of dual jet. In the present work, the most widely used five RANS-based turbulence models, that is, standard *k*-*ε*, RNG *k*-*ε*, realizable *k*-*ε*, standard *k*-*ω* and SST *k*-*ω* models, have been considered for fluid flow study of turbulent dual jet. Among these five turbulence models, the standard *k*-*ε*, RNG *k*-*ε* and realizable *k*-*ε* models are high Reynolds number (HRN) turbulence model, whereas the standard *k*-*ω* and SST *k*-*ω* models are low Reynolds number (LRN) turbulence model. In HRN *k*-*ε* turbulence models, the wall function is utilized for near-wall modelling which requires $y+$ value for first near wall grid points in the range of 30–100, whereas in LRN *k*-*ω* turbulence models, the near-wall integration method is used for modelling in the near-wall zone which requires few grids near the wall in the viscous sublayer with $y+$ value close to 1. In the present study, the standard wall function proposed by Launder and Spalding [8] is used for near-wall modelling in high-Reynolds number (HRN) turbulence models. All the numerical simulations have been performed using ANSYS FLUENT V. 17.2 commercial CFD code [9].

2 Mathematical Formulation

The fluid flow is assumed to be two-dimensional, incompressible and fully developed turbulent flow. Air is taken as working fluid and its thermo-physical properties are considered as constant. The turbulent flow field of dual jet is modelled with the help of Reynolds averaged continuity and momentum conservation equations, which can be written as:

Continuity equation:

$$\frac{\partial u_i}{\partial x_i} = 0 \tag{1}$$

Momentum equation:

$$\rho u_j \frac{\partial u_i}{\partial x_j} = -\frac{\partial p}{\partial x_i} + \frac{\partial}{\partial x_j}\left[\mu\left(\frac{\partial u_i}{\partial x_j} + \frac{\partial u_j}{\partial x_i}\right) - \rho\overline{u_i' u_j'}\right] \tag{2}$$

where u_i' and u_j' represent the fluctuating velocity components. The momentum equation is closed with the help of Boussinesq hypothesis. Boussinesq hypothesis relates the Reynolds stresses to the mean velocity gradients and can be given as:

Boussinesq hypothesis:

$$-\rho\overline{u_i' u_j'} = -\frac{2}{3}k\rho\delta_{ij} + \mu_t\left(\frac{\partial u_i}{\partial x_j} + \frac{\partial u_j}{\partial x_i}\right) \tag{3}$$

The details of transport equations of turbulent kinetic energy (*k*) and its dissipation rate (ε) for different HRN *k*-*ε* models and the transport equations of turbulent kinetic energy (*k*) and its specific dissipation rate (*ω*) for different LRN *k*-*ω* models are given in ANSYS FLUENT V. 17.2 theory guide [9].

3 Boundary Condition

The velocity inlet boundary condition is applied at inlet of both the offset jet and wall jet, which is characterized by uniform nozzle exit velocity u_0. The turbulent intensity at both the jets inlet is $I = 0.05$. The Reynolds number based on nozzle exit velocity u_0 and nozzle width w is $\rho u_0 w/\mu$. The no-slip and no-penetration boundary conditions are considered on the walls. The pressure inlet ($p_{\text{entrainment}} = p_{\text{a}}$) and pressure outlet ($p_{\text{outlet}} = p_{\text{a}}$) boundary conditions are employed at the upper top entrainment and the right outlet boundaries, respectively. The gradients of variables (u, v, k and ε) are taken as zero at the upper top entrainment and the right outlet boundaries.

4 Solution Methodology

The Reynolds averaged differential governing equations of flow and turbulence fields are solved using finite volume method proposed by Patankar [10]. The power law and second-order central difference schemes are employed for discretization of convective and diffusive terms, respectively. The SIMPLE [10] algorithm is used for coupling of mean velocity and pressure. A globally converged solution is obtained when the residuals of all variables are fallen below 10^{-6}. Domain size of $75w \times 60w$ is considered for the present study, where the width of nozzle w is taken as 0.0125 m.

5 Grid Sensitivity Test and Validation of Numerical Model

The grid sensitivity test has been performed by considering three sets of grid of sizes $190 \times 152, 230 \times 184$ and 270×216 for different HRN k-ε models, whereas another three sets of grid of sizes $370 \times 343, 410 \times 375$ and 450×407 have been considered for different LRN k-ω models in such a way so that the requirement of $y+$ value as per turbulence model is satisfied. Figure 2 shows the U-velocity profiles at the axial location of $X = 9$ for three sets of grid sizes considered. Since, the same number of grids with similar grid arrangements have been considered for each group of HRN k-ε models and LRN k-ω models. Thus, the grid sensitivity test of standard k-ε model from HRN k-ε models group and standard k-ω model from LRN k-ω models group are only presented here. The U-velocity profiles obtained from three sets of grids almost merge on each other for both the models under consideration, as shown in Fig. 2. However, to be on a safer side, a grid density of 230×184 is chosen for HRN k-ε models, whereas a grid density of 410×375 is selected for LRN k-ω models in the present numerical study.

To the best of authors' knowledge, no experimental work has been carried out for larger offset ratio of dual jet flow. Hence, the present numerical model has been

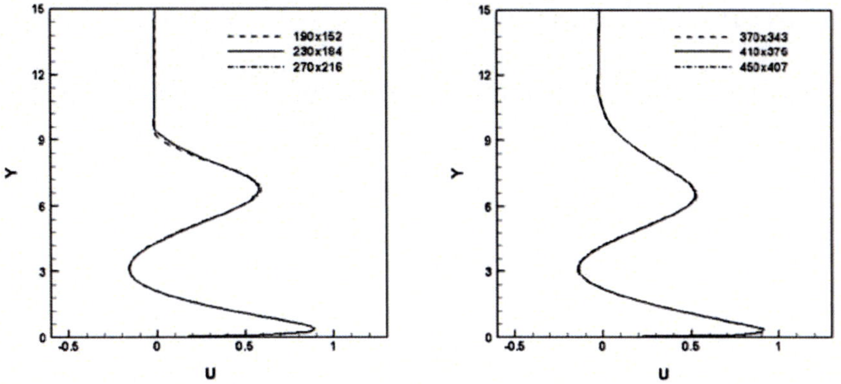

(a) Standard k-ε model. (b) Standard k-ω model.

Fig. 2 Grid sensitivity test showing spanwise distribution of U-velocity profile at $X = 9$

validated against the experimental results of Pelfrey and Liburdy [11] for Reynolds number 15,000 and offset ratio 7 of offset jet flow for all five turbulence models chosen for the comparative study. A good agreement has been found between present numerical and experimental results as shown in Fig. 3.

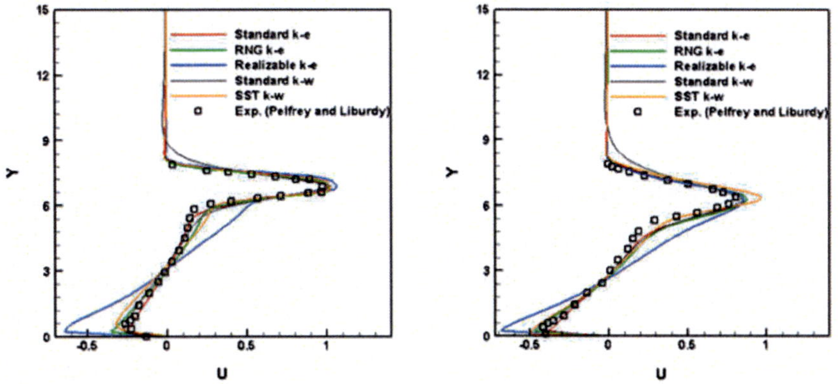

(a) $X = 3$ (b) $X = 6$

Fig. 3 Validation study showing comparison of U-velocity profile at $X = 3$ and $X = 6$ with the experimental results of Pelfrey and Liburdy [11]

6 Results and Discussion

The study of fluid flow behaviour of turbulent dual jet is more complex as it comprises the interaction of offset jet and wall jet, which results in the formation of rigorous converging, merging and combined regions. In the present study, the flow behaviour of dual jet is characterized by means of streamlines plot, U-velocity profile, velocity similarity profile, $U_{\max}$ profile, and so on. As mentioned earlier, there is a lack of detailed experimental work of dual jet flow for larger offset ratio. Therefore, some of the flow characteristics like streamlines, U-velocity profile, locations of upper and lower vortex centres ((UVC: X_{uvc},Y_{uvc}) and (LVC: X_{lvc},Y_{lvc})), merge point (MP: X_{mp},Y_{mp}) and combined point (CP: X_{cp},Y_{cp}) of dual jet flow have been plotted and compared against the numerical results of Kumar and Das [3] for Reynolds number 20,000 and offset ratio 9. Figure 4 presents the streamline plots with velocity magnitude contours of dual jet flow for all the five turbulence models under consideration. Two counter rotating vortices (upper vortex and lower vortex) are formed in the converging region. The vortex centres and merge point are identified at points, where

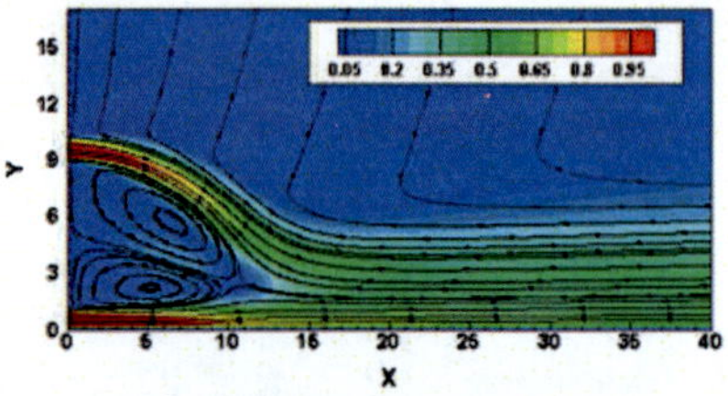

(a) Standard k-ε model.

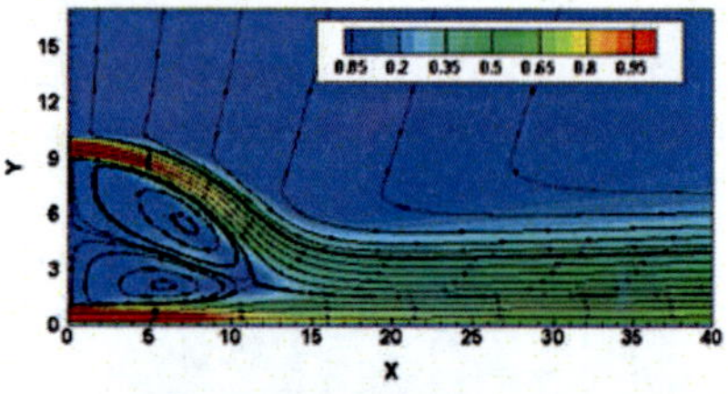

(b) RNG k-ε model.

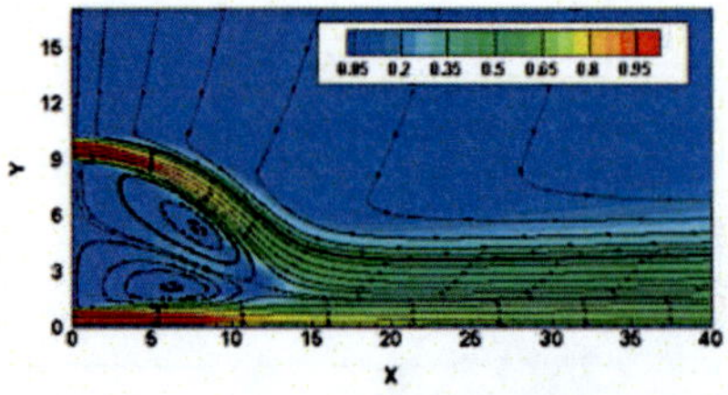

(c) Realizable k-ε model.

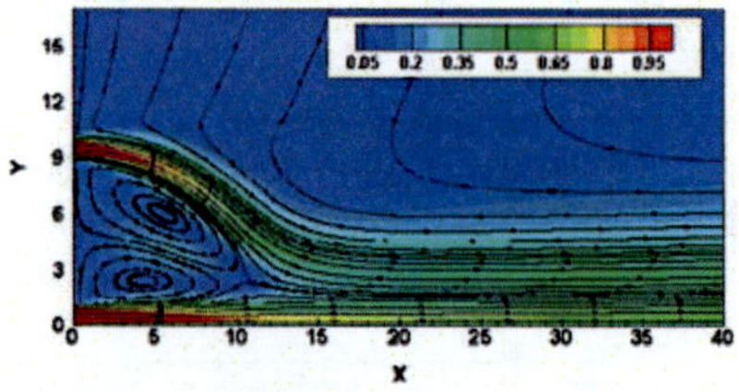

(d) Standard k-ω model.

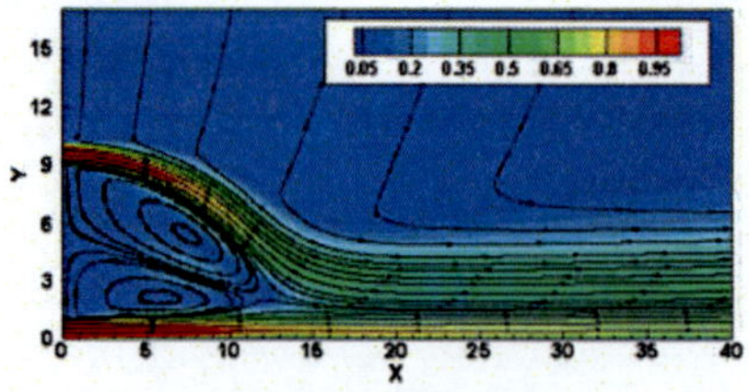

(e) SST k-ω model.

Fig. 4 Streamlines with velocity magnitude contours

Table 1 The locations of upper vortex centre (UVC: X_{uvc},Y_{uvc}), lower vortex centre (LVC: X_{lvc},Y_{lvc}), merge point (MP: X_{mp},Y_{mp}) and combined point (CP: X_{cp},Y_{cp}) of dual jet flow

Turbulence model	UVC (X_{uvc},Y_{uvc})	LVC (X_{lvc},Y_{lvc})	MP (X_{mp},Y_{mp})	CP (X_{cp},Y_{cp})
Standard *k*-*ε*	6.49, 5.65	5.16, 2.19	10.41, 2.41	17.78, 2.79
RNG *k*-*ε*	7.20, 5.48	5.85, 2.19	11.01, 2.34	18.19, 2.6
Realizable *k*-*ε*	7.85, 5.32	6.23, 2.16	11.29, 2.28	17.36, 2.6
Standard *k*-*ω*	5.50, 6.01	4.28, 2.40	10.22, 2.48	18.34, 2.62
SST *k*-*ω*	7.43, 5.43	5.81, 2.15	11.72, 2.20	19.13, 2.4
Kumar and Das [3] Predictions	6.49, 5.66	5.18, 2.18	10.46, 2.42	18.38, 2.78

the $U = 0$ and $V = 0$ contour lines intersect each other, whereas the combined point is located at extremity point of $\partial U/\partial Y = 0$ contour line. The locations of upper and lower vortex centres ((UVC: X_{uvc},Y_{uvc}) and (LVC: X_{lvc},Y_{lvc})), merge point (MP: X_{mp},Y_{mp}) and combined point (CP: X_{cp},Y_{cp}) are reported in Table 1. The observation of Table 1 reveals that predictions of standard *k*-*ε* model for identification of locations of vortex centres, merge point and combined point are matched very well with the results of Kumar and Das [3]. The largest converging region has been predicted by SST *k*-*ω* model, whereas the smallest converging region has been predicted by standard *k*-*ω* model, which can be decided by coordinates of merge point (X_{mp},Y_{mp}), from where the converging region ends and merging region starts with interaction of offset jet and wall jet.

Figure 5 depicts the spanwise distribution of *U*-velocity profile at various axial locations $X = 3$, 7, 11 and 15 of impingement wall. The predictions of various turbulence models for *U*-velocity profile at different axial locations are compared with the results of Kumar and Das [3] for Reynolds number 20,000 and offset ratio 9. It is observed that the predictions of standard *k*-*ε* model agreed very well with the results of Kumar and Das [3] as compared to other turbulence models under consideration as shown in Fig. 5.

The dual jet flow structure resembles to that of a single wall jet flow and shows the self-similarity behaviour in the combined region. Wygnanski et al. [12] have carried out a detailed experimental study on flow behaviour of wall jet at Reynolds number 19,000 and proposed the conventional outer-scaling method in which the dimensionless variables $Y/Y_{0.5}$ and U/U_{max} are used to plot the similarity solution, where the jet half-width ($Y_{0.5}$) is the dimensionless distance in the transverse direction (*Y*) at which $U = U_{max}/2$. In the present work, the self-similarity behaviour of dual jet flow is examined in the combined region, and numerical predictions of various turbulence models are compared against the experimental results of Wygnanski et al. [12] for Reynolds number 19,000 and offset ratio 9 and also with the experimental results of Wang and Tan [1] for Reynolds number 10,000 and offset ratio 2. Figure 6(a) shows the comparison of similarity profile at an axial location of $X = 60$ against the experimental results of Wygnanski et al. [12]. The similarity profiles obtained by

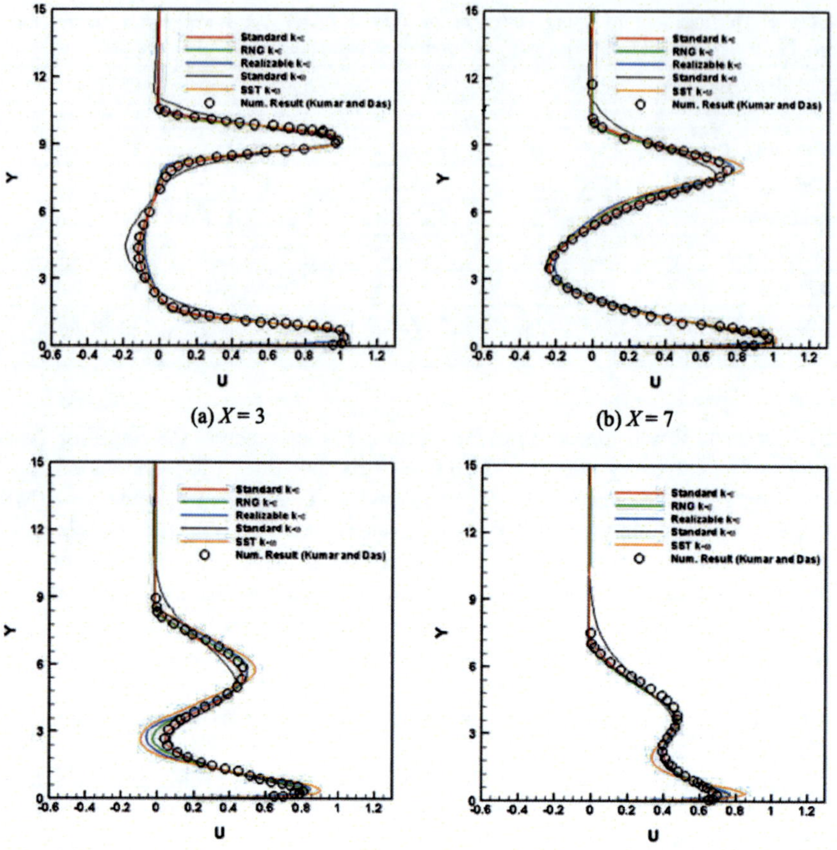

(a) $X = 3$

(b) $X = 7$

(c) $X = 11$

(d) $X = 15$

Fig. 5 Spanwise distribution of *U*-velocity profile at various axial locations

various turbulence models show a good agreement with the experimental results except in the outer shear layer region, where some deviations are observed, as shown in Fig. 6(a).

The numerical predictions of three HRN models, that is, standard *k-ε*, RNG *k-ε* and realizable *k-ε* are nearly the same and show a maximum deviation of approximately 17.64% with the experimental results of Wygnanski et al. [12], whereas the numerical predictions of LRN models, that is, standard k-ω and SST k-ω models show a maximum deviation of approximately 49.73 and 23%, respectively with the experimental results of Wygnanski et al. [12]. Figure 6(b) presents the comparison of similarity profile at an axial location of $X = 30$ against the experimental results of Wang and Tan [1]. The predictions of various turbulence models are almost similar and show a good qualitative agreement with the experimental results of Wang and Tan [1], as shown in Fig. 6(b).

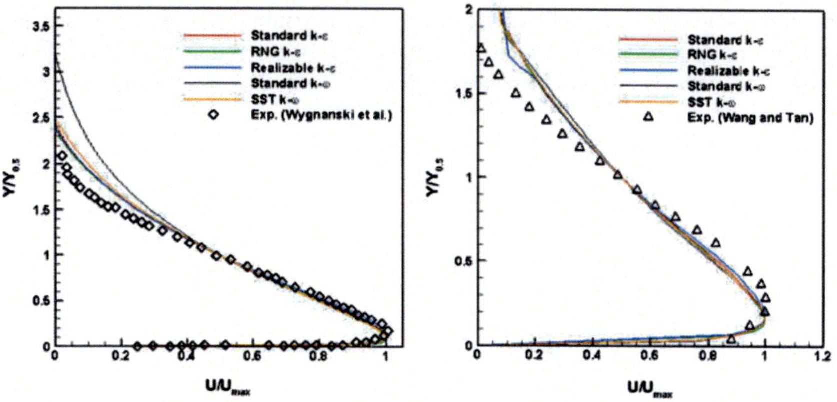

(a) $X = 60$ (b) $X = 30$

Fig. 6 Comparison of similarity profile at various axial locations with the experimental results

Figure 7 demonstrates the comparison of decay of local maximum axial velocity ($U_{\max}$) along X against the experimental results of Wang and Tan [1] and numerical results of Kumar [5] for Reynolds number 10,000 and offset ratio 2. The velocity $U_{\max}$ is initially constant in the potential core region where flow remains unaffected from the effect of viscosity. It then decays rapidly in the converging region and further decreases gradually in the merging and combined regions as shown in Fig. 7. The predictions of standard k-ε model show a good agreement with the results of Kumar [5], while show some deviation with the experimental results of Wang and Tan [1]. All the turbulent models predicted higher value for $U_{\max}$ as compared to

Fig. 7 Comparison of decay of local maximum axial velocity ($U_{\max}$)

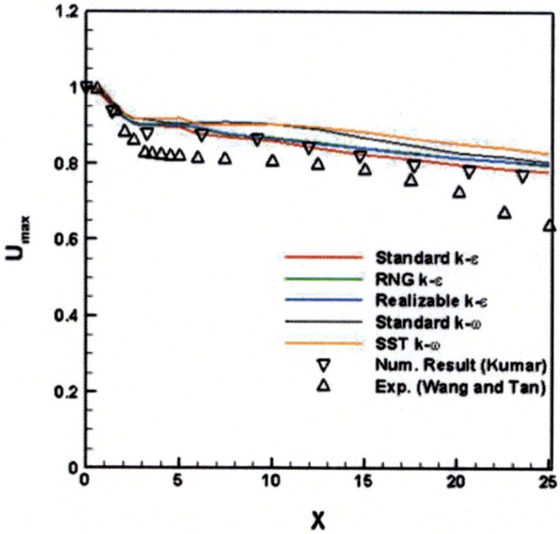

experimental results of Wang and Tan [1]. The maximum discrepancies between computational predictions and experimental results are approximately 23.8, 26.98, 28.57, 28.57 and 31.74% for standard k-ε, RNG k-ε, realizable k-ε, standard k-ω and SST k-ω models, respectively. The reason behind these discrepancies is the presence of periodic vortex shedding as mentioned by Kumar [5].

7 Conclusion

A detailed comparative study of fluid flow behaviour of turbulent dual jet comprising of an upper offset jet and a lower wall jet has been done considering different Reynolds averaged Navier–Stokes (RANS) based turbulence models. The standard k-ε, RNG k-ε, realizable k-ε, standard k-ω and SST k-ω turbulence models are considered for present comparative study of dual jet flow, and their relative performances are evaluated against the various experimental and numerical results available in the literature for a range of Reynolds number and offset ratios. Numerical predictions of dual jet flow characteristics are challenging because of the presence of complex converging, merging and combined regions that arise due to the interaction of upper offset jet and a lower wall jet. All the turbulence models are able to capture the basic flow structure and turbulence fields of dual jet, including two counter rotating vortices and three flow regions, that is, converging, merging and combined regions. Many significant information about the locations of vortex centres, merge point and combined point as predicted by different turbulence models are also reported in the present study. The SST k-ω model has predicted the largest converging region, whereas the standard k-ω model has predicted the smallest converging region. The predictions of standard k-ε model for identification of locations of vortex centres, merge point and combined point, U-velocity profile, velocity similarity profile and decay of U_{max} profile show the least deviation with various experimental and numerical results. The detailed comparative study reveals that the high Reynolds number (HRN) standard k-ε model performs better among the turbulence models considered and produces reliable results for two-dimensional fluid flow study of dual jet flow for the range of Reynolds number and offset ratios with less computational cost.

References

1. Wang XK, Tan SK (2007) Experimental investigation of the interaction between a plane wall jet and a parallel offset jet. Exp Fluids 42:551–562
2. Vishnuvardhanarao E, Das MK (2009) Study of the heat transfer characteristics in turbulent combined wall and offset jet flows. Int J Therm Sci 48:1949–1959
3. Kumar A, Das MK (2011) Study of a turbulent dual jet consisting of a wall jet and an offset jet. J Fluids Eng 133:1201–1211
4. Mondal T, Das MK, Guha A (2014) Numerical investigation of steady and periodically unsteady flow for various separation distances between a wall jet and an offset jet. J Fluids Struct

50:528–546
5. Kumar A (2015) Mean flow characteristics of a turbulent dual jet consisting of a plane wall jet and a parallel offset jet. Comput Fluids 114:48–65
6. Hnaien N, Marzouk S, Aissia HB, Jay J (2017) CFD investigation on the offset ratio effect on thermal characteristics of a combined wall and offset jets flow. Heat Mass Transf 53:2531–2549
7. Assoudi A, Said NM, Bournot H, Palec GL (2018) Comparative study of flow characteristics of a single offset jet and a turbulent dual jet. Heat Mass Transf 55:1109–1131
8. Launder BE, Spalding DB (1974) The numerical computation of turbulent flows. Comput Methods Appl Mech Eng 3:269–289
9. ANSYS FLUENT (2016) Theory guide vol 17.2. ANSYS Inc. USA
10. Patankar SV (1980) Numerical heat transfer and fluid flow. Hemisphere Publishing Corporation, New York
11. Pelfrey JRR, Liburdy JA (1986) Mean flow characteristics of a turbulent offset jet. Trans ASME J Fluids Eng 108:82–88
12. Wygnanski I, Katz Y, Horev E (1992) On the applicability of various scaling laws to the turbulent wall jet. J Fluid Mech 234:669–690

Fluid Structure Interaction Study of Damper for Swish and Rattle Noise Refinement

Swapnil S. Kulkarni, A. Satheesh, B. Ravi, and M. R. Saraf

1 Introduction

Structure noise perceived in shock absorber is mainly due to piston rod acceleration. Acceleration of piston rod changes due to opening and closing of valves in piston and base valve assembly. Considering the piston subassembly, this subassembly has to maintain the oil balance in chamber and needs to create the required pressure differential to meet the damping force characteristics. The fix orifice should act like a fulcrum about which other flexible orifices have to deflect. The outer diameter of fix orifice should be considered as critical parameter for pressure differential. Flow separation should be such that it should not generate the cavitation when disc deflection is restricted. The cavitation may result in drop in damping performance. The restriction to flow means poor energy dissipation, which results in the presence of acoustic energy in damper. Turbulence is generated by structural deformation of flexible orifice and these orifices can act as the obstacles for high-speed fluid flow which generates sound. This paper mainly addresses the reduction of velocity profile at fix and flexible orifices with the help of fluid structure interaction technique to reduce tendency of noise, and at the same time to generate desire pressure differential. Shu et al. [1] analyzed micro-process model of hydraulic shock absorber with abnormal structural noise. Abnormal noise and vibration of piston rod increased

S. S. Kulkarni · A. Satheesh (✉)
School of Mechanical Engineering, VIT, Vellore, Tamil Nadu 632014, India
e-mail: satheesh.a@vit.ac.in

S. S. Kulkarni · B. Ravi
Central Research and Development, Gabriel India Limited, Pune, Maharashtra 411501, India

S. S. Kulkarni · M. R. Saraf
Automotive Research Association of India (ARAI), Pune, Maharashtra 411038, India

M. Palanisamy et al. (eds.), *Theoretical, Computational, and Experimental Solutions to Thermo-Fluid Systems*, Lecture Notes in Mechanical Engineering,
https://doi.org/10.1007/978-981-33-4165-4_34

with increase in gap between piston rod head and oil. Improper fixing of orifice results in wider gap between piston and oil. This paper also addresses the damping coefficient of fixed and flexible orifices when oil passes through it. Benaziz et al. [2] showed the pressure flow rate relation for flexible orifice and mentioned that abrupt change in area and direction results in flow area less than 1000. The critical flow number handling transition from laminar to turbulent flow is also mentioned. Benaziz et al. [3] presented nonlinear dynamic analysis of shock absorber hydraulic spring valve in which the constant orifice section is analyzed. Oscillation frequency for three different values of constant orifices section is mentioned. The oscillation frequency decreases with increase in orifice section. Low damping of spring valve also results in unstable behavior and leads to high frequencies. High damping of spring valve does not show any oscillations. Bending modes of flexible orifice need to be studied to avoid the oscillation due to transient fluid flow. Satpute et al. [4] presented fluid flow model of fluid damper with spring loaded valve. Variable number of piston orifices and valve flow area is defined to get the pressure flow rate relations. Kulkarni et al. [5] presented coupled fluid structure interaction simulation of automotive shock absorber. Oil flows in small clearance between piston land and shim during low velocity and flexible orifice subjected to large displacement were analyzed. Also, they presented [6] nonlinear contact analysis of shock absorber shims using fluid structure interaction. Flexible orifices are considered with desired frictional effect to reduce fluttering of orifices as orifices are subjected to hydrodynamic fluid pressure. Contact condition between orifice and piston land is defined to get accuracy in FSI model. Kulkarni et al. [7] presented the influence of flexible orifice bending mode on damping force variation of hydraulic twin tube shock absorber. If the piston or base valve port positions are varied, then bending mode of orifice changes swish and rattling noise is produced by a series of pressure fluctuation generated in piston and base valve assembly of shock absorber. The pressure fluctuation at particular valving part is relatively small compared to mean value of pressure, and total pressure variation is accurately described by linear equation considering steady-state term. In case of large amplitude pressure variations, linearized wave equation no longer applies. The speed of propagation of disturbance depends on amplitude of disturbance. Wave steepening occurs as wave progresses by ultimately forming shock wave at the front of disturbance. Mehraby et al. [8] worked on numerical and analytical investigation in radiated noise by shock absorber. Rayleigh equation is used to calculate sound generated by surface of plate. The finite element method is employed to simulate the problem. Hou et al. [9] worked on shock absorber modeling and simulation based on modelica. In this work, a detailed model of shock absorber is established which contains rebound chamber, compression chamber, piston valve assembly and base valve assembly. B Gauduin et al. [10] described the rattle noise which is non-stationary and composed of short-time shocks. Succession of hydraulic chocks results in tapping noise. Alexander et al. [11] worked on NVH engineering of shock absorber modules. Structure-borne noise due to suspension attachment points is well described. Acoustic optimization of chassis components is achieved by reduced excitation amplitudes, shifted resonant frequency, and reduced noise transfer into vehicle interior. David Bogema et al. [12] presented noise path analysis process evaluation of automotive shock absorber transient noise. Transient noise referred to as chuckle or loose lumber is high-level noise. This noise transfers through top mounts or filtration units attached to damper.

Few works were initiated and reported in the literatures in the area of various structure and air-borne noise sources present in damper assemblies used in passenger car, and in-depth analyses of characterization of noise sources present in passenger car damper with the help of FSI have to be considered by the authors. Pressure flow rate relationship is obtained for orifice with fix and flexible sections.

Refer to double tube shock absorber model for noise and vibration analysis [2]. Flow rate equation is given by the following expression. For fixed orifice in piston

$$Q(\Delta P) = C_{\mathrm{d}}.A.\mathrm{sign}(\Delta P).\frac{\sqrt{2|\Delta P|}}{\rho} \tag{1}$$

For flexible orifice,

$$C_{\mathrm{d}} = C_{\mathrm{d,max}}\cdot\tanh\left(\frac{2l}{l_{\mathrm{c}}}\right) \tag{2}$$

$$C_{\mathrm{d}} = C_{\mathrm{d,max}}\cdot\tanh\left(\frac{D_{\mathrm{h}}}{\mu l_{\mathrm{c}}}\sqrt{\frac{2\Delta P}{\rho}}\right) \tag{3}$$

Refer to microprocess model of hydraulic shock absorber with abnormal structural noise [1]. By the theory of vibration, the differential equation of vibration of the piston head when oil passes through fixed orifice can be written as

$$\left\{m\ddot{x} + c\dot{x} + kx = sign(\dot{x}_r - \dot{x})f_d + c_f(\dot{x}_r - \dot{x})\right\} \tag{4}$$

$$c_f(\dot{x}_r - \dot{x}) \leqslant F_{cf}{}^{up} \tag{5}$$

where $F_{\mathrm{cf}}{}^{\mathrm{up}}$ represents critical damping force between the orifice and the piston. c_{f} represents damping coefficient as fluid passes through the orifice. By the theory of vibration, the differential equation of vibration of the piston head when oil passes through fixed and flexible orifice can be written as

$$\{m\ddot{x} + c\dot{x} + kx = \mathrm{sign}(\dot{x}_{\mathrm{r}} - \dot{x})f_{\mathrm{d}} + c_{\mathrm{c}}(\dot{x}_{\mathrm{r}} - \dot{x})\} \tag{6}$$

$$c_{\mathrm{c}}(\dot{x}_{\mathrm{r}} - \dot{x}) \leqslant F_{\mathrm{cf}}{}^{\mathrm{down}} \tag{7}$$

where $F_{\mathrm{cf}}{}^{\mathrm{down}}$ represents critical damping force as orifice closed.

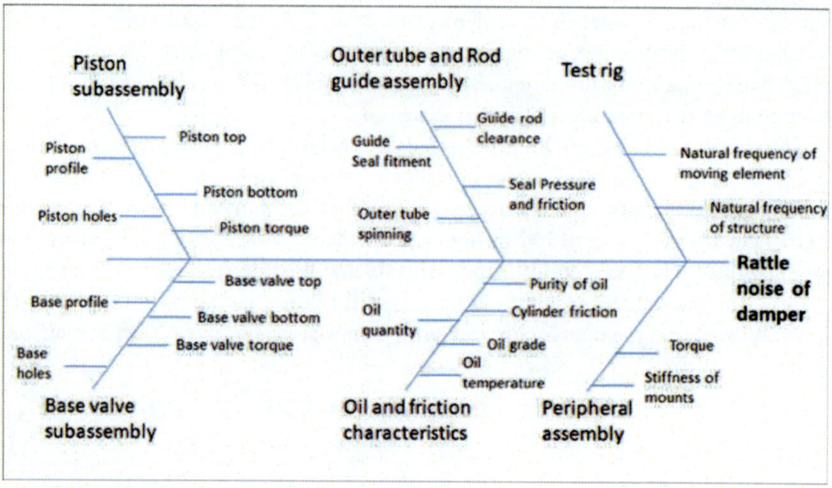

Fig. 1 Rattle noise of damper—Cause–effect diagram

Damping coefficient depends on bending modes of orifice. Uneven deflection of orifice results in high damping, which in turn increases the critical damping force and hence the piston vibrations.

In order to describe the causes of flow-induced noise, fluid–structure interaction should be performed in which nonlinear contact analysis of orifices should be taken into consideration. In this paper, we focused on fluid velocity and pressure differential across the piston. Fluid pressure and piston velocity characteristics are obtained using fluid–structure interaction (FSI) to get the desired damping performance.

2 ISHIKAWA Diagram for Rattle Noise of Damper

Based on the industrial experience on the present problem, **ISHIKAWA**—Cause–effect diagram is prepared to understand the potential sources present in damper which generates rattle noise (Fig. 1).

3 Methodology

Arbitrary Lagrangian–Eulerian (ALE) formulation is used for simulating damper oil and orifice structures. Contact algorithm has been considered for Lagrangian orifice structure subjected to variation in oil pressure and it has taken the nonlinear deflection characteristics and frictional effect between these orifices. Imposed velocity is applied considering the flow inlet as boundary condition. Flow domain is modeled completely where the nonlinear orifice deformation takes place. Computational domain is reduced to area of interest where the oil flow characteristics are analyzed. Comparatively fine mesh is prepared at oil and orifice interface. Effect on fluid pressure differential due to fix and flexible orifice has been studied. Flexible orifices are subjected to deflect while generating the pressure differential across chambers. Deflection mainly depends on orifice geometry, number of notches in fix orifice, size of piston ports, base valve ports and clamping orifice geometry on which flexible and fix orifices rested. Fluid pressure from the flow acts well on the surface area of flexible orifice that causes elastic bending of the orifices. Pressure differential and fluid velocity depend on bending modes of flexible orifice. Damping coefficient of fixed orifice should be lower than the critical damping force to avoid the swish and rattle noise. The bending mode coefficient of flexible orifice has been studied. Damping coefficient of flexible orifice should be lower than critical damping force. Figure 2 shows the finite element model of piston assembly works for rebound damping of dampers (Table 1).

Figure 3 shows the geometry and mesh model of piston ports. Comparatively, fine mesh is applied near the shim contact areas. Imposed velocity is applied as inlet boundary condition. Wall boundary condition is also applied to enclosure. Shim behavior has been studied by applying self-contact and friction value of 0.3. Figure 4 shows the deformation of shims. Self-contacts were defined in shims to get the deformation.

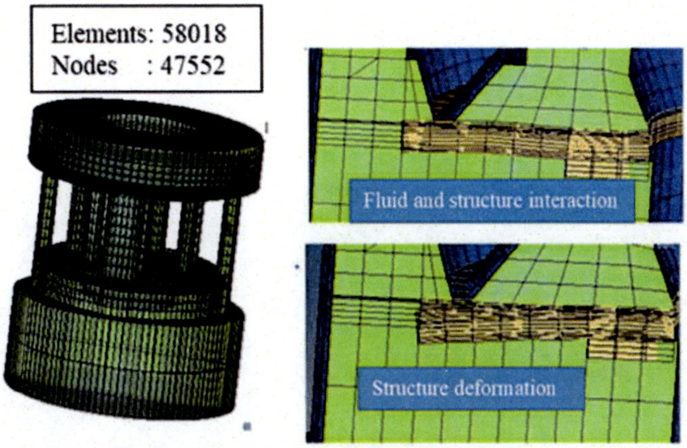

Fig. 2 FE model of piston assembly

Table 1 Material property and mesh details

Oil properties	Dynamic viscosity 9×10^{-6} kg/mm-S	Density 850×10^{-9} kg/mm^3	–
Structure properties	Young's modulus 2×10^5 N/mm^2	Poisson ratio 0.3	Density 8×10^{-6} kg/mm^3

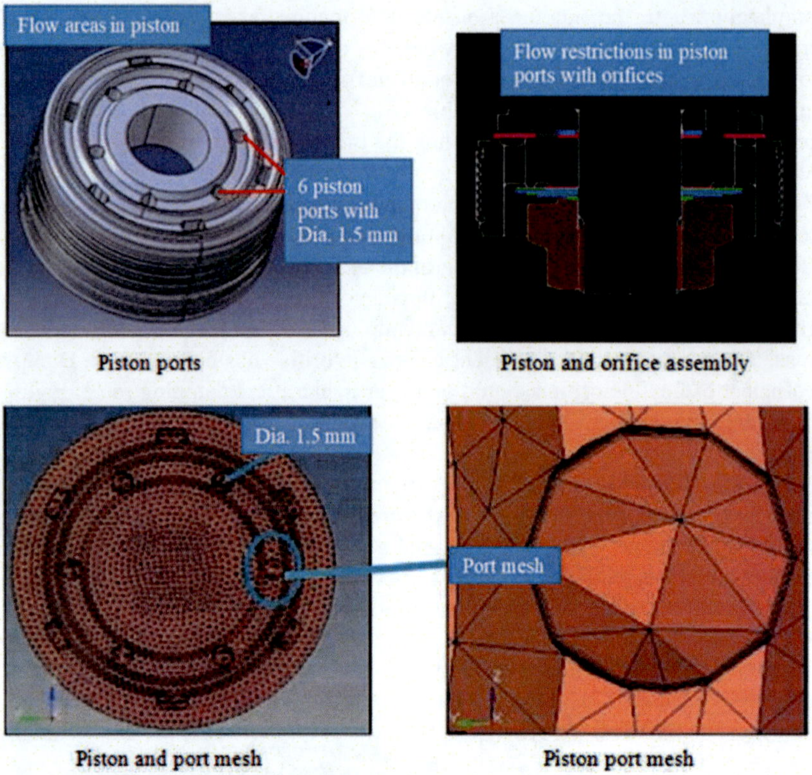

Fig. 3 Geometry and FE model of piston ports

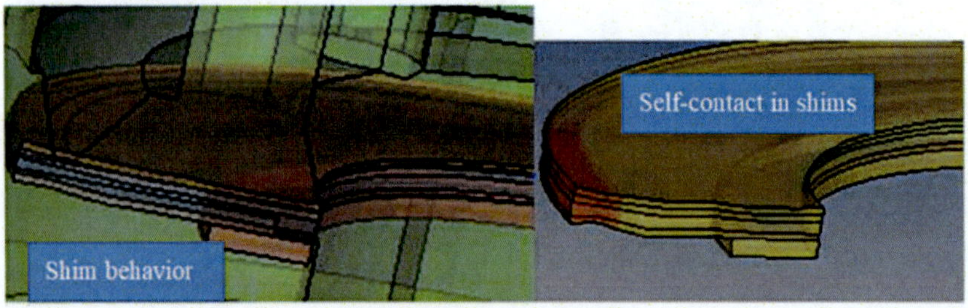

Fig. 4 Deformed FE model of flexible orifices

4 Results and Discussions

This paper presents numerical prediction of reduction of flow-induced noise using FSI technique. Three-dimensional unsteady flow simulations are carried out and compared with each other and also with fluid pressure and piston velocity characteristics. Oil flow velocity has been changed in fixed and flexible orifice to avoid the vibrations generated from piston assembly. Flow path has been altered by changing the number of notches in fixed orifice from 6 to 8 notches (keeping the same discharge area).

Figure 5 shows the velocity profile of fix orifice with six notches at piston velocity of 0.05 m/s. It shows the fluid velocity of 0.686 m/s at discharge side. Figure 6 shows the velocity profile of fix orifice with six notches at piston velocity of 0.05 m/s. It shows the reduction in fluid velocity from 0.68 to 0.51 m/s at discharge side.

Figure 7 shows the pressure differential and piston velocity graph of fix orifices. Reduction in pressure differential is observed with fix orifice with eight notches although the flow area of notches was the same. The flow path was altered due to increased notches. Higher pressure differential across piston velocity is observed in piston with six fix notches as compared to piston with eight fix notches. Higher pressure differential results in higher damping force. This will increase the vibration of piston body. Reduction of damping force in this low-frequency zone also helped in getting the desired comfort zone.

Figure 8 shows higher fluctuation in fluid pressure. This is observed in piston with six fix notches as compared to piston with eight fix notches. The higher pressure fluctuation of fluid around shims results in rattling behavior of shims. This phenomenon results in vibration of piston rod. The deflection of flexible orifice (blow off or

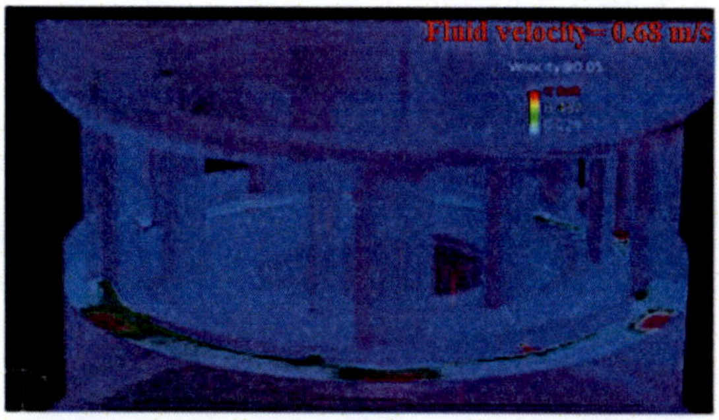

Fig. 5 Fix orifice with six notches

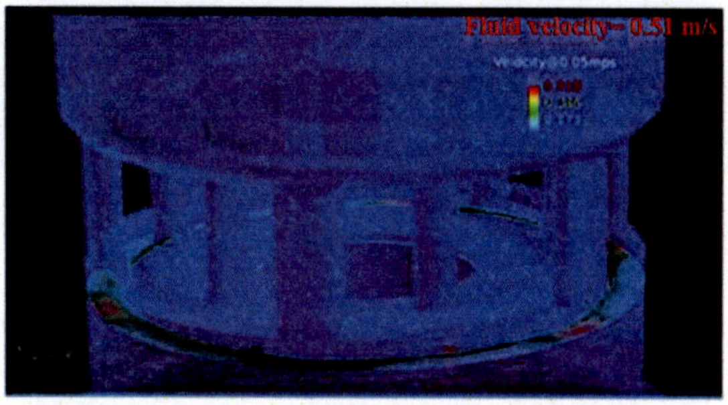

Fig. 6 Fix orifice with eight notches

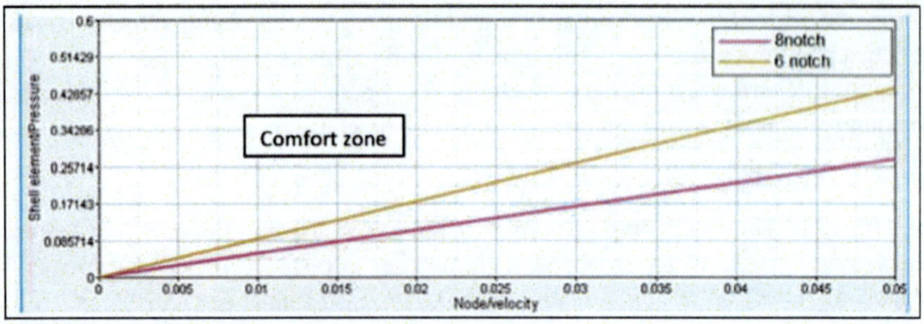

Fig. 7 Pressure differential and piston velocity

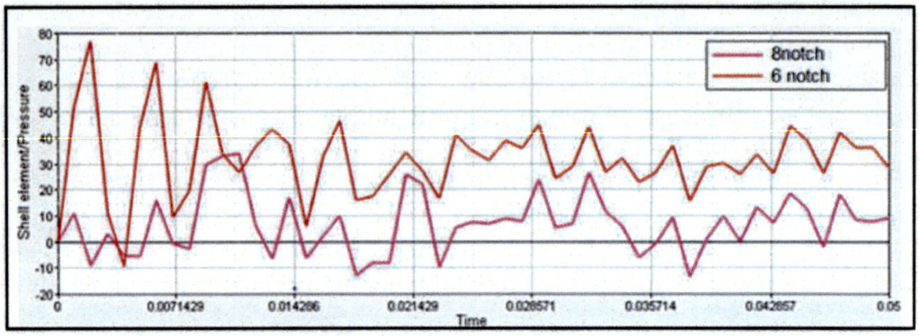

Fig. 8 Pressure differential and time

threshold point) occurs comparatively early with six notches results in higher pressure differential. Transition of flow from laminar to turbulent with six notches is not smooth compared to eight notches.

Figure 9 shows comparable fluctuation in fluid pressure with five and seven fix notches. The deflection of flexible orifice (blow off or threshold point) occurs comparatively at similar time. Transition of flow from laminar to turbulent is comparatively smooth.

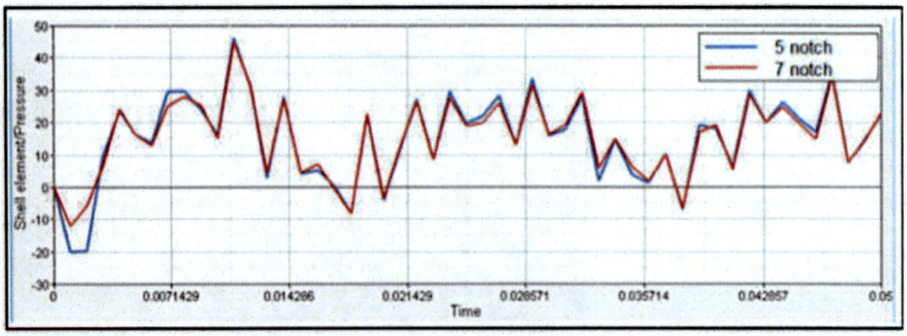

Fig. 9 Pressure differential and time

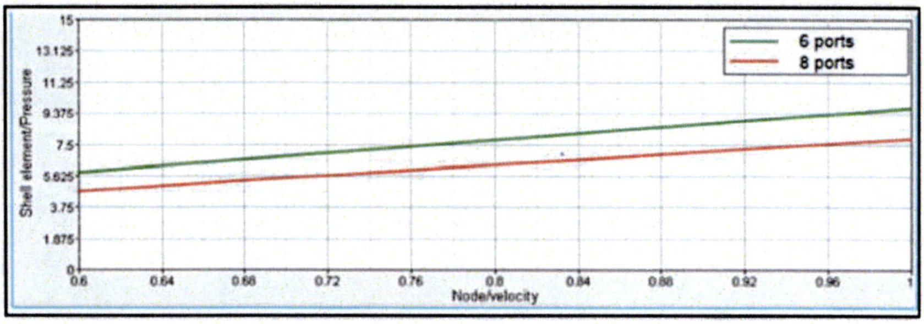

Fig. 10 Pressure differential and piston velocity

Fig. 11 Flow at piston port inlet

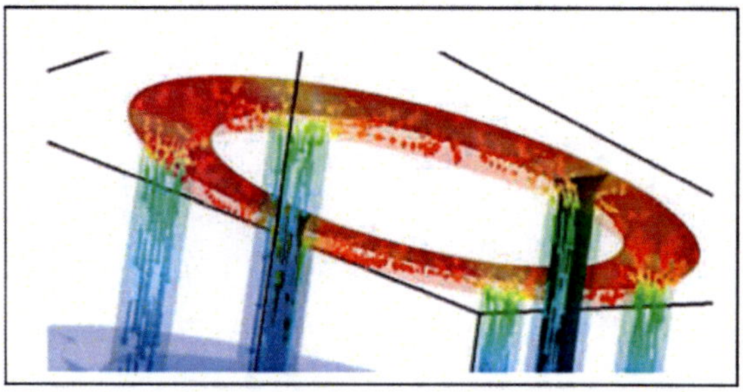

Figure 10 shows the pressure differential and piston velocity graph of fix orifices. Reduction in pressure differential is observed with piston with eight ports although the flow area of piston was the same. The flow path was altered due to increased ports. Figure 11 shows the flow at inlet. Flow is turbulent in nature. Flow recirculation (vortex shading) is also observed in this zone. Random flow is observed whereas turbulent intensity is also observed maximum. Figure 12 shows flow through piston ports where maximum jet velocity is observed in piston ports. Changing the shape and orientation of piston port helps to reduce the jet velocity and swish noise from damper. Figures 13 and 14 show the bending modes of flexible orifices subjected to fluid pressure. The bending modes of flexible orifices increases the sound pressure levels due to their vibrations. Uneven bending mode of flexible orifices led to high vibrations due to low damping and increase in pressure pulsation which resulted in rattling phenomenon as compared with even bending modes of flexible orifices.

5 Conclusion

Swish and rattle noise of passive hydraulic velocity sensitive damper is studied using fluid–structure interaction with contact nonlinearity approach. Full 360° damper piston assembly FSI model has been analyzed. Arbitrary Lagrangian–Eulerian (ALE)

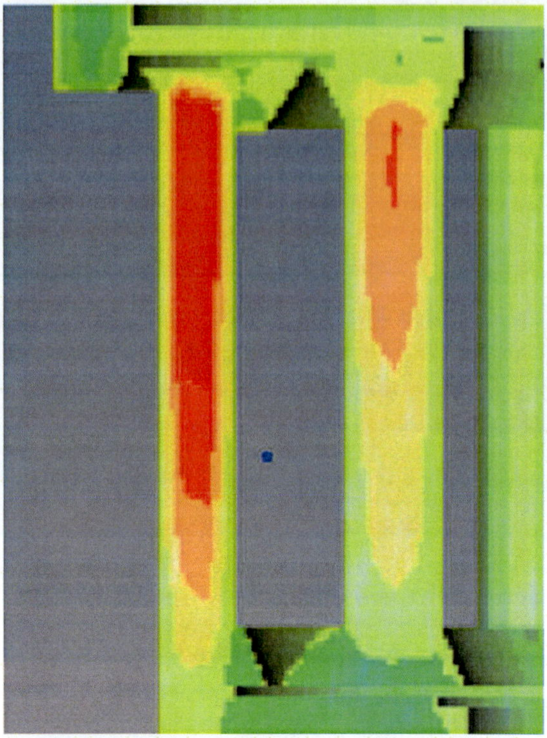

Fig. 12 Flow through piston port

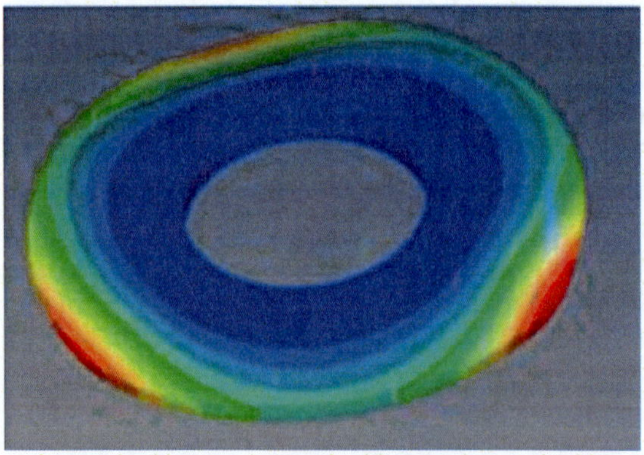

Fig. 13 Even bending mode

formulation is used for simulating damper oil and nonlinear orifice structures. Contact nonlinearities are also defined between orifice structures. Potential areas to reduce the uneven pressure fluctuation in automotive damper were investigated using FSI with contact nonlinearity approach. This work helped to avoid rattling noise due to vibration of piston. Potential areas to reduce the fluid velocity in certain regimes helped

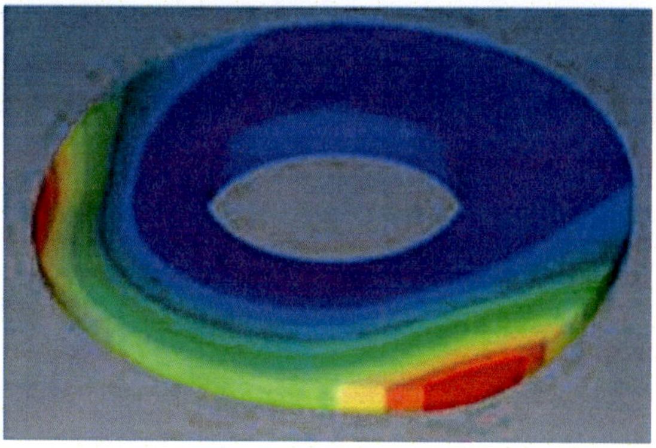

Fig. 14 Uneven bending mode

to reduce the swish noise phenomenon in damper. Fluid pressure–piston velocities were also investigated using FSI to achieve the desired damping performance along with acoustic refinement.

References

1. Shu H-Y, Zhang W-W, Feng Y (2008) Micro-process model of hydraulic shock absorber with abnormal structural noise, College of Mechanical Engineering, Chongqing University, Chongqing 400044, China. J Cent South Univ Technol 15:853–859. https://doi.org/10.1007/s11771-008-0157-x
2. Benaziz M, Nacivet S, Deak J, Thouverez F (2013) Double tube shock absorber model for noise and vibration analysis. SAE Int J Passeng Cars-Mech Syst 6(2013–01–1912):1177–1185
3. Benaziz M, Nacivet S, Thouverez F (2012) Nonlinear dynamic analysis of a hydraulic shock absorber spring valve. In: Proceedings of ISMA2012-USD
4. Satpute NV, Singh S, Sawant S (2013) Fluid flow modelling of a fluid damper with shim loaded relief valve. Int J Mech Eng 2(1):65–74. ISSN: 2319-2240
5. Swapnil K, Ravi B, Magdum M (2012) A coupled fluid structure interaction simulation of automotive shock absorber using Radioss block explicit solve. In: Altair hyperworks technology conference India 2012
6. Swapnil K, Ravi B, Magdum M (2013) Nonlinear contact analysis of automotive shock absorber shims using fluid structure interaction technique. In: Altair technology conference India 2013
7. Swapnil K, Ravi B, Magdum M (2014) Influence of shim bending mode on damping force variation of hydraulic twin tube shock absorber. SAE technical paper 2014-01-0045
8. Mehrabya K, Beheshtib H, Poursinab M (2013) Numerical and analytical investigation in radiated noise by a shock-absorber. Int J Eng 26(12):1525–1534. https://doi.org/10.5829/idosi.ije.2013.26.12c.13
9. Hou Y, Li L, He P, Zhang Y, Chen L (2011) Shock absorber modelling and simulation based on modelica. In: Proceedings 8th modelica conference, Germany, March 20–22
10. Gauduina B, Noela C, Meillierb J-L, Boussarda P (2008) A multiple regression model for predicting rattle noise subjective rating from in-car microphones measurements, euronoise acoustics08, Paris
11. Kruse A, Eickhoff M, Pagel J, Pöge D (2010) NVH engineering of shock absorber modules. SAE international 2010-01-0505, Published 04/12/2010
12. Bogema D, Goodes P, Apelian C, Csakan M (2009) Noise path analysis process evaluation of automotive shock absorber transient noise, SAE international 2009-01-20

A Numerical Model to Study the Fluid–Structure Interaction at a Bridge Pier

Kanchibhotla Hima Teja, G. Srivalli, and V. Balakrishna Murthy

1 Introduction

Over the last three decades, a number of bridges were spoiled due to the scour formation around piers. Scour is a term that is defined by a natural phenomenon which can be happened by the action of erosion of flowing water on the beds of sedimentary channels. In 2016, Abdullah [1] provided the definition of the scour as the pruning of the riverbed level by the erosion of water that results in a propensity to exhibit the foundations of structures like bridges, as given by Cheremisinoff (1987). As foundation is the most principal part in any structure, for the design and analysis of that one should be given the highest priority when compared to the remaining parts. The failure of a bridge is mainly encountered by three main causes, in which bridge scour occurs as the major one and others being overloading and collision. A serious damage could occur to the structure due to erosion of the soil bed by the local scour around bridge piers, and hence may lead to the destruction of the foundation. It was roughly evaluated that three-fifths of the bridge failures are recorded due to scour and other hydraulic-related issues.

In 2012, Arneson [2] quoted a study that was conducted in 1973 for reporting Federal Highway Administration (FHWA) of 383 bridge collapses due to flooding, one-fourth was of pier damage and three-fourths of abutment damage. Also, during the 1961–1976 period, in United States of America, 46 out of 86 major bridge failures were recorded due to scour around piers. This sequence of bridge failures was due to pier and abutment scour, which revealed the need of improving the better ways of securing bridges from the attacks of scour.

K. H. Teja (✉) · G. Srivalli · V. Balakrishna Murthy
V.R. Siddhartha Engineering College, Vijayawada, AP 520 007, India
e-mail: himateja711@gmail.com

M. Palanisamy et al. (eds.), *Theoretical, Computational, and Experimental Solutions to Thermo-Fluid Systems*, Lecture Notes in Mechanical Engineering,
https://doi.org/10.1007/978-981-33-4165-4_35

Local scour may occur in two ways: either as "clear water scour" or as "live-bed scour". The former defined as the removal of bed materials from the scour-hole but not again replaced by the flow of approach. And the latter is the continuous supply of scour hole with the sediment by the flow of approach till the equilibrium stage (the average quantity of sediment supplied into the scour hole is equal to the evicted amount of sediment from the scour hole) is attained. Under these equilibrium conditions, the depth of the local scour recurs periodically about an average value. The interaction between the bridge pier and the erodible soil around it is highly fusible. Perhaps, the fact is very limited success has been attained when trails are attempted to calculate scour numerically or computationally. The scour is mainly estimated by the pressure around the piers. The more the pressure withheld at the pier foundation, the more will be the depth of the scour. However, the physical models are the basic tools that are used for estimating the scour depths.

Tison [3] in 1961 fixed an obstruction in the water stream and concluded that the velocity of water is not only the influencing factor for scour formation but also the vertical component of the diving motion. It was proved that rounding off the edges of the pier leads to the lesser scour formation. Jueyi [4] in 2010 conducted various experiments, and conclusions drawn from these experiments indicated that as the flow velocity increases in all of the runs, that is, either under the case of same bed material and variable grain size of armour-layer or inverse. The equilibrium scour depth of the scour hole goes on increasing for both semi-elliptical and semi-circular abutments. The equilibrium depth of the scour hole is independent of both the armour-layered and bed-layered conditions.

Roy [5] in 2017 focussed on the shape of the pier to study the scour effect. Through various experiments, it has been identified that scouring rate is high at initial stage and at the backwater flow, and also it directly depends on the exposed open face of the pier nose which is in contact with the clear water. It was finally observed that oblong-shaped pier is the best among the three, which hinders the local scour by half of the maximum scour attained while using the rectangular pier. HEC-23 suggested a list of countermeasures for the scour formation around the piers of the bridge. Some of them are riprap, spurs, guide banks, gabions and so on. The main basic principles of any of these countermeasures are either reduction of vortex intensity or prevention of the increment in scour hole by stabilising the channel bed mechanically. Junhong [6] considered streamlining of the piers as a well-capable countermeasure for the formation of bridge scour by the reduction of vortex strength. Various CFD models had been developed for the evaluation of the viability of streamlining as a countermeasure. Five different test cases with various streamlining conditions are analysed and compared in this study. The results conclude that the vertical profiles of the side walls and pier nose have remarkable effect on the formation of vortex structures around the piers. It was also observed that the downward flow in front of the piers, intensity of the bed shear stress and vortex extent around the pier can be decreased with the help of side walls and pier nose which were streamlined. If in case of non-alignment of pier with the incoming flow, the countermeasure method of streamlining method may not be highly effective.

Bouabdellah [7] carried out numerical research on the effect of shape of the pier on scour formation using finite volume method. Through this numerical research a relation was discussed between the bed shear stress and the bridge pier shape which emerges as a scope for further experimentations to control the scour around the piers. Garde [8] described the principle of scour formation around the piers of bridges and summarised the methods for its prophecy. By using scour data in sandy beds, it was proved that the two methods by Melville Sutherland and Kothyari et al. gave almost similar accuracy and hence considered as the superior ones when compared with other methods like Lacey-Inglis, Laursen Toch and so on. Mosabbir in 2013 [9] performed an analysis for getting a list of variables that affect the scour depth around the bridge piers. It was concluded that the flow depth in channel, velocity of flow, shape of pier and approach flow velocity would directly affect the scour depth. It was also found that elliptical pier would amount to less scour depth when compared to the diamond-shaped pier. Kamini 2018 [10] studied experimentally the scour around the bridge pier in non-uniform sediments. The results of the experiments concluded that the scour depth is directly proportional to the velocity and inversely proportional to the sediment size. It was also concluded that the elliptical-shaped pier can be used as the best protective measure instead of others like circular and diamond.

Abdul-Hassan [11] conducted experiments on various shapes of piers to minimise the local scour and successfully concluded that the streamlined shape is considered as the best shape of piers which can reduce the maximum scour depth by three-fifths of the scour obtained when using the traditional rectangular pier. It has also been concluded from the results that the scour depth is directly proportional to the flow intensity. Muto in 2008 [12] discussed the local scour around a submerged pier by neglecting backwater effect. This study gave some empirical formulae relating the water depth, pier height and scour depth. Drysdale in 2008 [13] examined the effectiveness of an aerofoil-shaped pier in the reduction of downstream vortices and turbulence. In this experiment two piers, that is, one is aerofoil pier and the other is circular pier, were modelled and the results of the both were analysed. The intensity of the scour is minimal for the aerofoil pier when compared with the traditional pier, as the front of the pier is in fact a semi-circle in shape and has lesser vortex.

Boujia in 2017 [14] aimed to propose a technique of monitoring on the basis of first natural frequency of a pier-surrounded scour. This experiment gave some conclusions that the frequency will decrease significantly with the increase in the length which is exposed. Another major finding is that two rods with the same exposed length but different embedded lengths will amount to same frequency. Mohamed in 2018 [15] published a review article on the methods that are used to reduce the effect of scour on bridge pier. In this article, many methods have been discussed broadly. It was concluded that CFD multi-phase models like VOF, FVM and Fluent are the trending methods with less economy and high compatibility to calculate the scour around the bridge piers.

As conducting experiments on various flow conditions, pier shapes, bed conditions to study the scour effect is highly expensive, the researchers are adopting the computational methods. The present research work objectifies to develop a minimum

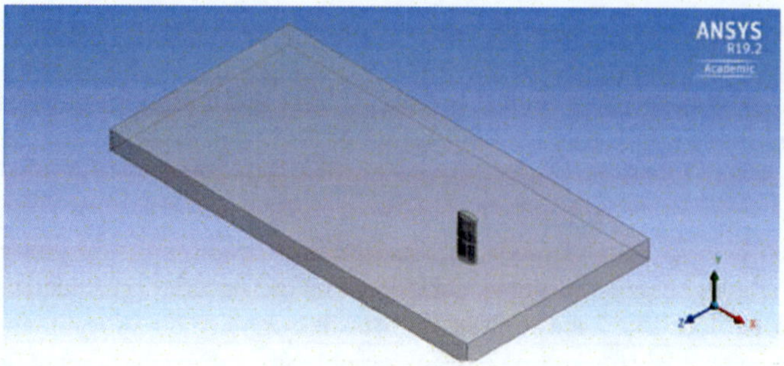

Fig. 1 Geometry of the problem

computational domain required to present the equivalence of pressure and velocity at the pier surface to study the scour parameter effectively.

2 Problem Description

In a trail of optimisation, a cuboid fluid domain of 40 m × 20 m × 2 m is initially taken and an elliptical pier (2 m × 1 m cross-section) is placed at three-fourths distance from the inlet with a height of 4 m, as shown in Fig. 1. A standard mesh with element size of 0.5 m has been generated with inlet, outlet, top wall, bottom wall and water surface as named selections. As the pier location shows symmetry conditions, one half of the geometry has been modelled. The selected domain is studied under laminar and transient conditions considering gravity effect with mass inflow rate of 79,856 kg/s.

Various meshes have been tested with an element size of 0.6, 0.5, 0.4, and 0.3 and 0.4 m has been finalised for the better convergence. Figure 2 shows the meshed geometry with an element size of 0.4 [16] indicating the named selections. The simulation is conducted by giving various time step sizes 1, 0.5, 0.25, 0.2, 0.125, 0.1. The maximum pressure at pier surface for various trails has been plotted in Fig. 3. From Fig. 3, 0.125 s has been considered as the best time step size for convergence.

3 Analysis of Results

3.1 Results for Initial Geometry

At the required time step, the velocity contour of the geometry is as shown in Fig. 4. Figures 5, 6 and 7 represent the d*u*/d*z* at inlet, *y*-component of velocity at water surface and pressure at pier surface, respectively.

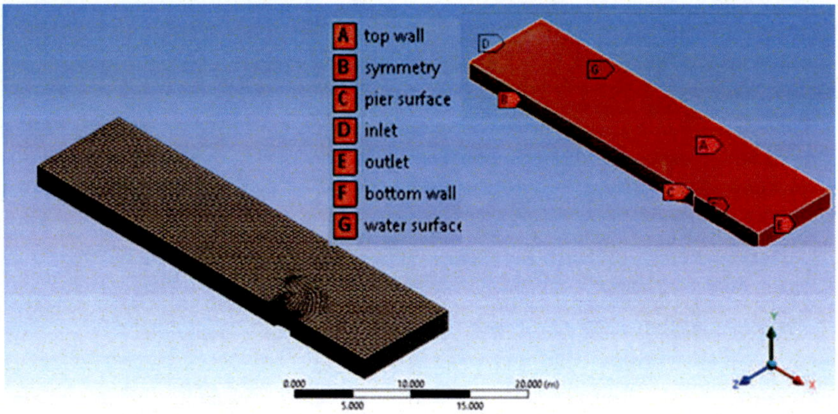

Fig. 2 Meshed geometry and named selections

Fig. 3 Variation of max. pressure at pier surface w.r.t. time step

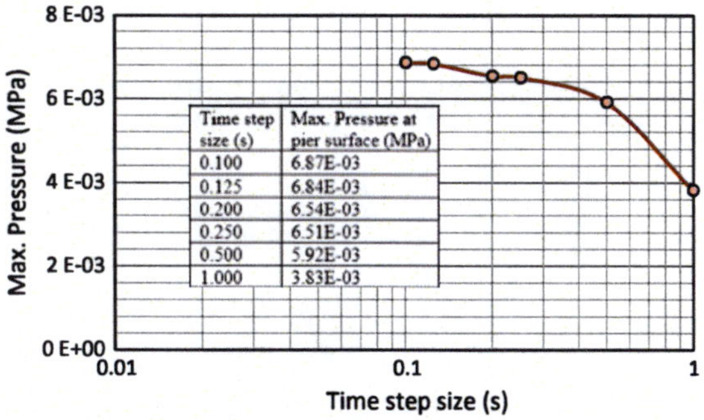

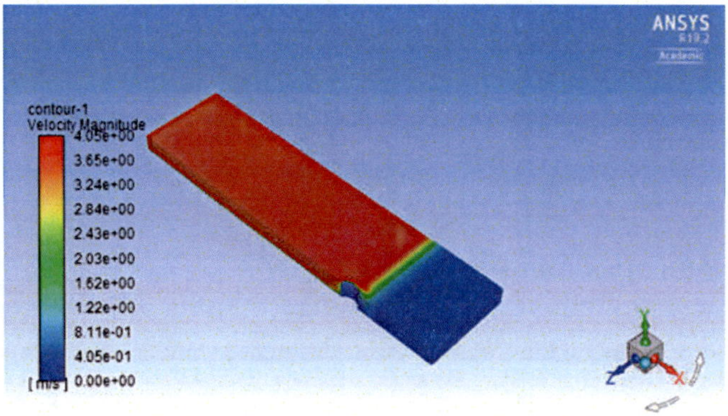

Fig. 4 Velocity contour

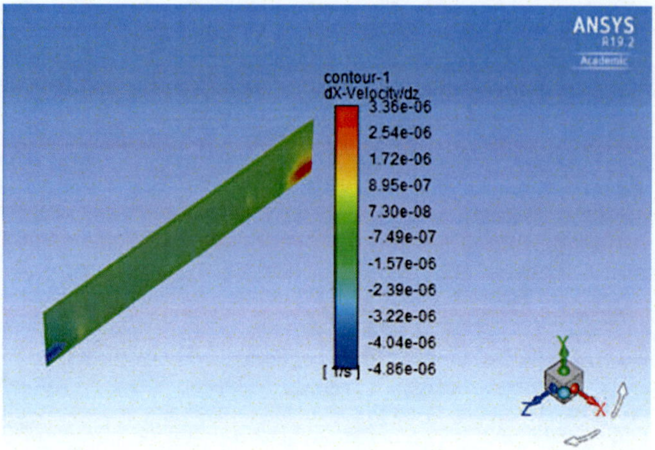

Fig. 5 d*u*/d*z* contour at the inlet

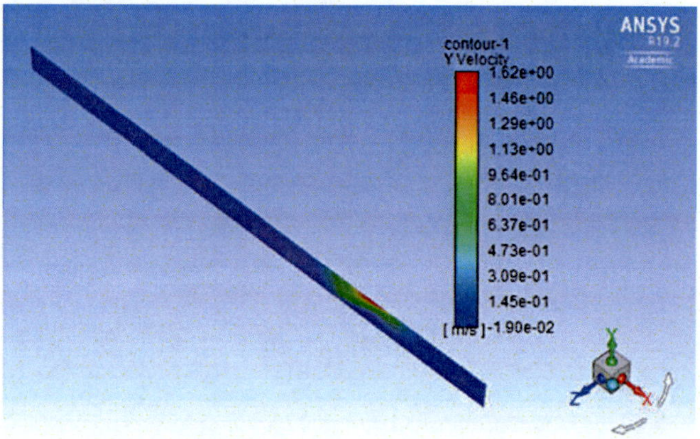

Fig. 6 *Y*-Velocity at the water surface

3.2 *Method of Optimising Geometry*

Planes are created along the z and x axes at different locations as shown in Figs. 8 and 9.

Figure 10 shows the variation of y-component of velocity with respect to width location. A linear drop can be observed up to 4 m of width followed by a constant value, exhibiting the propagation of disturbance up to 4 m width.

Figure 11 shows the variation of d(x-component of velocity)/dz with respect to length location. A nonlinear increase can be observed up to 22 m of length from pier

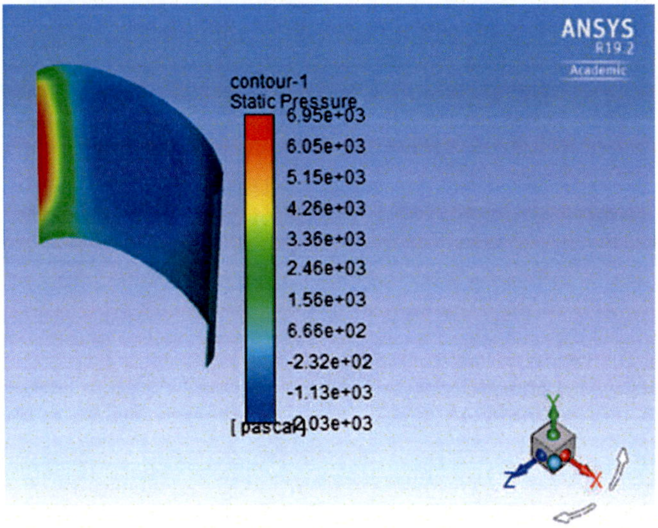

Fig. 7 Pressure at the pier surface

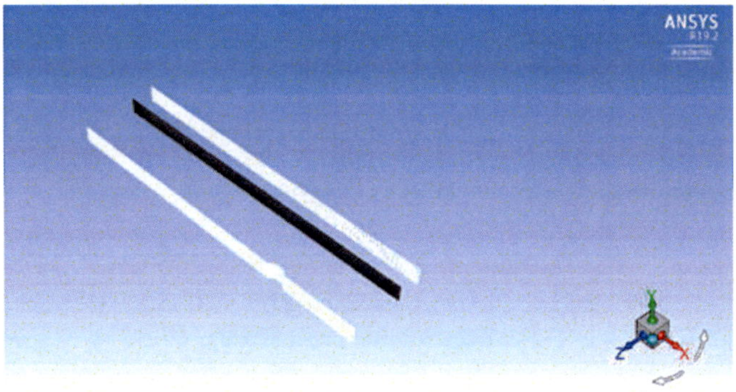

Fig. 8 Planes at different widths

followed by a negligible variation, exhibiting the propagation of disturbance up to 22 m of length location.

From the Figs. 10 and 11, it has been concluded that the geometry can be optimised up to 22 m × 4 m × 2 m.

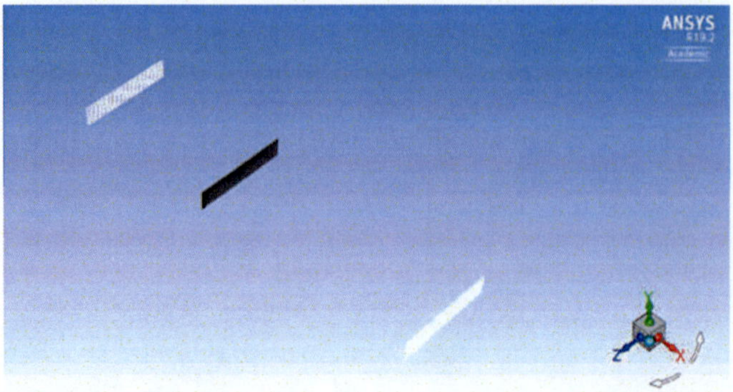

Fig. 9 Planes at different lengths

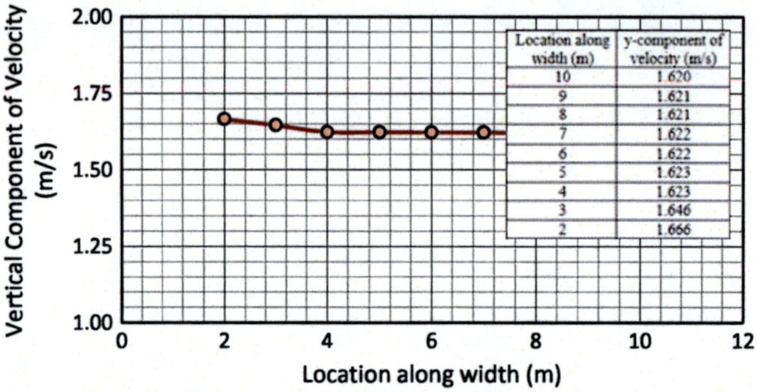

Fig. 10 Variation of *y*-component of velocity w.r.t. width location

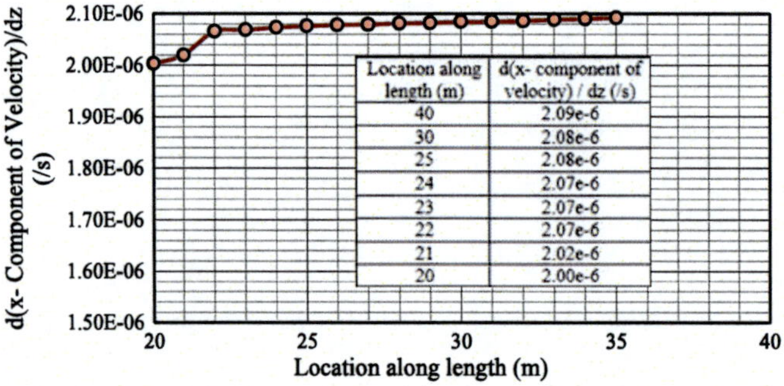

Fig. 11 Variation of d(*x*-component of velocity)/d*z* w.r.t. length location

3.3 Results for Optimised Geometry

Figure 12 shows the optimised geometry and Fig. 13 shows the velocity contour of it. Figures 14, 15 and 16 present the d*u*/d*z* at the inlet, *y*-component of velocity at the water surface and pressure at the pier surface of the optimised geometry, respectively.

The maximum pressure at pier surfaces of initial and optimised geometries is 6949.329 and 6894.878 Pa, respectively, and variation is 0.7% which can be neglected.

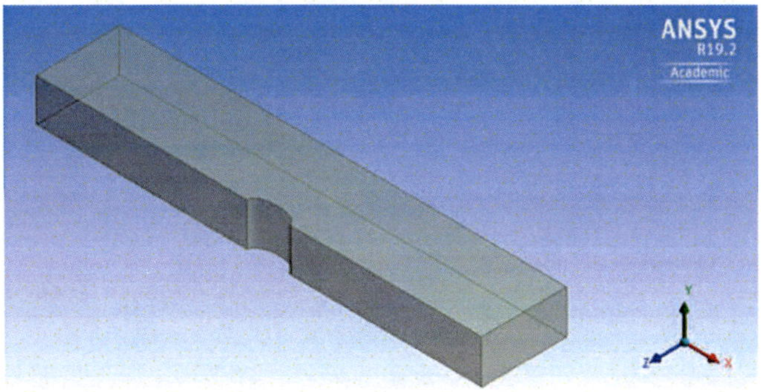

Fig. 12 Optimised geometry (21 m × 4 m × 2 m)

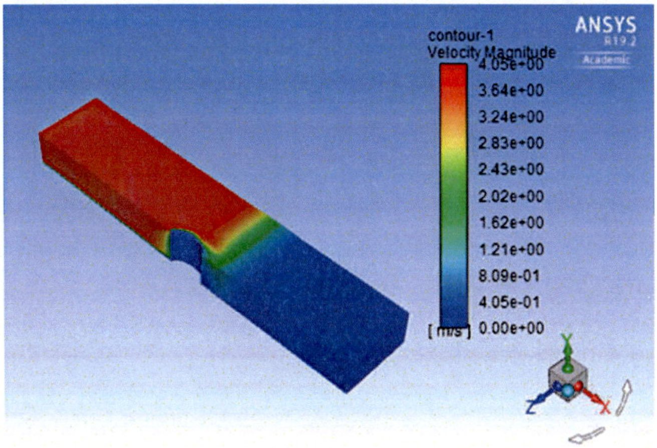

Fig. 13 Velocity contour

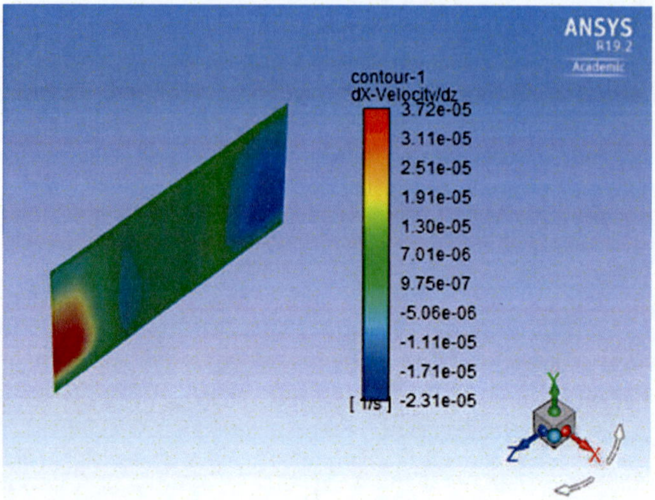

Fig. 14 d*u*/d*z* velocity at the inlet

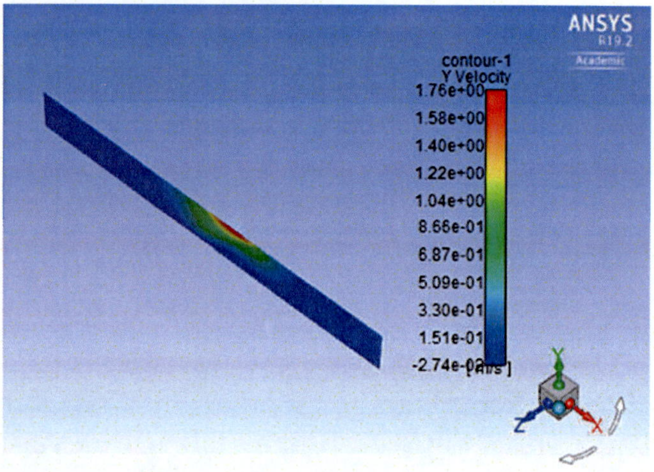

Fig. 15 *y*-velocity at the water surface

4 Conclusion

Effective utilisation of computational resources is very important to obtain a reasonably acceptable solution while applying the numerical methods such as CFD. The problem of fluid–structure interaction at a bridge pier is simulated in ANSYS-FLUENT software to present the ways to minimise the size of the computational

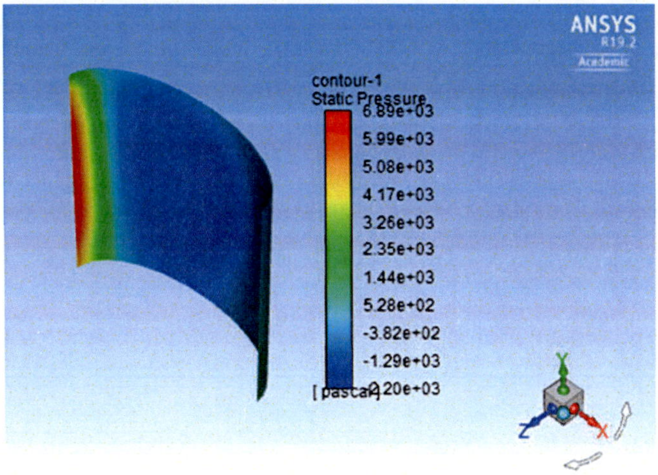

Fig. 16 Pressure at the pier surface

domain for a set of given parameters, such as mass flow rate, depth of water and pier shape. The size of the computational domain is minimised without affecting the interrupted zone near the pier and is verified with the initial model for maximum pressure at the pier surface.

References

1. Abdullah YA (2016) Experimental and theoretical investigations of scour at bridge abutment. J King Saud Univ Eng Sci 28(1):32–40
2. Arneson, LA (2012) Evaluating scour at bridges, 5th edition, report no: FHWA-HIF-12-003 HEC-18, Federal Highway Administration
3. Tison J (1961) Local scour in rivers. J Geophys Res 66(12):4227–4232
4. Jueyi S (2010) Clear-water scour around semi-elliptical abutments with armored beds. Int J Sedim Res 25:245–333
5. Roy C (2017) Effect of bridge pier geometry on local scouring. Int J Earth Sci Eng 2(2):374–377
6. Junhong L (2015) Streamlining of bridge pier as a scour countermeasure: a feasibility study, In: IFCEE, ASCE, pp 319–329
7. Bouabdellah G (2013) Numerical investigations of the bridge pier shape influence on the bed shear stress. Electron J Geotech Eng 5685–5698
8. Garde R (1998) Scour around bridge piers. PINSA 64(4):569–580
9. Mosabbir P (2013) An analysis of scoring effects on various shaped bridge piers. Brunei Darussalam J Technol Commer 7(1):29–42
10. Kamini B (2018) Effect of different shapes of bridge piers to minimize local scour. Int J Civil Eng Technol 9(1):628–638
11. Abdul-Hassan K (2016) Experimental study of bridge pier shape to minimize local scour. Int J Civil Eng Technol (IJCIET) 7(1):162–171
12. Muto Y (2008) Local scour around a submerged cylindrical pier. In: 4th international conference on Scour and Erosin, pp 180–185

13. Drysdale D (2008) The effectiveness of an aerofoil shaped pier in reducing down-stream vortices and turbulence. The University of Southern Queensland, Bachelor Degree Research project
14. Boujia N (2017) Modelling of a bridge pier subjected to scour. 10th international conference on structural dynamics, EURODYN, procedia engineering, vol 199, pp 2925–2930
15. Mohamed Z (2019) A review on the methods used to reduce the scouring effect of bridge pier. In: 2nd international conference on energy and power, ICEP 2018, energy procedia, vol 160, pp 45–50
16. ANSYS 19.2 Reference Manuals (2019)

Numerical Investigation on Performance Enhancement of Rectangular Microchannel by Using Quadrilateral Fins Embedded on the Channel Bed

Arindam Santra, Anirban Bose, and Arunabha Chanda

Nomenclature

A_{base}	Base area of heat sink (m^2)
A_c	Conjugated area (area between solid–fluid interfaces) (m^2)
$C_{p,w}$	Specific heat of water $(Jkg^{-1}K^{-1})$
D	Hydraulic diameter (m)
f	Friction factor
h	Average heat transfer coefficient $(Wm^{-2}K^{-1})$
k_w	Thermal conductivity of water $(Wm^{-1}K^{-1})$
k_s	Thermal conductivity of solid $(Wm^{-1}K^{-1})$
Nu	Nusselt number
p	Static pressure (Pa)
Re	Reynolds number
T	Temperature (K)
T_{in}	Inlet temperature (K)
T_m	Average mean temperature of fluid (K)
$T_{m,in}$	Bulk mean temperature at the inlet (K)
$T_{m,out}$	Bulk mean temperature at the outlet (K)
T_s	Temperature of solid (K)
T_w	Average wall temperature (K)
q''	Average heat flux $(W\,m^{-2})$

A. Santra (✉) · A. Chanda
Jadavpur University, Kolkata, India
e-mail: arindamsantra7@gmail.com

A. Bose
Meghnad Saha Institute of Technology, Kolkata, India

M. Palanisamy et al. (eds.), *Theoretical, Computational, and Experimental Solutions to Thermo-Fluid Systems*, Lecture Notes in Mechanical Engineering,
https://doi.org/10.1007/978-981-33-4165-4_36

u_{in} Inlet velocity (m/s)
μ_w Viscosity of water $\left(kg\,m^{-1}s^{-1}\right)$
ρ_w Density of water $\left(kg\,m^{-3}\right)$
$\vec{V}$ Velocity vector

1 Introduction

The microchannel heat sink has become crucial in modern times due to its high rate of cooling and high area to volume ratio. The electronic chips produce enormous amount of heat within very small area. Therefore, a rapid cooling within this small space is required to avoid the damage of electronic chip due to high temperature rise. The microchannel heat sink is used in place of earlier conventional heat sink for the purpose of cooling of integrated circuit. The concept of microchannel heat sink is established by Tuckerman et al. [1]. Wavy microchannel is experimented to enhance the heat transfer in terms of Nusselt number by Naphon et al. [2]. Heat transfer is enhanced because of flow disruption due to waviness. The effect of obstacles on heat transfer enhancement is studied by Meis et al. [3]. The obstacles are used as vortex promoters which enhance heat transfer inside microchannels. The effect of surface roughness on pressure drop and heat transfer in rough microchannel was numerically investigated by Zhang et al. [4]. Roughness elements persuade stronger recirculation and flow separation, due to which heat transfer is enhanced. The effect of waviness of microchannel on heat transfer and friction factor is experimentally and numerically investigated by Sui et al. [5]. The result shows that the wavy microchannel gives better heat transfer performance than the straight microchannel. The effect of geometrical configuration on heat transfer performance and fluid flow in a converging–diverging microchannel is numerically investigated by Ghaedamini et al. [6]. Heat transfer is increased because of increase in wall curvature and chaotic advection. The heat transfer performance in microchannel heat sink by introducing slanted passage in the channel wall between the adjacent channels in alternating orientation is numerically studied by Kappusamy et al. [7]. Due to the disruption in the hydrodynamic boundary layer, the average thermal boundary layer thickness is decreased. Therefore, the heat transfer performance is enhanced. The heat transfer performance of periodic converging–diverging microchannel of circular cross-section is numerically investigated by Chandra et al. [8]. Heat transfer is increased due to increase in fluid mixing and velocity jump in converging section. The characteristics of flow and heat transfer are investigated in microchannel with dimples by Xu et al. [9]. The transverse convection caused by dimple is very important to enhance the convection heat transfer under laminar flow. The effect of changing wavelength and amplitude in wavy microchannel is investigated by Lin et al. [10]. Heat transfer is increased due to formation of vortices because of curved wall.

2 Problem Description

In the present investigation, three types of rectangular microchannel are investigated: (i) Without fin, (ii) with inclined and straight fins embedded on channel bed and (iii) with vertical and staggered fins embedded on channel bed. The numerical simulation is done for these three models from Reynolds number 300–800. To investigate the performance of heat sink, two parameters, Nusselt number and friction factor, are analyzed for different microchannel. From Nusselt number, the heat transfer coefficient and from friction factor the pumping power can be estimated, which are the two main criteria for the performance of a heat sink.

2.1 Geometrical Description of Model

A single straight rectangular microchannel is studied for this investigation. The basic geometry of the channel for this investigation is taken from geometry of the model from Sui et al. [5]. The length of the channel is 25 mm, width is 206 μm, depth is 408 μm, and the thickness of bottom solid wall and width of solid wall on both sides is taken as 97 μm for this investigation. The dimensions of fin which is also used in this study are taken as width 200 μm, thickness 30 μm and height 200 μm. For the microchannel with fins, totally 24 number of fins are used, and the fins are placed in a manner of equal spaced. In case of inclined and straight fins, the inclination angle of the fin is taken as 45°. For this study, water is taken as coolant fluid and the solid substrate is made of copper. Different geometric models are shown in Figs. 1, 2 and 3.

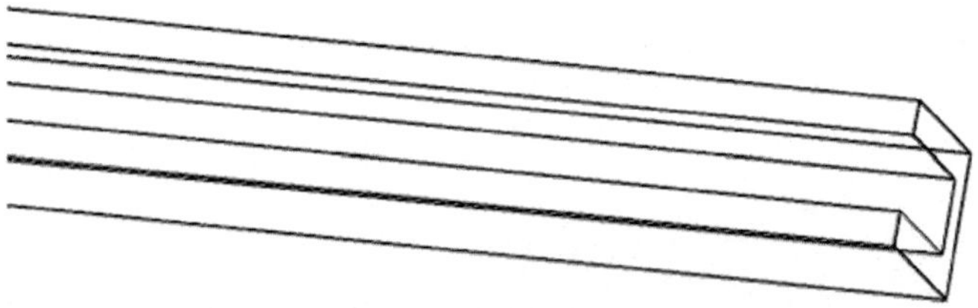

Fig. 1 Microchannel without fin

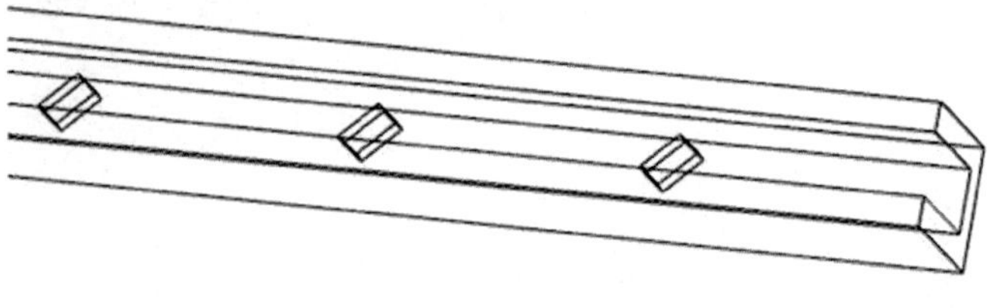

Fig. 2 Microchannel with inclined and straight fin

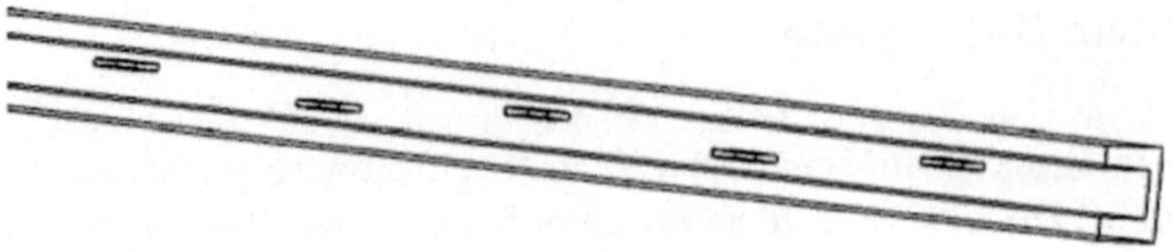

Fig. 3 Microchannel with vertical and staggered fin

3 Numerical Methods

For the present numerical study, Ansys Fluent 14.5 is used. For pressure discretization, standard scheme, pressure velocity coupling SIMPLE scheme is used. Second-order upwind scheme is used for solving momentum and energy equation. Conjugated heat transfer scheme is employed for the analysis of solid–fluid interfaces. The total computational domain is meshed in Ansys workbench. The mesh geometry of solid substrate of total computational domain for microchannel with vertical and staggered fins is shown in Fig. 4.

3.1 *Governing Equation*

A three-dimensional numerical simulation is done to analyze the performance of microchannel heat sink by employing the conjugated heat transfer model. The following assumptions are implemented: (1) laminar, (2) incompressible, (3) steady flow, (4) constant properties for solid and fluid, (5) no gravitational effect and (6) no heat loss and thermal contact resistance. As per these assumptions, the following governing equations can be written as:

Continuity equation for water:

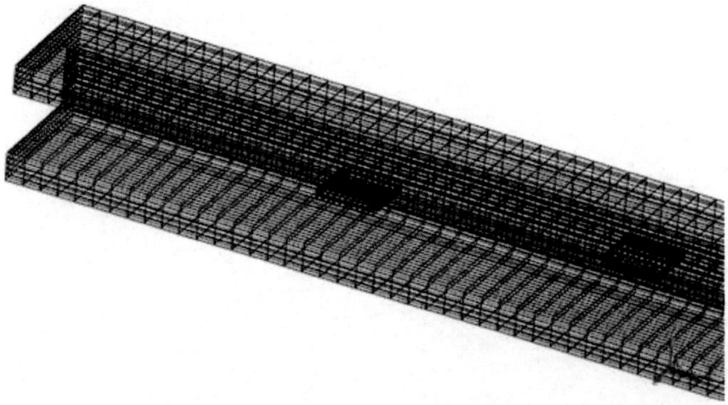

Fig. 4 Mesh geometry of microchannel with vertical and staggered fin

$$\nabla \cdot \vec{V} = 0 \tag{1}$$

Momentum equation for water:

$$\rho_{\mathrm{w}}\left(\vec{V} \cdot \nabla\right)\vec{V} = -\nabla p + \mu_{\mathrm{w}}\nabla^2\vec{V} \tag{2}$$

Energy equation for water:

$$\rho_{\mathrm{w}} C_{\mathrm{p,w}} \vec{V} \cdot \nabla T = k_{\mathrm{w}} \nabla^2 T \tag{3}$$

Energy equation for solid:

$$k_{\mathrm{s}} \nabla^2 T = 0 \tag{4}$$

3.2 *Boundary Conditions*

For this investigation, the inlet velocity is taken as constant and the inlet temperature is taken as 300 K. At the outlet, the outflow boundary condition is considered. At the bottom wall of microchannel heat sink, a constant average heat flux of 50 W/cm^2 is applied. At the solid–fluid interface, the temperature and heat flux are considered as continuous. The two outer sidewall is considered as periodic boundary and all other walls are considered as adiabatic.

At the inlet:

$$u = u_{\mathrm{in}}, v = 0, \ T = T_{\mathrm{in}}$$

At the outlet: outflow boundary condition

At the solid–fluid interface:

$$u = 0, v = 0, w = 0, T_{\mathrm{s}} = T \text{ and } \ k_s \nabla T_{\mathrm{s}} = k_{\mathrm{w}} \nabla T$$

Heat added to water from heat sink:

$$q = \rho_{\mathrm{w}} C_{\mathrm{p,w}} \, Q\left(T_{\mathrm{m,in}} - T_{\mathrm{m,\,out}}\right) \tag{5}$$

Average heat flux:

$$q'' = \frac{q}{A_{\mathrm{base}}} \tag{6}$$

Mean fluid temperature: average of the water inlet and outlet temperature

Average heat transfer coefficient:

$$h = \frac{q}{A_c(T_w - T_m)} \tag{7}$$

Average Nusselt number:

$$\mathrm{Nu} = \frac{hD}{k_w} \tag{8}$$

Reynolds number:

$$\mathrm{Re} = \frac{\rho_w u_{\mathrm{in}\,D}}{\mu_w} \tag{9}$$

Friction factor:

$$f = -\frac{(\mathrm{d}p/\mathrm{d}x)D}{0.5\ \rho_w u_{\mathrm{in}}^2} \tag{10}$$

4 Model Validation

The experimental data of straight rectangular microchannel (zero amplitude) heat sink experimented by Sui et al. [5] are used to validate the present model. For this purpose, the geometry and boundary conditions are taken as the same with the experiment. The heat sink in the case of experiment has 60 channels, but for validation purpose one channel is considered to minimize the computational duration. Therefore, the thickness of the two sidewalls taken as half of the actual thickness and the sidewalls is considered as periodic boundary condition for numerical simulation. The result of Nusselt number and friction factor for various Reynolds number from experimented data are used to validate the present numerical model. A comparison between the experimental and simulation results is shown in Figs. 5 and 6.

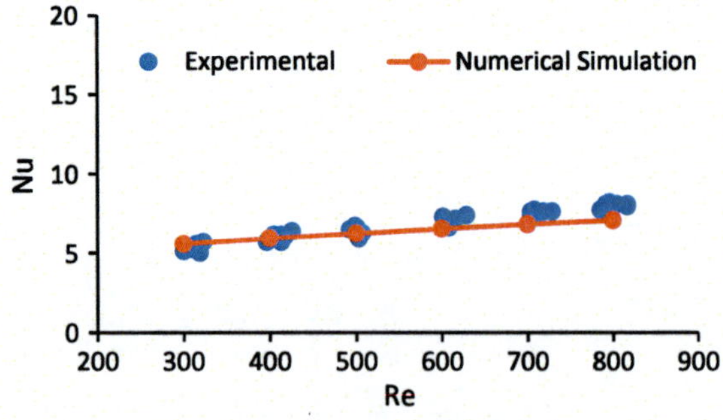

Fig. 5 Nusselt number versus Reynolds number

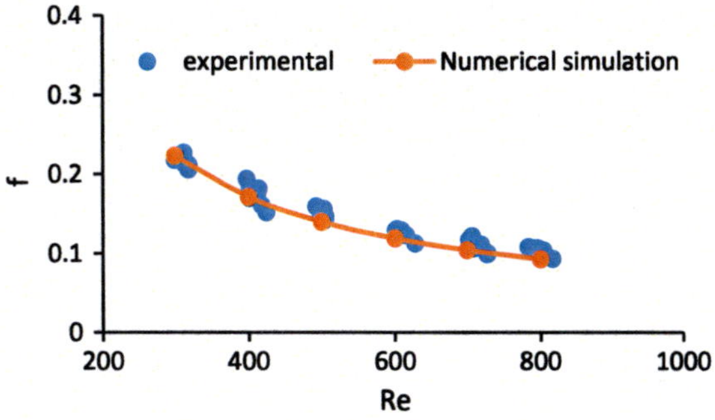

Fig. 6 Friction factor versus Reynolds number

5 Results and Discussion

It is clear from Fig. 7 that fins are increasing the heat transfer performance in terms of Nu. This is quite expected as fins acting as deflector or source of disturbance in the flow field are mixing the layers of fluid rapidly that ultimately increases the convective heat transfer coefficient. Fins are also increasing the heat transfer area and hence improve the heat transfer process. It is also observed in the figures that with the increase of Reynolds number, the advection rate increases and ultimately increases the Nusselt number. The addition of vertical and staggered fins heat transfer enhancement is significant in comparison to inclined and straight fins. The reason for this is flow disturbance in case of staggered fins is frequent and from both ends in the lateral direction, and ultimately mixing the flow at a greater rate in comparison to inclined and straight fins.

Figure 8 shows the penalty of improving heat transfer performance when fins are arranged in terms of friction factor and hence the pumping power. It has been

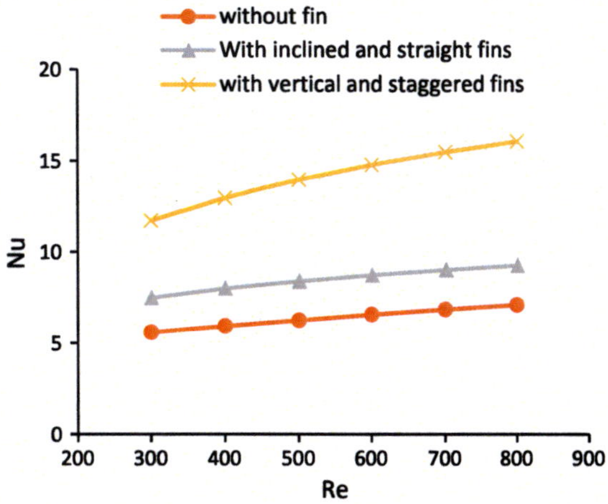

Fig. 7 Nusselt number versus Reynolds number

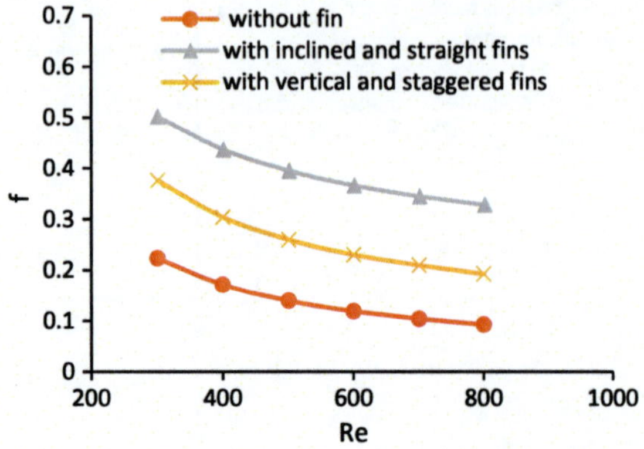

Fig. 8 Friction factor versus Reynolds number

observed with staggered arrangement of the fins that friction factor increment is less in comparison to straight arrangement. This is because of less form drag associated with the staggered arrangement of the fins in comparison to straight arrangement where flow separation area is larger along the width. Alignment of the large fin width is perpendicular to the flow direction and obstructing the flow more implies pressure drag is more; whereas in case of staggered arrangement, the fin width is small perpendicular to the flow direction, hence friction factor (considering both form drag and shear force) is smaller in comparison to inclined and straight arrangement.

It is obvious that with the implementation of fin the temperature distribution along the length of the channel at bottom wall (wall that is required to be cooled) gives better result, which is shown in Fig. 9 in comparison to the microchannel without fin. This is because of the high convective heat transfer, and effective fins are decreasing the thermal resistance to escape out the heat from the heat source from the bottom wall.

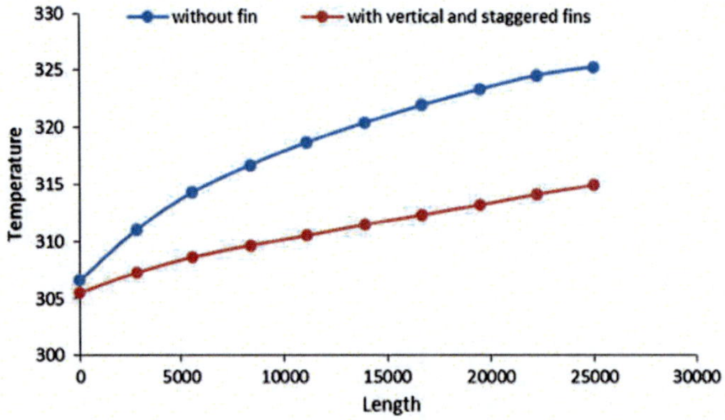

Fig. 9 Temperature distribution along length at bottom wall

6 Conclusion

From the present investigation we can conclude that the heat transfer can be enhanced by the implementation of fins embedded on channel bed. But simultaneously, friction factor is also increased, which in turn increases the pumping power. From the present study, it can be concluded that Nu is improved around 115% with moderate rise in friction factor in case of microchannel with vertical and staggered embedded fins with respect to the microchannel without fins. It can also be concluded that the microchannel with vertical and staggered fins gives the better heat transfer in terms of Nusselt number increasing around 65% with the reduction in friction factor around 30% than the microchannel with inclined and straight fins.

Acknowledgments The authors are thankful to the Department of Mechanical Engineering, Jadavpur University to carry out this research work.

References

1. Tuckerman DB (1981) Pease. High-performance heat-sinking for VLSI. IEEE Electron Dev Lett IEEE 2(1981):126–9
2. Naphon P (2007) Laminar convective heat transfer and pressure drop in the corrugated channels. Int Commun Heat Mass Transfer 34:62–71
3. Meis M, Varas F, Velázquez A, Vega JM (2010) Heat transfer enhancement in micro-channels caused by vortex promoters. Int J Heat Mass Transf 53:29–40
4. Zhang C, Chen Y, Shi M (2010) Effects of roughness elements on laminar flow and heat transfer in microchannels. Chem Eng Process 49:1188–1192
5. Sui Y, Lee PS, Teo CJ (2011) An experimental study of flow friction and heat transfer in wavy microchannels with rectangular cross section. Int J Therm Sci 50:2473–2482
6. Ghaedamini H, Lee PS, Teo CJ (2013) Developing forced convection in converging–diverging microchannels. Int J Heat Mass Transfer 65:491–499
7. Kuppusamy NR, Saidur R, Ghazali NNN, Mohammed HA (2014) Numerical study of thermal enhancement in micro channel heat sink with secondary flow. Int J Heat Mass Transf 78:216–223
8. Chandra AK, Kishor K, Mishra PK, Alam S (2015) Numerical simulation of heat transfer enhancement in periodic converging-diverging microchannel. Proc Eng 127:95–101
9. Xu M, Lu H, Gong L, Chai JC, Duan X (2016) Parametric numerical study of the flow and heat transfer in microchannel with dimples. Int Commun Heat Mass Transfer 76:348–357
10. Lin L, Zhao J, Lu G, Wang XD, Yan WM (2017) Heat transfer enhancement in microchannel heat sink by wavy channel with changing wavelength/amplitude. Int J Therm Sci 118:423–434

Investigation on Centrifugal Pump Performance for Various Nose-Cap Geometries

Neeta Amol Mandhare and K. Karunamurthy

1 Introduction

In today's world, one of the problems that are typically observed in centrifugal pumps is lower pressure head at the output for the same amount of mechanical power supplied at the input, especially at off-design conditions. There is a need to increase the net pressure generated at the output of the pump for the given amount of energy supplied. It is observed from the literature survey that the design modifications are carried out over impeller and casing and not in the suction region. Flow instabilities in the suction region (generally observed at part load or off-design conditions) include improper guidance of flow in the suction pipe and over the eye of impeller, unstable discharge(Q)–head(H) characteristics, flow recirculation and separation, stalling of blades and reverse flow from the impeller to suction region due to pressure fluctuations. Researchers have worked in many areas to overcome the flow randomness in centrifugal pump through different design modifications like impeller trimming [1–3], impeller rounding [4, 5], twisted blades [6], micro grooved impeller [7], drilled impeller [8], impeller with C, U and V cut [9], optimization of blade geometry [10–12] and design modifications over diffuser and casing like use of half-vanned diffuser [13], different volute design with same clearance between impeller and casing [14], rounding-off cut water edge [15], or radical changes which include designing new component like inlet guide vanes, impeller with splitter blades [16, 17] and volute with splitter blades [18, 19].

N. A. Mandhare (✉)
Pimpri Chinchwad College of Engineering, Pune, India
e-mail: na.624@rediffmail.com

K. Karunamurthy
Vellore Institute of Technology, Chennai, India
e-mail: karunamurthy.k@vit.ac.in

M. Palanisamy et al. (eds.), *Theoretical, Computational, and Experimental Solutions to Thermo-Fluid Systems*, Lecture Notes in Mechanical Engineering,
https://doi.org/10.1007/978-981-33-4165-4_37

From the extensive literature review carried out to explore the design modification efforts taken by researchers over the years, it is observed that the research related to design modifications over the suction region of the centrifugal pump is limited. Junaidi et al. [20] designed inlet guide vanes with different blade angles to guide the flow over the suction region at off-design conditions [20]. But this modification is not economical, and changing the angles of inlet guide vanes with a change in load conditions is very difficult. The main focus of research work carried out by authors is to reduce the flow instabilities over suction region through modification of hexagonal lock nut in the form of nose-cap of optimized profile and to increase the velocity of water before impeller, so that the net pressure head developed at the outlet increases for the same amount of mechanical power supplied at the input.

2 Matlab Programming

A standard program for quadratic exponential, power and linear curve generation is used. Output designed in CATIA is used to find equation of curve, as shown in Fig. 1. Model of nose-caps of different profiles generated is shown in Fig. 2.

$$\text{Equation of curve} = -1233.616037 \times x^2 + 113.109813 \times x + 4.108069 \quad (1)$$

Equation from exponential curves:

$$\text{Equation of curve} = 3.864371 \times \mathrm{e}^{11.767531} \times x \quad (2)$$

Equation from power curves:

$$\text{Equation of curve} = 12.739800 \times x^{0.228744} \quad (3)$$

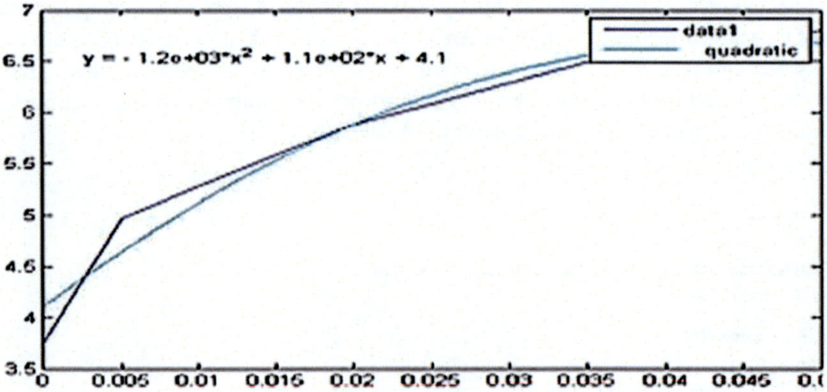

Fig. 1 Quadratic curve

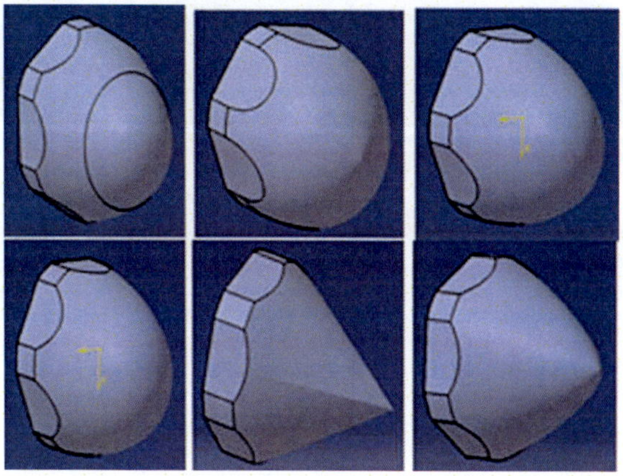

Fig. 2 Models of nozzle cap

Equation from linear curves:

$$\text{Equation of line} = 60.830000 \times x + 3.754977 \quad (4)$$

3 CFD Analysis and Selection of the Nozzle Cap

The main objective of work is to improve the performance of the centrifugal pump by use of a nose cone or a nozzle cap at the eye of the impeller. Before bringing this nozzle cap into actual use and testing it on the actual model, we need to validate the governing physics behind it.

3.1 Generation of Control Volume

After modelling the pump in CATIA, to perform fluid flow analysis it is required to generate the control volume on which the effect is to be observed. This was done using ANSYS ICEM 16, as shown in Fig. 3.

3.2 Meshing of Control Volume

The computational domain consists of a suction pipe, impeller, volute casing and outlet pipe. A moving reference frame technique is used where the impeller is placed in a rotating reference frame and the casing in a fixed reference frame. Tetrahedral

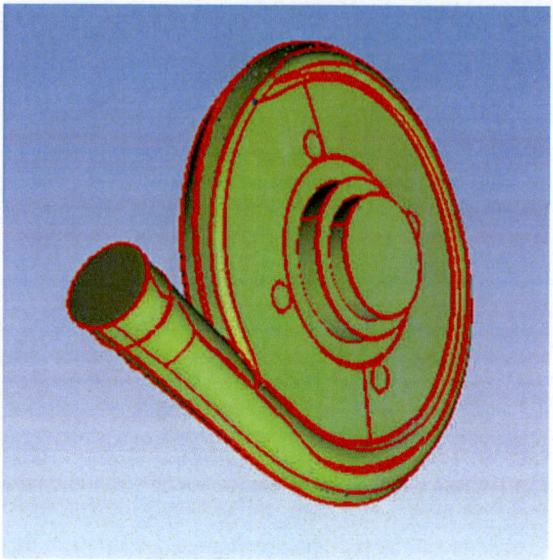

Fig. 3 Control volume with no defects

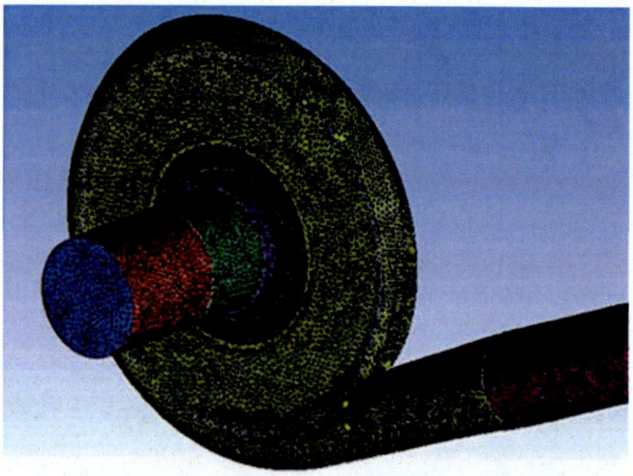

Fig. 4 Meshing of control volume

cells were used in all domains with prismatic boundary layer mesh used near the walls as shown in Fig. 4. The first layer thickness is 0.3 mm with a wall $y+$ less than 40. To ensure a smooth transition, 10 layers were used. For the impeller, the average mesh quality is 0.805, the aspect ratio is 5.524 and skewness is 0.263. For the casing, average mesh quality is 0.664, the aspect ratio is 24.213 and skewness is 0.283.

3.3 Selection of Turbulence Model

k–omega model is able to provide an analytical expression for the dissipation rate in the viscous sublayer. This is ideal for switching automatically from the near wall functions for coarse grids to near wall formulation for fine grids. This makes *k*–omega more flexible and robust. By using the shear stress transport (SST) function, the solver switches between the *k*–omega model in the inner region of the boundary layer and *k*–epsilon in the free shear flow, thus combining the strengths of both models.

3.4 Numerical Solution Method

The pressure and velocity contours as shown in Fig. 5 indicate an area average of

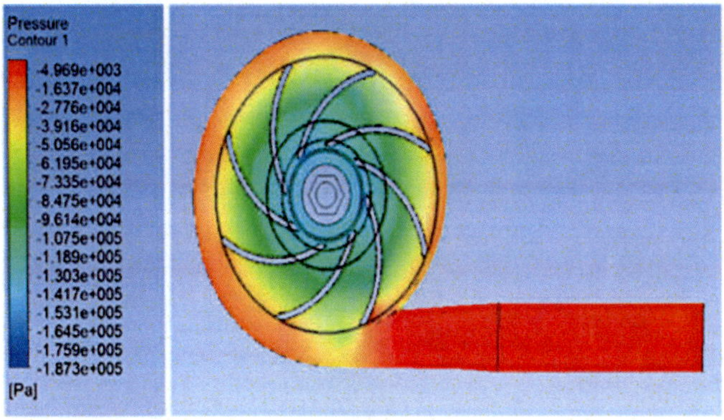

(a)

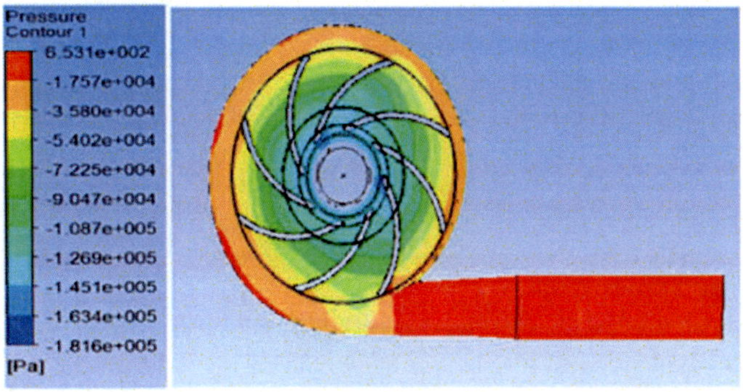

(b)

Fig. 5 **a** Pressure contour for hexagonal lock nut. **b** Pressure contour for nose-cap

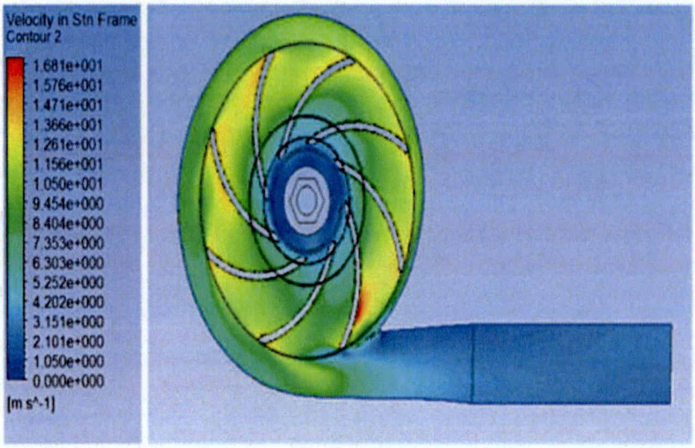

(a)

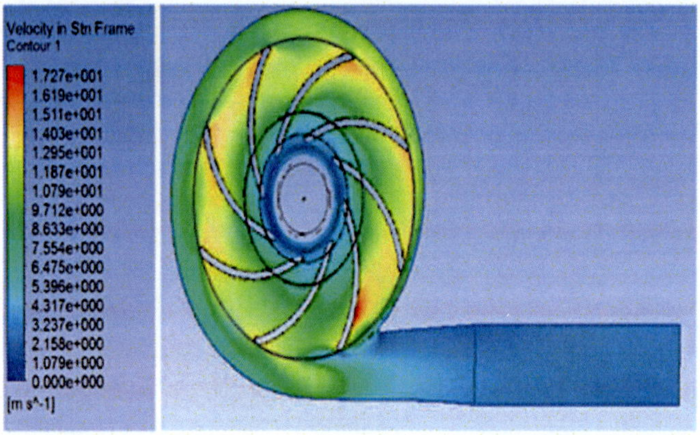

(b)

Fig. 6 **a** Velocity contour for hexagonal lock nut. **b** Velocity contour for nose-cap

the values of pressure and velocity throughout the cross-section of the pump. These contours typically indicate the values of pressure and velocity at different points inside the pump for different mass flow rates and speeds of the impeller.

3.4.1 Remarks from Pressure and Velocity Contours

i. From the pressure contours, as shown in Fig. 5, it is clear that the pressure is not uniformly distributed throughout the volute casing but has a relatively higher value when the water impinges on the wall of the volute while it issues out of the impeller blades.
ii. The velocity contours in Fig. 6 reveal that when the speed of the impeller is

increased the water is made to flow along with the blade profile due to higher amount of the centrifugal force acting on it. Therefore, the velocity of water near the wall of the blade is higher and tends to increase along the periphery of the impeller.

iii. As seen from the velocity contours though there is no any recirculation detected at the outlet pipe in most of the cases but the flow of water tends to stagnate after travelling a certain length of the outlet pipe, leading to a slight recirculation and backflow. But as the speed of the impeller is increased this tendency fades away leading to the flow of water covering the whole area of the outlet pipe.

iv. The wake region formed before the nozzle cap is slightly lesser than that formed in front of the shaft fitted without the nozzle cap.

v. From the pressure contours, the blue patches at the eye of the impeller tend to fade away at higher speeds of the impeller. This is correspondingly converted into the red patches at the wall of the volute casing at higher speeds of the impeller. This clearly indicates that the effect of the volute is best seen at the designed speed of the pump or at relatively higher speeds.

vi. In spite of rotational symmetry of the blades of the impeller, the pressure and velocity distribution is different for different blades as seen from the images.

vii. The pressure drops suddenly at the outlet of the volute casing and then increases to a higher value at the start of the outlet pipe. This indicates that there is a loss of pressure or valuable momentum of water at the end of the volute casing and it needs to be designed properly at the tongue to prevent this loss.

4 Experimental Setup

The experimental setup as shown in Fig. 7a and b consists of a centrifugal pump coupled to a belt-driven electric motor of 2 HP. The pump is of a low head, closed impeller, mixed flow, variable speed available at turbomachines laboratory of Pimpri Chinchwad College of Engineering, Pune, India. The impeller of the pump has an exit diameter of 0.214 m with eight backward curved blades and vaneless volute casing as shown in Fig. 3a. Inlet pipe diameter is of 2″ and exit pipe diameter is of 1.5″. The setup also has a water-collecting tank, gate valve and sophisticated measuring instruments like two pressure transducers with an accuracy of ±0.1%, data acquisition system, variable speed drive with the pulsation of speed around 5 rpm, laser tachometer, rotameter with an uncertainty of ±0.2% and an energy meter. The response time of the current transducer is lower than 0.2 μs.

Experimental results are compared with the simulation results as shown in Fig. 8a and b and the average % difference in head between simulation and experimental result is 4.23% and average difference in overall efficiency is 1.96%, which shows satisfactory comparison between experimental and simulation results.

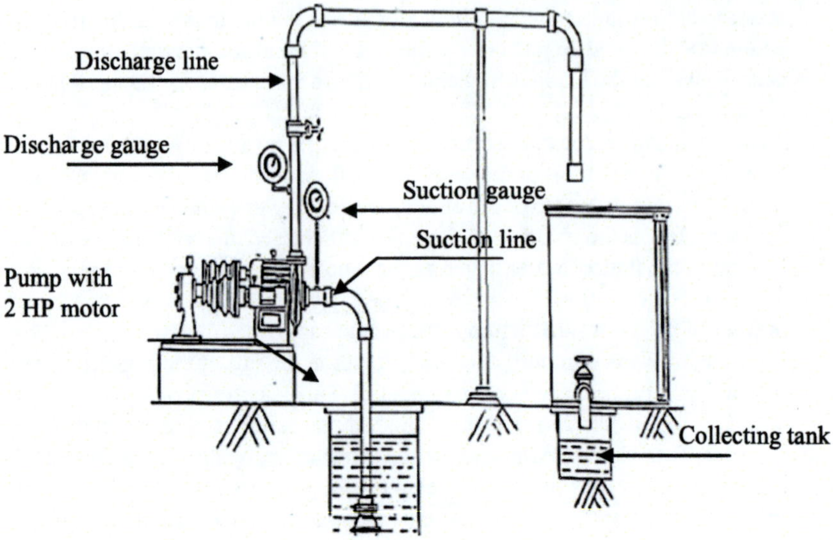

(a)

(b)

Fig. 7 **a** Block diagram of experimental setup. **b** Photograph of experimental setup

Fig. 8 a and b Comparison of experimental and simulation results for hexagonal nut

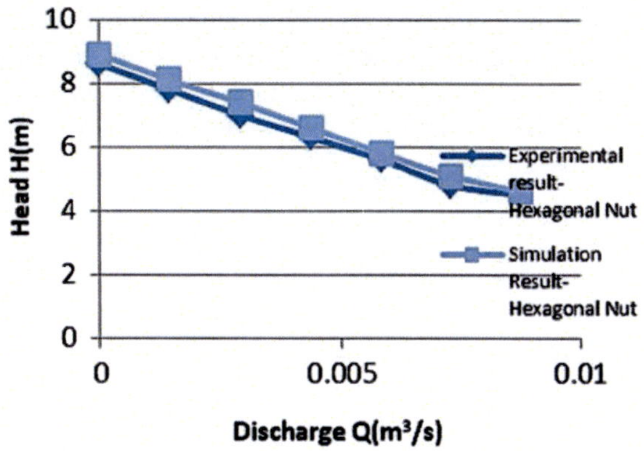

(a) Head (H) Vs Discharge (Q)

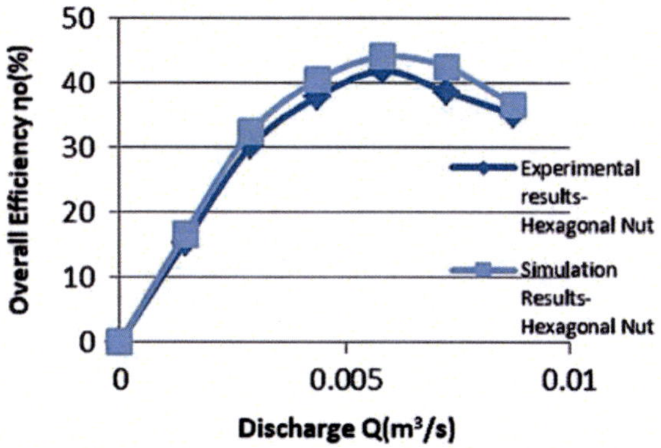

(b) Overall Efficiency (η_o) Vs Discharge (Q)

4.1 Conclusion

- The overall efficiency of the pump is on an average 2–3% higher for all the three nozzle caps as compared to the case when the nozzle cap is not attached as seen from the main characteristics (Graph 2).
- The pump was tested for 30 times without the nozzle cap. In all the cases the plots of main characteristics were more or less similar to the standard pump curves. But in case of the operating characteristics the curves deviated from the standard pump curves.
- It is clear from the curve of operating characteristics that most of the nozzle caps generate a head higher than the case without the nozzle cap.
- From the observations in the main and operating characteristics as well as the CFD results taken at random speeds of the impeller, three nozzle caps were selected for manufacturing and testing purposes.

- The average efficiency indicates how close the actual velocity rise is to the ideal velocity rise. It is highest in case of the linear and exponential profiles. The linear profile was eliminated due to flow separation at the end of the nozzle cap. The efficiency of the quadratic nozzle caps is more or less near the exponential ones.

References

1. World Pumps (1999) When trimming centrifugal pump impeller can save energy and increase flow rate
2. Šavar M, Kozmar H, Sutlović I (2009) Improving centrifugal pump efficiency by impeller trimming. Desalination 249:654–659. https://doi.org/10.1016/j.desal.2008.11.018
3. Yang S-S, Kong F-Y, Jiang W-M, Qu X-Y (2012) Effect of impeller trimming influencing pump as turbine. Comput Fluids 67:72–78. https://doi.org/10.1016/j.compfluid.2012.07.009
4. Singh P, Nestmann F (2010) Internal hydraulic analysis of impeller rounding in centrifugal pumps as turbines. Exp Therm Fluid Sci 35:121–134. https://doi.org/10.1016/j.expthermflusci.2010.08.013
5. Doshi A, Channiwala S, Singh P (2016) Inlet impeller rounding in pumps as turbines: an experimental study to investigate the relative effects of blade and shroud rounding. Exp Therm Fluid Sci 82:333–348. https://doi.org/10.1016/j.expthermflusci.2016.11.024
6. Chen H-X, He J-w, Liu C (2017) Design and experiment of the centrifugal pump impellers with twisted inlet. J Hydrodyn 29:1085–1088. https://doi.org/10.1016/S1001-6058(16)60822-3
7. Janusz Skrzypacz, Marcin Bieganowski (2017) The influence of micro grooves on the parameters of the centrifugal pump impeller. Int J Mech Sci. https://doi.org/10.1016/j.ijmecsci.2017.01.039
8. Skrzypacz J (2014) Investigating the impact of drilled impellers design of rotodynamic pumps on the efficiency of the energy transfer process. Chem Eng Process 87:60–67. https://doi.org/10.1016/j.cep.2014.11.006
9. Al-Qutub AM, Khalifa AE, Al-Sulaiman FA (2011) Exploring the effect of V-shaped cut at blade exit of a double volute centrifugal pump. J Press Vessel Technol 134:021301. https://doi.org/10.1115/1.4004798
10. World Pumps (2016) Design optimization of double volute suction pump
11. Abdelmadjid Chehhat, Mohamed Si-Ameur (2015) Blade exit angle impact on turbulent fluid flow and performance of Centrifugal pump using CFD, IEEE. ISSN: 2380-7393. https://doi.org/10.1109/IRSEC.2015.7455001
12. Farid Ayad Hassan, H. M. Abdalla, A. Abou El-Azm Aly (2017) Centrifugal pump performance enhancement by blade shape modification. In: ASME turbo expo 2017: turbomachinery technical conference and exposition. pp V02BT41A001. https://doi.org/10.1115/GT2017-63023
13. Zhu X, Li G, Jiang W, Lei Fu (2016) Experimental and numerical investigation on application of half vane diffusers for centrifugal pump. Int Commun Heat Mass Transfer 79:114–127. https://doi.org/10.1016/j.icheatmasstransfer.2016.10.015
14. Oliver Litfin, Antonio Delgado (2014) On the effect of volute design on unsteady flow and impeller volute interaction in a centrifugal pump. In: Proceedings of the ASME 2014 4th joint US-European fluids engineering division summer meeting FEDSM2014. https://doi.org/10.1115/FEDSM2014-21533
15. Morgenroth M, Weaver DS (1998) Sound generation by a centrifugal pump at blade passing frequency. J Turbomach 736–743. https://doi.org/10.1115/1.2841784
16. Yuan S, Zhang J, Tang YJ, Fu Y (2009) Research on the design method of the centrifugal pump with splitter blades. In: Proceedings of the ASME 2009 fluids engineering division summer meeting FEDSM2009, pp 107–120. https://doi.org/10.1115/FEDSM2009-78101

17. Ye L, Yuan S, Zhang J, Ye Y (2012) Effects of splitter blades on the unsteady flow of a centrifugal pump. In: Proceedings of the ASME 2012 fluids engineering summer meeting, FEDSM2012, pp 435–441. https://doi.org/10.1115/FEDSM2012-72155
18. Yan P, Chu N, Wu D (2017) Computational fluid dynamics-based pump redesign to improve efficiency and decrease unsteady radial forces. J Fluids Eng 139. https://doi.org/10.1115/1.4034365
19. Yan P, Wu P, Wu D (2015) High efficiency and low pressure fluctuation redesign of a centrifugal pump based on unsteady CFD analysis. In: Proceedings of the ASME-JSME-KSME 2015, joint fluids engineering conference, AJKFluids2015. https://doi.org/10.1115/AJKFluids2015-34110
20. Md. Abdul Raheem Junaidi, N.B.V. Laksmikumari, Mohd Abdul Samad (2015) CFD simulation to enhance the efficiency of centrifugal pump by application of inner guide vanes. In: 4th international conference on materials processing and characterisation (ICMPC 2015), 2 pp 2073–2082. https://doi.org/10.1016/j.matpr.2015.07.200

Determination of Coefficient of Contraction of Orifice with Variation of Geometrical Parameter

Santosh Kumar Panda and Alok Patra

1 Introduction

The mass flow rate of a flow in a pipe is measured by various flow-measuring devices, like orifice meter, nozzle and venturimeter. Orifice meter is a vital device to measure the flow with the principle of measuring the pressure drop in comparison to design, fabrication, reliability and cost-effectiveness. Orifice meters are used in petroleum, energy, nuclear, mining, chemical, process and food industries, power generation units, refrigeration apparatuses, oil wells, and pipelines where single phase and multi-phase with multiple fluid plays an important phenomenon. It is an alternate device for wellhead chokes, mechanical valves (rotary, screw, butterfly slide valves), chock valve, angle valve and so on. In the oil and gas industry, wellhead chokes are used to control the production from wells, avoid pressure fluctuations in the downstream of the choke and prevent formation damage due to excessive drawdown by providing the necessary backpressure on the reservoir. Single orifices or multiorifice and orifice with perforated plates are used to enhance the flow uniformity and mass distribution in downstream of manifolds and distributors. They are also used to enhance the heat-mass transfer in thermal and chemical processes by providing an extended surface to the increase the surface area. The objective of the present numerical study is for a flow analysis of an orifice to know the hydrodynamics analysis such as pressure profile by using single fluid (air and water) to determine pressure drop coefficient of discharge. The study is further extended by varying the geometry parameters of the orifice, like area ratio, thickness ratio and volume fraction.

A correlation is developed for wellhead chokes in a two-phase flow using P–V–T property relation with the variation of the geometrical parameter such as choke sizes, upstream pressure, gas–liquid ratios and oil API gravities of a 210 well [1].

S. K. Panda (✉) · A. Patra
National Institute of Science and Technology, Berhampur, Odisha, India
e-mail: santosh.panda@nist.edu

M. Palanisamy et al. (eds.), *Theoretical, Computational, and Experimental Solutions to Thermo-Fluid Systems*, Lecture Notes in Mechanical Engineering,
https://doi.org/10.1007/978-981-33-4165-4_38

The study concluded that the P–V–T property relation has minimum effect on the pressure output result. A study conducted an orifice experiment with single-phase flows, presented in their handbook and developed a correlation which is a function of Reynolds number (Re) and the geometrical parameters. The present research has been advanced to use orifice and multiorifice in the industry for replacing the wellheads for flow measurement and control. A theoretical study is carried out [2] for a sharp-edge orifice with two phases, gas–liquid flow analysis by considering interfacial shear force between the phases for the prediction of the single coefficient character of the pipe and orifice and ratio of the phase velocities with the effect of pressure ratio and gas–liquid weight ratio of the flow domain. Pressure drop along the length of the pipe and velocity profile at contraction of the orifice in the pipe were predicted experimentally by evaluating pressure loss coefficients, with the variation for wider range of Reynolds number and area ratio, contraction sharpness [3]. Two-phase flow analysis is considered [4] for a sharp-edged circular orifice for four different vapour to liquid density ratios with R-113 fluid with the variation of diameter ratios to determine the pressure drop. The predicted results are compared with the experimental study and five different proposed correlations. Compressible flow behaviour of air through knife-edge orifices, straight-bore orifices, rounded-entry nozzles and elliptical-entry nozzles of convergent nozzles with diameters ranging from 0.9 to 1.9 mm on a subsonic and critical-flow regions analysis experimentally [5] to determine discharge coefficient. A number of correlations are also reported for different types of orifice as a function of diameters, gas property, pressures and upstream temperatures. Two coefficients are defined [6] such as discharge coefficient and orifice-to-inlet ratio to calculate static and total pressure and the relation between these coefficients. Discharge coefficient and flow coefficient for a safety valve determine the effect of diameter ratio, size and valve of the flow. The difference between the coefficients also determines the safety valve at the inlet and at the orifice of the valve in a liquid flow. The valve stability is considered against the inlet pipe loss, diameter ratio effects and gravitational head for the determination of coefficients. A model of sudden expansion and sudden contraction [7] of flow through the thick and thin orifice plates for two-phase flow pressure analysis with the use of the reversible and irreversible losses through contractions and expansions is developed by taking R-113 as working fluid.

A two-phase flow analysis for thin and thick orifice is considered [8] for experimental study to evaluate local pressure drop. Single-fibre probes are used to measure the void fraction [9] distribution along the cross-sectional with a sudden area contraction for air–water as multiphase flow in a 2-D analysis. An experimental study is conducted for two-phase flow analysis [10] for orifice flow contractions and data of void fraction are collected at different locations of the pipe along the flow length with air–water as two-phase fluid combination to determine the pressure drop. An investigation is carried out experimentally [11] on the two-phase flow characteristics of a multiorifice valve to know the two-phase frictional multipliers, frictional pressure drop and void fraction with the variation of throat thickness–diameter ratios, and air and water flow rate. A numerical study is analysed [12] for thin and thick orifice with single and multiphase flow analysis with water as single-phase fluid and air–water for multiphase fluid in a horizontal pipe, to determine the local pressure drop, discharge coefficient and the two-phase local multiplier as a function of inlet velocity and the geometrical parameter of the orifice.

2 Theoretical Background

The orifice is classified as thin and thick orifice based on the thickness. If the length of the orifice is less than the expansion zone then the orifice is called thin, and if the expansion zone is shorter than the length of the orifice, it is called thick as described [2]. But [11] describes the ratio between the throat thickness and diameter (*s/d*) as a reference to define thick orifice is 1, so the *s/d* ratio is greater than 0.5 as shown in Fig. 1b. Figure 1a and b shows the geometrical models of thin and thick orifice.

Flow through a thin orifice contracts with a consideration of no losses of mechanical energy, to a vena-contract of area Ac that forms outside the restriction. In the downstream of the vena-contract, the flow expands in an irreversible process from the pipe wall to flow area *A*. According to Chisholm [2], for thin orifice the pressure drop,

$$\Delta P_{\mathrm{sp}} = \frac{\rho V^2}{2}\left[\left(\frac{1}{\sigma\sigma_{\mathrm{c}}} - 1\right)\right]^2 \tag{1}$$

$$\sigma_{\mathrm{c}} = \frac{1}{\left[0.639(1-\sigma)^{0.5} + 1\right]} \tag{2}$$

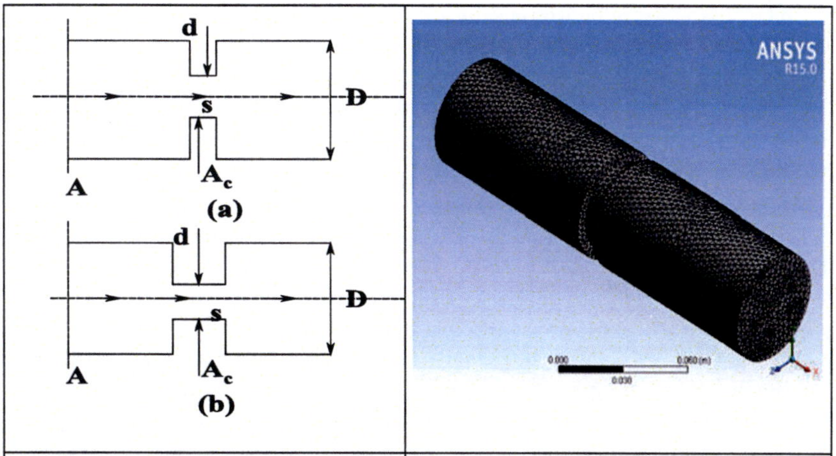

Schematic of orifice geometry (a) thin (b) thick	(c) Mesh topology

Fig. 1 Geometrical model

The local pressure drop can also be expressed as a function of orifice discharge coefficient C_d, [4, 13]

$$\Delta P_{sp} = \frac{\rho V^2}{2}\left[\left(\frac{1}{\sigma}\right)^2 - 1\right]\frac{1}{C_d^2} \tag{3}$$

$$\sigma_{c,thin} = \frac{1}{\sigma + \sqrt{1 - \sigma^2/C_d}} \tag{4}$$

3 Numerical Methodology

3.1 Solution Methodology

A pipe model is considered with a diameter of $D = 50$ mm, orifice diameter d is varying as per the required area ratio and the length of the cylindrical pipe is 200 mm for the numerical study, as shown in Fig. 1a. Three area ratios ($\sigma = (d/D)^2$) of 0.8, 0.6 and 0.4 are considered with a space ratio (s/d) from 0.125 to 0.75. The static pressure and velocity profile are predicted with varying Reynolds number, area and space ratio. A 3-D cylindrical coordinate mesh is created for the orifice by using commercial software. An unstructured mesh with good quality and aspect ratio has been used for flow domain analysis. Finite-volume method is chosen for grid discretization of the flow model domain. The uniform mesh size is created in circumferential direction with the help of O-grid but the size of the mesh is increasing in the radial direction by proper distribution for grid independent study and get accurate numerical result by minimizing the wall effect on the flow. The mesh topology of orifice domain is shown in Fig. 1c. The governing equation is chosen, such as continuity, momentum, and k–ε turbulent model. Grid independent study is carried out for better and accurate result.

3.2 Boundary Conditions

The present study works with single and multiphase problems. Mass inlet boundary condition is applied at the inlet, pressure outlet boundary condition is applied to the outlet and a no-slip and stationary wall is considered as boundary condition on the wall of the orifice pipe. Betterment of the results of k–ε model with enhanced wall treatment is used by considering the viscosity-affected near-wall region.

3.3 Numerical Solution

The governing equations of mass, momentum, and turbulent kinetic energy and dissipation rate of turbulent energy equation (k–ε model) for two-phase analysis volume fraction equation are used for solving the flow problem. The equations are discretized by using the finite volume technique for the control volume of the orifice which is used in commercial software. Good quality and aspect ratio are maintained in the mesh size to get better result. k–ε standard model is considered for the viscous model. Air and water are considered for a single-phase and single-fluid analysis separately. Water liquid and water vapour are used for multiphase analysis. For a good result, boundary conditions are used properly at the inlet, outlet and the wall, as mentioned in the boundary conditions. SIMPLE algorithm is used as pressure–velocity coupling. Least-square cell-based method is used as discretization scheme, and a standard scheme is used for pressure. A second-order upwind scheme is used for momentum, turbulent kinetic energy and turbulent dissipation rate. Under relaxation factors are used for pressure, density, body forces, momentum, turbulent kinetic energy, turbulent dissipation rate and turbulent viscosity as per the solution convergence criteria.

For solving multiphase problem, water–liquid is taken as primary phase and water vapour is considered as a secondary phase. Mixture multiphase model is considered with slip velocity to solve the multiphase problem. k–ε standard is considered as turbulent model for the viscous model. Mass inlet boundary condition with normal to boundary condition is chosen for the inlet condition. Turbulence specification method, intensity and viscosity ratio were considered in the turbulent model. The mass flow rate is mentioned as per the Reynolds number for primary and secondary phases. The outlet and wall boundary conditions are maintained the same as in single-phase flow. The solution methodology is also maintained the same as in single-phase flow. The required data are collected and graphs are plotted for the result analysis after achieving the converge criteria.

4 Results and Analysis

The numerical study is conducted with various parameters such as area ratio, space ratio and Reynolds number, for single and multiphase. The operating ranges of the analysis are area ratio (0.4, 0.6 and 0.8), space ratio (0.125–0.5) and for Reynolds number (5000–50,000). The outputs of numerical are predicted as pressure drop, static pressure, and velocity in single phase, pressure drop and contraction coefficient.

4.1 Single Phase

Static pressure profiles along the flow through orifice with varying Reynolds number are shown in Fig. 2a for air and Fig. 2b for water as a single-phase fluid. Figure 2 shows a sudden pressure drop along the flow around the orifice but as the flow progresses the pressure increases slightly in downstream section of the orifice and the flow continuous along the length. Figure 2 also predicts that with increase in Reynolds number the pressure drop increases because as Reynolds number increases the velocity increases and affects the pressure drop. The pressure drop is more in water as compared to air because of higher density.

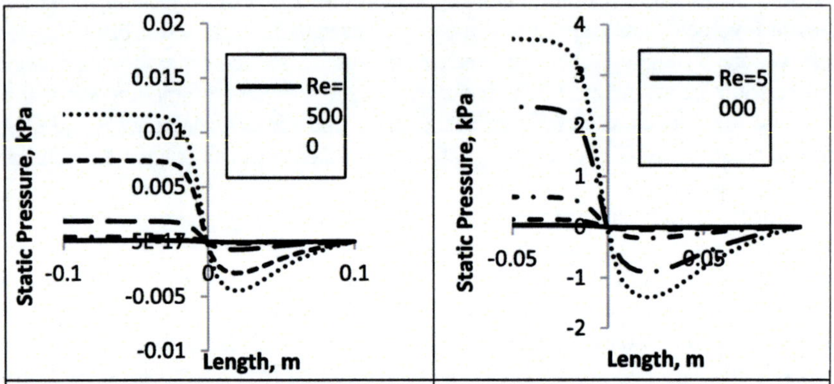

(a) Pressure profile for air (b) Pressure profile for water

Fig. 2 Static pressure profile along the flow through orifice with varying Reynolds number

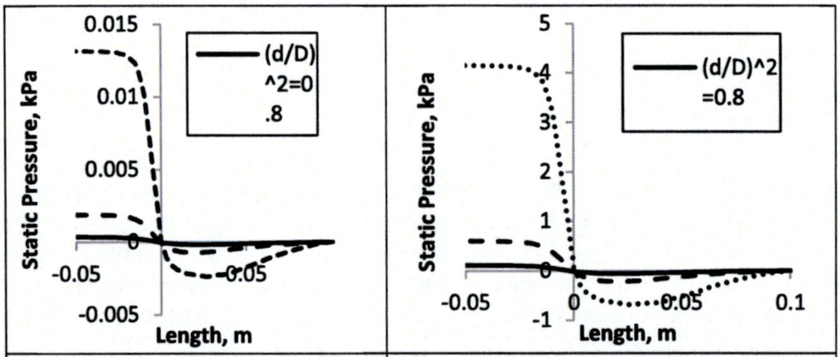

(a) Pressure profile for air (b) Pressure profile for water

Fig. 3 Static pressure profiles along the flow through orifice with varying area ratio

Figure 3 shows the static pressure drop along the flow through orifice with the variation of area ratio for air and water. As the area ratio decreases the more pressure drop occurs because of more blockages in the passages of the flow. Figure 4 shows the static pressure drop along the flow through orifice with the variation of space ratio. Two different static pressure profiles have been shown in upstream and downstream flow of orifice in Fig. 4a for air and Fig. 4b for water. In the upstream region the lesser space ratio shows maximum static pressure but in the downstream the pressure drop is more in lesser space ratio as compared to higher space ratio as represented in Fig. 4.

Velocity profiles along the flow with the variation of Reynolds number through the orifice are shown in Fig. 5a for air and Fig. 5b for water. Velocity of the fluid increases

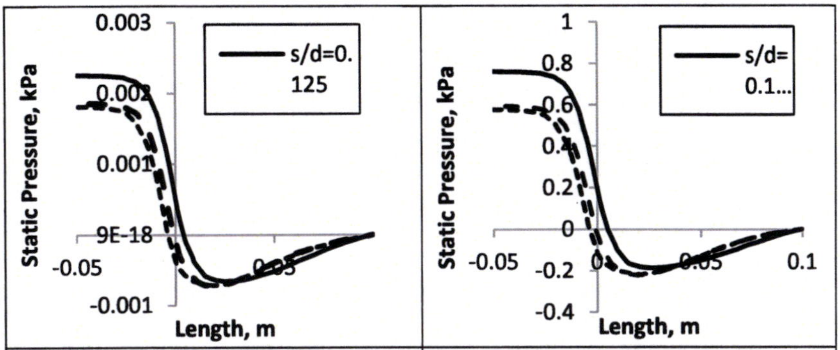

(a) Pressure profile for air | (b) Pressure profile for water

Fig. 4 Static pressure profile along the flow through orifice with varying space ratio

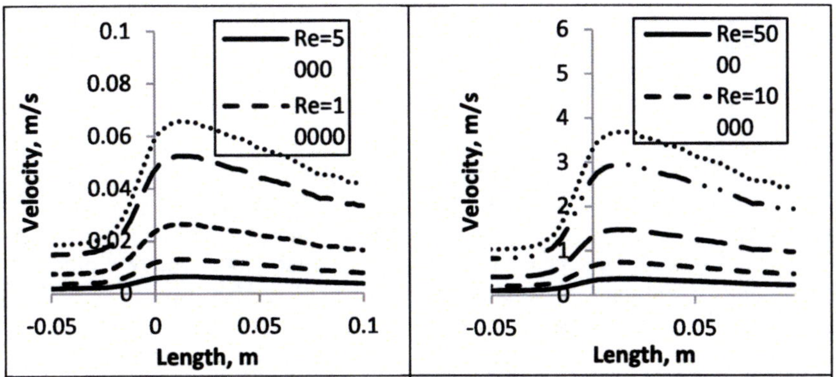

(a) Velocity profile for air | (b) Velocity profile for water

Fig. 5 Velocity profile along the flow through orifice with varying Reynolds number

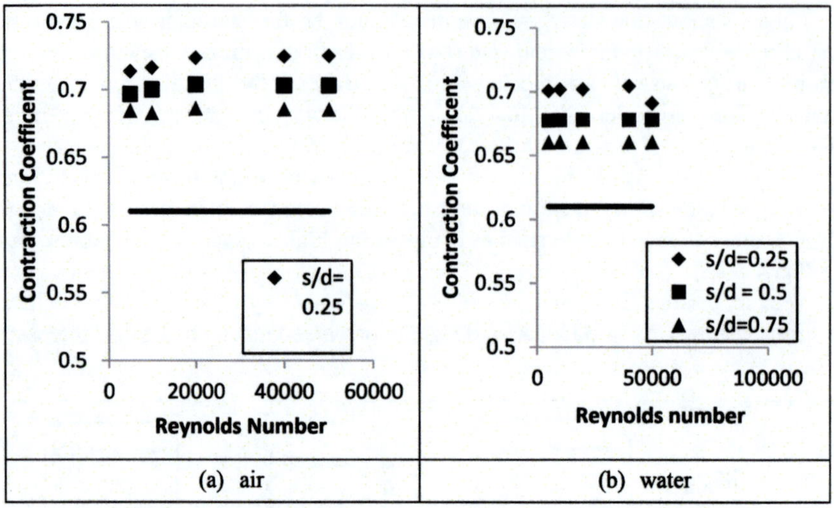

Fig. 6 Contraction coefficient for **a** air **b** water for an area ratio of 0.6

before the orifice because of reduction of the flow area due to the presence of orifice along the length of fluid flow, and the velocity also reduces after the orifice as flow progresses. The velocity is maximum for higher Reynolds number for a particular fluid as represented in Fig. 5. Figure 6a and b shows the contraction coefficient σ_c as a function of Reynolds number with the variation of area ratio and space ratio. The Reynolds number is varied from 5000 to 50,000. The predicted result of contraction coefficient data shows a good agreement with Chisholm formula as shown in Eq. (4). The contraction coefficients are shown for water and air separately as single-phase liquid with variation in space ratio for thin and thick orifice for an area ratio of 0.6. The area ratio for the orifice is high, so the recirculation zone is small which forms less pressure drop.

5 Conclusion

The numerical study is conducted with various parameters such as area ratio, space ratio, and Reynolds number to evaluate static pressure and velocity profile. The operating ranges of the analysis are area ratio (0.4, 0.6 and 0.8), space ratio (0.125–0.5) and for Reynolds number (5000–50,000). The outputs of numerical are predicted as pressure drop, static pressure, and velocity in single phase, pressure drop, slip ratio, two-phase multiplier and contraction coefficient. The static pressure shows maximum profile for the maximum Re, lesser area ratio and lesser space ratio. Similarly, the velocity profile gives better results as the increase in Re predicts better result for higher area ratio and lesser space ratio in single-phase flow. The numerical results are also compared with the available literature and correlation database, which gives better result.

References

1. Gassan H, Abdul-Majeed A, Maha RA (1991) Correlations developed to predict two phase flow through wellhead chokes. J Can Petrol Technol 30:47–55
2. Chisholm D (1967) Flow of incompressible two-phase mixtures through sharped edged orifices. Mech Eng Sci 9:72–78
3. Bullen PR, Cheeseman DJ, Hussain LA, Ruffel AE (1987) The determination of pipe contraction coefficients for incompressible turbulent flow. Int J Heat Fluid Flow 8:111–118
4. Lin ZH (1982) Two phase flow measurement with sharp edge orifices. Int J Multiph Flow 8:683–693
5. Kayser JC, Shambaugh RL (1991) Discharge coefficients for compressible flow through small-diameter orifices and convergent nozzles. Chem Eng Sci 46(7):1697–1711
6. Morris SD (1996) Liquid flow through safety valves: diameter ratio effects on discharge coefficients, sizing and stability. J Loss Prev Process Ind 9(3):217–224
7. Kojasoy G, Kwame MP, Chang CT (1997) Two-phase pressure drop in multiple thick and thin orifices plates. Exp Therm Fluid Sci 15:347–358
8. Fossa M, Guglielmini G (2002) Pressure drop and void fraction profiles during horizontal flow through thin and thick orifices. Exp Therm Fluid Sci 26:513–523
9. Bertola V (2004) The structure of gas-liquid flow in a horizontal pipe with abrupt area contraction. Exp Therm Fluid Sci 28:505–512
10. Fossa M, Guglielmini G, Marchitto A (2006) Two-phase flow structure close to orifice contractions during horizontal intermittent flows. Int Commun Heat Mass Transfer 33:698–708
11. Claudio A, Gioia F, Oladele B (2010) Two-phase flow characteristics in multiple orifice valves. Exp Therm Fluid Sci 34:1324–1333
12. Roul MK, Dash SK (2012) Single-phase and two-phase flow through thin and thick orifices in horizontal pipes. J Fluids Eng ASME 134:091301–091311
13. Grace HP, Lapple CE (1958) Discharge coefficient of small diameter orifices and flow nozzles. Trans ASME 10:639–647

Solution of Incompressible Navier–Stokes Equation Using Upwind Scheme

Banamali Dalai and Manas Kumar Laha

1 Introduction

Lid-driven cavity is a square cavity in which all the walls are stationary and fixed except top wall which is called lid. The lid moves toward the right with non-dimensional speed unit and the non-dimensional length of each side is unity. Initially, the cavity is filled with quiescent fluid. When the lid starts moving toward the right the fluid flow in the cavity sets up. To observe incompressible viscous flow patterns in that cavity, primitive variable formulation of conserved incompressible Navier–Stokes equation is solved using upwind scheme. As the Reynolds number changes, the flow pattern changes very large amount. Few literatures regarding the flow pattern in the cavity when the upwind scheme is incorporated on the convection terms are discussed below.

Shyy, Thakur and Wright [1] studied two-dimensional driven cavity flows with Reynolds numbers ranging from 100 to 3200. They compared the second-order accurate central differencing solution with third-order accurate upwind scheme on the convection term using SIMPLE algorithm in grid sizes 21 × 21 and 161 × 161 in a finite volume mesh. They observed that second-order accurate upwind scheme produces better result than the second-order accurate central difference scheme when compared with Ghia et al. [2]. Shyy and Thakur [3] solved Navier–Stokes equation in a lid-driven cavity using second-order accurate upwind scheme, higher-order accurate upwind scheme and QUICK scheme on the convection terms following the SIMPLE algorithm. They observed that second-order upwind scheme and QUICK

B. Dalai (✉)
Mechanical Engineering, CET BBSR, BPUT Odisha, Rourkela, India
e-mail: banamali.2000@gmail.com

M. K. Laha
Aerospace Engineering, Indian IIT, Kharagpur, West Bengal, India
e-mail: mlaha@aero.iitkgp.ernet.in

M. Palanisamy et al. (eds.), *Theoretical, Computational, and Experimental Solutions to Thermo-Fluid Systems*, Lecture Notes in Mechanical Engineering,
https://doi.org/10.1007/978-981-33-4165-4_39

scheme performs better than the first-order upwind scheme and second-order accurate central difference scheme and also observed that lesser number of grid points are required to converge the solution. Dalai and Laha [4] solved the Navier–Stokes equation in a three-dimensional lid-driven cavity using QUICK scheme on the convection terms following the SIMPLE algorithm in a finite volume mesh. They observed that the solution reaches unsteady state after Reynolds number 2000. Wu and Shao [5] computed the two-dimensional incompressible flow in a lid-driven cavity using multi-relaxation time (MRT) model in the parallel lattice Boltzmann–Bhatnagar–Gross-Krook method and compared the result with the solution obtained using single-relaxation time (SRT) model. They found that both the results were in good agreement with Ghia et al. [2] within the Reynolds number range 100–7500. The MRT model is superior than SRT model at higher Reynolds number. Dalai and Laha [6] studied two-dimensional flow in a lid-driven cavity using alternating direction implicit method and found that the method is good enough to get accurate solution when compared with other literatures [1, 5, 7, 8]. The objective in this study is to find the better result of the primitive variable formulation of Navier–Stokes equation using first-order and third-order accurate upwind schemes on the convection terms in a 2D lid-driven cavity and to find any sign of chaos from unsteady solution.

2 Formulations

The two-dimensional Navier–Stokes equation in primitive variable form is expressed as:

$$\frac{\partial u}{\partial t}+\frac{\partial (u^2)}{\partial x}+\frac{\partial (uv)}{\partial y}=-\frac{\partial p}{\partial x}+\frac{1}{\mathrm{Re}}\left(\frac{\partial^2 u}{\partial x^2}+\frac{\partial^2 u}{\partial y^2}\right) \tag{1a}$$

$$\frac{\partial v}{\partial t}+\frac{\partial (uv)}{\partial x}+\frac{\partial (v^2)}{\partial y}=-\frac{\partial p}{\partial y}+\frac{1}{\mathrm{Re}}\left(\frac{\partial^2 v}{\partial x^2}+\frac{\partial^2 v}{\partial y^2}\right) \tag{1b}$$

and the continuity equation is:

$$\frac{\partial u}{\partial x}+\frac{\partial v}{\partial y}=0 \tag{2}$$

Here Eqs. (1a) and (1b) are the x and y-momentum equations, respectively, where u, v and p are the x-component velocity, y-component velocity and pressure, respectively; Re is the Reynolds number of the flow. The boundary conditions for the lid-driven cavity are:

$$\text{At } x=0,\ u=v=0,\quad \text{at } y=0,\ u=v=0,\quad \text{at } x=1,\ u=v=0 \text{ and } y=1,\ u=1,\ v=0 \tag{3}$$

The equations are discretized using explicit scheme in a finite volume staggered grid mesh. Here first- and third-order accurate upwind scheme is applied on the convection terms. The application of first-order and third-order accurate upwind schemes is shown below:

First-order upwind scheme [8] at the 2D cell P along x-direction is:

$$\int_{x_i}^{x_{i+1}} \frac{\partial(u\varphi)}{\partial x}\mathrm{dx} = \left(\frac{(u\varphi)_e - (u\varphi)_w}{h}\right) \times h \times k; \quad u_e = \frac{u_P + u_E}{2} \text{ and } u_w = \frac{u_P + u_W}{2} \tag{4}$$

$$If\, u_e > 0,\ \varphi_e = \varphi_P \text{ and if } u_w > 0, \quad \varphi_w = \varphi_W \tag{4a}$$

$$If\, u_e < 0,\ \varphi_e = \varphi_E \text{ and if } u_w < 0, \quad \varphi_w = \varphi_P \tag{4b}$$

Third-order upwind scheme [8] at the 2D cell P along x-direction is:

$$\int_{x_i}^{x_{i+1}} \frac{\partial(u\varphi)}{\partial x}\mathrm{dx} = \left(\frac{(u\varphi)_e - (u\varphi)_w}{h}\right) \times h \times k; \quad u_e = \frac{u_P + u_E}{2} \text{ and } u_w = \frac{u_P + u_W}{2} \tag{5}$$

$$\text{If } u_e > 0, \quad \varphi_e = \frac{6}{8}\varphi_P + \frac{3}{8}\varphi_E - \frac{1}{8}\varphi_W \text{ and If } u_e < 0, \quad \varphi_e = \frac{6}{8}\varphi_E + \frac{3}{8}\varphi_P - \frac{1}{8}\varphi_{EE} \tag{5a}$$

$$\text{If } u_w > 0,\ \varphi_w = \frac{6}{8}\varphi_W + \frac{3}{8}\varphi_P - \frac{1}{8}\varphi_{WW} \text{ and If } u_w < 0,\ \varphi_w = \frac{6}{8}\varphi_P + \frac{3}{8}\varphi_W - \frac{1}{8}\varphi_E \tag{5b}$$

In Fig. 1, *P*, *E*, EE, *W* and WW are the central point of the cell, whereas *e*, *w*, ee and ww are the faces of the cell; '*h*' and '*k*' are the width and height of the two-dimensional cell; 'φ' is the flux which in this case are *u* and *v*. The fluxes at the faces of the cell along x-direction are evaluated as:

$$\varphi_e = \frac{\varphi_P + \varphi_E}{2}, \quad \varphi_w = \frac{\varphi_P + \varphi_W}{2}, \quad (u\varphi)_e = u_e\varphi_e \text{ and } (u\varphi)_w = u_w\varphi_w \tag{6}$$

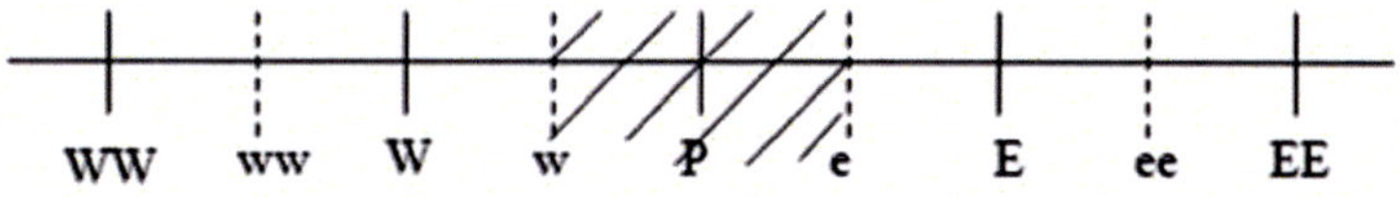

Fig. 1 1D computational domain along x-direction

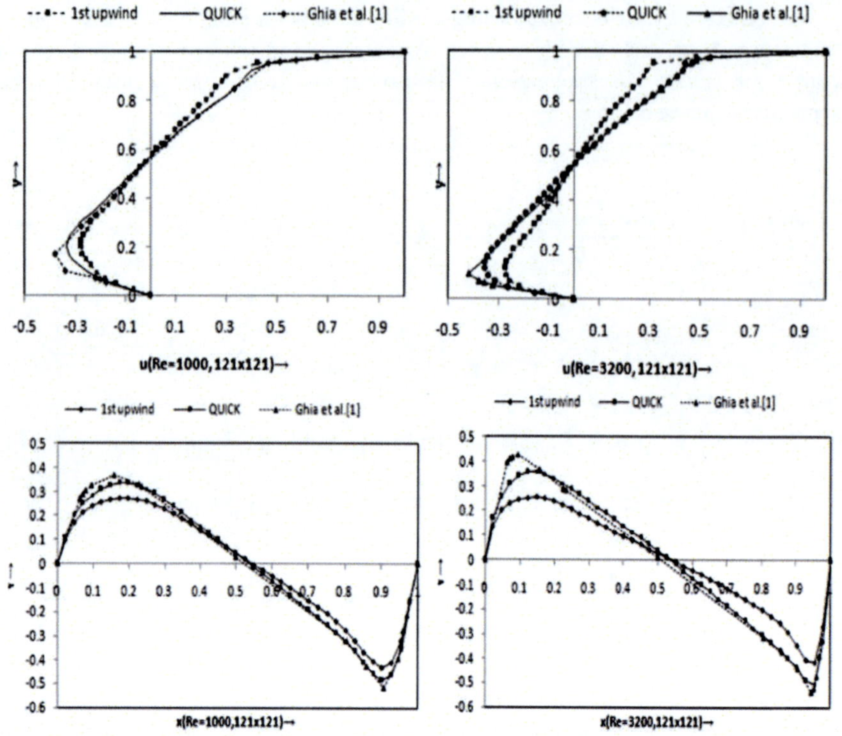

Fig. 2 *u* and *v*-velocity profiles along the center of the cavity at different Reynolds numbers

Similarly, the first- and third-order upwind scheme for $\partial(u\varphi)/\partial y$ can be evaluated along *y*-direction. The diffusion terms along *x*-direction are evaluated at cell P as:

$$\int_{y_i}^{y_{i+1}}\int_{x_i}^{x_{i+1}} \frac{\partial^2\varphi}{\partial x^2}\mathrm{d}x\mathrm{d}y = \frac{\varphi_E - 2.0\varphi_P + \varphi_W}{h^2} \times hk \tag{7}$$

Following similar manner, the diffusion terms along *y*-direction are also evaluated. The discretized equations are solved using SIMPLE [9] algorithm. The solution is said to be converged when the continuity equation (Eq. (2)) reaches up to order 10^{-18} (Figs. 2 and 3).

3 Convergence Details

Convergence criterion in primitive variable formulation using upwind scheme for two-dimensional lid-driven cavity is the satisfaction of continuity Eq. (3) up to the

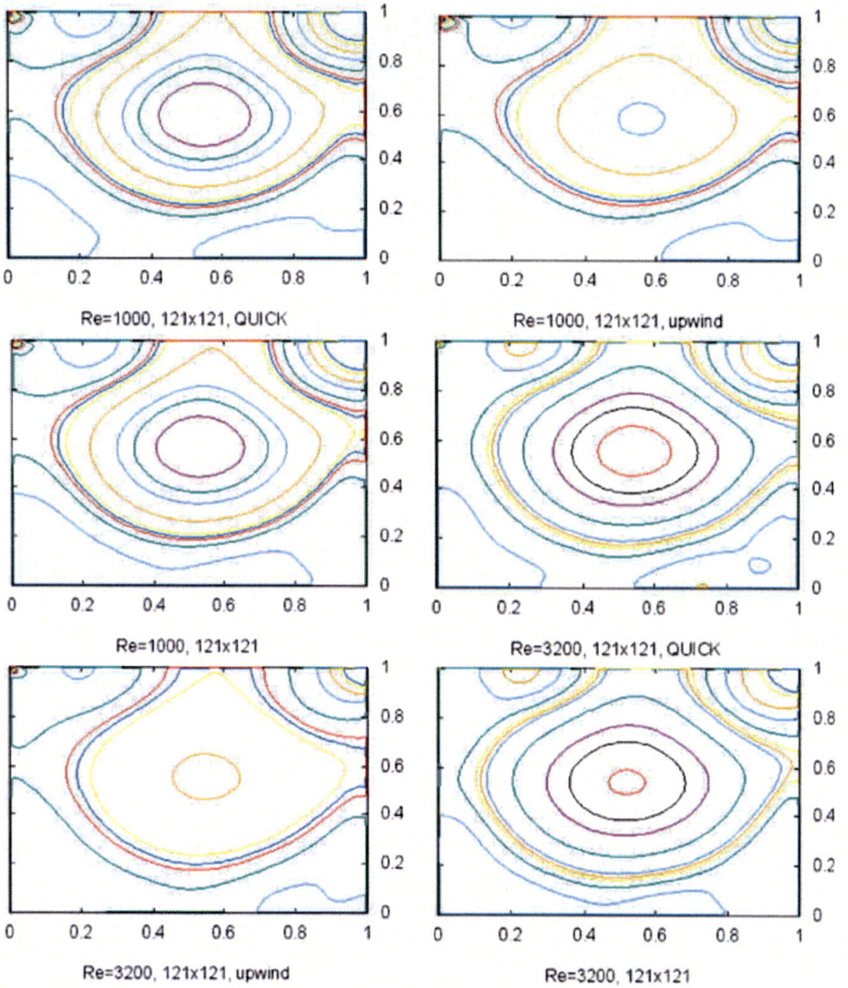

Fig. 3 Pressure contours at different Reynolds numbers

order 10^{-18}. First converged solution is obtained at Reynolds number 1000. Assuming this solution as initial value, the solutions for next higher Reynolds numbers are obtained. Similarly, the solutions at adjacent higher Reynolds numbers are obtained at an interval of 2500. The grid sizes used for this solution are 81×81, 101×101 and 121×121 in a range of Reynolds numbers from 1000 to 15,000.

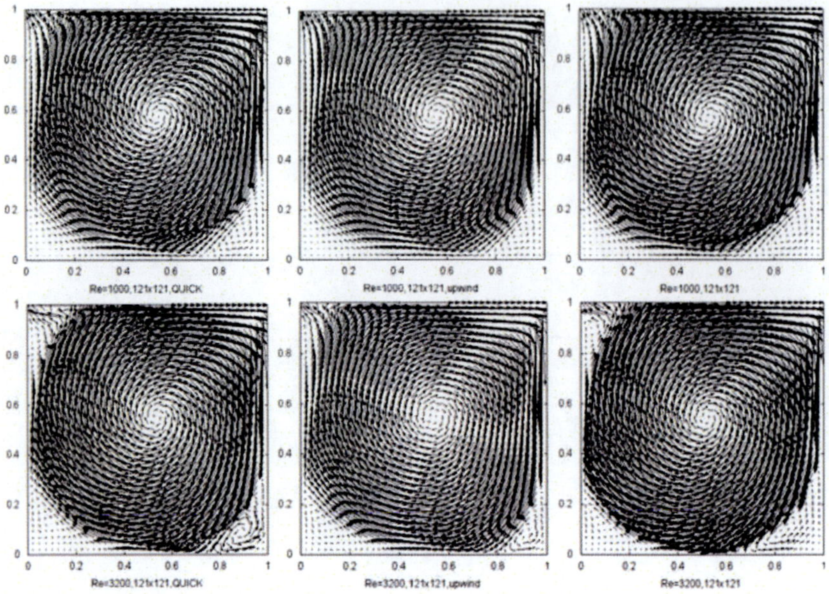

Fig. 4 Velocity vector plots at different Reynolds numbers

4 Validation Study

The computed results in the grid size 121 × 121 using first-order and third-order accurate upwind scheme are validated with the results of Ghia et al. [2] in the grid size 129 × 129 as shown in Fig. 2. The computed velocity profiles using first-order upwind scheme do not match exactly with Ghia et al. [2] but the computed results using third-order accurate upwind scheme match much better with Ghia et al. [2] as shown in Fig. 2, which shows third-order accurate upwind scheme (QUICK) result is much better than the first-order upwind scheme.

5 Results and Discussion

The converged solution is obtained up to Reynolds number 17,500 using first-order accurate upwind scheme in the grid size 121 × 121 and Re = 3200 using third-order accurate upwind scheme which is the steady-state solution. The solution at Reynolds number 20000 and 5000 becomes unsteady when first-order and third-order accurate upwind schemes applied, respectively, in the grid size 121 × 121. The pressure contours in the lid-driven cavity (Fig. 3) show that pressure distribution is uniform at the central region. Large variation of pressure is observed at the top right and left corner of the cavity in both upwind scheme and without upwind scheme which can be observed from the pressure contour plots as shown in Fig. 3. From Fig. 5, the pressure contour shows fluctuation near the lid which are unsteady solution at Reynolds number 5000 using QUICK scheme. On the other hand, if the pressure contours are observed at Re = 17,500 in Fig. 6 respectively, there is no fluctuation of pressure when first-order upwind scheme is applied. The velocity vector plot in Fig. 4 shows that as the Reynolds number increases the central primary eddy become larger in size and it becomes circular in nature. The vector plot also describes the development of secondary eddies at lower wall corners in both upwind schemes. As the Reynolds number increases there is development of unsteadiness in flow.

Pressure contour and velocity vector plot in Fig. 5 show that unsteady flow appears at Re = 5000 when third-order accurate upwind scheme (QUICK) is applied, whereas at the same Reynolds numbers the velocity vector plot does not show any sign of unsteadiness when first-order upwind scheme is applied on the convection terms.

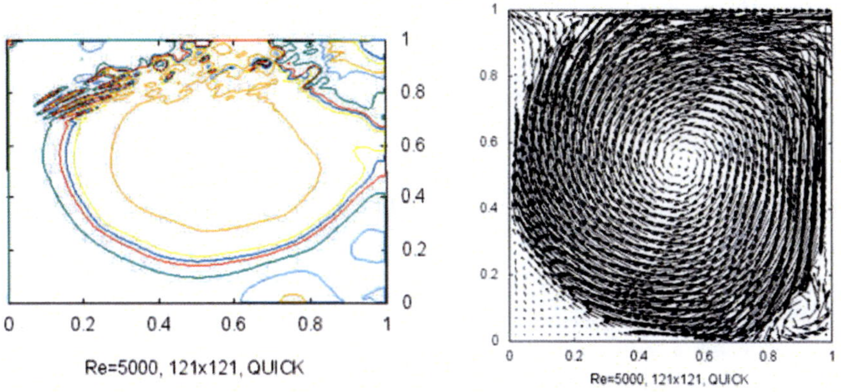

Fig. 5 Pressure contour and velocity vector plot at Re = 5000 using QUICK scheme

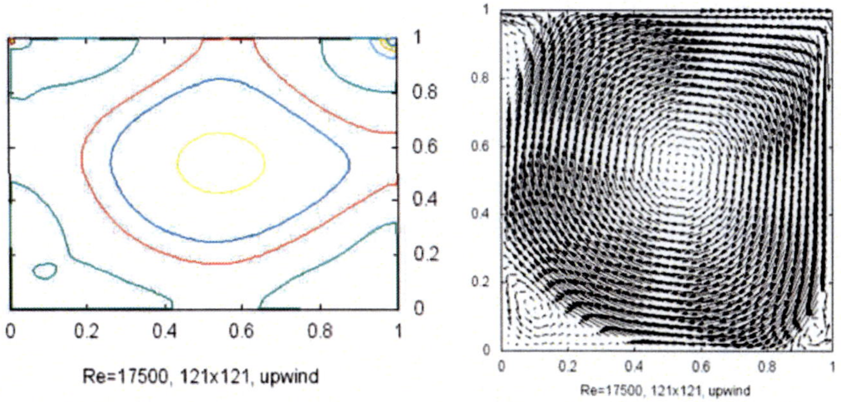

Fig. 6 Pressure contour and velocity vector plot at Re = 17,500 using upwind scheme

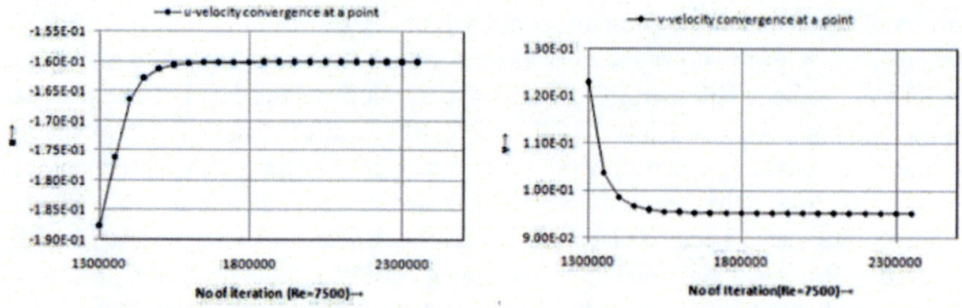

Fig. 7 Convergence history of *u* and *v* at Re = 7500 using first-order upwind scheme

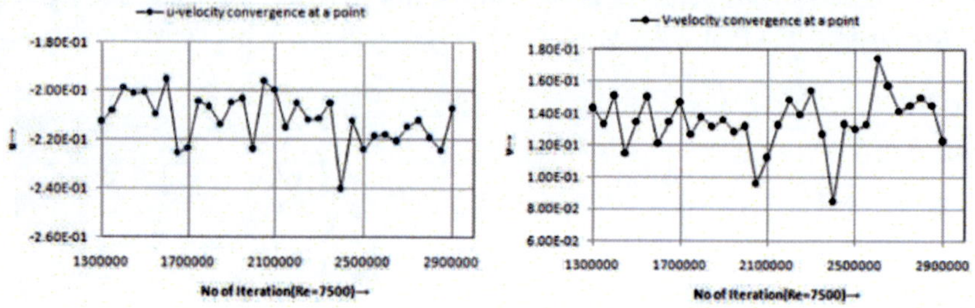

Fig. 8 Convergence history at a point at Re = 7500 using QUICK scheme

From Fig. 2, it is clear that QUICK scheme produces better accurate result than the first-order upwind scheme. So, the appearance of unsteadiness in the solution after Re = 3200 is not confirmed from the analysis of these two schemes. If it is assumed that QUICK scheme is accurate then the appearance of unsteadiness at Re = 5000 has reached to greater extent. If the grid size is made finer, again there may be appearance of chaotic flow at Re = 5000 because the lower right corner eddy and right and left corners of lid have been affected by the unsteadiness, which can be seen from Fig. 6.

Figure 7 shows the convergence history of *u* and *v* at a point (0.374, 0.2479) when first-order upwind scheme is applied on the convection terms at Re = 7500. It is observed that these are converged monotonically. On the other hand, if the third-order upwind scheme is applied on the convection terms using the same grid size, the properties do not converge, which are shown in Fig. 8 because the solution is unsteady at that Reynolds number.

6 Conclusions

First- and third-order upwind schemes are applied on the primitive variable form of the Navier–Stokes equation in a finite volume mesh. Following the SIMPLE procedure, the equations are solved in the two-dimensional lid-driven cavity. It is observed that flow is steady up to Re = 17,500 and 3200 using first-order and third-order upwind schemes, respectively. The unsteadiness appears at Re = 5000 using QUICK scheme which shows chaotic at Re = 10,000.

References

1. Shyy W, Thakur S, Wright J (1992) Second-order upwinding and central difference schemes for recirculating flow computation. AIAA J 30(4):923–932
2. Ghia U, Ghia KN, Shin CT (1982) High Reynolds number Solutions for incompressible flow using the Navier–Stokes equations and a Multigrid method. J Comput Phys 48:387–411
3. Thakur S, Shyy W (1993) Some implementational issues of convection schemes for finite volume formulations. Numer Heat Transfer Part B 24:31–55
4. Dalai B, Laha MK (2014) Incompressible viscous flows in two and three dimensional Lid-driven cavities. PhD Thesis Dissertation. Aerospace Engineering, IIT Kharagpur
5. Wu JS, Shao YL (2004) Simulation of lid-driven cavity flows by parallel lattice Boltzmann method using multi-relaxation-time scheme. Int J Numer Meth Fluids 46:921–937
6. Dalai B, Laha MK (2020) Numerical solution of steady incompressible flow in a lid-driven cavity using alternating direction implicit method. In: Recent advances in theoretical, applied, computational and experimental mechanics. Lect Notes Mech Eng 353–364. https://doi.org/10.1007/978-981-15-1189-9_28
7. Liao S-J, Mashayek F (2001) A multigrid approach for steady state laminar viscous flows. Int J Numer Meth Fluid 37:107–123
8. Versteeg HK, Malalasekera W (1995) An introduction to computational fluid dynamics. In: The finite volume method, 2nd edn. Pearson Education
9. Patankar SV (1981) Calculation of the flow field. In: Numerical heat transfer and fluid flow, 6.7, The SIMPLE algorithm. Hemisphere Publishing Corporation, pp 126

An Efficient Physically Adjusted Central Euler Solver

Souren Misra

1 Introduction

Numerical simulation of inviscid compressible fluid flows has been an active area of research in computational fluid dynamics (CFD) in the last four decades. The algorithms developed for solving the Euler equations of gas dynamics have a fascinating history, starting from the 1960s. The first successful numerical methods developed in the 1960s for solving the Euler equations, which are hyperbolic systems of conservation laws, were based on central discretization techniques and they did not take into consideration the directions of information propagation. The effort was focused mainly on devising stable schemes with reasonable accuracy in solving the above-mentioned convection equations. The algorithms which resulted from this effort are the popular Lax–Friedrichs method [1, 2], Lax–Wendroff method [3], two-step Lax–Wendroff method [4], MacCormack method [5] and Rusanov method [6]. A significant and popular addition to these central discretization methods was the scheme of Jameson, Schmidt and Turkel [7] in the early 1980s. The central schemes of 1960s and 1970s were the first successful algorithms for solving the hyperbolic systems of partial differential equations numerically. While the first-order accurate schemes in the above list were unsuitable for practical computations due to poor resolution of flow features (as a result of high dose of numerical diffusion), the second-order schemes generated oscillations or wiggles in capturing large gradients, and to suppress these oscillations, explicit numerical diffusion was often added. Since a theoretical insight about the right amount of numerical diffusion required to suppress oscillations was unavailable at that time, tuning parameters were introduced to adjust the numerical diffusion. A major criticism against the above central discretization methods was that the tuning parameters were not universal and were

S. Misra (✉)
National Institute of Science and Technology, Berhampur, Odisha 761008, India
e-mail: souren.misra@nist.edu

M. Palanisamy et al. (eds.), *Theoretical, Computational, and Experimental Solutions to Thermo-Fluid Systems*, Lecture Notes in Mechanical Engineering,
https://doi.org/10.1007/978-981-33-4165-4_40

problem-dependent. For this reason, upwind methods became more popular from 1970s.

Upwind discretization methods, in contrast to the above-mentioned central discretization methods, are based on recognizing the directions of information propagation and thus respect the hyperbolicity property of the governing equations by introducing splitting of the wave speeds into positive and negative parts and then suitably choosing the stencils in the discretization process. The upwind methods can be broadly classified into four major categories: (i) flux vector splitting methods, (ii) Riemann solvers (exact or approximate), (iii) kinetic (or Boltzmann) schemes and (iv) relaxation schemes.

The central discretization methods, which lost their popularity to upwind methods in 1980s, underwent a revival from the 1990s. The schemes of Nessyahu and Tadmor [8], Kurganov and Tadmor [9] initiated the development of central discretization methods which are free of Riemann solvers (thus retaining their simplicity) and yet are competent in accuracy. Jaisankar and Raghurama Rao [10] introduced a central Rankine-Hugoniot solver in which the coefficient of numerical dissipation is fixed based on the Rankine-Hugoniot (jump) condition. This algorithm, named as the Method of Optimal Viscosity for Enhanced Resolution of Shocks (MOVERS), can capture steady discontinuities exactly in a simple central discretization framework. The recently developed central schemes for various contexts are available in the website on Central Station, a collection of references on high-resolution non-oscillatory central schemes [11]. Most of these efficient central schemes are based on the local Lax–Friedrichs (LLF) method, which is the based on a minor reformulation of the original Rusanov method.

In the present work, we introduce a new Euler solver which is based on the LLF method. In smooth regions of the flow, the numerical dissipation is chosen as the predominant one of the relevant wave speeds (i.e. acoustic or fluid speed). The total density is used to distinguish the discontinuities from the smooth flow regions based on the fact that the total pressure and total density change across the discontinuities remain constant in the smooth flow regions. Further, the shock waves and the contact discontinuities are distinguished based on the jump in the static pressure, as there is no change of static pressure across the contact discontinuity. An exact expression available for the shock speed, derived from the shock relations, is used for fixing the numerical dissipation in the case of shocks. In the case of contact discontinuities, the numerical dissipation is fixed from the R–H condition, applied for each of the conservation equations, which leads to a unique scalar dissipation. The motivation is to retain the simplicity of the original LLF method and yet improve the accuracy compared to the parent scheme, by introducing the relevant physics of the flow into the discretization process such that steady discontinuities are captured exactly. This new central scheme, termed as physically adjusted central Euler (PACE) solver, is tested on some typical test problems for inviscid compressible flows in one and two dimensions to demonstrate its efficiency. The physics of the flow gives an idea of the domain of the exact wave speed for the smooth flow.

2 PACE: An Efficient Central Solver

The Euler solver presented in this work, the PACE scheme, is based on the local Lax–Friedrichs (LLF) method. Let us first consider the LLF method for one-dimensional system of conservation laws for inviscid fluid flows. The 1-D Euler equations can be written in conservation form as

$$\frac{\partial U}{\partial t} + \frac{\partial F}{\partial x} = 0 \quad (1)$$

where $U = (\rho, \rho u, \rho E)^{\mathrm{T}}$ is the vector of conserved variables and $F = (\rho u, \rho u^2 + p, \rho u E + p u)^{\mathrm{T}}$ is the inviscid flux vector. An explicit scheme in conservation form on a three-point stencil and with a piecewise polynomial approximation for conserved variable is given by

$$U_j^{n+1} = U_j^n - \frac{\Delta t}{\Delta x}\left(F_{j+\frac{1}{2}}^n - F_{j-\frac{1}{2}}^n\right) \quad (2)$$

where Δt and Δx are, respectively, time and space steps, j represents the cell centroids and $j \pm ½$ refers to the cell interfaces. In finite volume formulation, the differences among all the numerical schemes lie essentially in the definition of numerical flux $F_i = F_{j\pm1/2}$ evaluated at the cell interface. The interface flux for any stable scheme can be written in the generic form as

$$F_i = \frac{1}{2}(F_l + F_r) - D_i \quad (3)$$

where the subscripts l, r and i represent the left, right states and cell interface, respectively; the first term on the right-hand side (average flux) represents the central discretization of the flux terms and D_i (dissipative flux) represents the flux corresponding to numerical diffusion. The dissipative flux D_i can be defined as

$$D_i = \frac{1}{2}\alpha \Delta U \quad (4)$$

where α represents the coefficient of numerical diffusion and $\Delta U = U_r - U_l$. Each method differs from any other method in the coefficient of numerical diffusion, α. In the LLF method, α is defined as the local maximum of the eigenvalues of the flux Jacobian matrix for the left and right states. For the 1-D Euler equations, the flux Jacobian matrix $A = \partial F/\partial x$ has three eigenvalues, given by $\lambda_1 = u - a$, $\lambda_2 = u$ and $\lambda_3 = u + a$, where u is the fluid velocity and a is the speed of sound. Therefore, the coefficient of numerical diffusion in the LLF method is based on the maximum of the above three wave speeds, that is, $|u| + a$.

$$\alpha_{\mathrm{LLF}} = \max(|u_l| + a_l, |u_r| + a_r) \quad (5)$$

The LLF method is very simple, robust and is known to satisfy the entropy condition, thus generating physically correct weak solutions. The main drawback is the higher amount of numerical diffusion. In the PACE scheme, the coefficient of numerical diffusion of the LLF method is modified and is denoted as α_{p}. The dissipative flux D_i is modified by introducing the relevant physics of the flow and the cases of smooth flows, shock waves and contact discontinuities are treated separately.

2.1 *Criteria for Distinguishing Smooth Flows from Discontinuities*

It is well known from the gas dynamics that the total density and total pressure remain unchanged for isentropic flow, while across the discontinuities the entropy of the flow increases and the total density and total pressure decreases. In the present work, the change in total density is used as a guiding principle to distinguish discontinuities from the smooth flow. To distinguish further the contact discontinuities from the shock waves, the fact that the static pressure jumps across a shock wave remains unchanged but across a contact discontinuity is utilized.

The adapted criterion to recognize a shock wave moving toward the right is $(\Delta\rho_0)_{\mathrm{lr}}/\rho_{0_{\mathrm{r}}} > \theta$ and $|(\Delta p)_{\mathrm{rl}}|/p_{\mathrm{l}} < \eta_1$ and similarly for a leftward moving shock $(\Delta\rho_0)_{\mathrm{rl}}/\rho_{0_{\mathrm{l}}} > \theta$ and $|(\Delta p)_{\mathrm{lr}}|p_r < \eta_1$. The parameters θ and η_1 refer to some small tolerance values and are discussed further in the next subsections. The contact discontinuity is recognized by the jump of total density from left to right or from right to left with no change of static pressure. The criterion used for recognizing a contact discontinuity is $(\Delta\rho_0)_{\mathrm{lr}}/\rho_{0_{\mathrm{r}}} > \theta$ or $(\Delta\rho_0)_{\mathrm{rl}}/\rho_{0_{\mathrm{l}}} > \theta$ and $p_{\mathrm{l}} = p_{\mathrm{r}}$, where ρ_0 is the total density. $(\Delta\rho_0)_{\mathrm{lr}}$ is the decrease in total density across the interface and is defined by $\rho_{0_{\mathrm{l}}} - \rho_{0_{\mathrm{r}}}$. Similarly, $(\Delta\rho_0)_{\mathrm{rl}}$ is defined as $\rho_{0_{\mathrm{r}}} - \rho_{0_{\mathrm{l}}}$ and $(\Delta p)_{\mathrm{lr}}$ is defined as $p_{\mathrm{l}} - p_{\mathrm{r}}$

2.2 *Numerical Dissipation in Smooth Region*

For smooth flows, the numerical dissipation is chosen based on the predominant of the relevant speeds, as explained below. In supersonic flow, two clear zones are present: a zone of action and a zone of silence (see Fig. 3.2 in Shapiro [12]). In supersonic flow, all the information about the flow are obtained from the zone of action. In three dimensions, the zone of action is presented with the structure of a cone. Outside the Mach cone, in the zone of silence, no information of the flow is available. For subsonic flows, all the information of the fluid flow are present inside a sphere, the radius of which is determined by the sound speed. Thus, in the smooth flow regions, for supersonic flows, as $M > 1$ or $u > a$, the fluid velocity always dominates the sound speed. In that case, we can take the predominant wave speed as the coefficient of numerical diffusion and fix $\alpha = u$. For subsonic flows, as $M < 1$ or

$u < a$, the acoustic speed dominates the fluid velocity. Then, the predominant wave speed leads to the coefficient of numerical diffusion as the speed of sound and we can take $\alpha = a$. Thus, in the PACE scheme, the α_{p} is obtained from the predominant of the acoustic speed a or the fluid velocity u, for smooth regions of the flow. By this modification, the numerical diffusion of the PACE scheme is reduced for the smooth flow regions compared to the LLF method, as the coefficient of numerical diffusion will be $|u|$ or a which is less than $|u| + a$ chosen by the LLF method. Thus, we define the coefficient of numerical diffusion α_p for smooth regions as

$$\alpha_p = \max(\max(|u_{\mathrm{l}}|, |u_{\mathrm{r}}|), \max(a_{\mathrm{l}}, a_{\mathrm{r}})) \tag{6}$$

where $(u_{\mathrm{l}}, a_{\mathrm{l}})$ and $(u_{\mathrm{r}}, a_{\mathrm{r}})$ are fluid velocity and sound speed of the left and right sides of the interface. From the smooth flow computation, we can conclude that the wave speed at the interface for subsonic flow domain is $[0, a]$ and for supersonic flow the wave speed lies at $[a, u]$.

2.3 *Numerical Dissipation for Capturing Shock Waves*

The numerical diffusion for the case of a shock wave is obtained from the exact expression for the wave speed derived from the shock relations. The derivation of wave speed from the shock relations is derived in the appendix B of the thesis of Misra [13]. The numerical dissipation is fixed for left-running shock wave is given by

$$\alpha_{\mathrm{p}} = \left| |u_{\mathrm{r}}| - a_{\mathrm{r}}\sqrt{1 + \frac{\gamma+1}{2\gamma}\left(\frac{p_{\mathrm{l}}}{p_{\mathrm{r}}} - 1\right)} \right| \tag{7}$$

Similarly, for the right running shock wave

$$\alpha_{\mathrm{p}} = \left| |u_{\mathrm{l}}| - a_{\mathrm{l}}\sqrt{1 + \frac{\gamma+1}{2\gamma}\left(\frac{p_{\mathrm{r}}}{p_{\mathrm{l}}} - 1\right)} \right| \tag{8}$$

In higher dimension, the shock speed is obtained as presented in the following equation

$$\alpha_{\mathrm{p}} = \left| \left(|V_{\mathrm{l}}| - a_{\mathrm{l}}\sqrt{1 + \frac{\gamma+1}{2\gamma}\left(\frac{p_r}{p_l} - 1\right)} \right) \cos(\theta_V - \theta_i) \right| \tag{9}$$

where V, θ_V and θ_i are represented as $V^2 = u^2 + v^2$, $\theta_V = \tan - 1(v/u)$ and θ_{i} is the angle of the normal to the interface, respectively. In this work, the value of θ used is

1.0e−4 and the value of η_1 is taken as 0.18, which worked well for all the standard test cases presented.

2.4 Numerical Dissipation for Capturing Contact Discontinuities

For capturing contact discontinuities, we choose the numerical dissipation as the local maximum of the wave speeds (as in LLF method) for each of the conservation equations, and this procedure surprisingly leads to a unique scalar value of numerical dissipation. This procedure, though is different from the technique used by Jaisankar and Raghurama Rao [10] (who applied the LLF method to the energy equation as it contains the maximum possible information), leads to the same expression and further leads to exact capturing of contact discontinuities, as it leads to the satisfaction of Meng-Sing Liou's lemma [14]. The details of the derivation of the wave speed for each conservation law, leading to a unique value of scalar numerical dissipation based on the LLF method, are given in the appendix A of the thesis of Misra. For the case of a contact discontinuity, we then obtain

$$\alpha_p = \max(|u_l|, |u_r|) \tag{10}$$

In the present work, the parameter θ is chosen as 1.0e−4 and it worked satisfactorily for all the standard test cases presented in this paper. The higher order solutions are obtained with the usage of minmod limiter.

3 Results and Discussion

The PACE scheme is tested on some standard 1-D and 2-D test cases for Euler equations. The description of the initial and boundary conditions of 1-D benchmark problems are presented on: (i) 1-D stationary contact discontinuity in a shock tube, (ii) 1-D stationary normal shock, (iii) Sod's shock tube and (iv) Quasi 1-D nozzle flow.

PACE scheme resolves the 1-D stationary contact discontinuity and normal shock exactly. The results of both the test cases are presented in Fig. 1. The first- and second-order solutions (variation of density) of sod-shock test case are presented at time 0.01 s in Fig. 2. It is observed that there is improvement in the resolution of shock for the sod-shock test case with the PACE scheme compared to LLF. The quasi 1-D nozzle flow test case [12] in Fig. 3 and the results (variation of pressure) show that there are substantial improvements in capturing the shock. The wave speed and Mach number along the axis of nozzle are presented for LLF and PACE schemes in

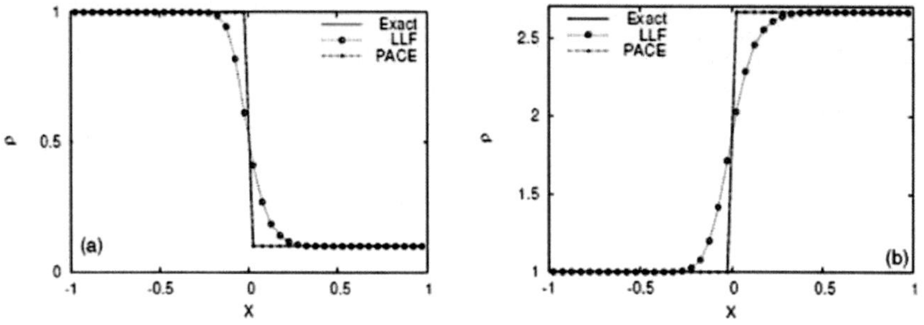

Fig. 1 Comparison of LLF and PACE schemes: **a** 1-D steady contact discontinuity in a shock tube and **b** 1-D steady normal shock test case

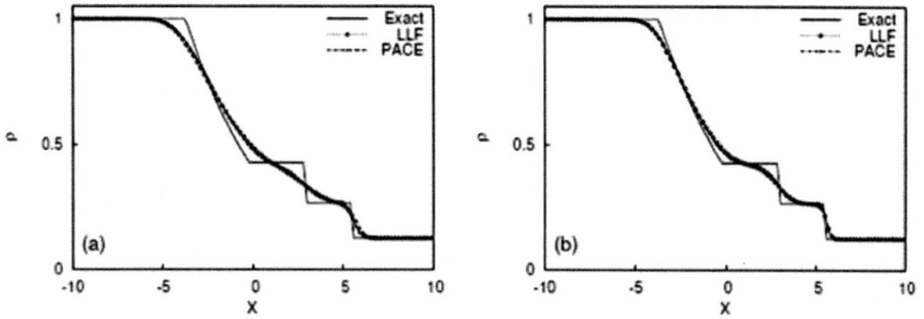

Fig. 2 Comparison of LLF and PACE schemes on 1-D Sod's shock tube with 100 grid points: **a** First-order and **b** second-order solutions

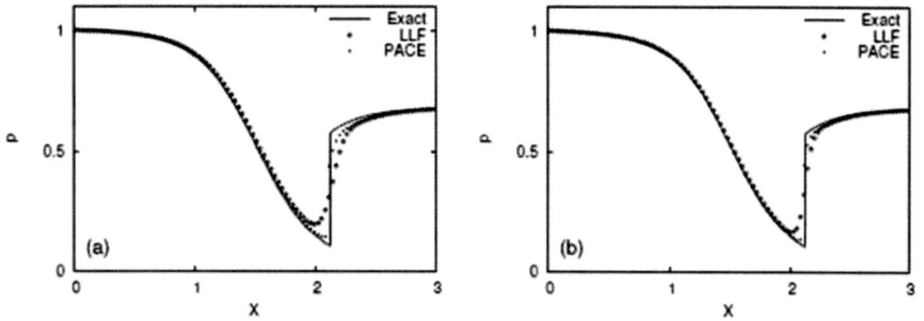

Fig. 3 Comparison of LLF and PACE schemes for quasi-1-D flow through a converging–diverging nozzle with 100 points: **a** First-order and **b** second-order solutions

Fig. 4. The wave speed of PACE scheme differs maximum at the sonic flow and is least toward the incompressible flow domain than the LLF scheme.

The PACE scheme is further tested on the following standard 2-D benchmark problems: (i) grid-aligned shear flow [15] and (ii) supersonic flow over a compression ramp of 15° in a channel [16].

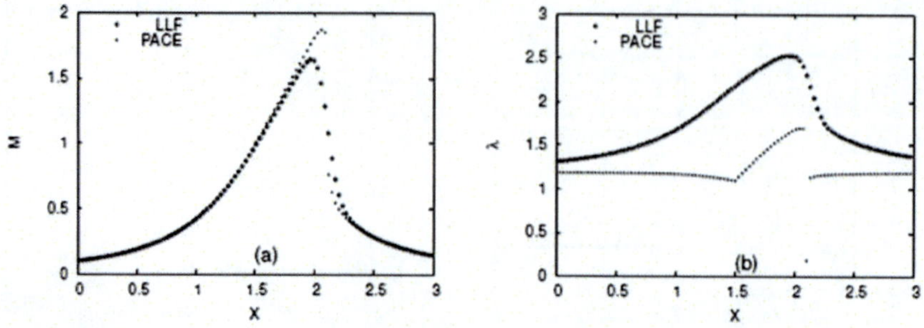

Fig. 4 Comparison of first-order LLF and PACE schemes: Variation of **a** Mach number and **b** wave speed along the axis of the quasi-1-D flow through a converging–diverging nozzle

The first-order and second-order results obtained with the PACE scheme have less numerical diffusion as compared to the LLF scheme in smooth as well as discontinuous regions. The results obtained with the first- and second-order accurate PACE are compared with those of LLF scheme in Figs. 5, 6 and 7. The first-order accurate LLF method is highly diffusive compared to that of PACE which captures the shock with less numerical dissipation. This is expected since the coefficient of numerical dissipation in LLF scheme is $|u| + a$, whereas for PACE it is $|u|$ or a in smooth regions, or based on shock relations for the shock waves, or just fluid velocity for contact discontinuities. The PACE captures the slip surface in the grid-aligned shear flow exactly. The results for the 15° wedge in the channel flow are presented in Figs. 6 and 7. The above test case illustrates the higher efficiency of the PACE scheme as compared to the LLF method.

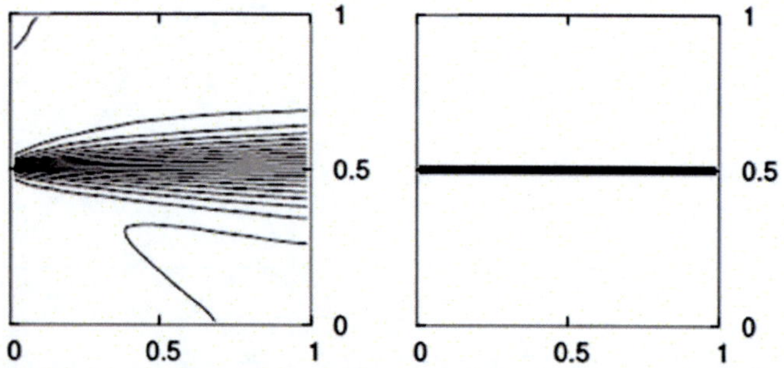

Fig. 5 Comparison of first-order LLF (left) and PACE (right) schemes: pressure contours (1.1:0.05:3.8) for ramp in a channel flow with grids: **a** 60 × 20, **b** 120 × 40 and **c** 240 × 80

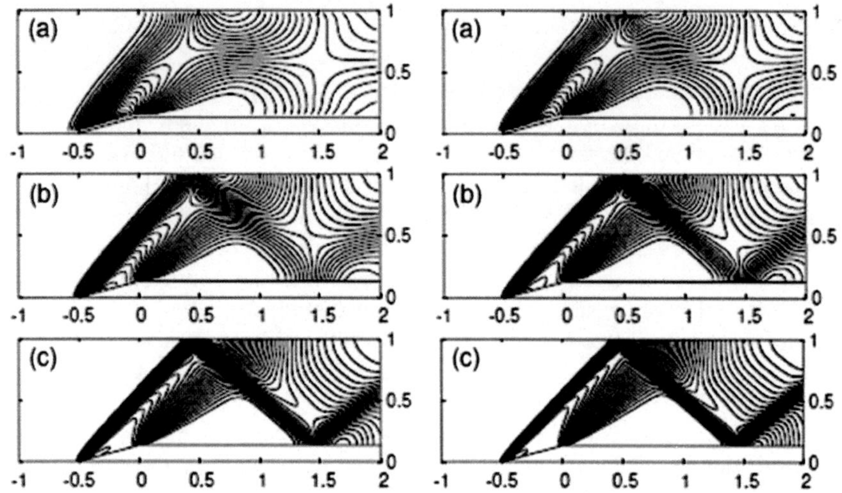

Fig. 6 Comparison of first-order accurate solutions with LLF (left) and PACE (right) schemes: pressure contours (1.1:0.05:3.8) for ramp in a channel flow with grids: **a** 60 × 20, **b** 120 × 40 and **c** 240 × 80

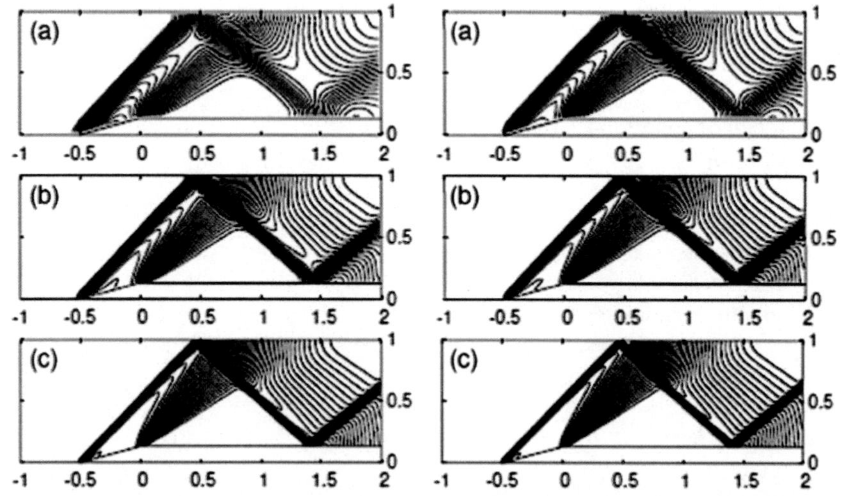

Fig. 7 Comparison of first-order LLF (left) and PACE (right) schemes: pressure contours (1.1:0.05:3.8) for ramp in a channel flow with grids: **a** 60 × 20, **b** 120 × 40 and **c** 240 × 80

4 Conclusion

A simple and efficient central solver, termed as PACE, is presented for solving the Euler's equations of inviscid compressible fluid flows. For smooth regions the coefficient of numerical diffusion is chosen as the maximum of the fluid velocity and the sound speed from the left and right states of the cell interface in a finite volume method. The maximum speed of the domain of the physical wave speed is used for computing the smooth flow computation. In this work, total density across an interface is the guiding criterion to distinguish the discontinuities from the smooth flow. The shocks are distinguished from contact discontinuities based on static pressure change, which must be zero across contact discontinuities. PACE scheme captures the grid-aligned steady contact discontinuity as well as the steady shock exactly. The numerical dissipation for each case is fixed based on the physically relevant wave speeds. PACE scheme is tested on several standard benchmark problems in one and two dimensions for Euler equations of gas dynamics, demonstrating its efficiency in capturing the flow features.

Nomenclature

A	flux Jacobian matrix
a	speed of sound
D	dissipative flux
F	invisid flux vector
u	fluid velocity
U	vector of conserved variables
ΔU	$U_r - U_l$
ρ	density
p	pressure
$\tilde{\rho}$	nondimensional density
$\tilde{V}$	nondimensional velocity
$\tilde{T}$	nondimensional temperature
$\tilde{A}$	nondimensional area
α	co-efficient of numerical diffusion
λ	Eigen value
γ	specific heat
φ	minmod limiter
θ_i	angle of the normal to the interface
θ_V	Direction of fluid velocity

Superscript

+	right running
−	left running

Subscript

0	total or stagnation state

j	cell centroid
$j + 1/2$	cell interface
l	left state
r	right state
i	cell interface
p	PACE
lr	moving towards right
rl	moving towards left
LLF	Local Lax-Friedrich
PACE	Physically Adjusted Central Euler

References

1. Lax PD (1954) Weak solutions of nonlinear hyperbolic equations and their numerical computation. Commun Pure Appl Math 7:159–193
2. Friedrichs KO (1954) Symmetric hyperbolic linear differential equations. Commun Pure Appl Math 7:345–392
3. Lax PD, Wendroff B (1960) Systems of conservation laws. Commun Pure Appl Math 13:217–237
4. Richtmyer RD, Morton KW (1967) Difference methods for initial value problems, 2nd edn. Wiley, New York
5. MacCormack RW (1969) The effect of viscosity in hypervelocity impact cretaring. AIAA paper no. AIAA-69-0354
6. Rusanov VV (1961) Calculation of interaction of nonsteady shock waves with obstacles. J Comput Math Phys 1:267–279
7. Jameson A, Schmidt W, Turkel E (1981) Numerical solution of Euler equations by finite volume methods using Runge–Kutta time stepping schemes. AIAA paper no. AIAA-81-1259
8. Nessyahu H, Tadmor E (1990) Non-oscillatory central differencing for hyperbolic conservation laws. J Comput Phys 87:408–463
9. Kurganov A, Tadmor E (2000) New high-resolution central schemes for nonlinear conservation laws and convection-diffusion equations. J Comput Phys 160:214–282
10. Jaisankar S, Raghurama Rao SV (2009) A central Rankine-Hugoniot solver for hyperbolic conservation laws. J Comput Phys 228:770–798
11. https://www.cscamm.umd.edu/centpack/publications/
12. Shapiro AH (1976) The dynamics and thermodynamics of compressible fluid flow, vol 1, Wiley, pp 50
13. Misra S (2014) Efficient numerical algorithms for nonlinear convection and convection-diffusion equations of fluid dynamics. PhD (Eng) Thesis. Department of Aerospace Engineering, Indian Institute of Science, Bangalore, India
14. Liou M-S (2001) Ten years in the making—AUSM family. AIAA paper no. AIAA-2001-2521; also NASA/TM 2001-210977
15. Chakravarthy SR, Harten A, Osher A (1986) Essentially non-oscillatory shock capturing schemes of arbitrary-high accuracy. AIIA Paper No. AIAA-86-0339
16. Manna M (1992) A three dimensional high resolution upwind finite volume Euler solver, von Karman Institute of fluid dynamics. Techn Note 180

Experimental Study on Vertical Axis Wind Turbine to Harness Wind Power from Rapidly Moving Railway Locomotives

Yendaluru Raja Sekhar, Muthuswamy Natarajan, Chalasani Chiranjeevi, Roy Sukanta, and Patil Yugandhar

1 Introduction

In recent decades, the use of renewable energy has considerably increased because of its abundant environment-friendly nature and government-supportive implications on the technical advancements. Particularly for tropical locations, energy generation from wind and solar power can be consistent and reliable, which can help us to meet the rising energy demand. However, for improved conversion efficiency, innovative harnessing methods as well as alternative technological designs are much sought. In this regard, wind velocities with satisfactory power density can be a suitable option, and it is considered to be one of the fastest-growing sources of clean energy worldwide, as stated by the Global Wind Energy Council [1]. One of the wind sources is the adjacent locations closed to moving railway locomotives. India has about 63,000 route km of railways and 14,300 trains running every day. Since these locomotives travel with velocities in the range of 60–120 km/h, they create high-velocity channelled wind source, which can generate power by placing wind turbines alongside the trains.

Based on the axis of rotation, the wind turbines are classified into horizontal axis wind turbines (HAWTs) and vertical axis wind turbines (VAWTs). Principally, the Savonius type VAWTs can generate higher torque even at lower wind velocity with considerable power capacity. Further, its operation and maintenance costs are relatively lesser as compared to HAWTs.

The design of Savonius-type VAWT was first patented by S. J. Savonius in 1929, followed by several experimental investigations in last few decades. In the available

Y. Raja Sekhar (✉) · M. Natarajan · C. Chiranjeevi · P. Yugandhar
School of Mechanical Engineering, Vellore Institute of Technology, Vellore 632014, India
e-mail: sekhar.rajay@gmail.com

R. Sukanta
Department of Mechanical Engineering, Curtin University, Sarawak, Malaysia

M. Palanisamy et al. (eds.), *Theoretical, Computational, and Experimental Solutions to Thermo-Fluid Systems*, Lecture Notes in Mechanical Engineering,
https://doi.org/10.1007/978-981-33-4165-4_41

literature, the value of turbine performance coefficients (*Cp*) were reported in the range of 0.15–0.38 [2]. Bethi and Laws [3] in their study mentioned the wind energy potential developed by trains when Savonius rotors were placed inside the tunnels. The new design of rotor configuration was studied and reported no drag on the train due to the presence of the turbine. They concluded to use the proposed method for powering electrical components and tunnel lighting. Mohamed [4] reported the impact of solidity on the self-stating capability of Darrieus VAWTs through numerical and experimental simulations, and reported improvement in H-rotor Darrieus turbine self-starting capability through a hybrid system using Savonius VAWTs. Darmawan and Winjaya [5] used micro wind electricity builders to power level crossing security gate to replace conventional energy usage. They considered fins for the rotor blades and power generation of 39.45 W at 3.9 m/s wind velocity, 278 rpm. Also, the total power of 394.5 W can be generated to charge 75 Ah/12 V battery for 22 h in a single device.

Roy and Saha [6] developed a two-bladed Savonius VAWTs for small-scale energy conversion and conducted a series of wind tunnel experiments, where they compared power and torque values with other blade designs such as semi-circular, semi-elliptic, Benesh and Bach types; and reported a gain of 34.8% in maximum power coefficient over conventional design. Based on the understanding from previous literature, it can be concluded that Savonius VAWT has better aerodynamic behaviour at low running speeds and can capture wind from any direction. Their flexible design, high torque at low speed make them ideal for harvesting intermittent wind energy produced by the trains. Hence, in this study, initial experiments were conducted to determine the possible wind velocity range from railway locomotives in which Savonius VAWT can work. Numerical simulations have been conducted with ANSYS CFX, followed by fabrication of experimental prototype, which were tested on a railway track to determine the actual power generation. Further, analysis of estimated annual power potential at different railway junctions in India is reported.

2 Numerical and Experimental Simulation

The project aims to use the wind energy produced by the trains on the railways, whose kinetic energy is left unused and can be recovered by installing wind turbines, as shown in Fig. 1a. A battery installed under the turbine assembly store DC power generated from wind, for later use to power signal control systems. Savonius rotor design was proposed as an optimum choice because of its advantage over other VAWTs. The theoretical evaluation of the proposed wind turbine design based on onsite wind velocity and the optimization of the rotor design is carried out using ANSYS CFX fluid analysis. Two-blade and three-blade rotor assembly was numerically analysed. An array of nine anemometers were placed on a PVC pipe network kept at a distance of 1.5, 2 and 2.5 m along the length from the horizontal reference and 3.0, 3.5 and 4.0 m in the axial reference from the railway track as shown in Fig. 1b. A safe distance of 3 m from the track was chosen, as reported by [3] and shown in

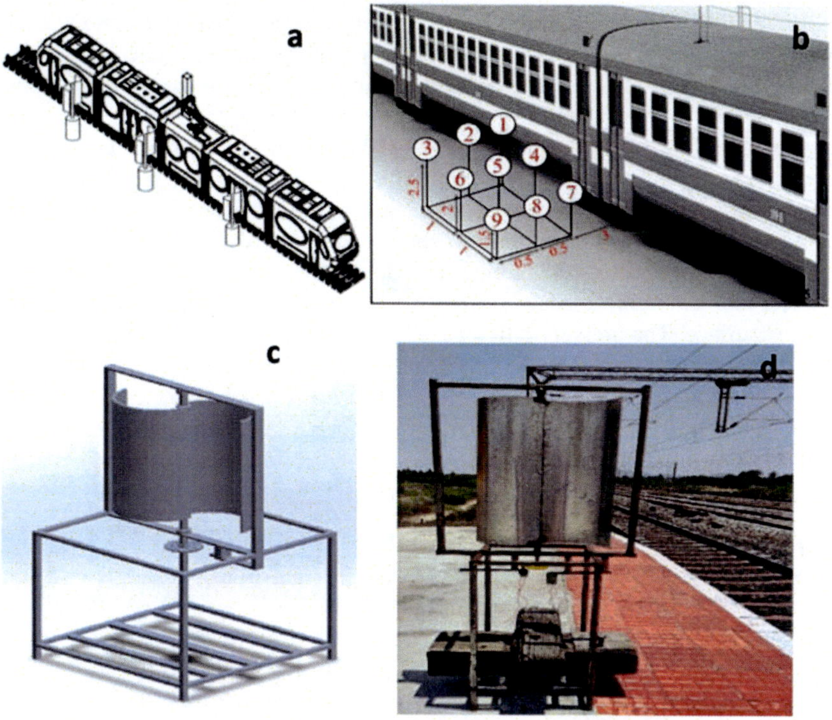

Fig. 1 **a** Proposed prototype system along the rail tracks. **b** Onsite wind velocity measurements using anemometers. **c** Fabricated VAWT. **d** Testing of VAWT prototype at site

Fig. 1b. The parameters and specifications of the Savonius rotor were adopted from the previous study of [7]. Wind measurements were measured for different trains such as Superfast and Loco trains plying in the section. The fabricated prototype has the dimensions of rotor swept area (*A*) of 0.464 m^2, rotor diameter (*D*) of 0.58 m, rotor height (*H*) of 0.80 m, chord length (*d*) of 0.29 m, overlap distance (*e*) of 0.03 m and blade thickness (*t*) of 0.001 m. The structural frame is fabricated using mild steel and the blade with aluminium.

3 Results and Discussion

Onsite measurements were taken during rapidly moving locomotive at a railway junction and wind velocities were recorded in the range of 2.0–6.0 m/s at location 2, as shown in Fig. 1b. Therefore, VAWT located at point 2 must generate maximum power output with high thrust 2.5 m height from the ground and at a horizontal distance of 3.5 m from the moving train. As reported in the literature, maximum rotor power delivered at point 2 was estimated by considering the power coefficient of VAWT as 0.37 for all the trains [8, 9]. Numerical simulations using ANSYS CFX were carried out by considering the onsite data to optimise rotor dimensions and blade design. Results of pressure and vector contours for the two-blade rotor are shown in Fig. 2a and b, respectively, to understand the distribution of wind. The maximum speed that hits the blade is represented by a red region, while the minimum speed of the wind gusts on the blades is represented by the blue region. The speed of gusts of wind was high in the returning blade [10], while it was low near the advancing blade. This indicates that the maximum wind contact surface was the returning blade, which rotates the rotor and sets the turbine in motion. The maximum value of the speed contour is 5 m/s in the concave surface, and the minimum value is 0.45 m/s in the convex surface. The pressure difference between the advanced and returning blade of the drag force, induced by the rotor, causes the blade to rotate. The average drag force for the value of gusts of 5 m/s is 8.63 N, while the drag force coefficient was 1.15. The drag force and the coefficient values were calculated for different wind gust speeds using CFD simulation.

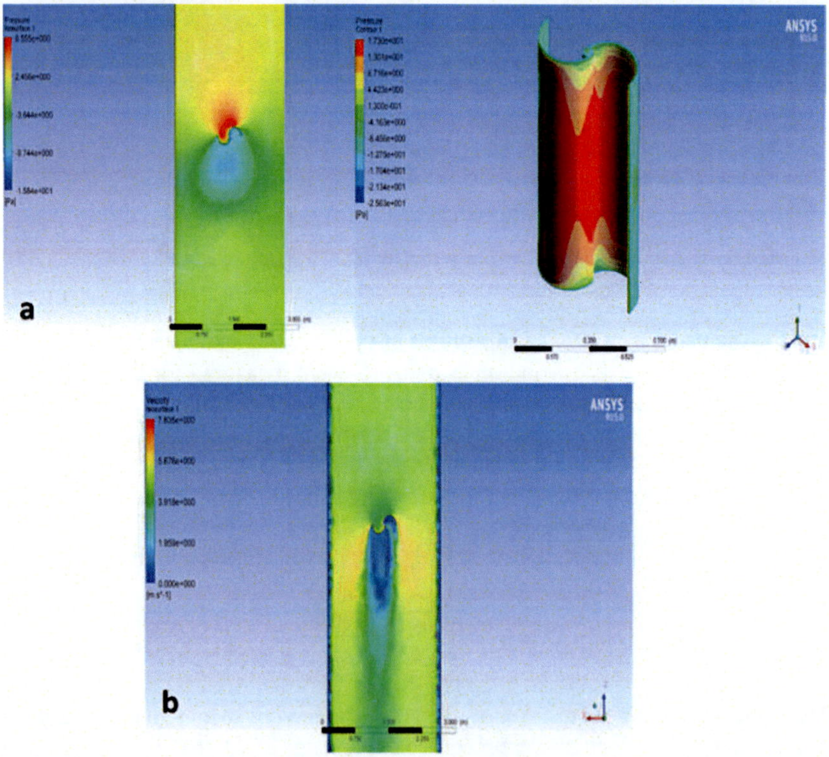

Fig. 2 **a** Pressure contours for two-blade rotor assembly. **b** Velocity contours for two-blade rotor assembly

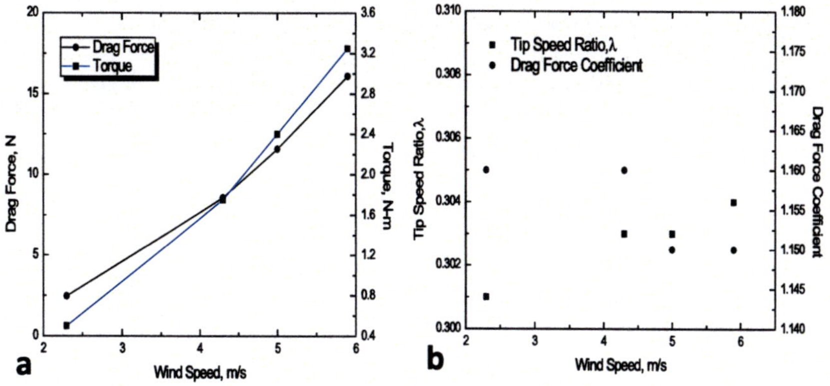

Fig. 3 **a** Variation of drag force and torque generated by rotor at different wind speed. **b** Comparison of tip speed ratio and drag coefficient at different wind speed

Though three-blade rotor assembly is considered for the study, better results were obtained using the two-blade design [11, 12]. For the two-blade rotor torque generated was higher by 1.2 times, and drag coefficient was lower by 0.5 times as compared to the three-blade rotor. Hence, the actual prototype with two-blade assembly was fabricated, which resulted in theoretical maximum generation at 5.9 m/s with rotor power and torque of 21.508 W and 3.297 N m at a tip speed ratio of 0.304, respectively, as shown in Fig. 3a and b. During no traffic, based on mean wind velocity of the region, the rotor was able to produce power 3.445 W @ 3.5 m/s wind speed having torque of 0.506 N m at a tip speed ratio of 0.301.

4 Conclusions

The results of the prototype confirm its feasibility and real-time power production. The practical results are supported by the theoretical and analytical results, as shown in Table 1. The maximum amount of energy produced by the prototype was calculated as 20.8 W. The actual efficiency of the prototype was found to be 31.2%. The analysis has been further extended to evaluate the energy produced for various busy stations in the country, as shown in Table 1, giving us an idea of the potential of this project. The number of systems assumed per km was 250 to obtain the results of Table 1. When implemented in large numbers and at multiple locations, it will be one of the innovative, viable solution marching toward the transition to sustainable and cleaner energy sources without posing much safety, efficiency and cost concern.

Table 1 Estimated annual power production at major railway junctions in India

S. No.	Name of the railway junction	No. of trains per day	Total power produced per rotor due to train gust (kWh/day/km)—High RPM	Total power produced due to ambient wind (kWh/day/km)—Low RPM	Total power from the rotor (kWh/day/km)
1	Howrah	600	4	1296	1300
2	New Delhi	350	4	2507.89	2511.89
3	Kanpur	230	2.5	3060	3062.5
4	Kaylan	180	2.5	405	407.5
5	Patna	173	1.25	3400	3401.25
6	Vijayawada	400	2.5	3060	3062.5
7	Allahabad	130	5	3420	3425
8	Itarsi	330	125	2812.5	2937.5
9	Vadodara	170	2	3685.5	3687.5
10	Lucknow	300	1.75	2925	2926.75
11	Chennai	111	1.5	5668.5	5670

References

1. Global Wind Energy Council (2019) Global wind report 2019, Brussels
2. Roy S, Saha UK (2013) Review of experimental investigations into the design, performance and optimization of the Savonius rotor. J Power Energy 227:528–542
3. Bethi RV, Laws P, Kumar P, Mitra S (2019) Modified Savonius wind turbine for harvesting wind energy from trains moving in tunnels. Renew Energy 135:1056–1063
4. Mohamed MH (2013) Impacts of solidity and hybrid system in small wind turbines performance. Energy 57:495–504
5. Darmawan A, Winjaya F (2019) Design of vertical axis wind turbine as an alternative of power supply in level crossing equipment. Indonesian Railway J 3
6. Roy S, Saha UK (2015) Wind tunnel experiments of a newly developed two-bladed Savonius-style wind turbine. Appl Energy 137:117–125
7. Sai S, Venkata J, Rao TV (2016) Design and analysis of vertical axis savonius wind turbine. Int J Eng Technol. e-ISSN: 0975-4024
8. Jon K, Fredrik B, Sandra E, Paul D (2011) Power coefficient measurement on a 12 kW straight bladed vertical axis wind turbine. Renew Energy 36:3050–3053
9. Çetin NS, Yurdusev MA, Özdamar A (2005) Assessment of optimum tip speed ratio of wind turbines. Math Comput Appl 10:147–154
10. Chen L, Chen J, Zhang Z (2018) Review of the Savonius rotor's blade profile and its performance. J Renew Sustain Energy 10:013306
11. Xiaoqing W, Dongxing W, Guozhong C (2012) The influence of blade angle on aerodynamic performance of small Savonius wind turbine. Power Energy 4
12. Hyman M, Ali MH (2019) Improved modelling for wind turbines on trains. In: 2019 IEEE power and energy society innovative smart grid technologies conference (ISGT), Feb 18, pp 1–5

Effect of Mach Number on the Rarefied Gas Flow Over a Forward-Facing Step

Deepak Nabapure, Arjun Singh, and K. Ram Chandra Murthy

1 Introduction

Forward-facing step (FFS) is among the most fundamental geometry that shows the impact of sudden expansion consisting of flow separation. FFS flow has various engineering applications such as flow around buildings, aircraft, diffusers and combustors. These types of flows exhibit separation and reattachment [1]. Various studies were performed on separation and reattachment to explore the impact of various parametric properties. The degree of fluid departure from the continuum assumption is described by the non-dimensional Knudsen number (Kn), which is the 'ratio of the mean free path (λ) to the characteristic dimension (L) of the system under consideration' [2]. The Kn categorises the flows into four regimes [3], namely continuum ($Kn \leq 0.001$), slip ($0.001 \leq Kn \leq 0.1$), transition ($0.1 \leq Kn \leq 10$) and free molecular ($Kn \geq 10$).

In the slip regime, the classical Navier–Stokes Fourier (NSF) equation does not hold. The Boltzmann equation improves the accuracy of modelling the non-equilibrium flows at the onset of the transition regime. However, during the last few years, direct simulation Monte-Carlo (DSMC) method articulated by Bird [4], which solves the Boltzmann equation stochastically, is gaining popularity due to its applicability in all regimes.

Aerospace vehicles during their re-entry phase are prone to intense aerodynamic loads; thus, a cautious heat shield design is required for these vehicles since they

D. Nabapure · A. Singh · K. Ram Chandra Murthy (✉)
High Performance Computing (HPC) Laboratory, Department of Mechanical Engineering, BITS Pilani, Hyderabad Campus, Hyderabad 500078, India
e-mail: rcmurthy@hyderabad.bits-pilani.ac.in

D. Nabapure
e-mail: deepak.nabapure@gmail.com

A. Singh
e-mail: arjun.sbhadwal@gmail.com

M. Palanisamy et al. (eds.), *Theoretical, Computational, and Experimental Solutions to Thermo-Fluid Systems*, Lecture Notes in Mechanical Engineering,
https://doi.org/10.1007/978-981-33-4165-4_42

operate at very high Mach numbers (often hypersonic). Although a smooth aerodynamic shape is preferable while designing such systems, however, discontinuities in the form of contour disruptions, such as steps, gaps or cavities, are present. Such discontinuities constitute increased heat flux or premature transition of laminar flow to turbulence. Therefore, to operate safely, careful design and analysis of such discontinuities are required. Such an analysis is carried out by various experimental as well as various numerical techniques.

Several experimental and numerical studies have been conducted to understand the impact of such discontinuities on aerodynamic properties of the vehicle. For FFS flows, the literature mostly deals with the experimental analysis. However, for this introduction, we shall discuss a few such studies.

Bogdonoff and Kepler [5] experimentally investigated the flow separation in the steps and shock-wave boundary layer interaction. Their work focused on continuum flows for a freestream $Ma = 3$. They showed that in the FFS, flow separation occurred at a pressure ratio ≈ 2. Rogers and Berry [6] experimentally investigated the supersonic flow in forward-facing steps. The investigation involved the flow of $Ma = 2$ with freestream pressure defined as 30, 50 and 70 μm Hg and having a thick laminar boundary layer. Their analysis involved variation of step heights ranging from 0.1 to 0.9 in. They concluded that the pressure rise was dependant on the (h/L) ratio. Pullin and Harvey [7] investigated 2D rarefied hypersonic flow around an FFS. They conducted numerical investigations to study the hypersonic flow of N_2 gas with a freestream $Ma = 22$. The results exhibited a rapid deceleration and compression in the flow near the step base. Grotowsky and Ballmann [8] studied the laminar flow at hypersonic speeds over both FFS and BFS employing the Navier–Stokes equations to simulate a flow with $Ma = 8$, $Re \approx 10^8$ and an altitude $Z = 30$km. Their computational results aligned with the experimental data from the literature. However, they pointed out significant differences in the wall heat flux. The probable cause for this could be the inherent difficulty in measuring it accurately. Leite and Santos [9] conducted numerical modelling of a 2D FFS flow. Their work focused on the hypersonic flow of Mach number roughly equal to 25 at an altitude of 70 km. Their work focused on flow in transition regime for three different step heights of 3, 6 and 9 mm. Their analysis showed a high compression ahead of the step face, analogous to continuum regime investigations. In parallel, Leite et al. [10] investigated the BFS past for different step heights and studied the flow characteristics.

From the literature survey, it is evident that there is limited literature about the rarefied flow past FFS in other regimes, which is addressed in the present study using DSMC. For numerical modelling, *dsmcFoam* [11] solver is used, modelled on the framework of OpenFOAM.

2 DSMC

Established by Bird [4] in the early 1960s, the DSMC technique is among the most popular and successful methods to model rarefied flows. The DSMC method is a probabilistic method, which solves the Boltzmann equation, and the method has been evolving subsequently. The DSMC currently maintains its significant role in addressing rarefied gas dynamics issues amid the emergence of alternative methods. It solves diverse problems involving high-altitude aerodynamics, laser ablation, vacuum techniques and rarefied plasma. The concept of this process is to independently calculate at every time step the movement and inter-molecular collision of a finite number of fictitious molecules separately, considering that each molecule constitutes a considerable number of actual molecules. The cells in the computing space should be less than $\lambda/3$ [12], where λ is the mean-free path (MFP) and the time step less than $\tau/3$, where τ is the mean time of collision [13]. The preliminary amount of '*n*' molecules is consistently allocated through the random function in the cells. The molecular collisions in the present study are through the use of variable hard sphere (VHS) [14] collision model with no time counter (NTC) [15] scheme. The DSMC technique is built on Boltzmann's equation [16] given below.

$$\frac{\partial f}{\partial t} + v \cdot \nabla_x f = \frac{1}{Kn} Q(f, f), \ x \in R^{\mathrm{dx}}, \ v \in R^{\mathrm{dv}} \tag{1}$$

where $f(t, x, v)$ denotes the gas density distribution function at a position x, velocity v, and time t, Kn is the Knudsen number, and the collision operator $Q(f, f)$ represents the binary collisions (Fig. 1).

3 Geometry, Grid and Freestream Conditions

Figure 2(a) shows the schematic of the 2D computational domain. The different geometrical parameters in terms of the step height '*h*' are given in Table 1. The boundary conditions are given in Table 2. The mesh used in the present investigation is shown in Fig. 2(b). The fluid considered is non-reacting air, containing 76.3% N_2 and 23.7% O_2. Freestream conditions employed are tabulated in Table 3. The study is carried out in the transition regime.

4 Validation

The present computational study is carried out using a *dsmcFoam* solver. The solver has been verified in our recent works for various geometries [19–25]. For the specific case of the transitional regime, we compare the findings of Leite et al. [10]. Leite et al. [10] studied the BFS flow in a transitional regime using the DSMC for $Ma = 25$. Figure 3 shows the tangential velocity u/U_∞ distribution at (a) $X = 10$ and (b) $X = 51$. Here X and Y represent the axial and perpendicular distance which is normalised by the MFP (λ). The outcomes show no considerable variation signifying

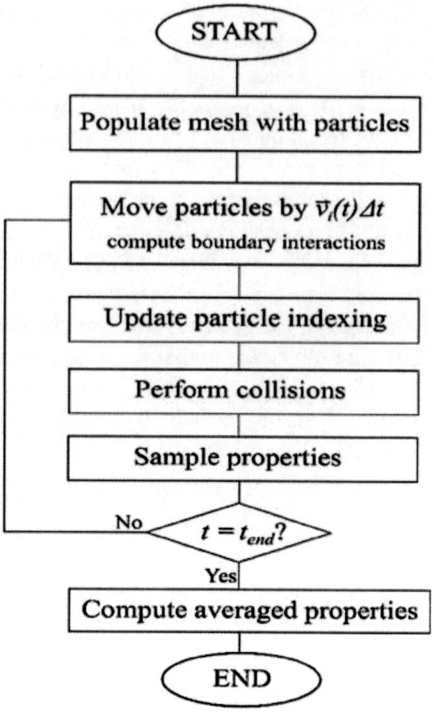

Fig. 1 Steps in the DSMC method adapted from Ref. [17]

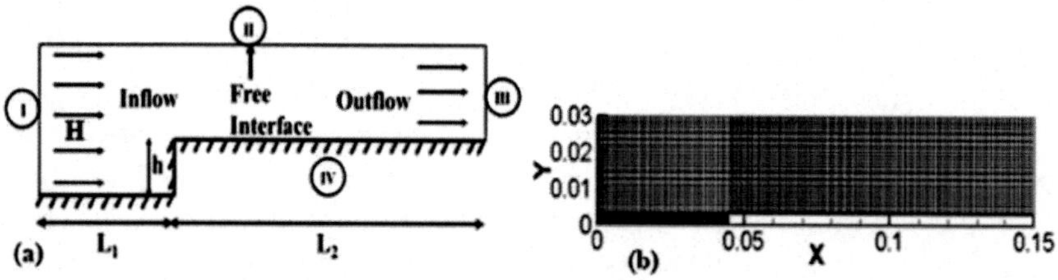

Fig. 2 **a** Computational domain. **b** Mesh for the present study

Table 1 Geometric parameters

Parameter	L_1	L_2	H	Step height (h)
Value	15 h	35 h	10 h	3 mm

Table 2 Boundary conditions

Surface	I	II	III	IV
Boundary condition	Inlet	Free interface	Outlet	Wall

Table 3 Freestream conditions [18]

Parameter	Value
Mach number (Ma)	0.5, 25
Wall temperature (T_w)	880 K
Freestream temperature (T_∞)	204 K
Pressure (p_∞)	1.582 N/m^2
Density (ρ_∞)	2.71×10^{-5} kg/m^3
Viscosity (μ_∞)	1.35×10^{-5} N s/m^2
Mean free path (λ_∞)	0.003 m
Number density (n_∞)	5.64×10^{20} m^{-3}
Knudsen number (Kn_h)	1 (Transition regime)

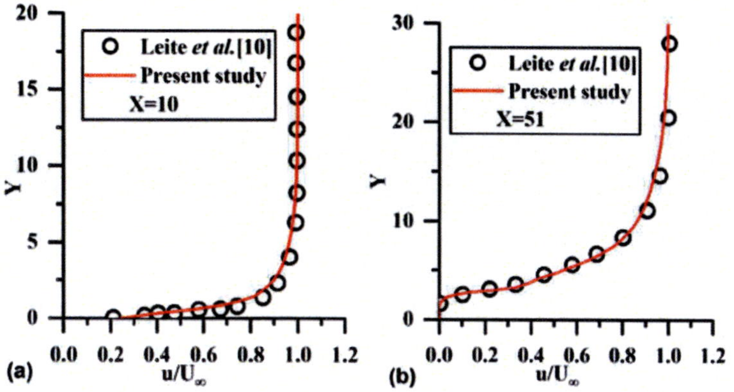

Fig. 3 Streamwise velocity (u/U_∞) distribution perpendicular to the surface of BFS for **a** $X = 10$, **b** $X = 51$

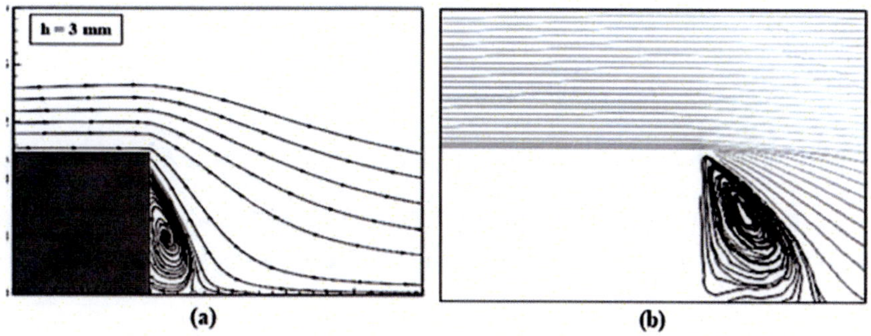

Fig. 4 Comparison of velocity streamlines for h = 3mm **a** Leite et al. **b** Present study

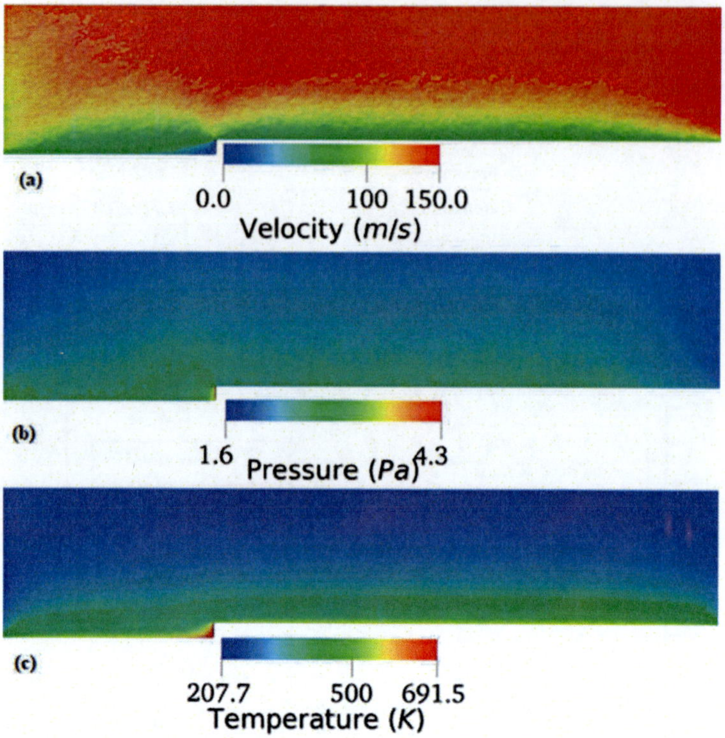

Fig. 5 **a** Velocity, **b** pressure, and **c** temperature contours for Ma = 0.5

the solver validation. Furthermore, the velocity streamlines shown in Fig. 4 also depict a close match with the published outcomes.

5 Results and Discussion

In this section, flow field properties for different Mach numbers (Ma $= 0.5, 25$) representing the subsonic and hypersonic flow are compared. The freestream velocity (U_∞) was 150 and 7149 m/s, respectively, for Ma $= 0.5, 25$. The flow field properties such as velocity contour, pressure contour and temperature contour are evaluated.

5.1 Rarefied Subsonic Flow (Ma = 0.5)

The velocity contour in Fig. 5a shows low velocity close to the wall, with increasing velocity away from the wall, due to boundary layer growth. This phenomenon is analogous to the continuum regime results, other than that the near-wall velocity is non-zero in the present study.

The pressure contour in Fig. 5a shows variations in pressure along the flow direction. The pressure changes in the transverse direction are more pronounced in the vicinity of the step, with regions of higher pressure near the walls and pressure reducing sharply away from the wall.

From the temperature contours in Fig. 5c, it is observed that the near-wall temperatures are of higher magnitude due to the viscous dissipation effects of the hypersonic flow and thermal boundary layer growth.

5.2 *Rarefied Hypersonic Flow (Ma = 25)*

For the rarefied hypersonic flow of Mach 25, the U_∞ was maintained at 7149 m/s.

Figure 6a shows that the velocity contour strongly matches the flat plate Blasius model found in the continuum regime. Equated to the subsonic flow, the boundary

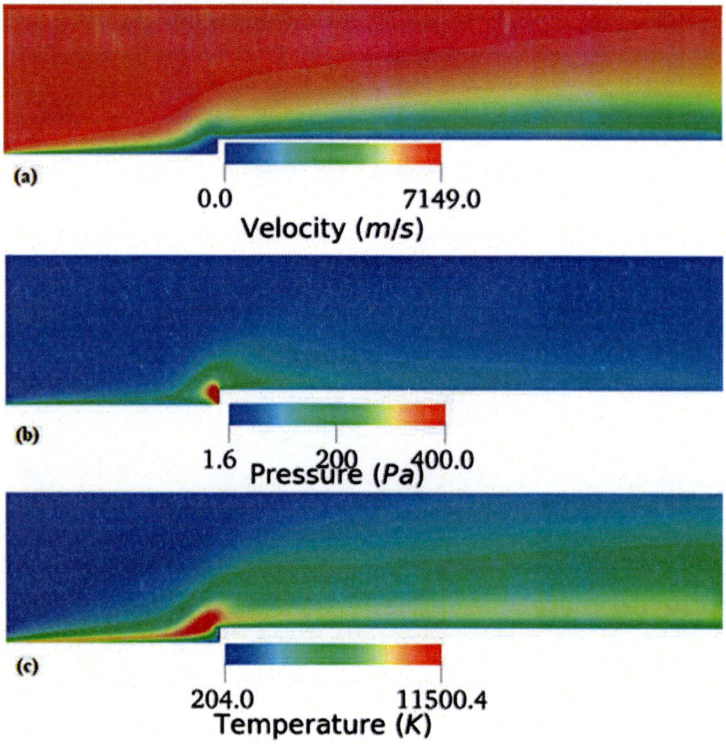

Fig. 6 **a** Velocity, **b** pressure and **c** temperature contours for Ma = 25

layer development in the hypersonic flow is far more prominent, along with higher thickness. Furthermore, near the vicinity of the step, a region of recirculation is also observed.

Figure 6b shows the pressure contour, which has high near-wall pressure and considerably lower outlet pressure. Near the step, there is pressure surge due to shock. Also, owing to compressibility and rarefaction effects, changes in pressure are observed. Downstream of the step, there is a minimal pressure drop towards the outlet.

Figure 6c shows the temperature contour with regions of high temperature near the wall due to the velocity effects of the hypersonic flow. The thermal boundary layer growth is also more pronounced compared to subsonic flow. Temperature magnitude is of order 10^4, which is higher than the imposed wall temperature.

6 Conclusions

The current research is centred on the DSMC analysis of the rarefied subsonic and hypersonic FFS flow. To analyse the fluid flow behaviour, properties such as velocity, pressure and temperature were used. From this research, we may draw the following conclusions:

1. The velocity contour show recirculation for hypersonic flow, whereas it is absent for subsonic flow.
2. The pressure contour shows even pressure distribution for subsonic flow, whereas it is uneven for hypersonic flow.
3. Velocity and shear effects on the temperature distribution are more profound in the hypersonic flow, and in the subsonic flow, the temperature variations are minimal.
4. The obtained conclusions help gain more insight in regards to the velocity and thermal boundary layer development in other rarefaction regimes.

References

1. Chen YT, Nie JH, Armaly BF, Hsieh HT (2006) Turbulent separated convection flow adjacent to backward-facing step—effects of step height. Int J Heat Mass Transf 49:3670–3680
2. Gavasane A, Agrawal A, Bhandarkar U (2018) Study of rarefied gas flows in backward-facing micro-step using direct simulation Monte Carlo. Vacuum 155:249–259
3. Mohammadzadeh A, Roohi E, Niazmand H, Stefanov S (2012) Myong RS (2012) Thermal and second-law analysis of a micro- or nanocavity using direct-simulation Monte Carlo. Phys Rev E Stat Nonlin Soft Matter Phys 85:056310
4. Bird GA (1994) Molecular gas dynamics and the direct simulation of gas flows. Oxford University Press, New York
5. Bogdonoff SM, Kepler CE (1955) Separation of a supersonic turbulent boundary layer. J Aeronaut Sci 22:414–430

6. Rogers E, Berry C (1965) Research at the NPL on the influence at supersonic speeds and low Reynolds numbers of thick laminar boundary layers. Rarefied Gas Dyn 1:574
7. Pullin DI, Harvey JK (1977) Direct simulation calculations of the rarefied flow past a forward-facing step. AIAA J 15:124–126
8. Grotowsky IMG, Ballmann J (2000) Numerical investigation of hypersonic step-flows. Shock Waves 10:57–72
9. Leite PHM, Santos WFN (2015) Computational analysis of the flow field structure of a non-reacting hypersonic flow over forward-facing steps. J Fluid Mech 763:460–499
10. Leite PHM, Santos WFN (2009) Direct simulation calculations of the rarefied hypersonic flow past a backward-facing step. In: 20th International congress of mechanical engineering, Brazil
11. Scanlon TJ, Roohi E, White C, Darbandi M, Reese JM (2010) An open-source, parallel DSMC code for rarefied gas flows in arbitrary geometries. Comput Fluids 39:2078–2089
12. Alexander FJ, Garcia AL, Alder BJ (1998) Cell size dependence of transport coefficients in stochastic particle algorithms. Phys Fluids 10:1540–1542
13. Hadjiconstantinou NG (2000) Analysis of discretization in the direct simulation Monte Carlo. Phys Fluids 12:2634–2638
14. Borgnakke C, Larsen PS (1975) Statistical collision model for Monte Carlo simulation of polyatomic gas mixture. J Comput Phys 18:405–420
15. Abe T (1993) Generalized scheme of the no-time-counter scheme for the DSMC in rarefied gas flow analysis. Comput Fluids 22:253–257
16. Cercignani C (1988) The Boltzmann equation and its applications. Springer-Verlag, New York
17. White C, Borg MK, Scanlon TJ, Longshaw SM, John B, Emerson DR, Reese JM (2018) dsmcFoam+: an OpenFOAM based direct simulation Monte Carlo solver. Comput Phys Commun 224:22–43
18. US Standard Atmosphere (1976). National Oceanic and Atmospheric Administration
19. Nabapure D, Murthy RC (2019) DSMC simulation of rarefied gas flow over a wall mounted cube
20. Nabapure D, Sanwal J, Rajesh S, Murthy KRC (2019) Investigation of subsonic and hypersonic rarefied gas flow over a backward facing step. In: J Phys: Conf Ser, IOP Publishing, p 012007
21. Nabapure D, Murthy KRC (2020) Simulation of flow in single and double-sided lid-driven square cavities by direct simulation Monte Carlo method. Therm Sci 24:3031–3045
22. Nabapure D, Kalluri RCM (2019) DSMC simulation of rarefied gas flow over a forward-facing step. In: ICTEA: international conference on thermal engineering
23. Nabapure D, Ram Chandra Murthy K (2021) DSMC investigation of rarefied gas flow over a 2D forward-facing step: effect of Knudsen number. Acta Astronautica 178:89–109
24. Nabapure D, Ram Chandra Murthy K DSMC simulation of rarefied gas flow over a 2D backward-facing step in the transitional flow regime: Effect of Mach number and wall temperature. In: Proceedings of the Institution of Mechanical Engineers. Part G: J Aero Eng 095441002095987
25. Nabapure D (2020) Investigation of rarefied open cavity flows in all rarefaction regimes using DSMC method. Bulletin Am Phys Soc

CFD Analysis of Twisted Gas Turbine Blade with Different Cooling Hole Geometries on Leading Edge

Moughbul Basha and Mithilesh Kumar Sahu

Abbreviations

BR	Blowing Ratio
LE	Leading Edge
PS	Pressure Side
SS	Suction Side
TIT	Turbine Inlet Temperature, K
T_m	Mainstream Temperature, K
T_c	Coolant Temperature, K

Symbols

η	Film Cooling Effectiveness
Ø	Cooling Hole Diameter
Re	Reynolds number
X	Geometric Distance, mm

M. Basha · M. K. Sahu (✉)
Department of Mechanical Engineering, Gayatri Vidya Parishad College of Engineering (Autonomous), Madhurawada 530048, India
e-mail: mithileshkumarsahu@gvpce.ac.in

M. Palanisamy et al. (eds.), *Theoretical, Computational, and Experimental Solutions to Thermo-Fluid Systems*, Lecture Notes in Mechanical Engineering,
https://doi.org/10.1007/978-981-33-4165-4_43

1 Introduction

Over a period of last two decades, the advances in the gas turbine engine technologies have become more sophisticated which lead to higher performance and lower maintenance costs. These new technologies are driven by the need to drive at higher operating temperatures (higher turbine inlet temperatures, TIT) to improve thermal efficiency, reduce engine emissions and more power output. The author reported in his previous works about performance variation of gas turbine cycles with varying operating condition and also investigated the exergoeconomic performance of cycle [1, 2]. To achieve higher TITs many sophisticated cooling technologies are used in which film cooling is one of the most effective methods of cooling the gas turbine blade. Many studies have been done on the surface of the flat plate to investigate the parameters that are influencing the performance of film cooling. Schmidt et al. [3] conducted experiments on a flat plate with holes for three compound angles. The film cooling effectiveness for round holes and diffused holes with compound angle of 60° for different flux variations is experimentally studied. It was concluded that the compounded angle holes with extended exits have enhanced distribution of coolant near the holes when compared to others. Chowdhury et al. [4] conducted different experiments on flat plate with three different leading edge models, comparing the two cooling hole configurations at three different blowing ratios on film cooling performance. The experiment concluded that film cooling effectiveness of radial-angle cylindrical holes increases with increasing blowing ratios. Shridhar Paregouda and Nageswara Rao [5] conducted experiments on flat surface and concluded that 3D CFD models with increased exit angles of cooling hole would increase the film cooling effectiveness and reduce the temperature of blade surface. Lingyu Zeng et al. [6] investigated the film cooling performance for three simplifications of blade, namely annular cascade vs. linear cascade, twist blade vs. straight blade, rotation vs. non-rotation. The geometric differences between the twist blade and straight blade coupled with rotation lead to the change of stagnation line at the leading edge. The film cooling at the leading edge depends on the main flow stagnation. When compared between twisted blade to straight blade there is amount of error due to geometry. The film cooling effectiveness of blade when rotating is high on the pressure side and low on suction side, because the coolant from cooling holes is subjected to Coriolis force pertaining to the pressure side when it is rotating. This results in more deflection of film over suction side due to the superposition of centrifugal force and Coriolis force. Hai-wang Li et al. [7] conducted experimental investigations on the effect of injection angle and blowing ratio on the leading edge film cooling on a rotating twisted gas turbine blade. For an injection angle of 30° and 45°, the average film cooling effectiveness increases as the blowing ratio increases in all regions. For an injection angle of 60°, it first increases and decreases as the blowing ratio increases from 0.5 to 2.0. Tommaso Bacci et al. [8] conducted experimental and CFD analysis on a highly loaded gas turbine. For clear view of heat transfer mechanisms to determine metal temperatures, infrared (IR) and conjugate heat transfer (CHT) were used. At the suction side, the temperature distribution is two-dimensional and is determined by the progressive thickening of the laminar boundary layer followed by a sudden transition to turbulent conditions. On the pressure side, the heat transfer is quite uniform. Therefore, the pattern of temperature distribution clearly indicates that it is influencing internal cooling performance. The increase in coolant flow and reduced flow of mainstream gases leads to lesser metal temperature. Azzi et al. [9] performed numerical analysis to study the effect of injection angles on the film effectiveness at the leading edge turbine blade. The injection angles are maintained at 25°, 30°, 35° and 45°. The injection hole with 25° has the optimal film coverage over the blade and the highest film cooling effectiveness. Gao et al. [10] conducted numerical study of leading edge holes (cylindrical) with five different compound angles on film cooling performance. It was noticed that as the blowing ratio increases the film cooling effectiveness increases but at highest blowing ratio 2.0, the film cooling effectiveness decreases. It was concluded that optimal blowing ratio is 1.4. Mithilesh Kumar Sahu and Sanjay have investigated the air film cooled complex gas turbine cycles for its exergoeconomic performance [11–14]. Liu et al. [15] investigated the film cooling and impingement cooling performance on the leading edge of turbine blade. It was found that film cooling effectiveness mainly depends on the blowing ratio parameter. Han et al. [16] have reported the cooling technology schemes of blade cooling. The book provides the detailed information on different kinds of cooling schemes and the effect of blowing ratio.

Based on the above literature it is concluded that several experimental and numerical investigations have been done on the surface of the flat plate to evaluate the parameters that are influencing the film cooling performance. The examination comprises numerical and experimental analyses on turbine blade with different radial cooling holes by varying the number of holes. Also, numerical and experimental analyses were conducted on turbine blade with cylindrical cooling holes at leading edge with different orientations (30°, 45° and 60°).

2 Modelling

The airfoil used in this work for model creation is Eppler Airfoil (EPPLER E1212MOD AIRFOIL). The turbine blade profile is twisted in nature with an angle of twist of 7 and is shown in Fig. 1. To achieve the objective, two-blade models are created with two different hole profiles. Model 1 is having cylindrical LE holes while Model 2 is created with tapered LE holes, as shown in Fig. 2.

Figure 3 presents the geometrical model of the present problem. Here mainstream gases (flue gases from combustion chamber) flow through an enclosure having dimension of 200 × 100 mm. The turbine blade is placed at a distance of 100 mm from inlet of the enclosure. The length of the enclosure is kept as 400 mm. The coolant air passage is from the bottom of blade through radial cooling holes. The mainstream gases interface with coolant at blade tip and also at leading edge of turbine blade. The mainstream gases have a temperature of 1273 K while coolant temperature is about

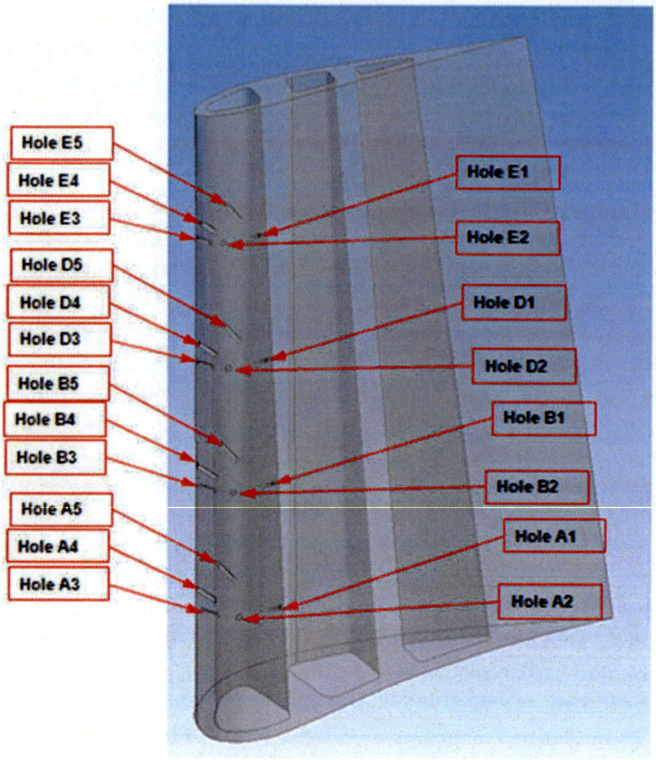

Fig. 1 Cooling holes of turbine blade

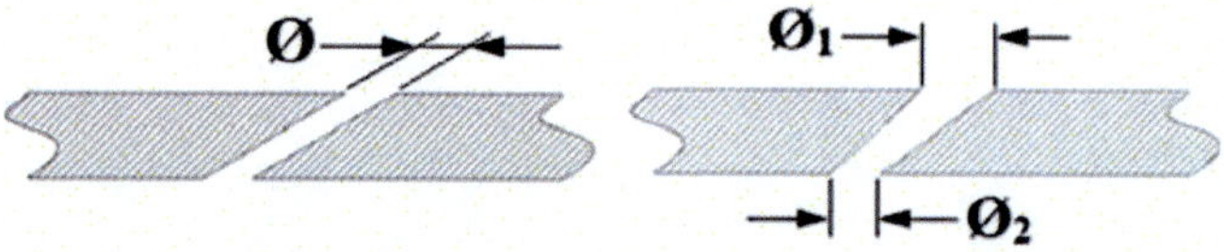

Fig. 2 Model 1 (cylindrical LE hole) and Model 2 (tapered LE hole) with 30° orientation

773 K. The mainstream Reynolds number considered is 200,000. In the analysis, all the enclosure walls are considered adiabatic and zero heat flux is applied to them.

3 Results and Discussion

The numerical investigations are performed on two models, namely Model 1 (cylindrical LE holes) and Model 2 (tapered LE holes) with four blowing ratios (0.9, 1.0, 1.1 and 1.2). Figures 4 and 5 illustrate the temperature profiles for Model 1 (cylin-

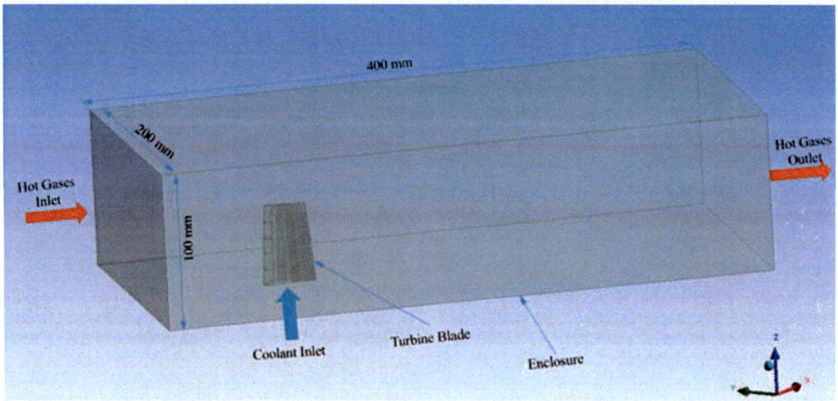

Fig. 3 Turbine blade with enclosure

drical LE holes) and Model 2 (tapered LE holes), respectively, with blowing ratio 1. The temperature profiles basically show when the mainstream gases of 1273 K flow over the blades; the parts of the blades which are covered with the film of coolant air have lesser temperature compared to the portions which do not have such film. So

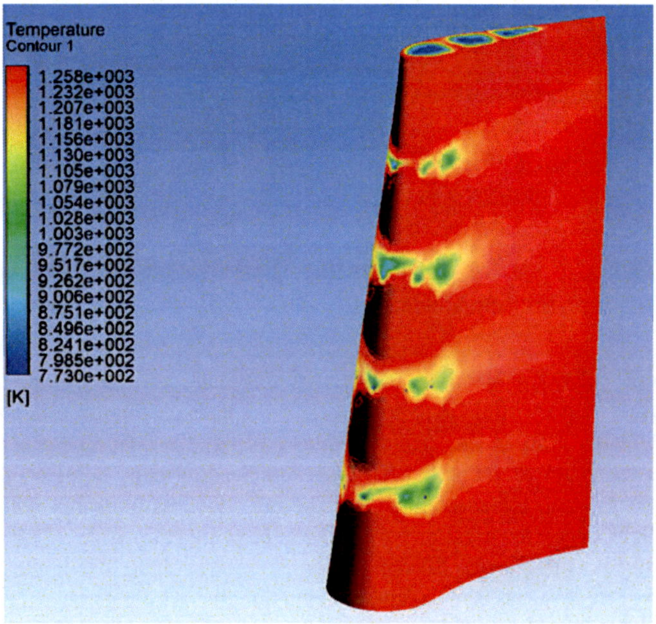

Fig. 4 Temperature distribution for model 1 (cylindrical LE holes) for blowing ratio (1)

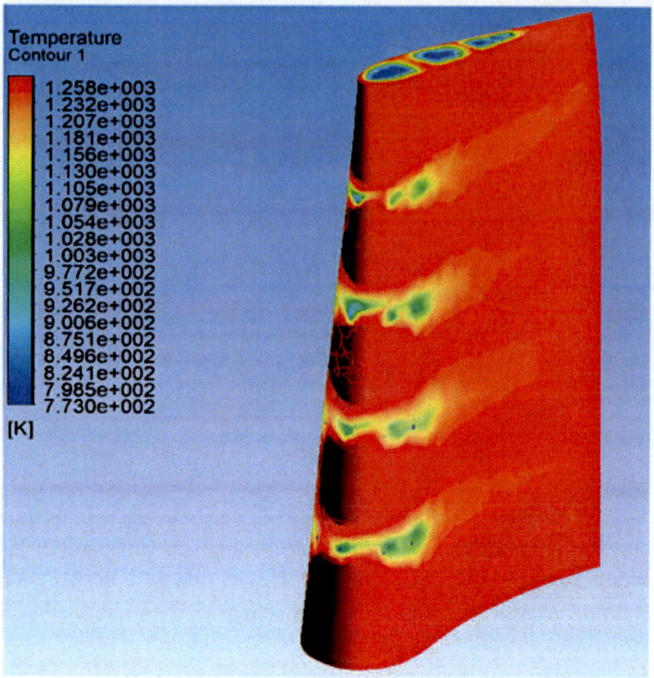

Fig. 5 Temperature distribution for model 2 (tapered LE holes) for blowing ratio (1)

to protect the turbine blades from higher thermal stresses, the leading edge of blade should be provided with more number of cooling holes to create more stable film, keeping in mind the physical strength of the blade also.

3.1 Effect of Blowing Ratio on the Span-Wise Film Cooling Effectiveness for Model 1

In this work, the compressed air is taken as a coolant, and the different blowing ratios considered for simulations are 0.9, 1.0, 1.1 and 1.2. Figure 6 illustrates the effect of blowing ratio on the span-wise film cooling effectiveness for Model 1. In the figure, three arrows indicate the position of leading edge holes on the blade. As the blowing ratio increases from 0.9, the effectiveness increases until 1.1 and there is slight decrease in effectiveness when blowing ratio is 1.2. This decrease in effectiveness is due to the lift off phenomenon of coolant. The highest peak of film cooling effectiveness is measured at stagnation line at hole number E3 (0.7333), which can be seen in Fig. 6. The effectiveness decreases at pressure side, which tends

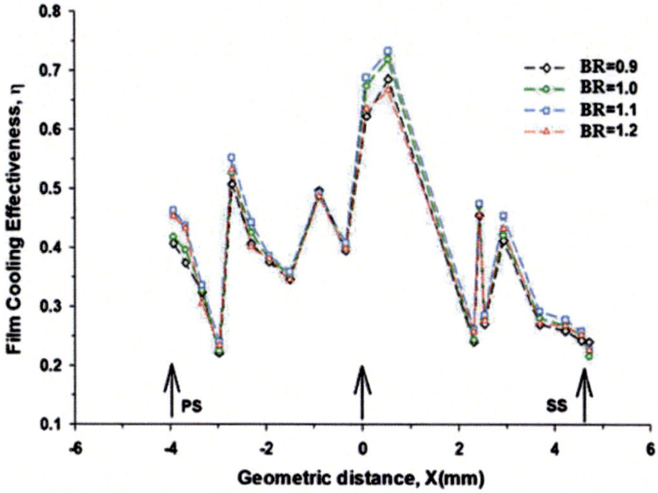

Fig. 6 Effect of blowing ratio on the span-wise film cooling effectiveness for model 1 (cylindrical LE holes)

to increase at stagnation, and then again decreases as it goes to the suction side. The effect of film cooling effectiveness is more at pressure side compared to suction side.

3.2 *Effect of Blowing Ratio on the Span-Wise Film Cooling Effectiveness for Model 2*

The effect of blowing ratio on the span-wise film cooling effectiveness for Model 2 is presented in Fig. 7. Model 2 (tapered LE holes) is also analysed for different blowing ratio ranging from 0.9 to 1.2 in step of 0.1. Figure 7 depicts the cooling effectiveness of different leading edge holes over the blade span. Results show that the effectiveness increases as blowing ratio increases from 0.9 to 1.1, and afterwards it shows fall in cooling effectiveness for increase in blowing ratio. As mentioned earlier that due to lift off phenomenon the formation of air film got affected and thus it results in decrease in film cooling effectiveness. In this case also the highest peak of film cooling effectiveness is measured at hole number E3 (0.7246). The cooling effectiveness increases as it moves from pressure side to stagnation line, and then starts decreasing as it moves towards the suction side. The figure also shows that effect of film cooling is more at pressure side compared to suction side.

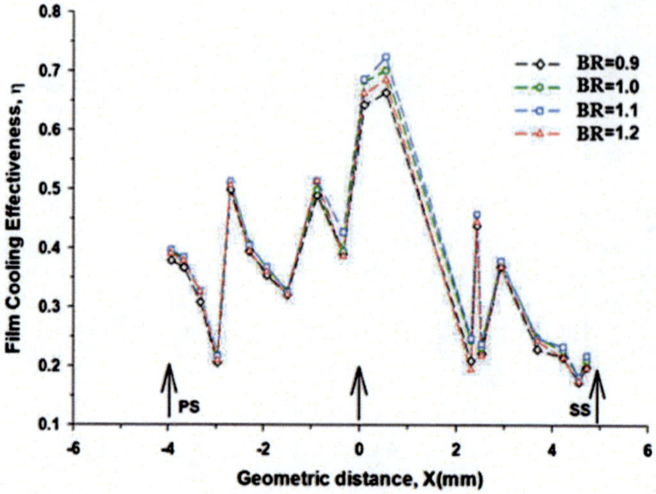

Fig. 7 Effect of blowing ratio on the span-wise film cooling effectiveness for model 2 (tapered LE holes)

4 Conclusions

Based on the results obtained and discussed in previous section, the important conclusions have been made and listed as follows:

- CFD analysis of twisted gas turbine blade has been carried out for two different models (cylindrical LE holes and tapered LE holes).
- Four different blowing ratios (0.9, 1.0, 1.1 and 1.2) were taken for analysis for air as coolant.
- Film cooling effectiveness of blade increases up to certain level of blowing ratio (here up to 1.1) and afterwards it starts decreasing.
- In this analysis it was found that Model 1(cylindrical LE holes) has better span-wise film cooling effectiveness than Model 2 (tapered LE holes).
- Considering E3 hole (highest effectiveness hole), Model 1 (cylindrical LE holes) has 1.2% more cooling effectiveness when compared to Model 2 (tapered LE holes) with air as coolant.

The discussed article is a helpful source of knowledge for both researchers working in this field and for gas turbine blade designers. The presented work helps them to select the right hole geometry, type of coolant and to identify the best blowing ratio for higher possible film cooling effectiveness of gas turbine blade with leading edge cooling holes.

Acknowledgements I express my gratitude to my M. Tech research supervisor, Dr. Mithilesh Kumar Sahu, Assistant Professor, Department of Mechanical Engineering, Gayatri Vidya Parishad

College of Engineering (Autonomous) Visakhapatnam for his continued support and encouragement during the course of this research work.

References

1. Sahu MK, Sanjay (2016) Investigation of the effect of air film blade cooling on thermoeconomics of gas turbine based power plant cycle. Energy 115:1320–1330
2. Sahu MK, Sanjay (2017a) Exergoeconomic investigation of power utility based on air film blade cooled gas turbine cycle. Appl Therm Eng 122:738–746
3. Schmidt DL, Sen B, Bogard DG (1996) Film cooling with compound angle holes: adiabatic effectiveness. ASME J Turbo Mach 118:807–813
4. Chowdhury NHK, Qureshi SA, Zhang M, Han J-C (2017) Influence of turbine blade leading edge shape on film cooling with cylindrical holes. Int J Heat Mass Transf 115:895–908
5. Paregouda S, Nageswara Rao T (2013) CFD Simulation on gas turbine blade and effect of hole shape on leading edge film cooling effectiveness. Int J Modern Eng Res 3(4):2066-2072
6. Zeng L, Chen P, Li X, Ren J, Jiang H (2018) Influence of simplifications of blade in gas turbine on film cooling performance. Appl Therm Eng 128:877–886
7. Li H-W, Han F, Zhou Z-y, Ma Y-W, Tao Z (2018) Experimental investigations of the effects of the injection angle and blowing ratio on the leading-edge film cooling of a rotating twisted turbine blade. Int J Heat Mass Transf 127:856–869
8. Baccia T, Gamannossib A, Mazzeia L, Picchia A, Winchlera L, Carcascia C, Andreinia A, Abbac L, Vagnolic S (2017) Experimental and CFD analysis of a highly-loaded gas turbine blade. Energy Procedia 126:770–777
9. Azzi A, Jubran BA (2004) Influence of leading edge lateral injection angles on the film cooling effectiveness of a gas turbine blade. Int J Heat Mass Transf 40:501–508
10. Gao WJ, Yue ZF, Li L et al (2017) Numerical simulation on film cooling with compound angle of blade leading edge model for gas turbine. Int J Heat Mass Transf 115:839–855
11. Sahu MK, Sanjay (2017b) Comparative exergoeconomic analysis of basic and reheat gas turbine with air film blade cooling. Energy 132:160–170
12. Sahu MK, Sanjay (2017c) Thermoeconomic investigation of power utilities: intercooled recuperated gas turbine cycle featuring cooled turbine blades. Energy 138:490–499
13. Sahu MK, Sanjay (2017d) Comparative exergoeconomics of power utilities: air-cooled gas turbine cycle and combined cycle configurations. Energy 139:42–51
14. Sahu MK, Sanjay (2018) Thermoeconomic investigation of basic and intercooled gas turbine based power utilities incorporating air-fi lm blade cooling. J Cleaner Prod 170:842–856
15. Liu Z, Ye L, Wang C, Feng Z (2014) Numerical simulation on impingement and film composite cooling of blade leading edge model for gas turbine. Appl Therm Eng 73:1432–1443
16. Han J-C, Dutta S, Ekkad SV (2013) In: Gas turbine heat transfer and cooling technology, 2nd edn, Taylor and Francis

Validation of Blade Failure of a Kaplan Turbine Under Adverse Conditions Using Numerical Analysis

Atul S. Tumane, K. Kumar, Abhijeet Kulkarni, R. A. Kubde, and S. Ajai

1 Introduction

Global renewable energy production is steadily increasing to meet the demands for clean and reliable energy. The International Hydropower Association (IHA) reports that renewable comprise 23% of the global electricity mix as of 2014, with 16% of the world's energy production coming from hydropower [1]. Kaplan turbines are exploited effectively worldwide under low head and high discharge conditions. The turbine displays reasonable performance during the range of operations [2]. Kaplan turbines are generally of double-regulated by which the adjustment of both guide vane and runner blade angle is synchronized to deliver a best efficiency point of above 90% efficiency at low head and larger flow rate environment. During the ranges of operation, the Kaplan turbines are subjected to both static and dynamic loads [3]. The discharge flows through the runner at the given net head and contributes to the static pressure head, whereas the rotor and stator interaction induces dynamic loads on the turbine. Research studies [4, 5] have shown that these loads have caused a high stress concentration at the root area of the blade and the control mechanisms.

During the operational conditions, some of the Kaplan turbine units are subjected to catastrophic failure due to fatigue stress, cavitation, mechanical or hydraulic unbalances because of defects in precise double-control mechanisms, silt erosion and so

A. S. Tumane · R. A. Kubde
Prof, Ram Meghe Institute of Technology and Research, Badnera, India

K. Kumar (✉)
Scientist 'C', Central Water Power Research Station, Pune, India
e-mail: krishnaswamy_kumar@yahoo.co.in

A. Kulkarni
Superintendent Engineer, Renewable Energy Centre, MAHAGENCO, Pune, India

S. Ajai
Scientist 'B', Central Water Power Research Station, Pune, India

M. Palanisamy et al. (eds.), *Theoretical, Computational, and Experimental Solutions to Thermo-Fluid Systems*, Lecture Notes in Mechanical Engineering,
https://doi.org/10.1007/978-981-33-4165-4_44

on. Cavitation is a most ubiquitous technical problem that causes material damage when cavity collapses [6, 7]. The secondary effects of these conditions develop a heavy noise and vibrations and leads to fatigue failures. Various experimental studies and their findings [8, 9] paved the ways for eliminating failures of the turbines and its parts. Some of the noted outcome and finding are cavitation-free runner by the way of effective coating; metallurgical advances, sediment management systems and erosion-free coating of runner blades and so on are preferred especially in the hilly regions. Life of a Kaplan turbine runner blades is around 30 years whereas the blades of hilly regions have been replaced in a short span of 2–3 years due to the erosion thinning of blades.

Another type of failure encountered by the hydro turbines is the cases of resonances. When the frequencies of the runner coincide with the dynamic pressure, frequencies of nature will substantially increase the blade vibration which will result in severe mechanical damages. In this paper, case studies of uncommon failure of a Kaplan turbine blades due to mechanical failures of 12 MW Dudh Ganga hydroelectric project Kolhapur in 2017 and 8 MW Surya hydroelectric project, Palghar in 2018 have been analyzed. The operations of the units were successfully put into generation mode after identifying and rectifying the root cause of fatigue failures. Periodical noise and vibrations of the units have been monitored and found to be satisfactory. In order to compliment and validate the cause of failures flow analysis was done using computational fluid dynamic (CFD) simulation under blade misalignment and muddy water flow conditions of Kaplan turbine blades.

2 Experimental and Damage Investigation of Kaplan Turbine

The case study of frequent blade failures of MAHAGENCO for 12 MW Dudh Ganga HE project is reported in November 2017 to CWPRS, Pune. The runner blades (Fig. 1) have been broke down twice in 2008 and 2016 since the commissioning of unit in 1999. The task of identifying the root cause of frequent blade failure is entrusted to

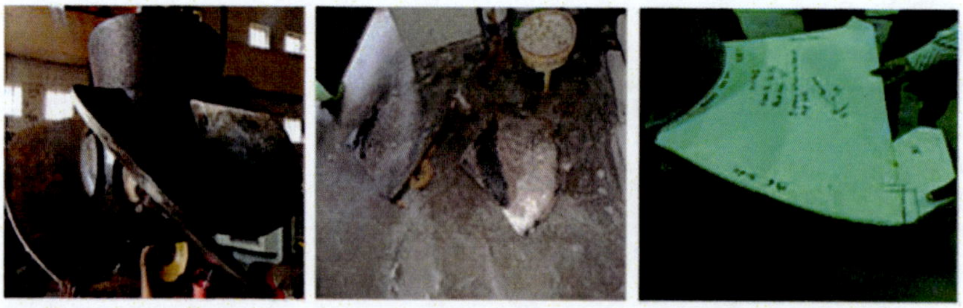

Fig. 1 Kaplan turbine blade

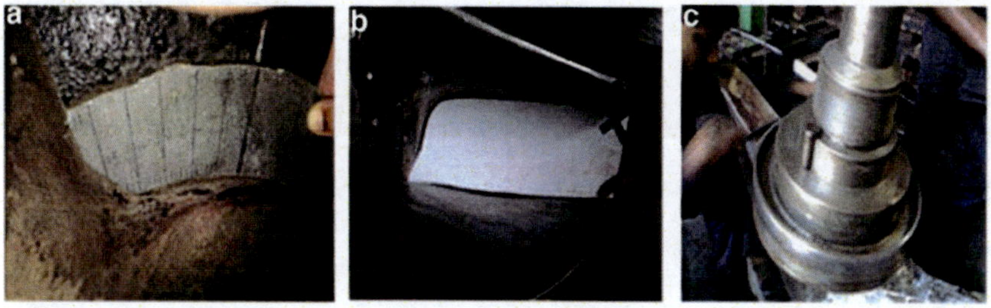

Fig. 2 **a** Template for blade angle, **b** Template for blade opening, **c** Off-centric position of dowel pin

CWPRS, Pune. Subsequently, team of officials conducted the initial investigations on the turbine units.

On initial findings, the unit was developing reasonable noise and vibrations as compared to another unit. Secondly, the unit was getting cavitation where the stream of water bubbles used to disperse heavily on the tail race section, and the unit power was fluctuating while connected to the power grid. Various templates have been prepared for operational range of governor control system of 11° to 33° blade angle positions at idle and full load conditions. Templates for blade opening (Fig. 2) at idle and full load positions are prepared to check the opening between the blades.

On verifying the runner blade angles and the opening of the blades revealed that the angle of two in number blades has been slightly off-centric with the adjacent blades. The openings of the off-centric blades also mismatched with the adjacent blades. On investigating the individual blades, the securing dowel pins of two in number blades have been off-centric with blade-control mechanisms. The off-centric position of the two blades has caused the churning effects on the blades and subsequently developed the cascade effects on the blades. This has developed the heavy noise and vibration and cavitation on the tail race sections.

On identifying the root cause of the frequent blade failures, the off-centric position of the two blades has been rectified (Fig. 3) by placing the sizing washers at the piston assemblies of the control mechanism. The angle of the individual blade angle and the opening positions are synchronized with governor control mechanisms.

In order to gain the generation loss due to the downtime for the availability of the new runner blades from OEM, the damaged blades have welded in place and the die penetration test and NDT have been carried out on the blades. The runner blades are assembled and aligned with generator for the operation of unit. The level of vibration measurement is periodically measured and the reading at various points is below the upper limit of 4.5 mm/s as per ISO 10816 [10]. The units have been generating the year targets of 27 MU in each year during the operation cycle during 2017–2018 and 2018–2019, respectively. Similarly, the turbine failure of 8 MW Surya HEP has been investigated and concluded that the one in number blade out of six blades has been sheared off because of the heavy impact of siltation due to heavy monsoon conditions.

Fig. 3 **a** Blade with hub assembly, **b** Piston block assembly, **c** Assembling of blades

Out of six blades, two in number blades have been replaced with newer ones. The runner blades are assembled after balancing and have been put into operation.

3 Numerical Simulation

This paper describes the numerical analysis of turbine blade failure for 12 MW Dudh Ganga HE project and 8 MW Surya hydroelectric project. We use computational fluid dynamics as a tool for numerical simulation of Kaplan turbine blade using flow 3-D software for Dudh Ganga unit. Figure 4 shows the computational fluid domain in which turbine rotor with draft tube is mentioned. The SST model is used for turbulent flow as an input boundary condition for 8 MW Surya HEP Turbine CFD simulation. The following equations describe the SST model for fluid flow pass through turbine [11, 12].

Kinetic Eddy viscosity term:

$$v_T = \frac{a_1 k}{\max(a_1 \omega, SF_2)} \tag{1}$$

Turbulence kinetic energy:

$$\frac{\partial k}{\partial t} + U_j \frac{\partial k}{\partial x_j} = P_k - \beta^* k_\omega + \frac{\partial}{\partial x_j}\left[(v + \sigma_k v_T)\frac{\partial k}{\partial x_j}\right] \tag{2}$$

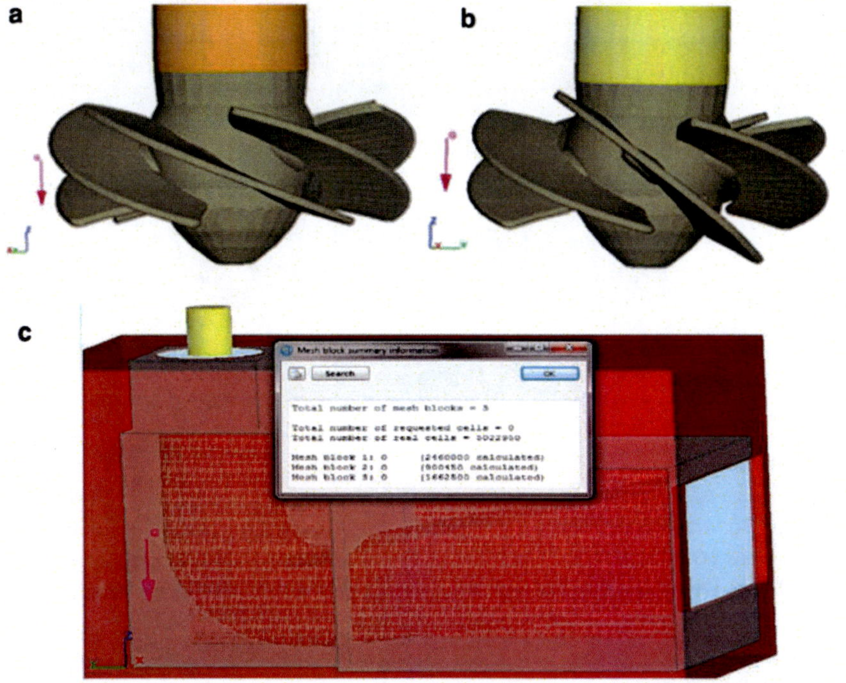

Fig. 4 **a** Normal blade position, **b** Tilted blade position, **c** Fluid flow path through runner and draft tube computational fluid domain

Specific dissipation rate:

$$\frac{\partial \omega}{\partial t} + U_j \frac{\partial \omega}{\partial x_j} = \alpha S^2 - \beta \omega^2 + \frac{\partial}{\partial x_j}\left[(v + \sigma_\omega v_T) \frac{\partial \omega}{\partial x_j} \right] + 2(1 - F_1)\sigma_{\omega 2} \frac{1}{\omega} \frac{\partial k}{\partial x_j} \frac{\partial \omega}{\partial x_j} \tag{3}$$

For validating the blade failure, numerical simulation is performed using computational fluid dynamics flow 3D software. Here the simulation is performed to check pressure fluctuation and severe turbulent zone. Figure 4 shows the geometry of turbine runner with normal and tilted blade position. Simulation is performed for both the position under the same boundary condition. Muddy water single-phase input condition is taken for inlet boundary condition. Total number of three mesh blocks is taken for computational domain.

Flow simulation provides pressure and velocity distribution profile through which we can do comparative analysis for normal and tilted blade position. Liquid–solid two-phase flow simulation is performed to check erosion rate on blade surface due to sediment particles in 8 MW Surya hydropower unit, Palghar Maharashtra. From visual inspection it is seen that heavy rain is accompanied by sediment particle which erodes the blade surface severely. Moreover, heavy rock stones with water make the

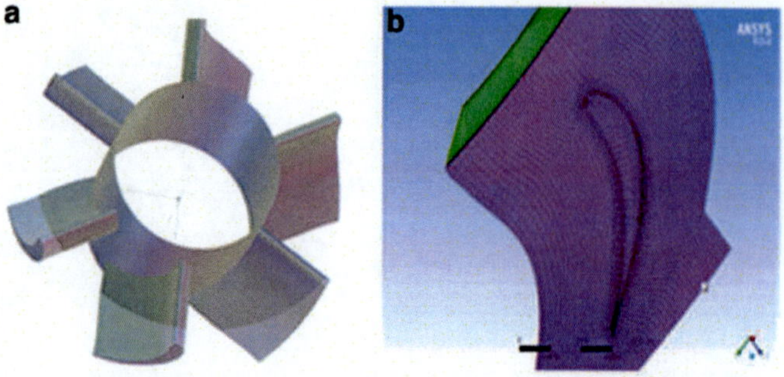

Fig. 5 **a** Kaplan turbine blade geometry. **b** Single runner blade flow passage meshing

surface weaker. To validate the above result simulation is performed on ANSYS Fluent. The computational domain includes runner passage flow which is generated in bladegen ANSYS software, as shown in Fig. 5. Meshing is done in turbo grid ANSYS software. In order to reduce the simulation time only single-blade flow passage has been taken into account.

Eulerian two-phase flow model is taken for simulation with discrete phase on. Quartz is taken as solid discrete particle with 0.001 m diameter and 0.01 kg/s flow rate. Erosion physical model is kept on to check erosion rate. The most often quoted expression of wear is

$$\text{Wear directly proportional to (velocity)}^{n} \quad (4)$$

The value of the exponent n depends on the flow conditions and material properties, but most of the values appear to be three. A bunch of more advanced equations estimating the erosion rate exist, but an accurate mathematical model is difficult to achieve, and the results may only be used as a qualitative estimate [13].

4 Results and Discussion

According to postprocessing result (Fig. 6) we observed that turbulent energy dissipation rate is maximum in tilted blade position as compared to normal blade position. Figure 7 shows that development and dissipation of turbulent energy is symmetrical to the axis of turbine rotation along the plane perpendicular to the flow direction in case of normal blade position. In the similar ground, Fig. 8 shows that turbulent energy is imbalanced to the turbine rotation in case of tilted blade position. The velocity flow of stream line is shown in Fig. 9. It shows no recirculation zone in case of normal blade position, but in case of tilted blade position, recirculation zone is predicted from the figure. The tilted blades induces high velocity fluctuations, pressure variations and asymmetric flow with respect to the axis of turbine rotation.

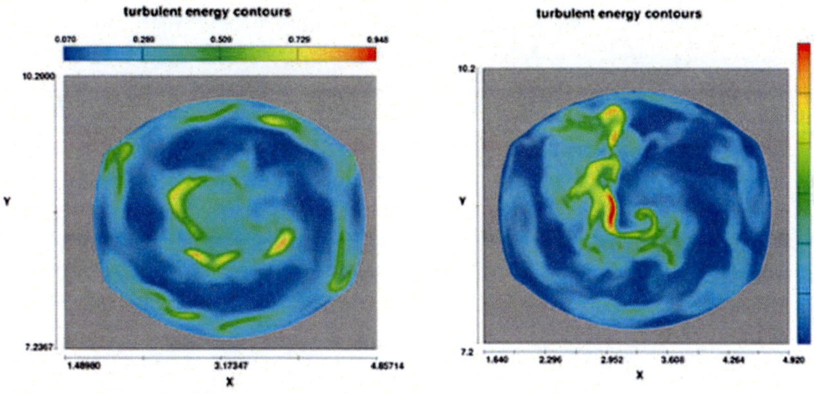

a) Normal Blade Position b) Tilted Blade Position

Fig. 6 Turbulent energy contours at the exit plane of the turbine impeller

The numerical results explain that imbalanced load on the turbine blade due to the adverse asymmetry flow caused by the tilted blade is the major reason for the blade failure. It was also evident that the severe water stream fluctuation at the tail race section was due to the uneven turbulent energy dissipation.

Postprocessing result of two-phase flow simulation shows that erosion rate is very severe in blade surface, which makes surface weaker and is one of the reasons for blade failure. When we analyze velocity contour and stream line, it shows maximum velocity distribution in horizontal location of blade, as shown in Fig. 10. One of the reasons for erosion failure is high velocity and high turbulence zone. As from visual inspection of the side, it is observed that flood flow condition arises due to heavy rain on 8 MW Surya hydropower unit of Palghar, Maharashtra. Blade failure occurs due to high fluctuation of pressure and velocity which creates high turbulence. Density and viscosity of water increases due to the muddy mix water which could cause high turbulence zone.

5 Conclusion

The root-cause analysis of frequent blade failure of 12 MW Dudh Ganga HEP is analyzed and the defect has been rectified by the proper positioning of spacing washer in the piston rod assembly in order to maintain streamlined opening and closing of blades for full range of operation. The governor and control mechanisms are synchronized to affect the precise double controls of turbines. The same has been validated by flow 3D simulation for the root analysis of the catastrophic failure. Similarly, a case of Surya HEP is analyzed by using Fluent solver to identify the failure of turbine blades due to the adverse operation conditions. The CFD simulations are

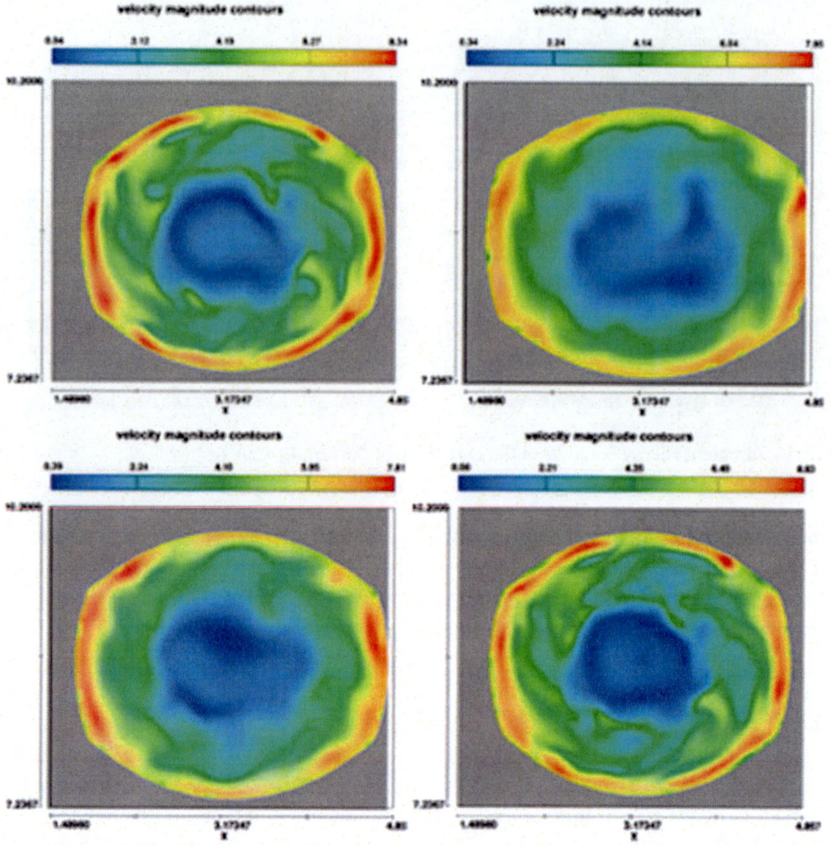

Fig. 7 Normal blade: development and dissipation of turbulent energy is symmetry to the axis of turbine rotation, along the plane perpendicular to the flow direction

reasonably complemented to the experimental investigations in identifying the major high fatigue failures of turbine blades.

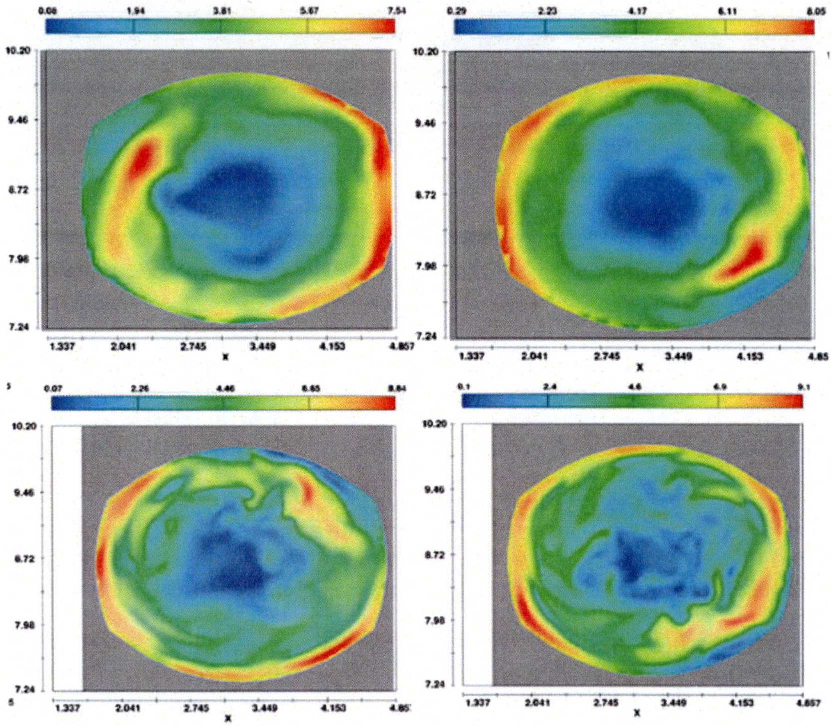

Fig. 8 Tilted blade: development and dissipation of turbulent energy is asymmetry to the axis of turbine rotation, along the plane perpendicular to the flow direction

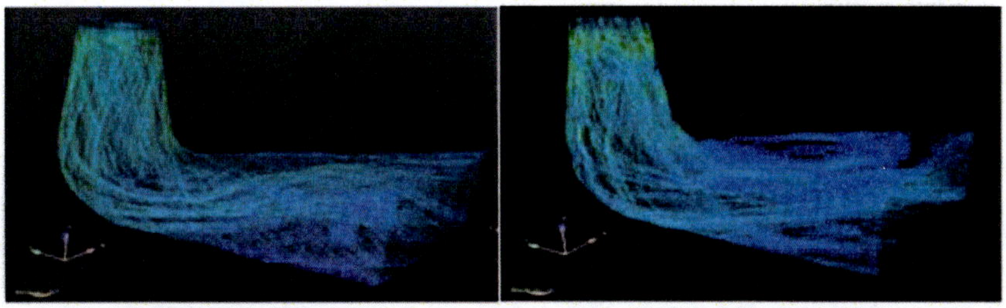

a)Normal blade position \ b) Tilted blade position

Fig. 9 Velocity flow stream line

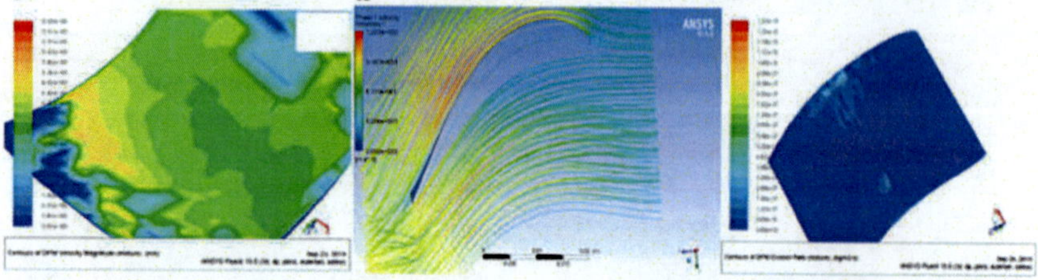

Fig. 10 **a** Velocity contour for mixture. **b** Velocity stream line. **c** Contour of DPM erosion rate

Acknowledgements The authors are grateful to Dr. (Mrs) V.V. Bhosekar, Director, CWPRS, Pune, for giving the guidance and motivation throughout the studies. The authors are also thankful to Dr. R G Patil, Scientist 'E' for his continuous support and guidance during the period of investigations at various stages.

References

1. Schellenberg G et al. (2017) Sedimentation and hydropower: impacts and solutions, Hatch
2. Estevez EE (2004) Comportament Dinàmic De Màquines Hidràuliques. Univ. Politèc, De Catalu-nya
3. Zhang M, David V, Carme Valenro, Agusquiza M, Egusquiza E (2019) Failure investigation of a Kaplan turbine blade. J Eng Failure Anal 97:690–700
4. Frunzaverde D, Campian V, Nedelcu D, Gillich G-R, Marginean G (2010) Failure analysis of a Kaplan turbine runner blade by metallographic and numerical methods. In: Proceedings 7th WSEAS international conference FLUID mechanics (FLUIDS'10), University Cambridge, UK, pp 60–67
5. Zhou L, Wang Z, Xiao R, Luo Y (2007) Analysis of dynamic stresses in kaplan turbine blades. Eng Comput 24:753–762
6. Khurana S, Navtej, Mann HS (2011) Effect of cavitation on hydraulic turbine–a review. Int J Current Eng Technol 2(1)
7. Brennen CE (2011) An introduction to cavitation fundamentals cavitation: turbo-machinery and medical applications, WIMRC FORUM 2011, University of Warwick, UK, 4th-6th July 2011
8. Escalera X, Egusquiza E, Avellan FM (2006) Detection of cavitation in hy-draulic turbines. Mech Syst Signal Process 20983–1007
9. Angulo M, Lucino C, Liscia S (2013) Experimental investigation of vibrations, noise and pressure fluctuations on a Kaplan turbine operating at on-cam cavitating conditions. In: 5th International workshop on cavitation and dynamic problems in hydraulic machinery, September 8–11, Lausanne, Switzerland
10. Evaluation of Machine Vibration by Measurements on non-rotating parts—Machine sets in hydraulic power generating and pumping plants, Multiple. Distributed through American National Standards Institute (ANSI) (23 August 2007)
11. Florian RM (1993) Zonal two equation for k-ω turbulent models for aerodynamic flows. In: 24th fluid dynamics conference, July 6–9, Orlando, Florida

12. Hellsten A (1997) Some improvements in Menter's k-ω SST model turbulent model. In: 29th AIAA fluid dynamics conference June 15–18, Albuquerque, NM
13. Truscott GG (1972) A literature survey on abrasive wear in hydraulic machinery. Wear 20(1):29–50

Determination of Nusselt Number Over Artificially Roughened Solar Air Heater Using Numerical Approach

Amit Kumar, Dheeraj Kumar, and Apurba Layek

Nomenclature

A	Fluid flow duct or pipe area, (m^2)
D	Hydraulic diameter size, (m)
e	Height of rib, (mm)
e/D_h	Relative roughness height ratio
h	Heat coefficient of flowing fluid, (W/m^2 K)
k	Thermal conductively, (W/mK)
P_i	Pitch, (mm)
P_i/e	Relative roughness pitch ratio
Re	Reynolds number (dimensionless)
V	Fluid velocity duct, (m/s)
Nu	Nusselt number (dimensionless)
W	Width, Duct (m)
ρ	The air density, (Kg/m^3)

1 Introduction

The efficiency on the basis of thermal aspect targeting the conventional type solar air heater was observed to be very less due to the lesser rate of heat transfer coefficient to the flowing stream and the absorber surface. This lower in heat transfer mainly due to the presence of viscous sub-layer present over the surface and creates

A. Kumar (✉) · D. Kumar · A. Layek
Department of Mechanical Engineering, National Institute of Technology, West Bengal, Durgapur 713209, India
e-mail: amit4310@rediffmail.com

M. Palanisamy et al. (eds.), *Theoretical, Computational, and Experimental Solutions to Thermo-Fluid Systems*, Lecture Notes in Mechanical Engineering,
https://doi.org/10.1007/978-981-33-4165-4_45

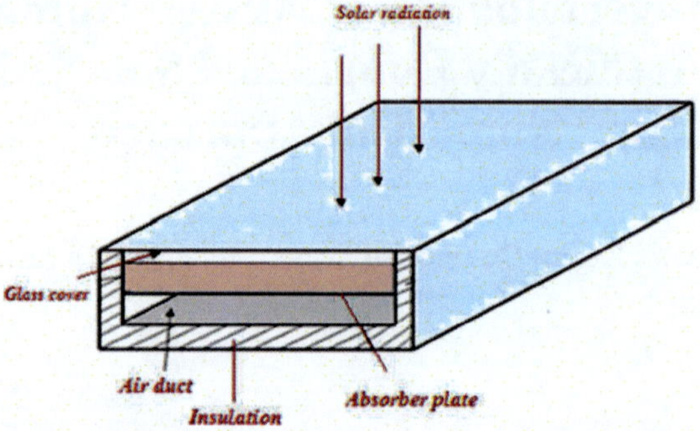

Fig. 1 Conventional-based solar air heater

thermal resistance to heat transfer convection. The conventional type solar air heater as represented in Fig. 1 mainly comprises of a surface plate where solar heat incidence on it and a transparent glass cover on the top surface in order to allow the direct incidence of solar radiation on the surface plate. This system has become more economical and reliable by advancing its convective heat transfer coefficient between the absorber plate and the flowing stream tends to improvise the thermal performance of a solar air heater. The presence of viscous sub-layer over the surface can be a disorder by a solid element also known as an artificial roughness has been placed over the surface which tremendously enhances the heat transfer rate of the flowing stream. The artificial roughness on the absorber surface accelerate the intensity of turbulence level while it creates an interruption in the flow direction leading to increase pumping power hence increases the frictional losses. The use of the computational approach has become one of the important tools to know the fluid flow problem and fluid behavior of roughened type solar air heater. The computational method nowadays used is the CFD technique and its application is widely used in fluid mechanics problem. Prasad and Mullick [1] found that transverse ribs on the surface lead to a higher transfer rate of heat and friction characteristics of the flowing fluid. Bhutan et al. [2] conducted an experimental analysis in order to get the best possible outcome of the thermo-hydraulic performance parameter (THPP) having roughness of protruded shapes using $(\text{CFD})_{\text{fluent}}$-based technique. Wang and Sunden [3] did the experimentation on the square duct to get the enhancement consequence on heat transfer and friction characteristics due to transversely placed ribs on the heated plate. Chaube et al. [4] studied various roughness geometries of a solar air heater system formulating a computational approach to get possible consequences on heat transfer improvement and friction factor wit minimum penalty of pressure drop. Kumar and Saini [5] used $(\text{CFD})_{\text{fluent}}$-based approach having arc-shaped ribs on the absorber surface of a solar air heater in order to get the heat transfer phenomenon and friction effect. Yadav and Bhagoria [6] observed on the basis of his $(\text{CFD})_{\text{fluent}}$-based numerical analysis on solar air heater and to get the most possible effect on heat transfer observed to be giving the best results as compared to the available data.

Using $(\text{CFD})_{\text{fluent}}$-based two-dimensional (2D) approach, the rectangular duct having square-shaped ribs on the heated plate has been investigated in the present analysis. The effects of using square-shaped ribs on Nusselt number and its maximum possible enhancement rate on flow structure have also been analyzed and the best possible results on the basis of its higher level of turbulent flow forced convection may be achieved with a maximum rate of heat transfer.

2 CFD Methodology

The methodology used nowadays is a computational fluid dynamics $(CFD)_{fluent}$-based numerical simulation technique in all types of fluid flow problems in order to understand the actual phenomenon of flowing medium. The Navier–Stokes equations as used for $(CFD)_{fluent}$-based analysis which indicates and shows its effect with respect to its pressure, temperature, velocity, and density depended on each other. In fluid dynamics, problems are generally governed by non-linear partial differential equations and extracted as per conservation law of mass, momentum, and energy equation. The simulation approach is generally used in design modification of a solar air heater achieving to get its optimum values on its design parameter. The complete process used very simple principle for the analysis which consists of grids, formation by using the generalized governing equations on the discrete elements and calculate the study of the solutions on velocity, temperature, pressure, and density of fluid medium consuming lesser time for the evaluation as well as its overall cost by reducing the experimental work [7]. The procedure generally adopted for solving the 2D problem of solar air heater comprises of square ribs on the heated surface discussed in the given sub-section.

2.1 Computational Domain

For such type of flow problem, it is important to simulate three-dimensional (3D) problem to 2D domain study in order to a reduction in computational time and cost. Yadav and Bhagoria [6] endorse that the 2D domain fluid study furnishes closer culmination to experimental results as compared to the 3D domain study. Hence, the present analysis focused on 2D domain study has been done to reduce computer memory and computational time. The complete guideline for the computational model design for the exploration of a solar air heater has been taken as suggested by ASHRAE standard 93–2003 [8]. It is a rectangular section comprised of three main parts named Inlet part, Test part, and Exit part as depicted in Fig. 2. The roughness geometry with different relative pitch distances for square-sectioned rib roughness element is represented in Fig. 3.

The solar air heater system with rib-roughened surface and its operating parameters for $(CFD)_{fluent}$ study are given in Table 1. The heat flux at a constant value of 1000 W/m^2 has been used on the top portion of the test surface and assuming that the complete wall material, absorber surface, and the roughness used are homogeneous

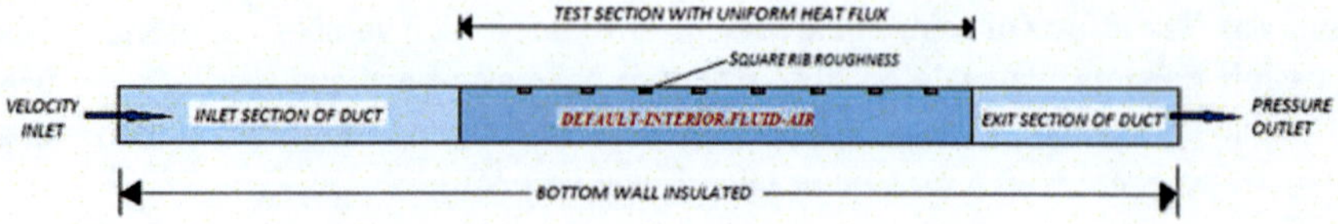

Fig. 2 2D Simulation domain

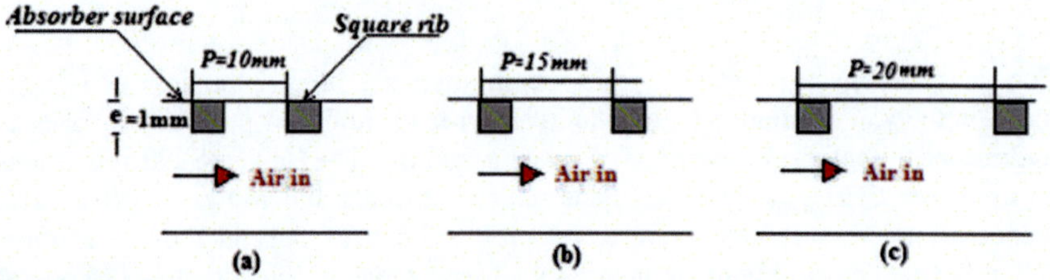

Fig. 3 Different relative pitch distance for square-sectioned rib roughness

Table 1 Square-shaped rib roughness and its operating parameters for $(\mathrm{CFD})_{\mathrm{fluent}}$ study

Relative pitch ratio, P_i/e	10,15,20
Heat flux, I	1000 W/m^2
Reynolds number	3800–18,000
Depth of duct, H	20
Width of duct, W	100
Duct hydraulic diameter, D_h	33.33
Roughness height ratio, e/D_h	0.030
Plate length, L_l	280 mm
Pitch length, P_i	10, 15, 20 mm

and isotropic throughout the system. The assumed condition for the analysis of a complete system of solar air heater and its outcome on heat transfer phenomenon for the dynamic fluid has been simulated by ANSYS FLUENT 16.2.

The noted assumptions for the exploration are as follows:

(1) 2D steady flow as well as fully developed.
(2) It is temperature-independent on thermal conductivity of the roughness material, wall, and absorber plate.
(3) Homogeneous and isotropic conditions throughout wall material, absorber plate, and roughness element.
(4) Density variation should be less and assumed to be incompressible.
(5) No-slip boundary condition.

2.2 Grid Generation

The uniform grid size has been taken for fluid flow distribution and thermal estimation of a solar air heater. The given computational domain generally consists of smooth meshing of triangular-shaped elements done by ANSYS ICEM CFD V16.2 software. The very high-quality mesh has a greater number of cells just adjacent to the heating surface so as to reconcile the turbulent boundary layer considered to be very fine comparing to the height of flowing field. The governing equation for $(\text{CFD})_{\text{fluent}}$ modeling involves the 2D appearance of continuity, incompressible Navier–Stokes, and the energy equation. The equations are represented as follows:

The continuity equations is as follows:

$$\frac{\partial u1}{\partial x1} + \frac{\partial v1}{\partial y1} = 0. \tag{1}$$

The momentum equation follows:

$$u1\frac{\partial u1}{\partial x1} + v1\frac{\partial u1}{\partial y1} = -\frac{1}{\rho}\frac{\partial p}{\partial x1} + v\left(\frac{\partial^2 u1}{\partial x1^2} + \frac{\partial^2 u1}{\partial y1^2}\right) \tag{2}$$

$$u1\frac{\partial v1}{\partial x1} + v1\frac{\partial v1}{\partial y1} = -\frac{1}{\rho}\frac{\partial p}{\partial y1} + v\left(\frac{\partial^2 v1}{\partial x1^2} + \frac{\partial^2 v1}{\partial y1^2}\right). \tag{3}$$

Energy equation is as follows:

$$u1\frac{\partial T}{\partial x1} + v1\frac{\partial T}{\partial y1} = \alpha + \left(\frac{\partial^2 T}{\partial x1^2} + \frac{\partial^2 T}{\partial y1^2}\right), \tag{4}$$

where υ defined as the kinematic viscosity and α defined as the thermal diffusivity.

2.3 Solver

ANSYS FLUENT v16.2 has been used for the numerical simulation in order to predict the behavior of the flowing fluid and its effect on heat transfer due to artificial roughness on the absorber surface. In this computational work, the top wall consists of uniform heat flux conditions, and the remaining three walls are kept insulated. The results thus obtained from the present numerical study are differentiated from the accessible experimental data. Out of different turbulence models, RNG k-ε as well as Realizable k-ε turbulence models are taken for the investigation purpose, and after evaluating all the results of the turbulence model thus comparing these results with empirical relation, the (RNG) k-ε turbulence model gives the finest results.

3 Data Reduction Approach

The Nusselt number and thermal performance thus obtained from the present analysis are used to evaluate the interpretation of solar air heater and show the heat transfer effectiveness within the collector.

The useful value of heat gain is represented by Eq. (5)

$$Q_u = mC_p(T_o - T_i) \tag{5}$$

The enhancement effect of heat transfer by using a roughness element can be achieved by using the Nusselt number (Nu).

$$N_u = \frac{hD_h}{k} \tag{6}$$

Friction factor represented by the following relations:

$$fr = \frac{(\Delta P/l)D}{2\rho V^2} \tag{7}$$

where D_h is known as the hydraulic diameter.

The thermal enhancement factor also called a THPP suggested by Webb and Eckert [9] is written by the Eq. (8)

$$THPP = \frac{(\mathrm{Nu}_r/\mathrm{Nu}_s)}{\left(\frac{fr}{fs}\right)^{1/3}} \tag{8}$$

where (N_{ur}/N_{us}) is known as the Nusselt number enhancement ratio.

The required value of Nusselt number for the smooth duct has been differentiated in Fig. 4 with corresponding correlation given by the Dittus–Bolter equation given by McAdams [10],

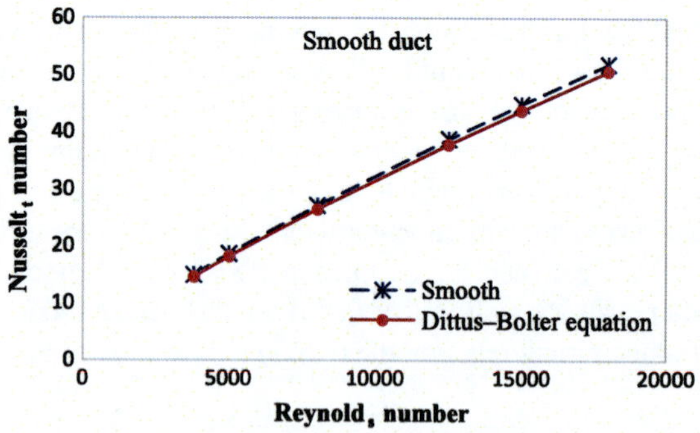

Fig. 4 Nusselt number versus Re for the smooth channel

$$N_{us} = 0.023\,\mathrm{Re}^{0.8}\mathrm{Pr}^{0.4} \tag{9}$$

4 Result and Discussion

The current work approaches toward investigation on the roughness element of square-type of ribs positioned on the absorber surface of a solar air heater and its effect on heat transfer phenomenon have been analyzed using $(\mathrm{CFD})_{\mathrm{fluent}}$-based simulation technique. The data which is obtained from the $(\mathrm{CFD})_{\mathrm{fluent}}$ model are compared to those with the available data of smooth channels working under the same conditions for evaluation purpose. In the case of fluid flow problem, the finest turbulence model taken for validation purposes are noted as RNG k-ε and Realizable k-ε model. Out of these model studies, it is observed that the evaluation of Nusselt number procured from RNG k-ε turbulence model inaugurated to be acceptable results with the experimental data of Yadav and Bhagoria [6] as depicted in Fig. 5 which clearly shows the contour image of turbulent intensity and the velocity magnitude of flowing fluid over the channel of a solar air heater are mainly due to the effect of ribs on the test surface.

For the solar air heater design, the Nusselt number is considered to be an important factor for heat transfer strengthening to the flowing fluid. Figure 6 depicts the action of relative roughness pitch on heat transfer at distinct values of Reynolds number clearly indicates that at P_i/e of 10 gives the ideal value of Nusselt number for all the values of Reynolds number; mainly, the fact deals that, at P_i/e of 10, the turbulence effect is more and, for furthermore increase of P_i/e values, Nusselt number reduces. The Nusselt number observed to be very low at low Reynolds number due to the existence of a viscous sub-layer very adjoining to the absorber plate and the flow is retarded by the roughness element that causes resistance to heat flow and thus lowers the heat transfer. Figure 7 depicts the variability of Nusselt number enhancement ratio versus Reynolds number for its optimum value obtained at P_i/e of 10 as fact concern that an average value of Nusselt number increases with the increased value of Reynolds number resulting in strengthening of turbulent intensity prior to increasing of turbulent kinetic energy as well as its dissipation rate. The optimum enhancement ratio for Nusselt number was observed to be 2.15 times to that of the smooth surface for its highest value of Reynolds number of 18,000 and initiated to be approximately good results with collected experimental details of Yadav and Bhagoria [6].

5 Conclusion

The effect of heat transfer and flowing stream phenomenon in a rectangular channel of a solar air duct consists of square-shaped ribs as a roughness structure on the absorber surface has been analyzed by the $(\mathrm{CFD})_{\mathrm{fluent}}$-based simulation technique.

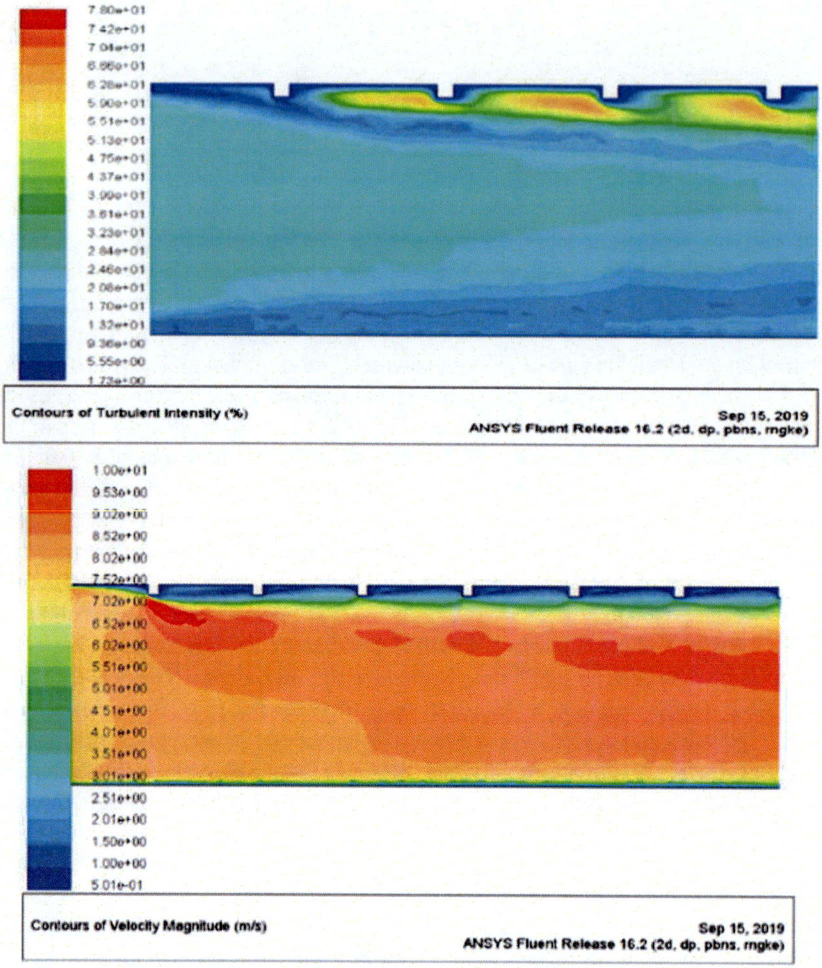

Fig. 5 The contour image of turbulent intensity and velocity magnitude

It is estimated that by varying the relative roughness pitch at different points of Reynolds number, Nusselt number found to improve as Reynolds number increases and attains its maximal value at relative roughness pitch (P_i/e) of 10 and decreases for furthermore increase in P_i/e value. The maximal Nusselt number observed to be 111.3 for a relative pitch ratio of 10 at its highest value of Reynolds number and the optimum value of Nusselt number enhancement ratio noted to be 2.15 times to that of the smooth surface for the considered parameters. The result thus obtained from the simulation technique was found to be in an acceptable range of assigned work and considered to be an important tool for predicting and analyzing the fluid stream behavior of a solar air heater system.

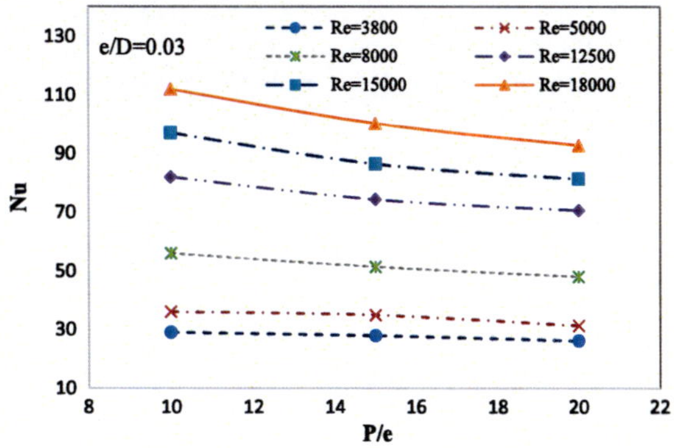

Fig. 6 Nusselt number versus P_i/e

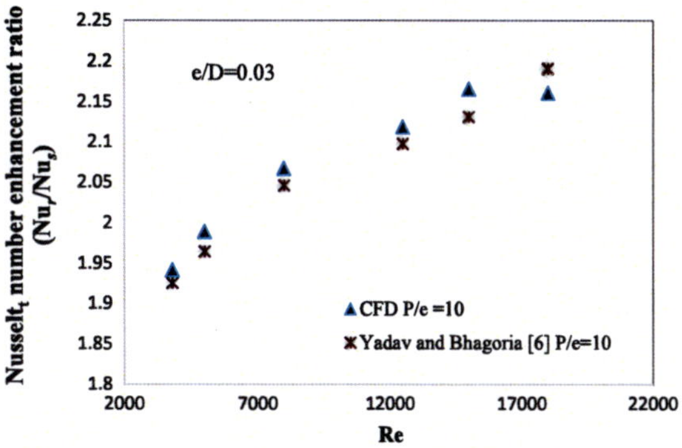

Fig. 7 Nu enhancement ratio versus reynold number

References

1. Prasad K, Mullick SC (1983) Heat transfer characteristics of a solar air heater used for drying purposes. Appl Energy 13:83–93
2. Bhushan B, Singh R (2012) Thermal and thermo hydraulic performance of roughened solar air heater having protruded absorber plate. Sol Energy 86(11):3388–3396
3. Wang L, Sunden B (2007) Experimental investigation of local heat transfer in a square duct with various shaped ribs. Heat Mass Transfer. 43:759–766
4. Chaube A, Sahoo PK, Solanki SC (2006) Effect of roughness shape on heat transfer and flow friction characteristics of solar air heater with roughened absorber plate. WIT Trans Eng Sci 53:43–51
5. Kumar S, Saini RP (2009) CFD based performance analysis of a solar air heater duct provided with artificial roughness. Renew Energy 34(5):1285–1291
6. Yadav AS, Bhagoria JL (2013) Numerical investigation of flow through an artificially roughened solar air heater. Int J Ambient Energy
7. Date AW (2005) Introduction to computational fluid dynamics. 1st edn. Cambridge University Press, New York
8. ASHRAE Standard 93 (2003) Method of testing to determine the thermal performance of solar collectors, american society of heating, refrigeration and air conditioning engineers, Atlanta, GA 30329

9. Webb RL, Eckert ERG (1972) Application of rough surface to the heat exchanger design. Int. J. Heat Mass Transfer 15(9):1647–1658
10. McAdams WH (1942) Heat Transmission. Mc Graw-Hill, New York

Unsteady Wake Dynamics Past a Triangular Cylinder at Incidence with a Downstream Semi-circular Cylinder at Re = 100

Akram Hossain Middya, Amartya Samanta, Chitrak Mondal, Sandip Sarkar, and Nirmal Kumar Manna

1 Introduction

Vortices arise in nature and technology in a large range of sizes. Several investigations have been done so far for problems regarding vortex dynamics. Vortex-induced vibration occurs anytime when a sufficiently bluff body is exposed to a fluid flow that produces vortex shedding at, or near, a structural natural frequency of the body. Various examples of designing using vortex dynamics include flow inside cooling towers, design of tubular and pin type heat exchanger, flow around chimney stacks and offshore structure, and so on. Vortex-induced vibration (VIV) is probably the single most important design issue for steel centenary risers, particularly for high current locations.

Earlier, various researches have been done on flow past a single cylinder only and this itself gives such rich vortex dynamics and that too shows different behavior when we change the Reynolds number. Recent studies on vortex dynamics past a single bluff body have been carried out by Verma et al. [1], where they showed mixed convective flow and heat transfer characteristics past a triangular cylinder. Here observations were made for flow characteristics for different blockage ratio. Another research was done by Biswas et al. [2] which demonstrates the vortex shedding process behind a heated circular cylinder in a cross flow at low Reynolds number of 10° ≤ Re ≤ 45° under the influence of thermal buoyancy. Here, they observed that the steady separated flow becomes unsteady periodic when thermal buoyancy is absent. Now, complexity further increases if the flow takes place past two bluff bodies, and thus nowadays more importance is given to it in this field. Now the

A. H. Middya · A. Samanta · C. Mondal · S. Sarkar (✉) · N. K. Manna
Department of Mechanical Engineering, Jadavpur University, Kolkata 700032, India
e-mail: sandipsarkar.mech@jadavpuruniversity.in

M. Palanisamy et al. (eds.), *Theoretical, Computational, and Experimental Solutions to Thermo-Fluid Systems*, Lecture Notes in Mechanical Engineering,
https://doi.org/10.1007/978-981-33-4165-4_46

vortex generated due to one cylinder affects the dynamics of the cylinder downstream. Earlier researches were done by taking circular cylindrical geometries, but now observations are also made by taking different geometries. Simulations were performed by Meneghini et al. [3] for both isolated cylinder as well as two cylinders in tandem arrangement. They observed that the drag on downstream cylinder becomes less negative as the distance between the two increases. Also, as the gap is increased the amplitude of lift on downstream cylinder becomes of the same magnitude as of the isolated cylinder. Sisodia et al. [4] performed numerical simulations for flow past a semicircular cylinder and square cylinder in tandem arrangement. They performed the simulations at low Reynolds number ranging from $10° \leq \text{Re} \leq 45°$ for various angle of incidences and also show the effect of thermal buoyancy. They varied the Richardson number from $0 \leq \text{Ri} \leq 2$. They computed certain parameters like lift coefficient, drag coefficient, moment coefficient, and average Nusselt number. They observed that the vortex shedding frequency increases with increase in Reynolds and Richardson number, and also at low frequency the vortex starts shedding due to given thermal buoyancy. They also observed that the Nusselt number increases with increase in Reynolds and Richardson number. Bearman and Wadcock [5] did numerical simulation for the flow around two circular cylinders that are placed in a plane normal to the free stream. The spaces between the two cylinders range from 0.1D–1D at a Re no. of 2500. They observed that at very small gap the drags of the cylinders are less than the sum of the drag of the cylinder in isolation. They also observed that there are two vortex streets when the gap is more than 1D, and when the cylinders are very close, only single vortex street is observed.

2 Physical Problem and Numerical Methodology

Figure 1 shows the schematic diagram of the domain with the boundary conditions. In a two-dimensional regime, bluff bodies are in tandem arrangement. As clear from the figure there is an equilateral triangle with each side equal to the semicircle diameter, d and both of them being separated by a distance of $X = 2\text{d}$. The walls of both semicircle and triangular cylinders are maintained at a constant temperature, T_w and are exposed to a constant incoming free stream jet of velocity U_∞. Air is chosen as the working fluid with $\text{Pr} = 0.71$. The distance of the upstream to the end of the semicircle is $a = 8\text{d}$, and the downstream length is $b = 37\text{d}$. The total cross-sectional height is $h = 48\text{d}$ with the centroid of the triangle, and the center of the semicircle aligned at a same horizontal level with a distance of 24d from the vertical slip boundaries. The distance separating these two walls have been chosen as 2d because of some importance according to some researchers' previous observation on vortex shedding, for the specific range of parameters considered in the present problem.

The acceleration due to gravity (g) acts perpendicular to incoming fluid flow direction, that is, along negative Y direction. In accordance to the present scenario the flow

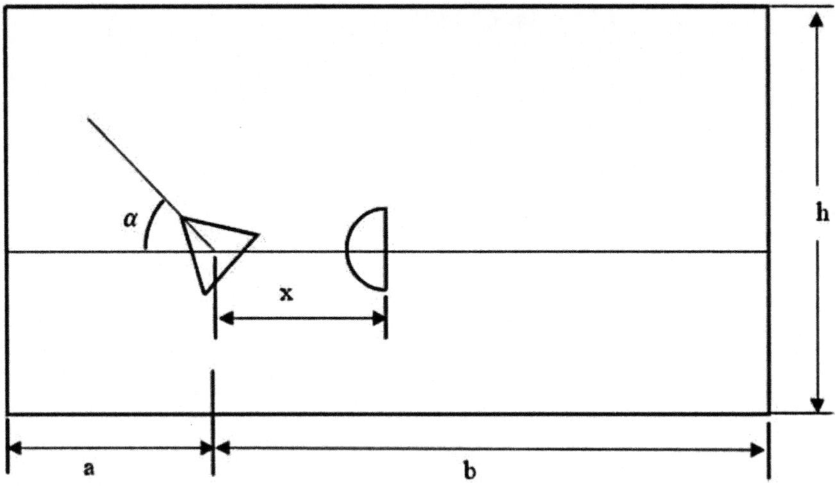

Fig. 1 Geometry of the two-cylinder tandem arrangement

air (air) is considered as unsteady, laminar, and incompressible with constant thermo-physical properties. The flow here is considered to be two-dimensional as the three-dimensional flow is predominant only when the value of Reynolds number exceeds its critical value of 180. Based on these assumptions as discussed, the governing equations of continuity, momentum, and energy conservation in non-dimensional form can be expressed as

$$\frac{\partial u}{\partial x} + \frac{\partial v}{\partial x} = 0 \tag{1}$$

$$\frac{\partial u}{\partial t} + u\frac{\partial u}{\partial x} + v\frac{\partial u}{\partial y} = -\frac{\partial p}{\partial x} + \frac{1}{\mathrm{Re}}\left(\frac{\partial^2 u}{\partial x^2} + \frac{\partial^2 u}{\partial y^2}\right) \tag{2}$$

$$\frac{\partial u}{\partial t} + u\frac{\partial u}{\partial x} + v\frac{\partial u}{\partial y} = -\frac{\partial p}{\partial x} + \frac{1}{\mathrm{Re}}\left(\frac{\partial^2 v}{\partial x^2} + \frac{\partial^2 v}{\partial y^2}\right) \tag{3}$$

$$\frac{\partial u}{\partial t} + u\frac{\partial \theta}{\partial x} + v\frac{\partial \theta}{\partial y} = \frac{1}{\mathrm{Re}.Pr}\left(\frac{\partial^2 \theta}{\partial x^2} + \frac{\partial^2 \theta}{\partial y^2}\right) \tag{4}$$

Here, $u = U/U_\infty$, $x = X/D$, $y = Y/D$, $p = P/\rho U_\infty^2$, $\theta = (T - T_\infty)/(\mathrm{T_w} - \mathrm{T_\infty})$, Reynolds no. $\mathrm{Re} = \rho U_\infty D/u$, Prandtl no. $\mathrm{Pr} = \mu c_p/k$. Here D is the length scale.

2.1 Boundary and Initial Conditions

The governing equations (Eqs. 1–4) are solved by using the following boundary conditions in dimensionless form:

(i) At inlet: uniform inlet velocity of fluid along the x-direction along with constant inlet temperature.

$$u = 1,\ v = 0,\ \theta = 0 \tag{5}$$

(ii) At outlet: flow is assured to be fully developed and zero heat flux condition along downstream

$$\partial u/\partial x = 0, \partial v/\partial x = 0, \partial \theta/\partial x = 0 \tag{6}$$

(iii) At the top and bottom slip boundary walls: adiabatic walls

$$v = \partial u/\partial x = 0, \partial \theta/\partial x = 0 \tag{7}$$

(iv) At the semicircular and triangular walls: no-slip condition for flow with constant wall temperature

$$u = 0,\ v = 0,\ \theta = 1 \tag{8}$$

2.2 Drag and Lift Coefficients

The drag force coefficient has two components, that is, the viscous ($C_{\mathrm{d}v}$) and the pressure force ($C_{\mathrm{d}p}$) coefficient and the net drag force coefficient is the summation of the two force coefficient components as

$$C_D = C_{\mathrm{d}v} + C_{\mathrm{d}p} = \frac{F_D}{\frac{1}{2}\rho U_{\infty}^2 D} \tag{9}$$

Similarly, the lift force also has two force components and can be expressed by the relationship

$$C_L = \frac{F_L}{\frac{1}{2}\rho U_\infty^2 D} \tag{10}$$

Here, F_L and F_D represent the lift and drag forces, respectively.

2.3 Strouhal Number

The periodic flow that is being observed is mainly due to vortex shedding. The non-dimensional frequency of vortex shedding is expressed in terms of a number known as Strouhal number (St) expressed as:

$$St = \frac{fD}{U_\infty} \tag{11}$$

2.4 Nusselt Number

The heat transfer coefficient between the triangular and the semicircular walls with that of the surroundings is described by the Nusselt number, and the Nusselt number (Nu) is expressed as:

$$\text{Nu} = -\frac{\partial \theta}{\partial n} \tag{12}$$

Here, n is the direction perpendicular to the triangular and semicircular surface. However, this Nusselt number is the local Nusselt number and to find the average value of Nusselt number we need to integrate the local Nusselt numbers over the entire length, that is

$$\text{Nu}_{\text{avg}} = \frac{1}{l}\int_0^l \text{Nu} \cdot \text{d}l \tag{13}$$

Here, l is the dimensional length along these wall surfaces.

2.5 Mesh Generation

Now, we will check out the details on how to create the mesh as shown in Fig. 2. Clearly the mesh is in the form of a plus sign with smaller and finer divisions within

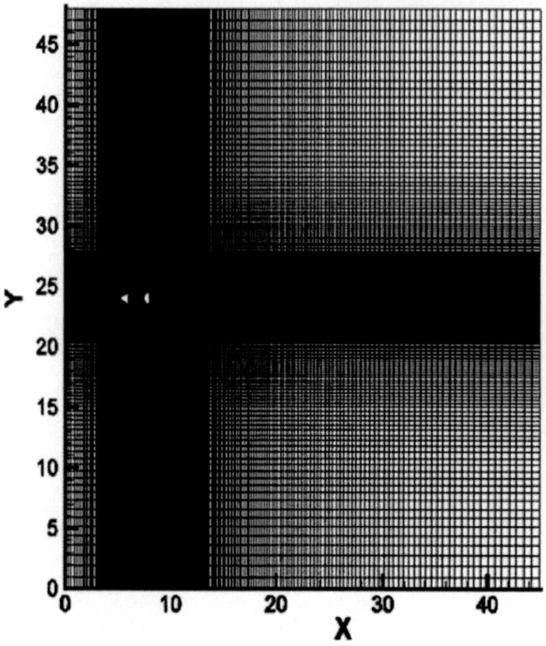

Fig. 2 Actual mesh generation

that region and the divisions getting more and more broad and coarse as we go far away from it.

3 Results

Figure 3a displays the instantaneous streamlines, vorticity, and isotherms, respectively, for different angle of incidence varying between $0° \leq \alpha \leq 105°$ and for Re = 100. An illustrated examination of all these leads us to the following set of conclusions. The fully separated zone starts to develop at the downstream of the cylinder. The formation of the disturbance for the upstream triangular cylinder is interrupted by the downstream semicircular cylinder. It should be noted that the instantaneous separation points separate with time due to vortex shedding cycle. With variation in α the separation points are seen to move along the surface of triangular cylinder. So, due to interference of wake of triangular cylinder separation, bubbles are seen to form at the right side of the semicircular cylinder. At $\alpha = 0°$ the separation point on the triangular cylinder is located at the middle and as α is increased the separation point shifts toward the upward corner till $\alpha = 60°$. As α is increased the point moves toward the bottom. It is also seen that the recirculation bubble disappears as α is increased up to 75°. It is also noticed that the maximum curvature will be at $\alpha = 105°$ due to flow fluctuations.

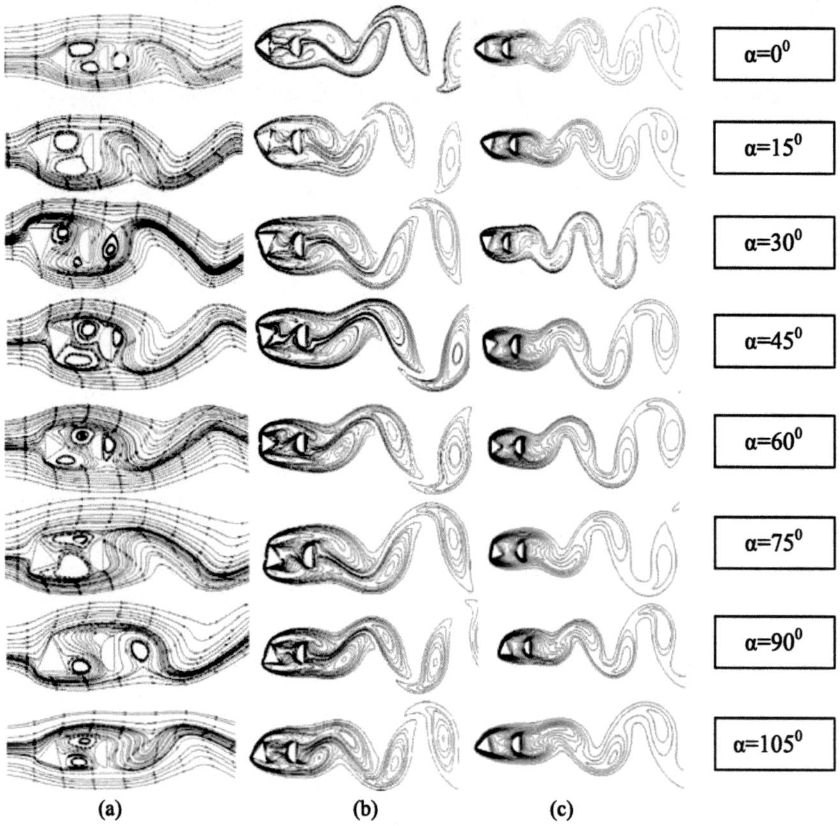

Fig. 3 Contours of **a** Streamlines **b** Vorticity **c** Isotherm

Figure 3b shows the vorticity contours originating from the two-cylinder surfaces till the far wake. The vortex shedding occurs for all angle of incidence at Re = 100. It is observed that the length of vortex strands that originates from the two tandem cylinders increases in length as α is increased.

Figure 3c shows the isotherm contours which look similar to the vorticity contours. We observe that the isotherms in close proximity to each cylinder acquire the shape similar to that of the cylinder. The isotherms originating from the cylinders will interact with one another and cause significant distortion of the isotherms because of the rotation of the triangular cylinder. It is seen that the detachment length of the isotherms decreases with increasing value of α. It is also observed that the detachment length of the thermal energy blobs in isothermal contours is more in comparison to the vortex strands in vorticity contours.

3.1 C_D For Semicircular and Triangular Cylinder

As can be observed from Fig. 4 for the semicircle, the C_D value first decreases from $\alpha = 0°$ to $\alpha = 15°$ in negative direction and then increases gradually till $\alpha = 60°$ in negative direction. At $\alpha = 15°$ due to formation of recirculation zone at downstream to semicircular cylinder; the pressure downstream is low and so creates a positive drag. But as the angle of incidence is further increased the recirculation zone in the upstream of semicircular cylinder increases and thus generates a negative drag which increases in negative direction till $\alpha = 60°$. When angle of incidence is further increased the recirculation zone in the upstream of semicircular cylinder decreases and thus the drag upstream decreases and thus drag downstream increases. Now for the triangle as seen from Fig. 5, on increasing the angle of incidence from $\alpha = 0°$ the C_D increases gradually till $\alpha = 60°$. This can be explained as follows: at $\alpha = 0°$ the front stagnation point is located at the center of the front face. A recirculation zone which corresponds to a low-pressure region is also formed between the two cylinders. Now as the angle of incidence is increased the recirculation zone seems to increase and thus the pressure downstream decreases due to which the drag downstream increases. As the angle of incidence is further increased beyond $\alpha = 60°$, the recirculation region decreases and thus the drag force downstream also decreases.

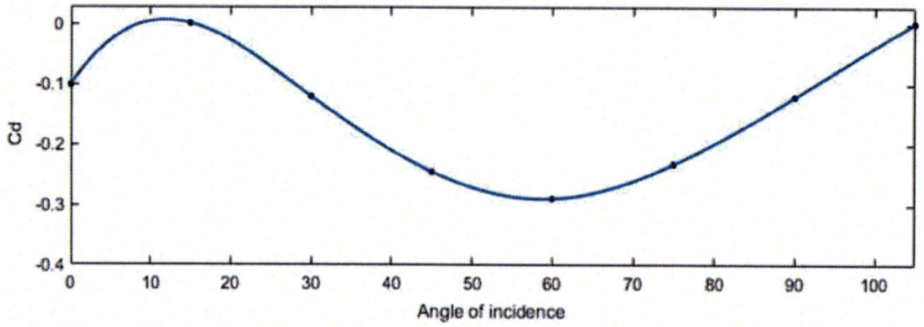

Fig. 4 C_D for semicircular cylinder

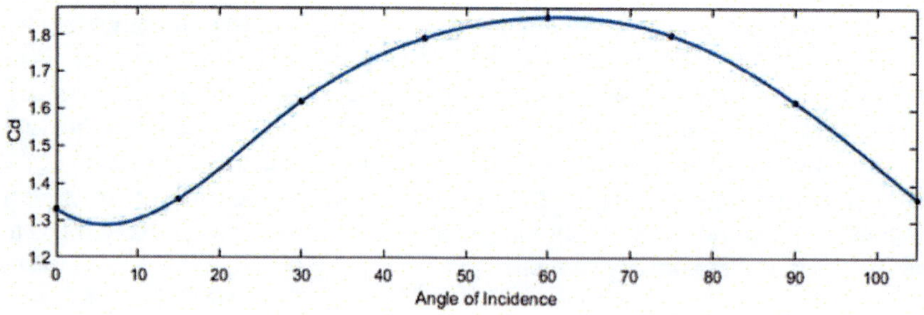

Fig. 5 C_D for triangular cylinder

3.2 C_L for Semicircular and Triangular Cylinder

Figures 6 and 7 depict the variation of lift coefficients with α for the semicircular and triangular cylinders, respectively, at Re = 100. Here the graph is plotted by taking the rms values which actually tells us about the magnitude of C_L values ignoring the direction at which lift acts. The general trend that is being observed after looking at the curve for semicircle is that it is periodic in nature and has a symmetry about $\alpha =$ 60°. The global maxima of this curve is at $\alpha =$ 60° and minima at $\alpha =$ 0°. Figure 9 shows the variation of lift coefficient of the triangular surface with angle of incidence α. This curve is also similar to the one shown above as it is also periodic and having symmetry about $\alpha =$ 60°. The maximum values are obtained at $\alpha =$ 30° and $\alpha =$ 90°, whereas the global minima is obtained at $\alpha =$ 0°. If we observe the contours, we can clearly see that at $\alpha =$ 0° the upward and the downward direction forces are almost equal making the magnitude of force very near to zero. Now as we further increase the value of α the magnitude of either force keeps on decreasing making the overall lift value to increase up to $\alpha =$ 30° where the vertical force component becomes maximum. After that, further we see that either of the force value increases making

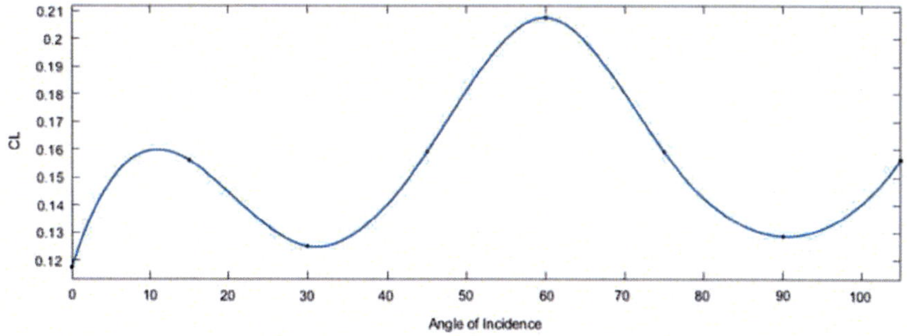

Fig. 6 C_L for semicircular cylinder

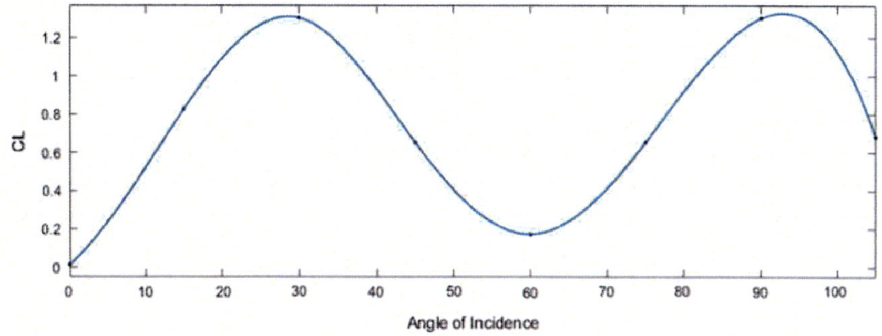

Fig. 7 C_L for triangular cylinder

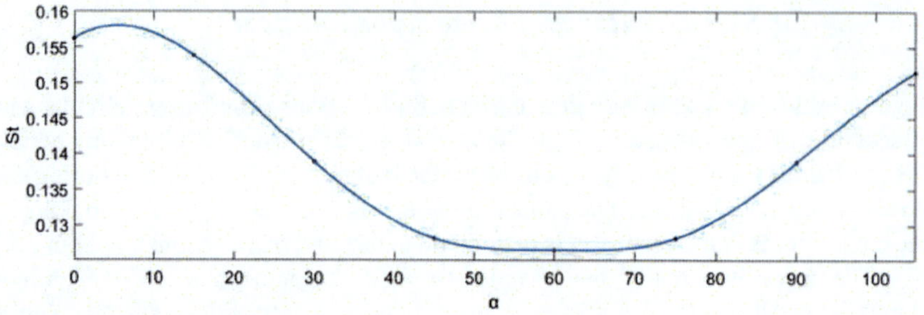

Fig. 8 *St* for semicircular cylinder

the overall value to decrease and at $\alpha = 60°$ both the force components magnitude become nearly equal making it a local minima.

3.3 *St for Semicircular Cylinder*

In dimensional analysis Strouhal number (*St*) is a dimensionless number describing oscillating flow mechanisms. The Strouhal number can be important when analyzing unsteady, oscillating flow problems. The Strouhal number represents a measure of the ratio of the inertial forces due to the unsteadiness of the flow or local acceleration to the inertial forces due to changes in velocity from one point to another in the flow field. The numerical value of Strouhal number is found from the lift curve for semicircular cylinder at saturation state. Figure 8 depicts the variations of Strouhal number for different angle of incidences (α). As can be seen from the curve the Strouhal number decreases with increasing α and attains global minimum at $\alpha = 60°$. It is because in that region the vortices stay attached to the cylinder surface for a longer duration due to which there is delay in separation, thus reducing the strength of vortex shedding.

3.4 *Average Nu for Semicircular and Triangular Cylinder*

The average Nusselt number is calculated by taking the average of the local Nusselt number values over the surface of the cylinder. Figures 9 and 10 show the variation of average Nusselt number to the angle of incidence α for both the semicircular and the triangular cylinders, respectively, at Re = 100. It is seen that in case of triangle the value of average Nusselt number decreases up to $\alpha = 60°$ after an initial increase at the start for very small value of α. After that, the value increases steadily for the rest of the curve. The reduction in the average Nusselt number with increasing values of α till 60° is attributed to the interference between the two cylinders with increasing

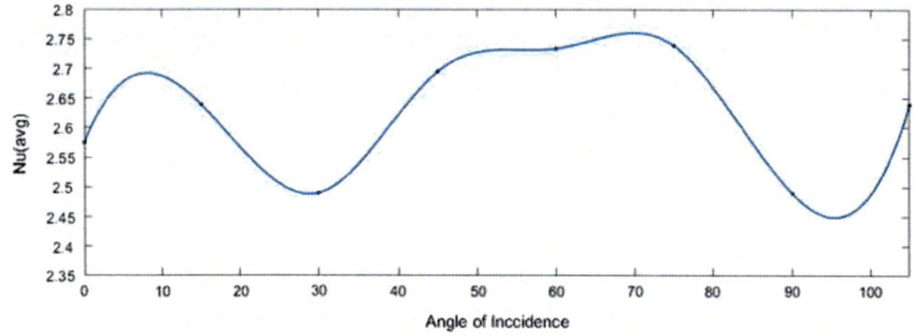

Fig. 9 Nu_{avg} for semicircular cylinder

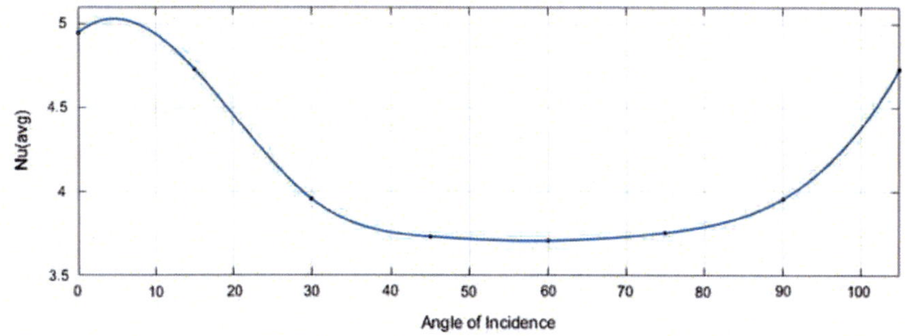

Fig. 10 Nu_{avg} for triangular cylinder

separation. On further increasing the values of α the alteration in the thermal boundary layer thickness takes place increasing the values of average Nusselt number (Nu_{avg}) since then. For the semicircle the values of Nu_{avg} are comparatively lower which can be attributed to the fact that higher convective heat transfer from the triangular cylinder is directly facing the incoming stream. The semicircular cylinder facing the heat stream later results in comparatively lower values. For the semicircular cylinder the Nu_{avg} increases and then decreases to attain local minima at $\alpha = 30°$ and the value further increases and after that attains the global maxima at $\alpha = 70°$. Then the value further decreases to attain the global minima at $\alpha = 95°$ and then increases to $\alpha = 105°$.

4 Conclusion

In this paper a numerical study is performed to study the fluid flow and heat transfer characteristics flowing past a triangular and semicircular cylinder. Here air is being considered as the working fluid and the results are obtained for Re = 100 with the

Table 1 Nomenclature

C_D = Coefficient of drag C_L = Coefficient of lift D = Projected with of triangular cylinder (m) f = Frequency of vortex shedding (Hz) F_D = Drag force acting on cylinder (N/m) F_L = Lift force acting on cylinder (N/m) g = Gravitational acceleration due to gravity(m/s^2) k = Thermal conductivity $\mathrm{Nu_L}$ = Local nusselt number $\mathrm{Nu_{avg}}$ = Average nusselt number Re = Reynolds number $\left[\frac{\rho U_\infty D}{\mu}\right]$ Pr = Prandtl number $\left[\frac{\mu c_p}{k}\right]$ St = Strouhal number $\left[\frac{fD}{U_\infty}\right]$	T_w = Temperature of cylinder wall (K) T_∞ =Temperature of free stream (K) t = Flow time(s) U_∞ = Free stream velocity (m/s) U, V = Velocity in x, y directions u, v = Non-Dimensional velocity x, y = Co-ordinate axis α = Angle of incidence μ = Dynamic viscosity of fluid (Pa.s) Υ = Kinematic viscosity of fluid (m^2/s) θ = Dimensionless temperature $\left(\frac{T-T_\infty}{T_w-T_\infty}\right)$ ρ = Density of fluid (kg/m^3)

fluid striking a triangular cylinder kept at different angles α, varying from 0° to 105°. The parameters such as coefficient of drag (C_D), coefficient of lift (C_L), Strouhal number (St) and average Nusselt number ($\mathrm{Nu_{avg}}$) are computed for varying angles of incidence. Now as observed for Re = 100, the vortex shedding takes place for all values of α. The time average drag coefficient curve was periodic for both the semicircle and the triangle and similar about $\alpha = 60°$. In case of lift coefficient, the rms values are taken into consideration but still the property of being similar about α= 60° persist even in this case of both cylinder surfaces. However, the average Nusselt number curves for the triangular and semicircular cylinders are dissimilar, with the previous one attaining minima at α= 60° and the latter one attaining maxima at the same angle (Table 1).

References

1. Verma N, Dulhani JP, Dalal A, Sarkar S, Ganguly S (2015) Effect of channel confinement on mixed convective flow past an equilateral triangular cylinder. J Heat Transfer 137:121013
2. Biswas G, Sarkar S (2009) Effect of thermal buoyancy on vortex shedding for flow past a circular cylinder in cross flow at low reynolds number. Int J Heat Mass Transf 52:1897–1912
3. Meneghini JR, Saltara F, Siqueira CLR, Ferrari JA (2001) Numerical simulation of flow interference between two circular cylinders in tandem and side by side arrangement. J Fluids Struct 15:327–350

4. Sisodia SS, Sarkar S, Saha SK (2017) Fluid flow and mixed convective heat transfer around a semi-circular cylinder at incidence with a tandem downstream square cylinder in cross flow. Int J Thermal Sci 121:13–29
5. Bearman PW, Wadcock AJ (1973) The interaction between a pair of circular cylinders normal to a stream. J Fluid Mech 61:499–511